Practical Antenna
Handbook

About the Authors

Joseph J. Carr was a military electronics technician and the author of several popular electronics books, including McGraw-Hill's *Secrets of RF Design*, and *Old Time Radios! Restoration and Repair.* He wrote a monthly column for *Nuts & Volts* magazine.

George W. (Bud) Hippisley, W2RU, earned his BSEE degree from MIT and was formerly chief operating officer for a major supplier of electronics to the cable TV industry. A longtime active amateur radio operator who has won or ranked nationally in many competitive on-the-air operating events, he has given talks on the basics of antennas and ionospheric propagation to radio clubs and other groups for more than 35 years.

Front cover: Two towers support the author's mono-band Yagi antennas for 40, 20, 15, 6, and 2 meters. The taller tower is shunt-fed on 160 meters and supports one end of an 80-meter dipole; the shorter tower supports a 40-meter dipole. *In the foreground:* a section of open-wire transmission line feeding the 80-meter dipole. *Out of view:* Beverage wires for low-band receiving and a triband Yagi for 10, 15, and 20 meters.

Practical Antenna Handbook

Joseph J. Carr
George W. Hippisley

Fifth Edition

New York Chicago San Francisco
Lisbon London Madrid Mexico City
Milan New Delhi San Juan
Seoul Singapore Sydney Toronto

The McGraw·Hill Companies

Cataloging-in-Publication Data is on file with the Library of Congress.

McGraw-Hill books are available at special quantity discounts to use as premiums and sales promotions, or for use in corporate training programs. To contact a representative please e-mail us at bulksales@mcgraw-hill.com.

Practical Antenna Handbook, Fifth Edition

4567890 QFR/QFR 0198765

ISBN   978-0-07-163958-3
MHID   0-07-163958-6

The pages within this book were printed on acid-free paper.

Sponsoring Editor
   Judy Bass

Acquistions Coordinator
   Bridget Thoreson

Editorial Supervisor
   David E. Fogarty

Project Manager
   Virginia Carroll, North Market Street Graphics

Copy Editor
   Virginia Carroll

Proofreaders
   Virginia Landis, Sue Miller, and Stewart Smith

Production Supervisor
   Pamela A. Pelton

Composition
   North Market Street Graphics

Art Director, Cover
   Jeff Weeks

Contents

V

Appendices

Preface

My paternal grandfather was born and raised in Newfoundland, on the outskirts of St. John's, its capital city. As a young man, he came to the United States in 1903 to find work. To his dying day he understood very little about radio or television beyond the simple mechanics of adjusting the volume control or changing the station. Yet he became directly responsible for my nearly lifelong love affair with "wireless" communications when he gave me an AM broadcast band clock radio for my thirteenth birthday. It was my very first radio receiver, and, although my grandfather thought I would prize the alarm on it, the "magical" feature I came to love instead was the Sleep knob. Now I could climb into bed and fall asleep to the sounds of faraway stations, my parents and I secure in the knowledge that the radio would turn itself off.

As I gained familiarity with my newfound window on the world, I marveled at the distant stations I could hear on certain nights, and soon I was keeping a paper diary, or *log*, of broadcast station call letters I was able to pluck from the cacophony of late-night radio waves. Without knowing the name of my obsession, I had become a broadcast band *DXer*! ("DX" is radio shorthand for "long distance".)

At about the same time, my next-door neighbor obtained his amateur radio license, and soon I was spending occasional evenings in his basement, sitting silently by his side while he made two-way contacts with like-minded hobbyists around the world. The seed was sown, and with the help of the local radio club I obtained my own "ham radio" license before I turned 14. I had plunged headlong into a passionate love affair with a technology that has shaped my life ever since, from my choice of college major to my choice of career and even my choice of where to live on this planet.

But it wasn't until I was in the midst of preparing new material for this fifth edition of the *Practical Antenna Handbook* that I realized my grandfather had another, albeit indirect, connection with the world of radio I had so quickly taken to.

Picture the world in the fall of 1901: The only way to travel between North America and Europe is by giant steamship; the Wright brothers haven't yet made their first flight on the sands of the North Carolina Outer Banks; there is no such thing as radio or TV broadcasting; it will be more than a half century before the first artificial satellite beams its telemetry data down to Earth; and it will take three-quarters of a century for the Internet to be born. In 1901 the only way to communicate across the sea is by sending mail on the oceangoing steamships or by *wire-line telegraphy*, using the relatively new undersea cables that had been laid in the late 1800s.

In Europe, meanwhile, young inventor Guglielmo Marconi is having scant success convincing governments and corporations of the value of his recently discovered ability to signal across towns and villages without wires. Discouraged with his native Italy's lackluster response to his experiments, he has come to England, where he has received encouragement from that government—especially after his success in communicating across the English Channel.

But something more is needed to capture the public's attention. So Marconi boards a ship and, with some of his assistants, heads to Newfoundland after setting up a transmitting station overlooking the Atlantic near the westernmost point of England, at Poldhu, Cornwall. On December 11, 1901, Marconi and his North American team hoist a wire with a large kite launched from Signal Hill at the entrance to the bay at St. John's, and on the following day Marconi reports hearing the letter "S"—the prearranged code letter being sent by other assistants back at Poldhu—in the headphones of his receiving apparatus! As they say, the rest is history.

Did my grandfather ever meet Marconi? Was he aware of the transatlantic experiments being conducted within a few miles of his home? Did he visit Signal Hill to see what all the fuss was about? I'll probably never know. But it's fun to speculate.

In distinct contrast to water waves and sound waves, or to the Internet, which moves information from place to place largely via optical fibers and other physical media, radio communication takes place with no supporting medium at all! Electromagnetic waves launched at the transmitting site travel essentially forever, until intercepted at a receiving site. The task of efficiently converting transmitter output power into radio waves capable of traveling into the infinite reaches of space or of picking up those waves for amplification and conversion to audible sound waves by my clock radio falls to a class of devices known as *antennas.*

It is no exaggeration to say that without antennas, there would be no radio. Thus, to design radio or television equipment, to be an amateur radio operator or shortwave listener, or to successfully install a wireless Internet service for a group of users, a *practical* understanding of antenna basics and basic antennas is a necessity—hence the title of this book.

In the decade since the fourth edition of the *Practical Antenna Handbook* first appeared, entirely new classes of specialized antennas—including *medium-frequency* (MF) directional receiving antennas for limited spaces and budgets—have come into existence and are in daily use by thousands of radio amateurs and shortwave listeners around the world. For many older antenna designs, theories of antenna operation originating in crude experiments from as long as a century ago have been either confirmed or supplanted by new theories and practices based on modern analytic techniques, coupled with tests conducted with new generations of test equipment, and backed up by extensive computer modeling using inexpensive and readily available software that until just a few years ago was found only in the laboratories of universities and large corporations.

Scientific and technical information transfer has also been upended by the Internet. Today, numerous online discussion groups (variously known as *reflectors* and *listservs*) allow experts to report and critique recent antenna findings in near real time while hundreds or thousands of other professionals and rapt hobbyists "listen in".

In putting together the fifth edition, every chapter of the *Practical Antenna Handbook* has been reviewed and revisions made wherever appropriate. Occasionally, antenna designs or topics that are too esoteric or arcane for the overview of basic antenna types this book represents have been winnowed out to make room for new material of broader or more topical interest. In many chapters, new explanations of fundamental concepts are the result of developing and refining effective (i.e., "lightbulb") answers to the most-asked questions during antenna talks I have long given at radio club meetings and conventions. Similarly, the use of the half-wave dipole as a conceptual building block for "quick and dirty" analysis of many other antenna types found throughout this book has been well-received over the years.

With few exceptions, the use of math in this book is kept at or below high school level. Appendix A has been added to give readers a foundation for understanding the *decibel*, since it is essential to the discussion of antenna performance, and to provide a summary of a few specific trigonometric right triangle and sine-wave relationships necessary for some of the explanations at various places in the text. Perhaps the only "advanced" math concept in the book is the use of the imaginary number operator j to simplify and clarify the manipulation and display of complex numbers.

Two structural changes have been made to the *Practical Antenna Handbook*:

- The order of presentation of the material has been revised to collect related topics in groups of chapters and to help the book as a whole flow logically from the most basic concepts and antennas to more complicated configurations and techniques.

- The fourth edition of the *Practical Antenna Handbook* included a CD with supplemental information and spreadsheets on it. Today it is far more practical to commit this material—and more—to a Web site, where it can be updated and augmented, as appropriate, over the lifetime of the book. Go to www .mhprofessional.com/carr5 for:

  o Tables of worldwide geographic coordinates and antenna dimensions vs. frequency

  o Supplier updates

  o Author's blog

  o Additional photographs and schematics

  o Links to tutorials and specialized calculators

Before I became captivated by radio, my goal had been to be a journalist. Throughout my adult life I have never lost that love of writing, and during my career I have occasionally had the good fortune to be able to marry the two interests in a few brief work assignments. Ordinarily, writing a book about antennas would represent nirvana for me, except that I cannot lose sight of the fact that I have this opportunity only because Joe Carr passed away shortly after completing the fourth edition. So I hope you will find this fifth edition a worthy expansion and updating of Joe's fine work over the years, and I ask your indulgence as I dedicate it to my grandfather, Archie Munn Hippisley, who started me on this path many years ago.

GEORGE W. (BUD) HIPPISLEY, W2RU
Old Forge, New York

Acknowledgments

As with other fields, in engineering and the sciences virtually all advances in knowledge build upon accomplishments and results that are the work of others. Thus, an overview volume such as this must necessarily draw upon and reflect the collective wisdom of the larger community of contributors to the *state of the art*. To identify all such individuals would require a second book, even larger than this one.

However, certain people who provided specific guidance, inspiration, and resources should not go unrecognized. In particular:

Guy Olinger, K2AV; Jim Lux, W6RMK; Tom Rauch, W8JI; Jim Brown, K9AY; and Chuck Counselman, W1HIS—exemplary members of a cadre that is advancing the collective state of the art in radio communications technology by freely disseminating and sharing their considerable knowledge, analytic skills, and experimental results with others via the Internet;

W. L. Myers, K1GQ, for a near-lifetime of stimulating dialog and many kindnesses, and for introducing me to *cocoaNEC,* antenna modeling software for the Macintosh by Kok Chen, W7AY;

Roy Lewallen, W7EL, the author of *EZNEC,* and Dan Maguire, AC6LA, for *NEC-2* modeling guidance;

Professors Amir Fariborz (SUNY Institute of Technology) and Orlando Baiocchi (University of Washington/Tacoma Institute of Technology) for making electromagnetics interesting and a lot more fun during my second "tour" around the EM classroom; and

James Rautio, AJ3K, author of the first antenna modeling software (*Annie*) I ever used, who has leveraged his unparalleled appreciation of James Clerk Maxwell's accomplishments into concrete projects to help preserve Maxwell's legacy for generations to come.

In amateur radio, those who provide assistance to newcomers to the hobby are known as *Elmers.* Some of us can point to a single such Elmer; others, to many. Long ago, I was fortunate to have many, virtually all of whom have since passed on. In the area of antennas, one deserves posthumous recognition:

Carl Herrling, ex-K2TVT, who helped a 13-year-old neophyte radio amateur and budding author construct and erect his very first antennas—including a three-wire dipole for 80 meters!

On the publishing side of this project I must mention:

Tom Nutt-Powell, an author in an entirely unrelated field, whose one-sentence motivational "mantra" (that I can't repeat here) kept me on track when it seemed as though I would never find the right words—or, for that matter, *any* words;

Ginny Carroll, whose efficiency, competence, and shared love of the English language made the final stages of the review and editing process a joy; and

Judy Bass, who enthusiastically carried the ball for this project within McGraw-Hill, and who worked with me to develop a schedule we all could live with.

Finally, my entire family has been enthusiastic beyond belief, but two members in particular deserve specific praise:

Nathan Dougall, my stepson, whose gift to me of the fourth edition of this book started this whole project, even as he embarked on his own book; and

Linda, my wife, who was instantly supportive at the onset of this project, and who has since administered the loving but firm prodding I periodically needed.

Bud Hippisley

PART I

Background and History

CHAPTER 1

Introduction to Radio Communications

Useful communication by radio has been with us for the entire twentieth century and shows no sign of abating in the twenty-first. Originally determined to be mathematically "possible" by the Scottish inventor James Clerk Maxwell in the mid-1860s, radio waves remained science fiction until 1887, when a German, Heinrich Hertz, was first to convert Maxwell's theoretical equations into laboratory experiments that demonstrated and confirmed their existence. Italian Guglielmo Marconi then picked up the ball and through the 1890s took the first productive steps toward making radio communication a popular and commercial success. Along the way, Nikolai Tesla from Serbia, Alexander Fessenden in Canada, Alexander Popov in Russia, and Lee De-Forest in the USA were also early pioneers in advancing the state of the art. Thus, the story of radio's inception is as international as the wireless long-distance communications it ultimately engendered.

By the turn of the century "wireless telegraphy" (as radio was called then) was sparking (pun intended) the imaginations of countless people across the world. Radio communications began in earnest, however, when Guglielmo Marconi successfully demonstrated wireless telegraphy as a commercially viable technology. The "wireless" aspect so radically changed communications that the word is still used in many countries of the world to denote all radio communications even though today in the United States the term *wireless* has come to be associated almost entirely with *WiFi*—the short-range, very short wavelength method of interconnecting computers and other digital devices. Until Marconi's successful transatlantic reception of signals near the end of 1901, and despite his earlier success at spanning the English Channel, wireless was widely seen as a neighborhood or cross-town endeavor of limited usefulness. Of course, although ships close to shore or each other could summon aid in times of emergency, the ability to communicate over truly long distances was absent. But all that changed on that fateful December day in Newfoundland when the Morse letter "S" tickled Marconi's ears.

Wireless telegraphy was immediately pressed into service by shipping companies because it provided an element of safety that had been missing in the pre-wireless days. In fact, some early shipping companies advertised that their ships were safer because of the new wireless installations aboard. It was not until 1909, however, that wireless telegraphy actually proved its usefulness on the high seas. Two ships collided in the foggy Atlantic Ocean and were sinking. All passengers and crew members of both ships were in imminent danger of death in the icy waters, but radio operator Jack Binns became the first man in history to send out a maritime distress call. His call for help in Morse code was received and relayed by nearby ships, allowing another vessel to come to the aid of the stricken pair of ships.

All radio communication prior to about 1916 was carried on via telegraphy (i.e., the on-off keying of a radio signal in the Morse code). But one night in 1916 there was more magic. Radio operators and monitors up and down the Atlantic seaboard—from the Midwest to the coast and out to sea for hundreds of miles—heard something that must have startled them out of their wits: crackling out of the "ether," amid the whining of Alexanderson alternators and the *zzzchht* of spark-gap transmitters, came a new sound—a human voice. Engineers and scientists at the Naval Research Laboratory in Arlington, Virginia, had transmitted the first practical *amplitude-modulated* (AM) radio signals. Earlier attempts (prior to 1910) had been successful as scientific experiments but did not employ commercially available equipment.

Although radio activity in the early years was unregulated and chaotic, today it is quite structured and (relatively) heavily regulated. At the international level, radio is coordinated by the International Telecommunications Union (ITU) in Geneva, Switzerland, based on treaties signed at World Administrative Radio Conferences (WARCs) held every few years. Of course, each sovereign country retains control of its own internal radio regulations. In the United States, for instance, all matters pertaining to radio communications are handled by the Federal Communications Commission (FCC), headquartered in Washington, D.C.

Amateur radio has grown from a few thousand "hams" prior to World War I to more than 900,000 today, about one-third of them in the United States. Amateurs were ordered off the air during World War I, and the hobby almost did not make a comeback after the war. There were, by that time, many powerful commercial interests greedily coveting the frequencies used by amateurs, and they almost succeeded in keeping post-war amateurs off the air. But the amateurs won the dispute—at least partially.

In those days, the frequencies with wavelengths longer than 200 m (i.e., 20 to 1500 kHz) were the ones considered valuable for communications. The cynical attitude attributed to the commercial interests regarding amateurs was "Put 'em on two hundred meters and below . . . they'll never get out of their backyards there!" But there was a surprise in store for those commercial operators, because the wavelengths shorter than 200 m are in the high-frequency region that we now call *shortwaves*. Today, those shortwaves are well-known for their ability to support communications over transcontinental distances, but in 1919 that ability was not suspected by anyone.

One of the authors of this book once heard an anecdote from an amateur radio operator "who was there". In the summer of 1921 this man owned a beautiful, large, wire "flattop" antenna array for frequencies close to 200 m on his family's farm in southwestern Virginia. Using those frequencies he was accustomed to regularly communicating several hundred miles into eastern Ohio and down to the Carolinas. Returning home at Thanksgiving of 1921 during his first semester at college, he noticed that his younger brother had replaced the long flattop array with a short dipole antenna. Confronting his brother over the incredible sacrilege, he was told that they no longer used 150 to 200 m, but were using 40 m instead. Everyone "knew" that 40 m was useless for communications over more than a few blocks, so (undoubtedly fuming) the older sibling took a turn at the key. He sent out a "CQ" (general call) and was answered by a station with a call sign similar to "8XX". Thinking that the other station was in the eighth U.S. call district (at the time covering West Virginia, Ohio, Michigan, and western New York), he asked him to relay a message to a college buddy in Cincinnati, Ohio. The other station replied in the affirmative, but suggested that ". . . you are in a better position to reach Cincinnati than I, I am *French* 8XX." (Call signs in 1921 did not yet include any system of national prefixes, such as we have today.) The age of international amateur communications had arrived! And with it came a new problem: National identifiers in call signs became necessary (which is why all U.S. call signs begin with K, W, N, or portions of the A block).

During the 1930s, radio communications and broadcasting spread like wildfire as technology and techniques improved. World War II became the first war to be fought with extensive electronics. Immediately prior to the war, the British developed a new weapon called RADAR (*RAdio Detection And Ranging*; now spelled *radar*). This tool allowed them to see and be forewarned of German aircraft streaming across the English Channel, planning to strike targets in the United Kingdom. The German planes were guided by (sophisticated for their times) wireless highways in the sky, while British fighters defended the home islands by radio vectoring from ground controllers. With night fighters equipped with the first *centimeter* (i.e., microwave) radar, the Royal Air Force was able to strike the invaders accurately—even at night. The first kill occurred one dark, foggy, moonless night when a *Beaufighter* closed on a spot in the sky where the radar in the belly of the plane said an enemy plane was flying. Briefly thinking he saw a form in the fog, the pilot cut loose a burst from his quad mount of 20-mm guns

FIGURE 1.1   This AM/FM broadcast tower bristles with two-way antennas.

slung in the former bomb bay. Nothing. Thinking that the new toy had failed, the pilot returned to base—only to be told ground observers had reported that a German *Heinkle* bomber fell from the overcast sky at the exact spot where the pilot had his ghostly encounter!

Radio, television, radar, and a wide variety of services are available today under the general rubric "radio communications and broadcasting". Increasing numbers of families now obtain their radio and television broadcasts via small *dishes* aimed at satellites in geosynchronous orbit 23,000 mi out in space. Inside, these same families network their home computers, audio/video equipment, and printers with wireless *nodes*. Amateur operators are able to communicate worldwide on low power and have even launched their own "OSCAR" satellites. And, of course, no summary of present-day radio communications would be complete without mentioning the ubiquitous cell phone.

Whether visible or hidden inside the consumer's device, antennas are required at both ends of the circuit for each of these communications systems. Thus, the antenna is arguably one of the most important parts of the receiving and/or transmitting station (Fig. 1.1). That is what this book is all about.

PART II

Fundamentals

CHAPTER 2

Radio-Wave Propagation

To intelligently choose the right antenna for our purposes we first need an understanding of how radio waves get from one place to another. Intuitively, radio signal propagation seems similar to light propagation; after all, light and radio signals are both electromagnetic waves. But the propagation of radio signals is not the simple matter that it seems at first glance: Simple inverse square law predictions, based on the optics of visible light, fall down drastically at many radio frequencies because other factors come into play—especially in the vicinity of earth.

In the *microwave* region of the spectrum, for instance, atmospheric pressure and water vapor content of the air through which a *terrestrial* electromagnetic wave moves become more important than for visible light. Clearly, those differences do not exist when comparing light and radio waves in free space. Similarly, near-earth microwave propagation differs from that of the lower VHF and HF bands. But perhaps the most amazing difference is found in the *medium-frequency* (MF) and *high-frequency* (HF) portions of the electromagnetic spectrum, where solar ionization of the upper reaches of the atmosphere provides a *refracting* layer that acts much like a mirror at these frequencies and hence is capable of supporting long-distance "skip" communications essential to such uses of radio as international broadcasting and amateur *DXing*.

This chapter examines radio propagation phenomena so that you have a better understanding of what an antenna is expected to accomplish and which design parameters are important to ensure the communications results that you desire.

Radio Waves

Today it is well recognized that radio signals travel in a wavelike manner, but that fact was not always so clear. It was well known in the first half of the nineteenth century that wires carrying electrical currents produced an *induction field* surrounding the wire that was capable of exerting a force over short distances. It was also known that this induction field is a magnetic field, and this knowledge formed the basis for the invention of electrical motors. But it was not until 1887 that German physicist Heinrich Hertz succeeded in demonstrating the ability to transmit and receive radio signals between two separate sets of equipment in his laboratory. In so doing, he confirmed experimentally Maxwell's theoretical work of a quarter century earlier predicting the existence of previously unimagined radio signals that, like light, were *electromagnetic waves*.

Like the induction field, the electromagnetic wave is created by an electrical current. Unlike the induction field, however, the electromagnetic wave requires a *changing* electric current. And once launched, the *radiated field* further differs from the induction field in that it is self-sustaining and no longer depends on the existence of a conductor or a current for its path or its amplitude. Instead, once it leaves the conductor, the radiated field propagates through space according to the universal equations governing wave motion of all kinds.

Wave propagation is easily visualized with a water-wave analogy. Although not a perfect analogy, it serves to illustrate the point. Figure 2.1 shows a body of water into which a ball is dropped (Fig. 2.1A). When the ball hits the water (Fig. 2.1B), it displaces water at its point of impact and pushes a leading wall of water away from itself. The ball continues to sink (Fig. 2.1C) and the wave propagates away from it until the energy is dissipated. Although Fig. 2.1 shows the action in only one dimension (a side view), the actual waves propagate outward in all directions, forming concentric circles when viewed from above.

The wave produced by a dropped ball does not last forever but, rather, is *damped*—i.e., it will decline in amplitude on successive crests until the energy is dissipated (by friction between adjacent water molecules) and the wave ceases to exist (Fig. 2.1D). To

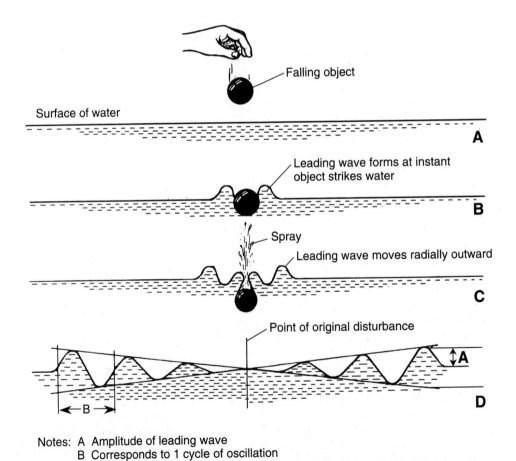

Notes: A Amplitude of leading wave
        B Corresponds to 1 cycle of oscillation

FIGURE 2.1   A ball dropped into water generates a wavefront that spreads out from the point of original disturbances.

make the analogy to radio waves more realistic, the wave must be generated in a continuous fashion, as shown in Fig. 2.2A: A ball is dipped up and down at a cyclic rate, successively reinforcing new wave crests on each dip. New waves continue to be generated at point A as long as the ball continues to oscillate up and down. The result is a *continuous wave train* spreading out through the water.

Perhaps the most important point to note about the water-wave analogy is that careful observation of individual water molecules (if that were possible) a small distance from the ball would show little or no net *horizontal* movement or displacement of any specific molecule sideways across the surface of the water. Nonetheless, a clearly visible wavefront travels outward from the point where the ball was dropped. Similarly, the motion of a radio wave does not require that electrons travel with it. To the contrary, we shall see in our discussion of basic antennas that maximum radiation occurs at right angles to the direction of travel of electrons along a wire antenna!

There are two fundamental properties of all waves that are important to our study of radio waves: *frequency (f)* and *wavelength (λ)*.The frequency is the number of oscillations (or cycles) in the wave's amplitude per unit of time. Since the fundamental unit of time is universally accepted to be the second, frequency is often expressed as the number of *cycles per second* (cps). If the time between successive positive peaks at any single point in space is 0.5 s, as suggested by Fig. 2.2A, exactly two complete amplitude cycles of the wave occur in 1 s, so the frequency of the wave created by the oscillating ball is 2 cps. Note that we can make the measurement from consecutive observable points elsewhere on the waveform, such as negative peaks, zero crossings, etc.

At one time, frequencies (along with the frequencies of other electrical, mechanical, and acoustical waves) were expressed in cycles per second but, in honor of Heinrich Hertz, the unit of frequency was renamed the *hertz* (Hz) many years ago. Since the units are equivalent (1 Hz = 1 cps), the wave in Fig. 2.2 has a frequency of 2 Hz.

Because the frequencies best suited for efficient radio propagation are so high, they are usually expressed in kilohertz (kHz = 1000s of Hz), megahertz (MHz = 1,000,000s of Hz), or gigahertz (GHz = 1,000,000,000s of Hz). Thus, the frequency of WBZ in Boston, operating near the mid-

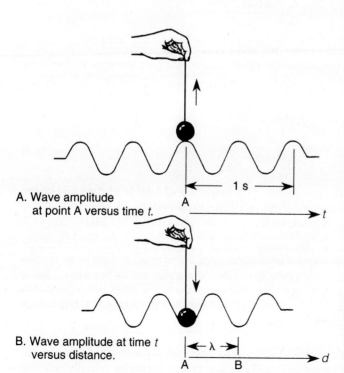

A. Wave amplitude at point A versus time *t*.

B. Wave amplitude at time *t* versus distance.

Figure 2.2 A bobbing ball on a string demonstrates continuous wave generation. (A) Wave amplitude at point A versus time *t*. (B) Wave amplitude at time *t* versus distance *d*.

dle of the AM broadcasting band, can be properly expressed as 1,030,000 Hz, 1030 kHz, or 1.03 MHz, all of which are equivalent and equally correct. While radio receiver tuning dials in North America are usually calibrated in kilohertz or megahertz, in Europe and the rest of the world it is not uncommon to find radio dials calibrated in meters, the universal unit of wavelength, as well as in frequency. In radio work, wavelength is expressed in *meters* or its subunits, such as when talking about *millimeter waves*.

The *wavelength* of any wave is the physical distance between adjacent corresponding points of a spatial waveform, measured at the exact same time. In Fig. 2.2B the wavelength (λ) is drawn as the distance between successive negative peaks in any direction from the source, but we could also measure the wavelength from the distance between successive positive peaks, or between any two corresponding features on successive waves as we move outward from the source in a straight line.

Wavelength is proportional to the reciprocal of the frequency. The relationship is given by $f\lambda = v$, where f is the frequency in Hz, λ is the wavelength in meters, and v is the *velocity of propagation* in meters per second (m/s). In each case, v depends on certain characteristics of the medium (water, air, etc.). In water at room temperature, the velocity of a sound wave is typically 1482 m/s, or about 4861 ft/s—just a bit less than a mile per second. In air, however, sound waves travel at about 343 m/s, or 1125 ft/s. In contrast, radio waves in free space propagate at the speed of light—approximately 300,000,000 m/s (or 186,000 mi/s) in free space and almost exactly the same in the earth's atmosphere. For historical reasons, the lowercase letter c is usually used to represent the speed of light. Thus:

$$f = \frac{c}{\lambda} = \frac{300,000,000}{\lambda} \tag{2.1}$$

where f is in hertz (Hz), λ is in meters (m), and c is in meters per second (m/s).

For convenience, these equations are often recast with frequency in kilohertz or megahertz:

$$f(\text{kHz}) = \frac{300,000}{\lambda} \tag{2.2}$$

$$f(\text{MHz}) = \frac{300}{\lambda} \tag{2.3}$$

Thus, a wavelength of 300 m corresponds to a frequency of 1 MHz (the middle of the AM broadcast band), while a wavelength of 10 m corresponds to 30 MHz (just above what we call the amateur 10-m band, which lies between 28.0 and 29.7 MHz).

NOTE *When a wave of any type—be it a mechanical wave, sound wave, or radio wave—passes from one linear medium to another, only its wavelength changes, not its frequency. Thus, a radio signal having a wavelength of 40 m in free space may have a wavelength of, say, only 30 m in one of the solid-dielectric transmission lines discussed in Chap. 4. A 1-kHz tone in the air will still be a 1-kHz tone in water.*

The place where the water-wave analogy falls down most profoundly is in the implied need for a *medium* of propagation. Water waves propagate by actually moving water molecules; that is, water is the *medium* through which the wave propagates. Similarly, sound travels to your ears by actually vibrating the molecules of the air or water between you and the guitar string or other source of the mechanical vibration. In a vacuum there is no sound! Maxwell's equations, on the other hand, make it clear that electromagnetic waves can exist in laboratory vacuum jars and in free space. Neither Maxwell nor other scientists of the late 1800s could conceive of the "action at a distance" in a vacuum predicted by his equations, so they invented a hypothetical medium called the *ether* (or *æther*) for transporting all electromagnetic waves, including radio waves and light, from one place to another. Despite a profoundly important experiment conducted in the late nineteenth century by American physicists Michelson and Morley, who conclusively disproved the existence of an ether, scientists, engineers, the press, and the lay public all continued to refer to (and probably believed in the need for) such a medium until at least the 1920s. Even today, radio enthusiasts still refer to the "stuff" out of which radio waves arrive as the "ether". The term, however, is merely an archaic, linguistic echo of the past.

The Electromagnetic Wave: A Brief Review

The study of electromagnetic wave radiation and transmission is typically a semester-long, calculus-intensive upper-class college course, necessarily preceded by a semester of basic electromagnetic field theory to develop a facility with Maxwell's equations. Here we shall limit ourselves to summarizing some of the key issues of electromagnetism central to our understanding of how antennas work. In other words, our goal here is to help you develop an intuitive "feel" for the generation and propagation of electromagnetic fields in the radio portion of the electromagnetic spectrum without subjecting you to advanced math (or the cost of a college degree).

- Radio waves are the result of accelerated electrical charges.

Acceleration is, by definition, the rate of change of a body's velocity, and velocity is the rate of change of a body's position. Electrons traveling at a constant velocity along a wire create a constant current in that wire. Thus, even though stationary electric charges or constant (or direct) currents may create static or slowly changing electric or magnetic fields, they do not generate radio waves.

Any alternating current in a wire or other conductor has the potential to radiate, but a few hurdles have to be overcome. For one thing, the acceleration undergone by electrons in response to the application of an alternating voltage is proportional not only to source amplitude but also to the frequency of the source. Thus, high-frequency circuits radiate a lot more easily than low-frequency ones. Further, accelerated charges must also be hosted by a physical structure that promotes EM radiation by forcing, if you will, the electric and magnetic fields to leave the vicinity of the radiating structure. (That's what distinguishes an efficient antenna from a lumped circuit component or a transmission line, and why antenna geometry is so important.)

- Electromagnetic radiation as we know it depends for its existence on a *finite* speed of light.

Even though the speed of light (*c*) is extremely high, it's still a finite number. It takes, for instance, about 8 minutes for light emitted from the surface of our sun to reach us (93,000,000 mi/186,000 mi/s = 500 s, or about 8 minutes). The radiated wave from an accelerated electron exists as a real entity, carrying its own energy from source to receiver, only because there is a finite time delay to a distant point before the pre- and postacceleration fields from the electron reach that point.

- Radio waves are fundamentally different from static or quasi-static charge fields and induction fields.

Many of us have experienced a variety of high school experiments helping us to visualize the strength and spatial distributions of electric and magnetic fields. (Iron filings on a sheet of paper with a bar magnet underneath, for instance.) Beyond helping us understand that invisible "entities" called *fields* exist, these quasi-static fields have little in common with radio waves. Specifically:

- Electrostatic charge fields are radially directed outward from the charge sources, like needles from a porcupine's body. Static magnetic fields wrap around a straight wire carrying a constant current. A static electric field can exist without a corresponding magnetic field, and vice versa. In contrast, a radio wave requires the simultaneous existence of linked time-varying electric and magnetic fields at right angles to each other.

- In the absence of radiation, electric field lines must start and end on charged bodies. But the field lines for radio waves become detached from their source and travel through free space closed upon themselves.

- An antenna is a physical conducting structure deliberately designed to facilitate or "encourage" the departure into space of *linked* time-varying electric and magnetic fields originating in the accelerated charges (resulting from time-varying currents) on the conductor. Sometimes poorly designed or laid-out circuits become *unintentional radiators*.

- A radio wave propagates outward from the radiator with its electric and magnetic fields at right angles to the direction of wavefront motion and to each other. In "techie-speak", radio waves propagating through free space are *transverse electromagnetic* (TEM) *waves*, consisting of mutually perpendicular *electric* and *magnetic fields* (see Fig. 2.3) varying and traveling together in synchronism.

- The strength of static and quasi-static electric and magnetic fields falls off as $1/r^2$, where r is the distance from the source to a receiver at a distant point. In sharp contrast, the strength of a TEM radio wave falls off according to $1/r$. In other words, we can detect the radio wave from accelerating charges much, much farther away than we can detect the effects of the static charge distributions (such as by observing iron filings around a magnet).

- Radio waves are exactly like light—whether *visible, infrared* (IR), or *ultraviolet* (UV)—and other free-space radiation except for their frequency. Radio waves occupy the lowest frequencies of the entire electromagnetic (EM) spectrum; hence, they have much longer wavelengths than the other forms of radiation we encounter.

- Radio waves are real, not just artificial or abstract concepts to help our visualization. Radio waves carry energy with them, expanding outward in

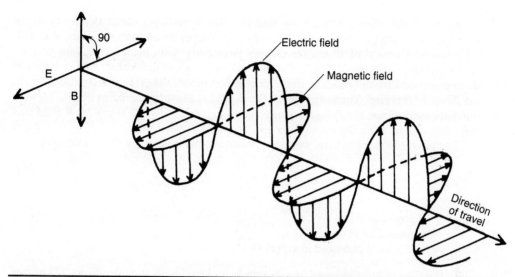

FIGURE 2.3   A transverse electromagnetic (TEM) wave consists of electric and magnetic fields at right angles to each other and to the direction of wave propagation.

all directions from the source. When a radio wave encounters a receiving antenna, it delivers energy to that antenna and to any associated load. If the load impedance includes a real, or resistive part, real power is dissipated in it—power that can only have come from the arriving wave!

Radio-Wave Intensity

In free space, a radio wave expanding outward over an ever-growing spherical surface centered on the source (accelerated electrons on a conductor) neither gains nor loses energy. Thus, with the passage of time and distance, the original amount of energy is spread over a larger and larger surface (the wavefront). A helpful analogy is to visualize the process of blowing up a balloon. The total energy in the wavefront is analogous to the total rubber in the skin of the balloon. Of course, as you blow into the balloon, the balloon expands and the skin becomes thinner. Similarly, as the radio wavefront occupies a larger and larger spherical surface, the power per unit area of the surface anywhere on the sphere is smaller and smaller.

Thus, even though there are no obstacles in free space to dissipate the energy in the wavefront, if we could measure the amplitude of the passing wavefront at points progressively farther from the source we would see that it decreases with distance from the source. In common radio lingo, we say the radio wave is *attenuated* (i.e., reduced in apparent power) as it propagates from the transmitter to the receiver, but this can be misleading unless we understand that the so-called attenuation we are referring to is not the same as when we attenuate a signal in a resistive voltage divider. Because a fixed amount of power in the wavefront is spread over a larger and larger surface area, the amplitude of the wave is inversely proportional to the total area of the sphere's surface. That area is, of course, proportional to the *square* of the radius of the surface from the source, so radio waves at all frequencies suffer losses according to the *inverse square law*. Let's take a look at that phenomenon.

Electric fields (E-fields) are measured in units of volts per meter (V/m), or in the subunits millivolts per meter (mV/m) or microvolts per meter (μV/m). Thus, if a vertically oriented E-field of 10 mV/m crosses your body from head to toe, and you are about 2 m tall, a voltage of (2 m) × (10 mV/m), or 20 mV, is generated. As a radio wave travels outward from its source, the electric field vector falls off in direct proportion to the distance traveled. The reduction of the E-field is linearly related to distance (i.e., if the distance doubles, the E-field voltage vector halves). Thus, a 100-mV/m E-field that is measured 1 mi from the transmitter will be 50 mV/m at 2 mi.

The power in any electrical system is related to the voltage by the relationship

$$P = \frac{V^2}{R} \tag{2.4}$$

where P = power in watts (W)
R = resistance in ohms (Ω)
V = electrical potential in volts (V)

In the case of a radio wave, the R term is replaced with the impedance (Z) of free space, which happens to be 377 Ω. If the E-field intensity is, for example, 10 V/m, then the *power density* of the signal is

$$P = \frac{(10 \text{ V/m})^2}{377\Omega} = 0.265 \text{ W/m}^2 = 26.5 \text{ mW/m}^2 \tag{2.5}$$

The power density, measured in watts per square meter (W/m²) or subunits thereof, falls off according to the square of the distance, for the "stretched balloon" reasons described previously. This phenomenon is shown graphically in Fig. 2.4. Here, you can see a lamp emitting a light beam that falls on a surface (A), at distance L, with a given intensity. At another surface (B) that is $2L$ from the source, the same amount of energy is

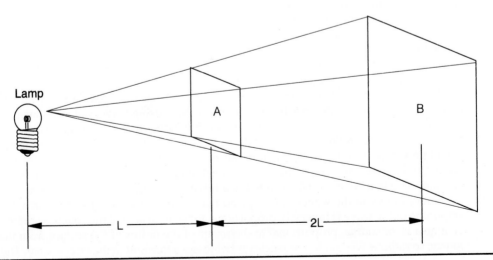

FIGURE 2.4   Wave propagation obeys the inverse square law.

distributed over an area (B) that is *four times* as large as area A. Thus, the power density falls off according to $1/d^2$, where d is the distance from the source. This is called the *inverse square law.*

An advancing wave can experience additional reductions in amplitude when it passes through matter. In this case, true dissipative attenuation may indeed occur. For instance, at some very high microwave frequencies there is additional path loss as a result of the oxygen and water vapor content of the air surrounding our globe. At other frequencies, losses originating in other materials can be found. These effects are highly dependent on the relationship of molecular distances to the wavelength of the incident wave. For many frequencies of interest, the effect is extremely small, and we can pretend radio waves in our own atmosphere behave almost as though they were in the vacuum of free space. Many materials, such as wood, that are opaque throughout the spectrum of visible light are fundamentally transparent at most, if not all, radio frequencies.

Isotropic Sources

In dealing with both antenna theory and radio-wave propagation, a totally fictitious device called an *isotropic source* (or *isotropic radiator*) is sometimes used for the sake of comparison, and for simpler arithmetic. You will see the isotropic model mentioned several places in this book. An isotropic source is a very tiny spherical source that radiates energy equally well in all directions. The radiation pattern is thus a perfect sphere with the isotropic antenna at the center. Because such a spherical source generates uniform output in all directions, and its geometry is easily determined mathematically, the signal intensities at all points can be calculated from basic geometric principles. Just don't forget: Despite all its advantages, there is no such thing in real life as an isotropic source!

For the isotropic case, the average power in the extended sphere is

$$P_{av} = \frac{P_t}{4\pi d^2} \tag{2.6}$$

where P_{av} = average power per unit area on surface of spherical wavefront
$\quad\quad P_t$ = total power radiated by source
$\quad\quad d$ = radius of sphere in meters (i.e., distance from radiator to point in question)

The *effective aperture* (A_e) of a receiving antenna is a measure of its ability to collect power from the EM wave and deliver it to the load. Although typically smaller than the surface area of a real antenna, for the theoretical isotropic case, $A_e = \lambda^2/4\pi$. The power delivered to the load is then the incident power density times the effective collecting area of the receiving isotropic antenna, or

$$P_L = P_{av} A_e \tag{2.7}$$

By combining Eqs. (2.6) and (2.7), the power delivered to a load at distance d is given by

$$P_L = \frac{P_t \lambda^2}{(4\pi)^2 d^2} \tag{2.8}$$

where P_L = power to load

λ = wavelength (c/f) of signal

From these expressions, we can then derive an expression for ordinary free-space path losses between an isotropic transmitter antenna and a receiver antenna (see App. A for an explanation of logarithms):

$$L = 10 \log \left(\frac{P_t}{P_L} \right) \tag{2.9}$$

or, by rearranging to account for individual terms:

$$L = [20 \log d] + [20 \log F] + k \tag{2.10}$$

where L = path loss in decibels (dB)

d = path length

F = frequency in megahertz (MHz)

k is a constant that depends on the units of d as follows:

k = 32.4 if d in kilometers

= 36.58 if d in statute miles

= 37.80 if d in nautical miles

= –37.87 if d in feet

= –27.55 if d in meters

The radiated sphere of energy gets ever larger as the wave propagates away from the isotropic source. At a great distance from the center, a small slice of the advancing wavefront is essentially a flat plane, as in Fig. 2.5. (This situation is analogous to the apparent flatness of the prairie in the midwestern United States, even though the surface of the earth is really spherical.) If you could see the electric and magnetic field vectors, they would appear to be at right angles to each other in the flat-plane wavefront.

The polarization of an EM wave is, by definition, the direction or orientation of the electric field with respect to some agreed-upon reference. In most cases, the reference is the earth's surface, although that may well be meaningless for antennas in outer space. In

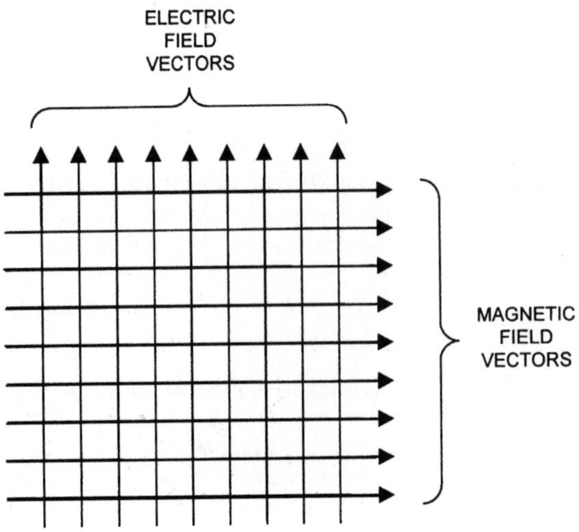

ELECTRIC
FIELD
VECTORS

MAGNETIC
FIELD
VECTORS

Figure 2.5 A spherical wavefront far from the source can be treated as a plane wave with E- and H-fields at right angles to each other.

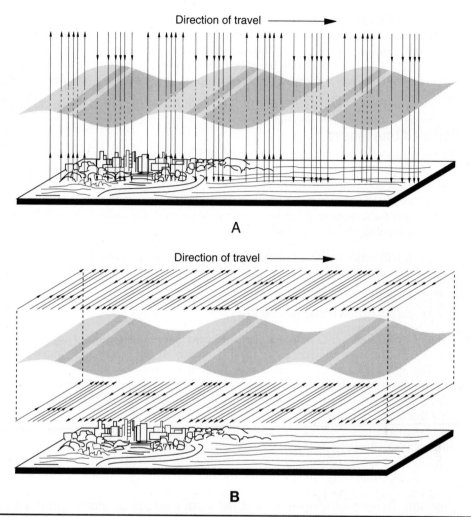

FIGURE 2.6 Wave polarization is determined by the direction of the electric field lines of force. (A) Vertically polarized electromagnetic wave. (B) Horizontally polarized wave.

the case of small mobile antennas, the reference is usually the closest large body of conducting material—such as the roof of a car. Figure 2.6A is an example of vertical polarization because the E-field is vertical with respect to the earth, the assumed reference plane. If the fields are interchanged (as in Fig. 2.6B), the EM wave is said to be horizontally polarized.

As mentioned earlier in this chapter, an EM wave travels at the speed of light, designated by the letter c, which is about 300,000,000 m/s (or 186,000 mi/s). To put this velocity in perspective, a radio signal originating on the sun's surface will reach earth in about 8 minutes. The velocity of the wave slows in dense media, but in air the speed is so close to the "free-space" value of c that the same velocity is used for both air and the near-perfect vacuum of outer space when solving most practical problems. In pure water, which is much denser than air, the speed of radio signals is about one-ninth that of the free-space speed. (Contrast this with sound waves, which travel almost four times faster in water than in air!) This same phenomenon shows up in the basis for the *velocity factor* (v_F) of transmission lines. In foam dielectric coaxial cable, for example, the value of v_F is 0.80, which means a signal propagates along the line at a speed of 0.80c, or 80 percent of the speed of light.

The Earth's Atmosphere

Electromagnetic waves do not need an atmosphere in order to propagate, as Marconi and his contemporaries would have immediately realized if they could have witnessed today's data communications between NASA's ground stations and their interplanetary probes in the near vacuum of outer space. But when a radio wave does propagate in the earth's atmosphere, it interacts with it. In general, the radio wave will be subject to three potential effects as a result:

- It will suffer increased attenuation (relative to normal free-space path loss over a comparable distance).

- Its path may be bent or redirected.

- Its polarization may be altered.

All of these effects vary with frequency. In addition, different regions of the atmosphere, which consists largely of oxygen (O_2) and nitrogen (N) gases, play differing roles:

The atmosphere (Fig. 2.7) consists of three major regions: *troposphere, stratosphere,* and *ionosphere*. The boundaries between these regions are not very well defined, and they change both *diurnally* (over the course of a day) and seasonally. Keep in mind, also, that the distinctions between these regions are a matter of definition of gas densities and behaviors by groups of scientists, not the result of some fundamentally profound differences in the gases that inhabit each.

The troposphere occupies the space closest to the earth's surface, extending upward to an altitude of 6 to 11 km (4 to 7 mi—roughly the cruising altitude of most jet airliners). The temperature of the air in the troposphere varies with altitude, becoming considerably lower than temperatures at the earth's surface as altitude increases. For example, a +10°C surface temperature could exist simultaneously with –55°C at the upper edges of the troposphere.

The stratosphere begins at the upper boundary of the troposphere (6 to 11 km) and continues upward to the beginning of the ionosphere (≈50 km). The stratosphere is called an *isothermal region* because the temperature in this region is relatively constant across a wide range of altitudes.

The ionosphere occupies the region between altitudes of about 50 km (31 mi) and 300 km (186 mi). It is a region of very low gas density because it is the portion of our atmosphere where the earth's gravity has the least "pull" on individual molecules. Beyond the ionosphere, our gravity system is so weak that any air molecules that stray there ultimately drift off into space, never to be seen again.

Of particular interest to our study of high-frequency propagation is the fact that a mixture of cosmic rays, electromagnetic radiation (including ultraviolet light from the sun), and atomic particle radiation from space (most of these from the sun also)—all impinging on the outermost edges of our atmosphere—has sufficient energy to strip

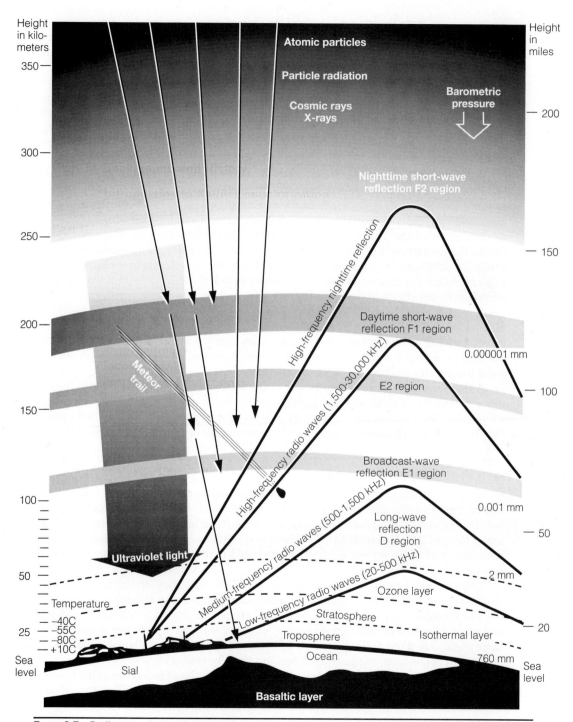

FIGURE 2.7 Radio waves in the atmosphere and some external radiation sources.

electrons away from the gas molecules of the ionosphere. These freed electrons are *negative ions*, while the O_2 and N molecules that lost the electrons become *positive ions*. The density of the air is quite low at those altitudes, so each free electron can travel a long distance before bumping into a positive ion, at which point they neutralize each other's electrical charge by *recombining*.

Ionization does not occur at lower altitudes—i.e., in the troposphere and stratosphere—partly because much of the incoming radiation is blocked before it can reach that far and partly because the air density is so much greater at lower altitudes that the positive and negative ions are more numerous and closer together, and recombination occurs rapidly.

Because a large percentage of the total radiation "raining" on our atmosphere comes from the sun, ionization levels in the ionosphere vary with the time of day, with the season (since the earth is farther from the sun during the northern hemisphere's summer), and with the solar radiation levels, both short term and long term.

Ionization of the upper level of the earth's atmosphere by radiation from space causes the ionosphere to have electrical characteristics not shared by the lower levels. While the details are beyond the scope of this book, the net effect is to set up the possibility of electrical interaction between the ionosphere and radio waves that reach it. Not surprisingly, the effects of the interactions are frequency dependent.

A second effect of radiation from space is to alter the characteristics of bands of magnetism (so-called magnetic belts) encircling our globe. When these bands are disturbed by excessively high cosmic radiation, they alter the magnetic fields surrounding the earth, causing disruption of normal EM propagation. As with ionization effects, some frequencies are affected more than others.

As we shall see, the ionization of our upper atmosphere is a major factor in the extreme variability of medium- and high-frequency radio-wave propagation. It is, quite simply, the "stuff" of magic for those of us who have been mesmerized by the unpredictability of long-distance terrestrial radio communications on those bands.

EM Wave Propagation Phenomena

If you have ever studied optics, you know that the path and polarization of visible light can be modified through *reflection, refraction, diffraction, and dispersion.* Radio waves (which, like visible light, are electromagnetic waves) can be affected the same way. Figures 2.8A and 2.8B illustrate some of the wave behavior phenomena associated with both light and radio waves. All four effects listed here play important roles in radio propagation.

Reflection and refraction are shown in Fig. 2.8A. *Reflection* occurs when a wave strikes a denser reflective medium, such as when a light wave in air strikes a glass mirror. The incident wave (shown as a single ray) strikes the interface between less dense and more dense media at a certain *angle of incidence* (a_i), and is reflected at exactly the same angle, called the *angle of reflection* (a_r). Because these angles are equal, a light beam or radio signal undergoing pure reflection can often be traced back to its origin.

If the incoming light wave arrives at certain other angles of incidence at the boundary between media of two different densities, *refraction* is the result. The amount and direction of the change are determined by the ratio of the densities between the two media. If Zone B is much different from Zone A, the bending is pronounced. In radio systems, the two media might be different layers of air with different densities. It is pos-

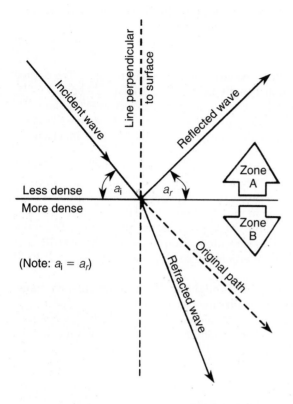

FIGURE 2.8A   Reflection and refraction phenomena.

sible for both reflection and refraction to occur in the same system. Indeed, more than one form of refraction might be present. More on these topics later.

Diffraction is shown in Fig. 2.8B. In this case, an advancing wavefront encounters an opaque object (e.g., a steel building). The shadow zone behind the building is not simply perpendicular to the wave but takes on a cone shape as waves bend around the object. The "umbra region" (or diffraction zone) between the shadow zone ("cone of silence") and the direct propagation zone is a region of weak (but not zero) signal strength. In practical situations, signal strength in the cone of silence rarely reaches zero. Also, it is not uncommon for a certain amount of reflected signal scattered from other surfaces to fill in the shadow a little bit. The degree of diffraction effect seen in any given case is a function of the wavelength of the signal relative to the dimensions of the object and the object's electromagnetic properties.

Dispersion can be seen in the spreading of a beam of white light into its separate colors, such as happens with a prism. It is a direct result of the angle of refraction for a

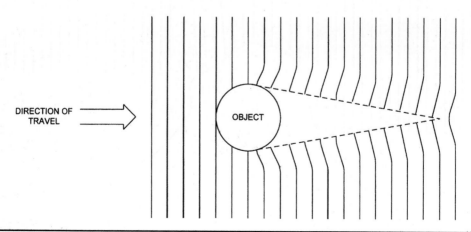

FIGURE 2.8B   Diffraction phenomena.

medium varying with frequency. We will have occasion to discuss dispersion further when we discuss modes of *fading* over propagation paths.

Diffraction Phenomena

Electromagnetic waves *diffract* when they encounter a radio-opaque object. The degree of diffraction and the resulting harm or benefit are frequency related. Above 3 GHz, wavelengths are so small (approximately 10 cm) compared to object sizes that large attenuation of the signal occurs. In addition, beamwidths (a function of antenna size compared with wavelength) tend to be small enough above 3 GHz that blockage of propagation by obstacles is much more effective.

Earlier in this chapter, large-scale diffraction around structures (such as buildings) was discussed in conjunction with Fig. 2.8B, a top-down view of diffraction effects in the horizontal plane. But there is also a diffraction phenomenon in the vertical plane. Terrain or man-made objects intervening in the path between UHF microwave stations (Fig. 2.9A) cause diffraction, along with the concomitant signal attenuation. There is a minimum clearance required to prevent severe attenuation (more than 20 to 30 dB, say) from diffraction. Calculation of the required clearance comes from Huygens-Fresnel wave theory.

Consider Fig. 2.9B. A wave source *A*, which might be a transmitter antenna, transmits a wavefront to a destination *C* (receiver antenna). At any point along path *A-C*, you can look at the wavefront as a partial spherical surface (B_1-B_2) on which all wave rays have the same phase. This plane can be called an *isophase plane*. You can assume that the d_n/d_h refraction gradient over the height extent of the wavefront is small enough to be considered negligible.

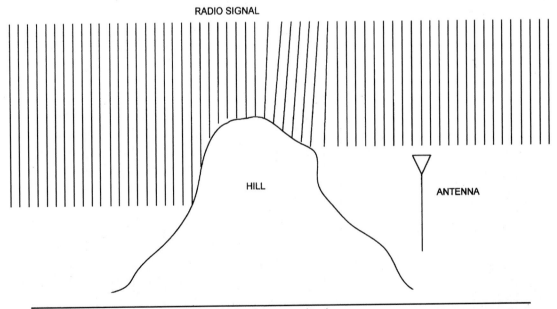

FIGURE 2.9A   Terrain masking of VHF and higher-frequency signals.

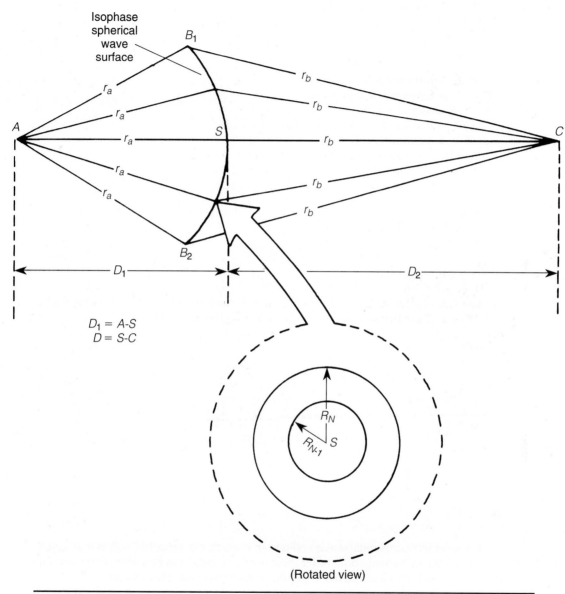

FIGURE 2.9B   Fresnel zone geometry.

Using ray tracing, we see rays r_a incoming to plane $[B_1\text{-}B_2]$ and rays r_b outgoing from plane $[B_1\text{-}B_2]$. The signal seen at C is the algebraic sum of all rays r_b. The signal pattern will have the form of an optical interference pattern with wave cancellation occurring between r_b waves that are a half wavelength apart on $[B_1\text{-}B_2]$. The ray impact points on plane $[B_1\text{-}B_2]$ form radii R_n called *Fresnel zones*. The lengths of the radii are a function of the frequency and the ratio of the distances D_1 and D_2 (see Fig. 2.9B). The general expression is

$$R_n = M \sqrt{\frac{N}{F} \left(\frac{D_1 D_2}{D_1 + D_2} \right)} \qquad (2.11)$$

where R_n = radius of nth Fresnel zone
F = frequency in gigahertz
D_1 = distance from source to plane B_1-B_2
D_2 = distance from destination to plane B_1-B_2
N = an integer $(1, 2, 3 \ldots)$
M = constant of proportionality equal to:
   17.3 if R_m is in meters and D_1, D_2 are in kilometers
   72.1 if R_m is in feet and D_1, D_2 are in statute miles

If you first calculate the radius (R_1) of the critical first Fresnel zone, you can calculate the nth Fresnel zone from

$$R_n = R_1 \sqrt{n} \qquad (2.12)$$

Example 2.1   Calculate the radius of the first Fresnel zone for a 2.5-GHz signal at a point that is 12 km from the source and 18 km from the destination.

Solution

$$R_1 = M \sqrt{\frac{N}{F} \left(\frac{D_1 D_2}{D_1 + D_2} \right)}$$

$$= (17.3) \sqrt{\frac{1}{2.5} \left(\frac{12 \times 18}{12 + 18} \right)}$$

$$= (17.3) \sqrt{(0.4)(7.2)}$$

$$= (17.3)(2.88)$$

$$= 29.4 \text{ m}$$

For most terrestrial microwave systems an obstacle clearance of 0.6 R_1 is required to prevent diffraction attenuation under most normal conditions. However, there are conditions in which the clearance zone should be more than one Fresnel zone.

Propagation Paths

There are four major propagation paths for a radio wave emanating from a terrestrial source: *surface wave, space wave, tropospheric,* and *ionospheric.* The ionospheric path is a major factor in medium-wave (MW) and HF propagation, but serves primarily only as a potential source of loss for VHF, UHF, or microwave propagation to/from space-based antennas. Ionization and recombination phenomena in the ionosphere add to the noise level experienced at VHF, UHF, and microwave frequencies. In satellite communications, there are some additional trans-ionospheric effects.

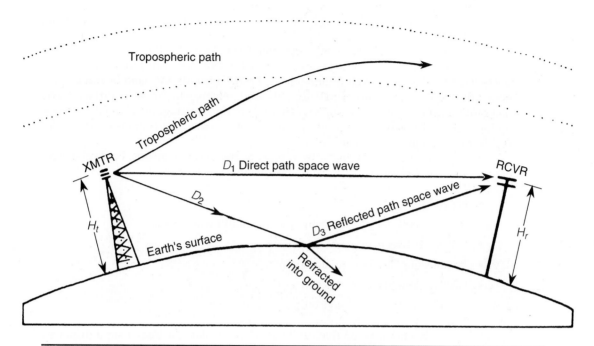

FIGURE 2.10 Space-wave propagation.

The space wave and surface wave are both *ground waves,* but they behave differently enough to warrant separate consideration. The surface wave travels in direct contact with the earth's surface and it suffers a severe frequency-dependent attenuation caused by continual contact with, and absorption by, the earth.

The space wave is also a ground-wave phenomenon, but it is radiated from an antenna a wavelength or more above the surface. No part of the space wave normally travels in contact with the surface; VHF, UHF, and microwave signals are usually space waves. There are, however, two components of the space wave in many cases: *direct* and *reflected* (see Fig. 2.10).

The tropospheric wave is lumped with the direct space wave in some texts, but it has unique properties in certain practical situations. The troposphere is the region of the earth's atmosphere between the surface and the stratosphere, or about 4 to 7 mi above the surface. Thus, most forms of ground wave propagate in the troposphere. But because certain propagation phenomena (caused mostly by weather conditions) occur only at higher altitudes, tropospheric propagation should be differentiated from other forms of ground wave.

Ground-Wave Propagation

The *ground wave,* naturally enough, travels along the ground, or at least in close proximity to it. There are three basic forms of ground wave: *space wave, surface wave,* and *tropospheric wave.* The space wave does not actually touch the ground. As a result, space-wave attenuation as a function of distance in clear weather is about the same as in free space (except above about 10 GHz, where H_2O and O_2 absorption increases dramatically). Of course, above the VHF region, weather conditions add attenuation not found in outer space.

The *surface wave* is subject to the same attenuation factors as the space wave, but it also suffers ground losses. These losses are caused by ohmic resistive losses in the conductive earth. Bluntly stated, the signal heats up the ground! AM band broadcast stations utilize ground-mounted vertically polarized antennas to radiate a surface wave that provides good signal strength throughout their local coverage area but which relies on these ground losses to avoid interference to other broadcast stations on the same frequency in communities farther away.

Surface-wave attenuation increases rapidly as frequency increases. For both forms of ground wave, reception is affected by the following factors:

- *Wavelength*
- *Height* of both the receiving and the transmitting antennas
- *Distance* between antennas
- *Terrain* along the transmission path
- *Weather* along the transmission path
- *Ground losses* (surface wave only)

Figure 2.11 is a nomograph that can be used to calculate the line-of-sight distances in miles over a spherical earth, given a knowledge of both receiving and transmitting antenna heights. Similarly, Figs. 2.12A and 2.12B show power attenuation with frequency and distance (Fig. 2.12A), and power attenuation in terms of field intensity (Fig. 2.12B).

Ground-wave communication also suffers another difficulty, especially at VHF, UHF, and microwave frequencies. The space wave is like a surface wave, but it is radiated from an antenna at least a wavelength above the earth's surface. It consists of two components (see Fig. 2.10 again): the *direct* and *reflected* waves. If both of these components arrive at the receiving antenna, they will add algebraically (more accurately, *vectorially*) to either increase or decrease signal strength. There is always a phase shift between the two components because the two signal paths have different lengths (i.e., D_1 is less than $D_2 + D_3$). In addition, there may possibly be a 180-degree (π radians) phase reversal at the point of reflection (especially if the incident signal is horizontally polarized). If the overall phase shift experienced by the reflected wave as a result of path length differences and polarization shifts is an odd multiple of 180 degrees at the receiving antenna, as shown in Fig. 2.13, the net signal strength there will be reduced (*destructive interference*). Conversely, if the direct and reflected signals arrive in phase, the resulting signal strength is increased (*constructive interference*). The degree of cancellation or enhancement will, of course, depend on the relative amplitudes of the two signal components as they arrive at the receive antenna.

The loss of signal over path D_1 can be characterized with a parametric term n that is defined as follows:

$$n = \frac{S_r}{S_f}$$

(2.13)

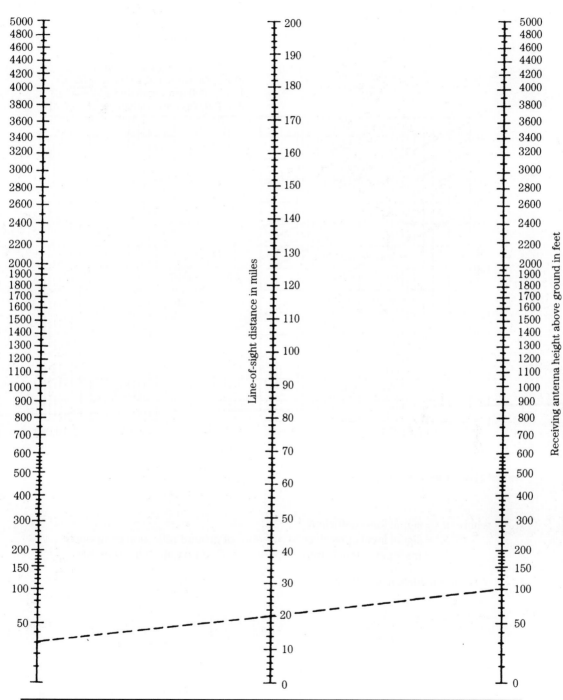

FIGURE 2.11   Nomograph of line-of-sight transmission distance as a function of receiving and transmitting antenna heights.

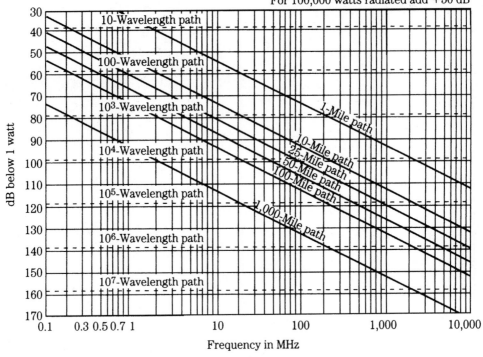

FIGURE 2.12A Power in a free-space field (normalized to 1 W).

where n = signal loss coefficient
   S_r = signal level at receiver in presence of ground reflection component
   S_f = free-space signal strength over path D_1 if no reflection takes place

You can calculate n as follows:

$$n^2 = 4\sin^2\left(\frac{2\pi h_t h_r}{\lambda D_1}\right) \tag{2.14}$$

or

$$n = 2\sin\left(\frac{2\pi h_t h_r}{\lambda D_1}\right) \tag{2.15}$$

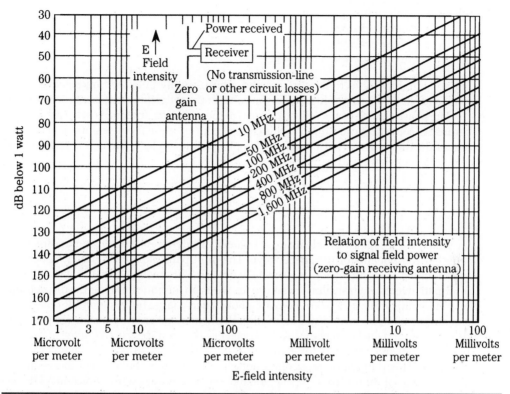

FIGURE 2.12B   Relation of field strength to signal field power.

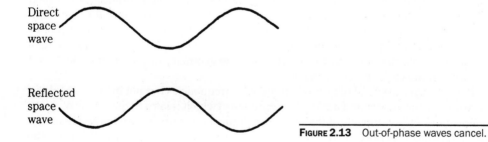

FIGURE 2.13   Out-of-phase waves cancel.

where h_t and h_r are the heights of the transmitting and receiving antennas, respectively.

The reflected signal contains both amplitude change and phase change. The phase change is typically π radians (180 degrees). The amplitude change is a function of frequency and the nature of the reflecting surface. The *reflection coefficient* can be characterized as

$$\gamma = pe^{j\phi} \tag{2.16}$$

where γ = reflection coefficient
 p = amplitude change
 φ = phase change
 j = imaginary operator ($\sqrt{-1}$)

For smooth, high-reflectivity surfaces and a horizontally polarized microwave signal that has a shallow angle of incidence, the value of the reflection coefficient is close to –1.

The phase change of the reflected signal at the receiving antenna is at least π radians because of the reflection. Added to this change is an additional phase shift that is a function of the difference in path lengths. This phase shift can be expressed in terms of the two antenna heights and path length:

$$s = \pi + \left(\frac{4\pi h_t h_r}{\lambda D_1} \right) \tag{2.17}$$

Multipath reception problems exist because of interference between the direct and reflected components of the space wave. The multipath phenomenon that is perhaps most familiar to many readers is *ghosting* of over-the-air analog television signals. Some multipath events are transitory in nature (as when an aircraft flies over the direct transmission path), while others are permanent (as when a large building or hill reflects the signal). In mobile communications, multipath phenomena are responsible for reception dead zones and picket fencing. A *dead zone* exists when destructive interference between direct and reflected (or multiply reflected) waves drastically reduces signal strengths. This problem is most often noticed at VHF and above when the vehicle is stopped; the solution is to move the antenna one half wavelength (which at VHF and UHF may be a matter of a few inches). *Picket fencing* occurs as a mobile unit moves through successive dead zones and signal enhancement (or normal) zones; it is hard to describe via written word, but it sometimes sounds like a series of short noise bursts.

At VHF, UHF, and microwave frequencies, the space wave is limited to so-called line-of-sight distances. The horizon is theoretically the limiting distance for communications, but the radio horizon is actually about 15 percent farther than the optical horizon (Fig. 2.14). This phenomenon is caused by refractive bending in the atmosphere around the curvature of the earth, and it makes the geometry of the situation look as if the earth's radius is ⁴⁄₃ its actual radius.

Refraction occurs at VHF through microwave frequencies but not in the visible light spectrum because water and atmospheric pressure (which relates to the effects of atmospheric gases on microwave signals) become important contributors to the phenomenon. The *K factor* expresses the degree of curvature along any given path, while the

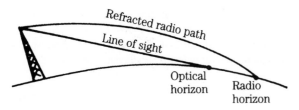

FIGURE 2.14   Radio horizon versus optical horizon.

index of refraction n measures the differential properties between adjacent zones in the atmosphere.

The K factor, also called the *effective earth's radius factor,* is determined by the relationship of two hypothetical concentric spheres, both centered on the earth's center. The first sphere is an idealization of the earth's surface, which has a nominal radius $r_o = 3440$ *nautical miles* (nmi) or 6370 km. The second sphere is larger than the first by the curvature of the signal *ray path*, and has a radius r. The value of K is then approximately

$$K = \frac{r}{r_o} \tag{2.18}$$

A value of $K = 1$ indicates a straight path (Fig. 2.15); a value of $K > 1$ indicates a positively curved path (refraction); and a value of $K < 1$ indicates a negatively curved path (subrefraction). The actual value of K varies with local weather conditions, so one can expect variation not only between locations but also seasonally. In the Arctic regions, K varies approximately over the range 1.2 to 1.34. In the "lower 48" of the United States, K varies from 1.25 to 1.9 during the summer months (especially in the south and southeast), and from 1.25 to 1.45 in the winter months.

The index of refraction n can be defined in either of two ways, depending on the situation. When a signal passes across boundaries between adjacent regions of distinctly different properties (as occurs during temperature inversions, for instance), the index of refraction is the ratio of the signal velocities in the two regions. In a homogeneous region n can be expressed as the ratio of the free-space velocity c to the actual velocity υ in the atmosphere:

$$n = \frac{c}{\upsilon} \tag{2.19}$$

At the surface, near sea level, under standard temperature and pressure conditions, the value of n is approximately 1.0003, and in homogeneous atmospheres it will decrease by 4×10^{-8} per mile of altitude. The units of n are a bit cumbersome in equations,

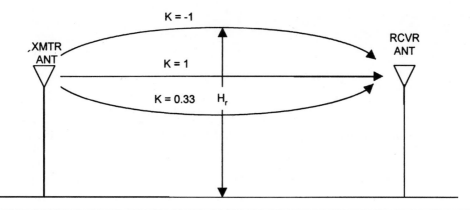

FIGURE 2.15   Refraction changes path length.

so the UHF and microwave communities tend to use a derivative parameter, N, called the *refractivity of the atmosphere*:

$$N = (n-1) \times 10^6 \tag{2.20}$$

N tends to vary from about 280 to 320 and, since n varies with altitude, so will N. In nonhomogeneous atmospheres (the usual case), these parameters will vary approximately linearly for several tenths of a kilometer. All but a few microwave relay systems can assume an approximately linear reduction of n and N with increasing altitude, although airborne radios and radars cannot. There are two methods for calculating N:

$$N = \left(\frac{77.6}{T}\right)\left(P + \frac{4810 e_s H_{rel}}{T}\right) \tag{2.21}$$

and

$$N = \frac{77.6}{T}\left(\frac{3.73 \times 10^5 e_s}{T^2}\right) \tag{2.22}$$

where P = atmospheric pressure in millibars (1 torr = 1.3332 mbar)
T = temperature in kelvins
e_s = saturation vapor pressure of atmospheric water in millibars
H_{rel} = relative humidity expressed as a decimal fraction (rather than a percentage)

Ray path curvature (K) can be expressed as a function of either n or N, provided that the assumption of a linear gradient d_n/d_h holds true:

$$K = \frac{1}{\left(1 + \frac{r_o d_n}{d_h}\right)} \tag{2.23}$$

or

$$K = \frac{1}{\left(1 + \frac{d_N / d_h}{157}\right)} \tag{2.24}$$

For the near-surface region, where d_n/d_h varies at about 3.9×10^{-8} m, the value of K is 1.33. For most terrestrial microwave paths, this value ($K = \frac{4}{3} = 1.33$) is called a *standard refraction*, and is used in calculations in the absence of additional data. For regions above the linear zone close to the surface, you can use another expression of refractivity:

$$N_a = N_s e^{-C_e(h_r - h_t)} \tag{2.25}$$

where N_1 = refractivity at 1 km altitude
h_r = height of receive antenna
h_t = height of transmit antenna
$C_e = L_n (N_s/N_1)$
N_a = refractivity at altitude
N_s = refractivity at earth's surface

For models close to the surface, use the geometry shown in Fig. 2.16A, where distance d is a curved path along the surface of the earth. But because the earth's radius r_o is about 4000 statute miles and thus very much larger than any practical antenna height h, the simplified model of Fig. 2.16B can be used. The underlying assumption, of course,

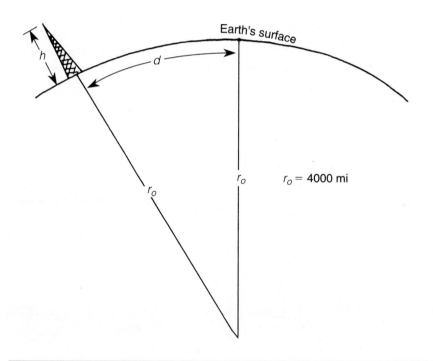

FIGURE 2.16A Geometry for calculating radio line-of-sight distances.

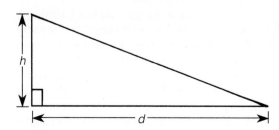

FIGURE 2.16B Simplified geometry.

is that the earth has a *radio radius* equal to about ⅔ (K = 1.33) of its *physical radius,* as discussed previously.

Distance d is found from the expression

$$d = \sqrt{2r_o h} \qquad (2.26)$$

where d = distance to radio horizon in statute miles
  r_o = radius of earth in statute miles
  h = antenna height in feet

Accounting for all constant factors, the expression reduces to

$$d = 1.414 \sqrt{h} \qquad (2.27)$$

all factors being the same as defined previously.

Example 2.2   A radio tower has a UHF radio antenna that is mounted 150 ft above the surface of the earth. Calculate the radio horizon (in statute miles) for this system.

Solution

$$d = 1.414 \ (h)^{1/2}$$
$$= (1.414)(150 \ ft)^{1/2}$$
$$= (1.414)(12.25)$$
$$= 17.32 \ mi$$

For other units of measurement:

$$d \ (nmi) = 1.23 \sqrt{h} \qquad (2.28)$$

and

$$d \ (km) = 1.30 \sqrt{h} \qquad (2.29)$$

Surface Waves

The surface wave travels in direct contact with the earth's surface, and it suffers a severe frequency-dependent attenuation from absorption by the ground.

The surface wave extends to considerable heights above the ground level, although its intensity drops off rapidly at the upper end. The surface wave is subject to the same attenuation factors as the space wave but, in addition, it suffers ground losses. These losses are caused by ohmic (resistive) losses in the conductive earth and by the dielectric properties of the earth. In short, the signal heats up the ground. Horizontally polarized waves are not often used for surface-wave communications because the earth tends to short-circuit the E-field component. (A perfectly conducting plane has no voltage between any two points on it; hence, no E-field can exist on the plane.) For verti-

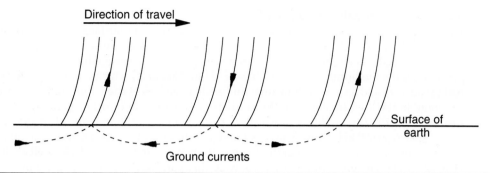

FIGURE 2.17   Distortion of vertically polarized electric field by lossy ground resistance.

cally polarized waves, however, the earth offers electrical resistance to the E-field and returns currents to subsequent waves (Fig. 2.17). The conductivity of the soil determines how much energy is returned. Table 2.1 shows the typical conductivity values for several different forms of soil, as well as saltwater and freshwater.

Type of Soil	Dielectric Constant	Conductivity (Siemens/Meter)	Relative Quality
Saltwater	81	5	Best
Freshwater	80	0.001	Very poor
Pastoral hills	14–20	0.03–0.01	Very good
Marshy, wooded	12	0.0075	Average/poor
Rocky hills	12–14	10	Poor
Sandy	10	0.002	Poor
Cities	3–5	0.001	Very poor

TABLE 2.1   Sample Soil Conductivity Values

The wavefront of a surface wave is tilted because of the losses in the ground that tend to retard the lower end of the wavefront (also shown in Fig. 2.17). The tilt angle is a function of the frequency, as shown in Table 2.2.

Frequency (kHz)	Tilt Angle Ratio	Earth/Seawater (Degrees)
20	207	4.3/0.021
200	104	13.4/0.13
2,000	64	32.3/0.5
20,000	25	35/1.38

TABLE 2.2   Tilt Angle as a Function of Frequency

Ground-Wave Propagation Frequency Dependence

The frequency of a radio signal in large measure determines its surface-wave behavior. In the low-frequency (LF) band (30 to 300 kHz), ground losses are small for vertically polarized signals, so medium-distance communication (up to several hundred miles) is possible. In the medium-wave (MW) band (300 to 3000 kHz, which includes both the AM broadcast band and the 160-m amateur band), distances of a few hundred miles are common. In the high-frequency (HF) band, ground losses are typically greater, so the surface-wave distance reduces drastically. It is possible, in the upper end of the HF band (3000 to 30,000 kHz), for surface-wave signals to die out within a few dozen miles. This phenomenon is often seen in the 15- and 10-m amateur radio bands, as well as the 11-m (27-MHz) citizens band. Stations only 20 mi apart may not be able to communicate, but both can talk to a third station across the continent via ionospheric skip!

Tropospheric Propagation

The troposphere is the portion of the atmosphere between the surface of the earth and the stratosphere (or about 4 to 7 mi up). Some older texts group tropospheric propaga-

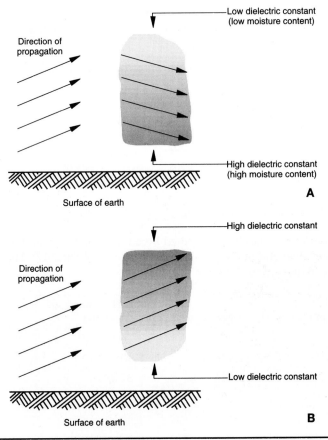

Figure 2.18 Refraction in the troposphere.

tion with ground-wave propagation, but modern practice requires separate treatment. The older grouping overlooks certain tropospheric propagation phenomena that simply don't happen with space or surface waves.

Refraction is the mechanism for most tropospheric propagation phenomena. The dielectric properties of the air, which are set mostly by the moisture content (Fig. 2.18A and B), are a primary factor in tropospheric refraction. Recall that refraction occurs in both light-wave and radio-wave systems when the wave passes between media of differing density. Under that situation, the wave path will bend in approximate proportion to the difference in densities.

Two general situations are typically found—especially at UHF and microwave frequencies. First, because air density normally decreases with altitude, the top of a beam of radio waves typically travels slightly faster than the lower portion of the beam. As a result, those signals refract a small amount. Such propagation provides slightly longer surface distances than are normally expected from calculating the distance to the radio horizon. This phenomenon is called *simple refraction,* and was discussed in a preceding section.

A special case of refraction called *superrefraction* occurs in areas of the world where warmed land air flows out over a cooler sea (Fig. 2.19). Typically these areas have deserts adjacent to large bodies of water: The Gulf of Aden, the southern Mediterranean, and the Pacific Ocean off the coast of Baja California are examples. Communications up to 200 mi at VHF/UHF/microwave frequencies are frequently reported in such areas.

The second form of refraction is weather-related. Called *ducting,* this form of propagation (Fig. 2.19) is actually a special case of superrefraction. Evaporation of seawater causes temperature inversion regions—i.e., layered air masses in which the air temperature is greater than in the layers below it—to form in the atmosphere. (*Note:* Air temperature normally decreases with altitude, but at the boundary with an inversion region it increases.) The inversion layer forms a "duct" that acts like a waveguide. In Fig. 2.20, the distance D_1 is the normal *radio horizon* distance, and D_2 is the distance over which duct communications can occur.

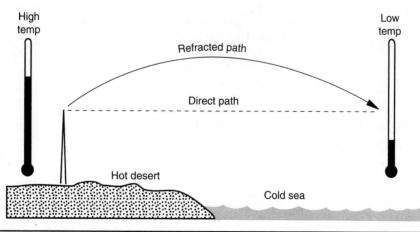

FIGURE 2.19 Superrefraction phenomena.

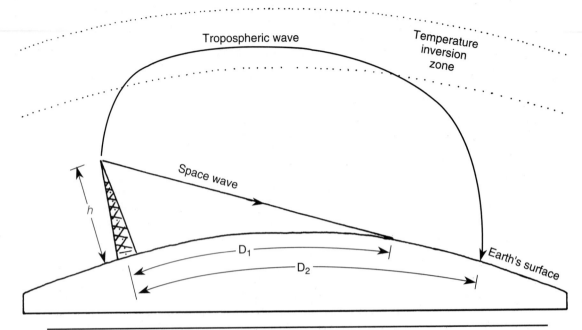

FIGURE 2.20   Ducting phenomenon.

Ducting allows long-distance communications from lower VHF through micro-wave frequencies, with 50 MHz being a practical lower limit and 10 GHz being an ill-defined upper limit. Airborne operators of radio, radar, and other electronic equipment sometimes report ducting at even higher microwave frequencies.

Antenna placement is critical for successfully using ducting as a propagation medium. Both the receiving and the transmitting antennas must be either (1) inside the duct physically (as in airborne cases) or (2) able to propagate at an angle such that the signal gets trapped inside the duct. The latter is a function of antenna radiation angle. Distances up to 2500 mi or so are possible through ducting. Certain paths where frequent ducting occurs have been identified:

- Great Lakes to Atlantic seaboard
- Newfoundland to Canary Islands
- Florida to Texas across the Gulf of Mexico
- Newfoundland to the Carolinas
- California to Hawaii
- Ascension Island to Brazil

Another condition is noted in polar regions, where colder air from the land mass flows out over warmer seas (Fig. 2.21). Called *subrefraction*, this phenomenon bends EM waves away from the earth's surface—thereby *reducing* the radio horizon by about 30 to 40 percent.

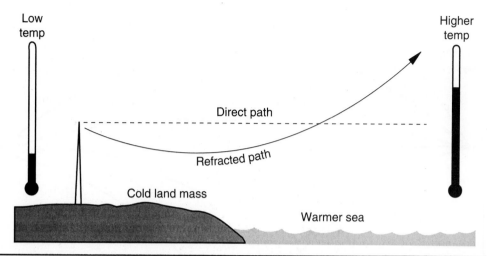

FIGURE 2.21   Subrefraction phenomena.

All tropospheric propagation dependent upon air-mass temperatures and humidity shows *diurnal* (i.e., over the course of the day) variation caused by the local rising and setting of the sun. Distant signals may vary 20 dB in strength over a 24-hour period. These tropospheric phenomena explain how TV, FM broadcast, and other VHF signals can propagate great distances—especially along seacoast paths—some of the time, while being weak or undetectable at others.

Fading Mechanisms

Fading is the name given to the perceptible effects of variations over time in received amplitude and/or phase. These variations are assumed to originate in changing propagation effects between the transmitter and the receiver, rather than as a result of changes in transmitter output power. Some of the most common sources include weather-related path loss, multipath anomalies, vehicles and other objects in motion, and variations in the height and density of the ionospheric layer of the atmosphere.

You might not expect line-of-sight radio relay links to experience fading, but they do. Fading does, in fact, occur, and it can reach levels of 30 dB in some cases (20 dB is relatively common). In addition, fading phenomena in the VHF-and-up range can last several hours, with some periods of several days (although very rare) reported.

Generally, fading is the result of varying path loss and/or phase shift on at least one path between the transmitting (TX) and receiving (RX) antennas. In Fig. 2.22, Ray *A* represents the direct path that, ideally, is the only path between source and destination. But other paths are also possible. Ray B is an example of energy from the same transmitting antenna traveling through an elevated layer or other atmospheric anomaly capable of providing refraction or subrefraction of the transmitted wave. Ray C adds signal via a ground reflection path, and Ray D is an example of yet another wave from the transmitting antenna arriving at the receiver via subrefraction fading.

All three of Rays B, C, and D will have amplitudes and phases that differ from those of Ray A. One or more may, for instance, be totally out of phase with the received voltage from Ray A. Unless one or more of them varies over time, fading is not likely to be

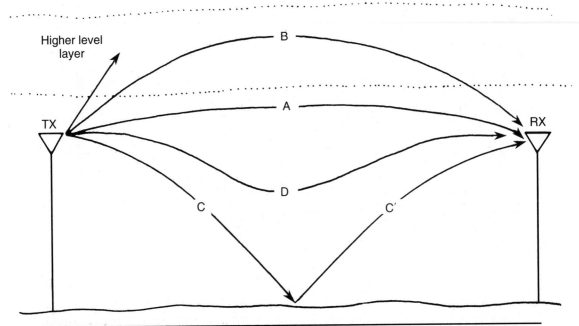

FIGURE 2.22   Multiple signal paths between transmitter and receiver.

present. But each path (including that of Ray A itself) is subject to fading. A moving vehicle midway between the transmit and receive sites may interrupt path C or C′ during its travels. In a more complex example, while the moving vehicle is in the ground reflection path, it may create a new and different geometry for reflection if the incoming signal hits its roof and reflects off in the direction of the receive antenna! Paths A, B, and D are all subject to weather-related changes: a gust of wind, a temperature or pressure gradient moving through the region, raindrops falling, etc.

In general, these mechanisms are frequency-sensitive, so a possible countermeasure is to use *frequency diversity*. In many cases, *hopping* over a 5 percent frequency range will help eliminate or reduce the fading. If other system constraints and/or local spectrum usage prevent a 5 percent range, try for at least 2 or 3 percent.

Over oceans or other large bodies of water *fair weather surface ducting* is sometimes encountered. These ducts form in the midlatitudes, starting about 2 to 3 km from shore, up to heights of 10 to 20 mi, with wind velocities in the 10- to 60-km/h range. The resultant *power fading* is due to the presence of the duct along with surface reflections (see Fig. 2.23). Power fading alone can occur when there is a superrefractive duct elevated above the surface. The duct has a tendency to act as a waveguide and focus the signal (Fig. 2.24). Although the duct shown is superrefractive, subrefractive ducts are also possible.

Microwave communications paths above about 10 GHz suffer increasingly severe attenuation caused by water vapor and oxygen in the atmosphere. Figure 2.25 shows the standard attenuation in decibels per kilometer for microwave frequencies. Note that in addition to a general upward slope to the graph there are strong peaks at several

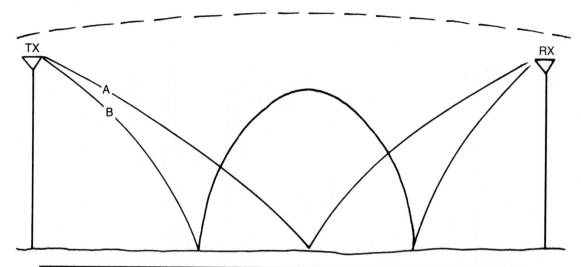

Figure 2.23   Multihop interference.

specific frequencies. Selecting a system frequency within any of these ranges will cause poor communications or will require a combination of more transmit power, better receiver sensitivity, and better antennas at both receiving and transmitting locations. Further, the curves of Fig. 2.25 are based on certain assumed standard conditions. Rain and other forms of airborne precipitation can severely increase the attenuation of signals beyond the amounts shown in the chart.

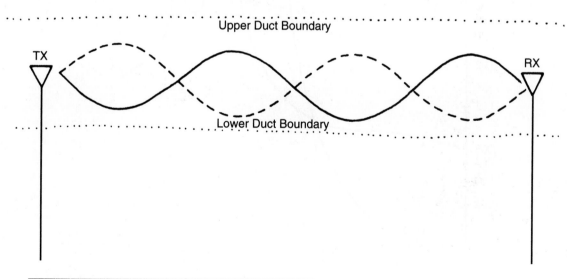

Figure 2.24   Duct focusing.

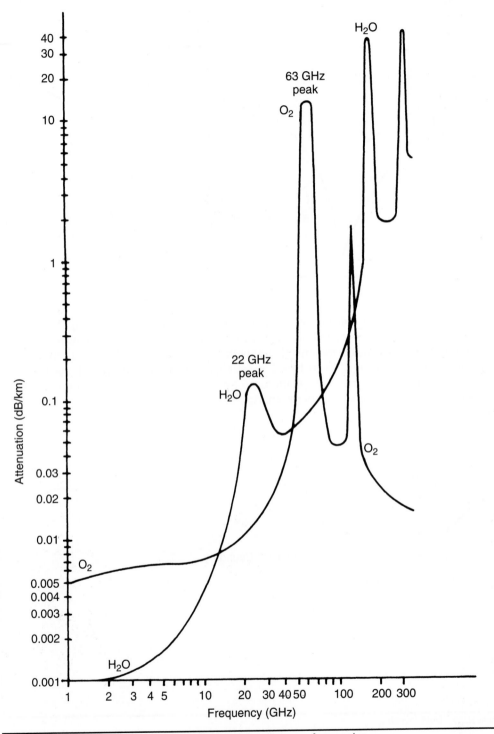

FIGURE 2.25   Atmospheric absorption of radio signals at microwave frequencies.

Ionospheric Propagation

Following on the heels of seminal discoveries by Hertz, Marconi, and other pioneering radio scientists of the late 1800s, the early 1900s found experimenters from all walks of life constructing simple radio transmitting and receiving equipment. By 1909, radio clubs had been formed at a few colleges around the United States, and by 1914 the American Radio Relay League (ARRL), dedicated to these amateur experimenters, had been founded. Commercial, government, and amateur stations shared the airwaves in willy-nilly fashion until around 1912 when the United States began to regulate radio. In those days, "good DX" might mean the ability to communicate between, say, Boston and New York City, and the prevalent assumption was that the longer wavelengths were superior for extending those distances. Forced to shift from wavelengths as long as 1000 m (well below the bottom of today's AM broadcast band) to 200 m (roughly around 1500 kHz, near the top of the AM broadcast band) by the new regulations, amateurs grumbled a bit but dug in and went back to figuring out how to attain better and better DX distances. As the distances increased, however, the experts were finding it harder and harder to explain how it was possible.

World War I brought a halt to all amateur transmissions, and when the war was over the U.S. government was inclined to keep it that way forever. But the ARRL convinced Congress and various federal agencies of the merits of radio amateurs, and in the fall of 1919 amateurs returned to the airwaves. In 1921 and again in 1922, the ARRL sponsored transatlantic receiving tests, and in both events too many stations were heard for it to be a fluke: 30 or so the first year and 10 times that number the following year!

Spurred on by these encouraging results, the ARRL and others began to investigate what it would take to convert these successes into *two-way* communication across the ocean. Despite the almost unanimous opinion of the best scientists of the day that the shorter wavelengths would prove to be useless, some hardy experimenters plunged below 200 m and, within a year, had shown that the shorter the wavelength, the better the DX results!

Textbooks and magazines of the era were rife with theories to explain these totally unexpected results. Ultimately, the "correct" theory turned out to be one originally advanced in 1901, when Arthur Kennelly of the United States and Oliver Heaviside of England independently proposed the existence of a "reflecting" layer circling the earth some miles above it. Soon named the *Kennelly-Heaviside layer*, it is now known as the *E layer* (Fig. 2.26) of the ionosphere.

Thus, it is the ionosphere that makes intercontinental radio communications possible on the MF, HF, and lower VHF bands (between roughly 0.3 and 70 MHz). Included in the range of frequencies for which ionospheric propagation is possible are the AM broadcast band, many traditional maritime frequencies, amateur bands from 160 to 6 m, and the first few channels (2, 3, and 4) of the analog broadcast television spectrum. True, in some cases (especially at the upper end of the range) "possible" does not mean "probable", but the potential is there during sunspot maxima or during brief periods of unusual solar activity. The author recalls the winter of 1957–58, when he saw firsthand a neighboring amateur make two-way contact with six continents in one evening on the 6-m band (50 MHz), using a 5-W AM transceiver and a three-element 6-m beam dangling from the I-beam running through his *basement*!

Kennelly and Heaviside described the unseen layer as a *reflecting* layer; in truth, the sparseness of the ionization and the variation of ionization with altitude make

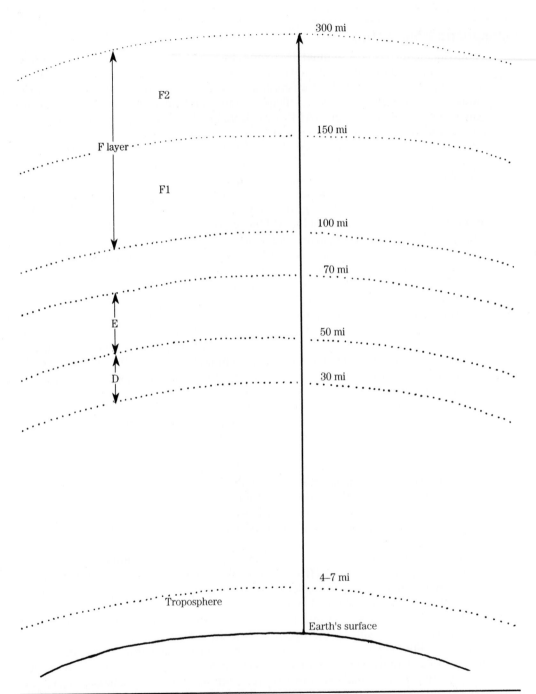

FIGURE 2.26 Classification of the earth's atmosphere for radio propagation.

it a *refracting* layer. But, as can be seen from Fig. 2.27, the refracting action of the layer beginning at B-C can be approximated by pretending there is a reflecting surface at altitude C'.

The mirror analogy is imperfect in another way, as well. If you stand directly in front of a mirror, you are able to see your image clearly. But the ability of the ionospheric layer to return a perfectly vertical signal back to earth is frequency dependent. That is, above a certain frequency, called the *critical frequency*, or f_c, signals sent straight up continue on into outer space! If, on the other hand, transmitted signals hit the layer at an angle to the vertical (Fig. 2.28), at some transmitted takeoff angle (α_r) the transmitted signal no longer escapes from the ionosphere but remains within it. This is called the *critical angle*, and it is different for every frequency.

Most important, takeoff angles smaller than the critical angle will be refracted from the layer and returned to earth. Thus, we see from Fig. 2.28 that there is a range of takeoff angles for the transmitted signal that will be "reflected" back to earth at some distance from the source. This is the great magic of the ionosphere.

Conversely, at takeoff angles greater than the critical angle, the transmitted wave either stays within the ionospheric layer or continues on into space. In other words, for angles between the critical angle and 90 degrees (pure vertical incidence), the iono-

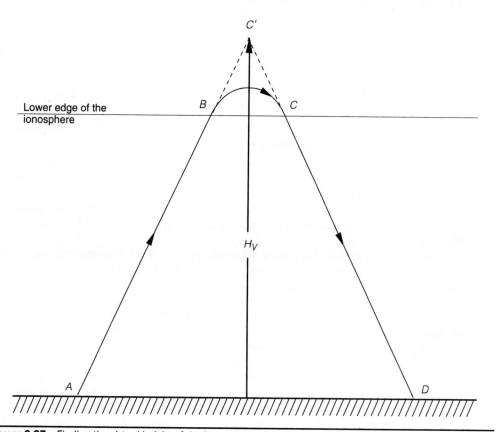

FIGURE 2.27 Finding the virtual height of the ionosphere.

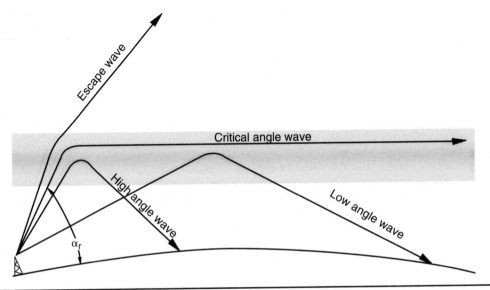

FIGURE 2.28   Sky-wave propagation as a function of antenna radiation angle.

sphere does not return the transmitted signal to earth. In short, it does not support earth-to-earth communications links at this frequency. The distance on the earth's surface from the transmitting antenna out to the closest point of returned signal from the ionosphere is called the *skip distance*, and all locations closer to the transmitter than that distance are said to be *inside the skip zone*.

The Sun's Effect on the Ionosphere

Several sources of energy contribute to ionization of the upper atmosphere. Cosmic radiation from outer space creates some degree of ionization, but the majority is caused by solar energy. Events on the surface of the sun sometimes cause the radio mirror to seem to be almost perfect, and this situation makes spectacular propagation possible. At other times, however, solar disturbances (Fig. 2.29A) can disrupt radio communications for days at a time.

There are two principal forms of solar energy that affect shortwave communications: *electromagnetic radiation* and *charged solar particles*. Most of the radiation is above the visible spectrum, in the ultraviolet and x-ray/gamma-ray region of the spectrum. Because electromagnetic radiation travels at the speed of light, solar events that release radiation cause changes to the ionosphere about eight minutes later. Charged particles, on the other hand, have a finite mass and thus travel at a considerably slower velocity, requiring two or three days to reach earth.

Multiple sources of both radiation and particles exist on the sun. *Solar flares*, for instance, can release huge amounts of both radiation and particles. These events are unpredictable and sporadic. Some layers of the sun rotate with a period of approximately 27 days, causing specific sources of radiation to face the earth once every 27 days. Thus, many components of solar radiation levels and the corresponding ionospheric activity tend to repeat every 27 days, as well.

FIGURE 2.29A   Solar event affecting radio propagation on the earth.

At some frequencies, solar and galactic noise establish a practical lower limit on the reception of weak signals. Solar noise can also affect radio propagation and act as a harbinger of changes in propagation patterns. Solar noise can be demonstrated by using an ordinary radio receiver and a directional antenna, preferably operating in the VHF/UHF regions of the spectrum (150 to 152 MHz frequently is used). Simply aim the antenna at the sun on the horizon at either sunset or sunrise—a dramatic change in background noise will be noted as the sun rises or sets across the horizon.

Sunspots

Sunspots are arguably the most well known contributor to the sun's effect on our ionosphere. Sunspots (Fig. 2.29B) can be as large as 70,000 to 80,000 mi in diameter and often appear in clusters. The number of sunspots typically visible on the face of the sun on any given day is quite variable in the short term, but as shown in Fig. 2.29C exhibits a fairly consistent and pronounced long-term cycle over a period of approximately 11 years. This is a very rough number; actual cycles since 1750 (when records were first kept) have varied from 9 to 14 years. The sunspot number is reported daily as the statistically massaged *Zurich smoothed sunspot number,* or *Wolf number.* The monthly smoothed sunspot number (SSN) is based on a weighted average of a 12-month period centered on the month being reported. Therefore, it is not possible to know what the SSN is for a given month until a half year later! Nonetheless, a century of observations has confirmed the strong relationship between the SSN and general levels of ionospheric radio propagation. The lowest monthly SSNs calculated during the entire radio era hovered around 1.5 in mid-1913 and did not go that low again until mid-2008. The highest recorded monthly SSN was 201 (in March of 1958). Daily sunspot numbers vary "all over the map" and can remain at zero for weeks on end (as they did throughout much of 2008 and 2009) or rise to values of 150 percent or more of the corresponding monthly SSNs. NOAA's National Geophysical Data Center Web site (www.ngdc.noaa .gov/stp/SOLAR) is an excellent source of background information on the methodology involved.

For many decades, the smoothed sunspot number was the best (and perhaps the only) available indicator of ionospheric propagation potential. With advances in science and technology, and especially with the availability of satellites, balloons, and other extraterrestrial probes, other measures now augment SSNs. One such indicator is the *solar flux index* (SFI). This measure is taken in the microwave region (at a wavelength of 10.2 cm, or 2.8 GHz), at 1700 UTC at Ottawa, Canada. The SFI is reported in an over-the-air announcement hourly by the National Institute for Standards and Technology (NIST) radio stations WWV (Fort Collins, Colorado) and WWVH (Maui, Hawaii) and is also available from many additional sources via the Internet. The calculated SSN is roughly proportional to the measured SFI, but the two use different scales (the SFI has a minimum value of around 60 when the SSN is at zero) and the relationship is not perfect, as can be seen by examining a scatter diagram of hundreds or thousands of data points collected over at least two consecutive 11-year cycles.

Two short-term propagation forecasting tools, the *A-index* and the *K-index*, relate to how "unsettled" the ionosphere is and are included in the NIST announcements. These, and many other monitoring and analysis tools are available; readers are encouraged to

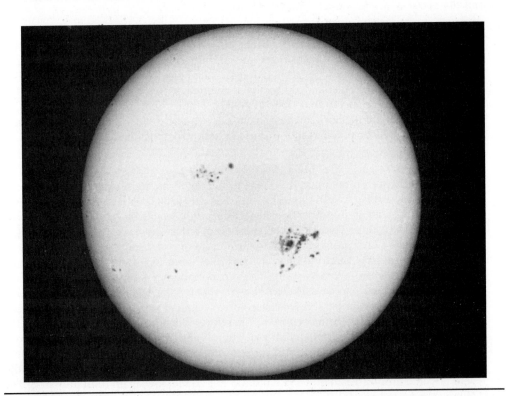

FIGURE 2.29B   Sunspots.

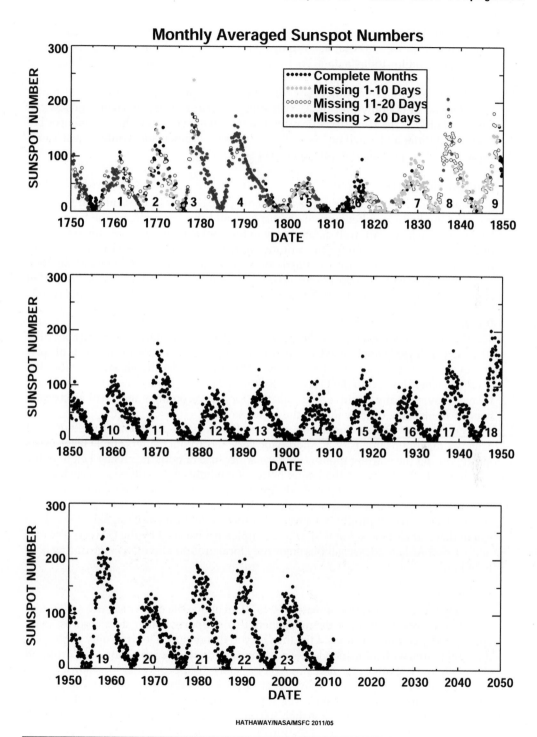

FIGURE 2.29C   Monthly averaged sunspot numbers, 1750–2011.

search the Internet using terms such as "sunspot numbers" and "solar cycle 24". (Appendix B includes a bountiful list of propagation-related Web sites.) As this is written, the current data for today's date, partway into solar cycle 24, are: SFI = 88; SSN = 51; A = 12; K = 2.

The ionosphere offers properties that affect radio propagation at different times. Variations occur not only over the 11-year sunspot cycle, but also diurnally and seasonally. Obviously, if the sun affects propagation in a significant way, then differences between nighttime and daytime, as well as between summer and winter, must cause variations in the propagation phenomena observed.

Ionospheric Layers

The ionosphere is divided, for purposes of radio propagation studies, into multiple layers. Like the boundary between the ionosphere and the stratosphere, the boundaries between these layers are well defined only in textbooks. However, even there you will find varying heights given for the altitudes of the layers above the earth's surface. In reality, they don't have sharply defined boundaries but, rather, blend one into another. Thus, the division of the ionosphere into layers is quite arbitrary. These layers (Fig. 2.26) are designated D, E, and F (with F being further subdivided into F1 and F2 sublayers).

D Layer

The D layer is the lowest layer in the ionosphere, at approximately 30 to 50 mi above the surface. This layer is not ionized as much as the higher layers because all forms of solar energy that cause ionization are severely attenuated by the higher layers before the energy reaches down into the D layer. Also, the D layer is much denser than the E and F layers, causing any positive and negative ions to recombine and form electrically neutral atoms more quickly than in the higher layers.

The extent of D-layer ionization is proportional to the elevation of the sun, so it achieves maximum intensity at midday or very shortly thereafter. The D layer exists mostly during the warmer months of the year because of both the greater height of the sun above the horizon and the longer hours of daylight. As might be expected, therefore, the D layer almost completely disappears after local sunset, although some observers have reported sporadic incidents of D-layer activity for a considerable time past sunset. The D layer typically is a strong absorber of medium-wave signals (to such an extent that signals below 3 or 4 MHz are completely absorbed by the D layer). Because of this, most AM broadcast stations enjoy only local ground-wave coverage during daylight hours.

E Layer

The E layer exists from approximately 50 to 70 mi above the earth's surface and is considered the lowest region of the ionosphere that supports, rather than absorbs, ionospheric radio communications. Like the D layer, this region is ionized only during the daylight hours, with ionization levels peaking at midday. The ionization level drops off sharply in the late afternoon and almost completely disappears after local sunset.

For most of each year, the E layer is absorptive and does not reflect radio signals. During the summer months, however, E-layer propagation occurs frequently. A phenomenon called *short skip* (i.e., less than 100 mi for medium-wave and 1000 mi for short-wave signals) occurs in the E layer during the summer months and in equatorial regions at other times, as well.

Sporadic E propagation is another phenomenon associated with the E layer. This mode is made possible by scattered zones of intense ionization in the E-layer region of the ionosphere. Sporadic E varies seasonally, and some believe it results from solar particle bombardment of the layer. Sporadic E propagation affects the upper HF and lower VHF region. It is observed most frequently in the lower VHF spectrum (30 to 150 MHz), but it is also sometimes observed at higher frequencies. Skip distances on VHF can reach 500 to 1500 mi on one hop—especially in the lower VHF region (including the 6-m band).

F Layer

The F layer of the ionosphere is the primary support for long-distance shortwave communications. This layer is located between 100 and 300 mi above the earth's surface. Unlike the lower layers, the air density in the F layer is low enough that ionization levels remain high all day and decay slowly after local sunset. Minimum levels are reached just prior to local sunrise. Because of its height, the F layer can support skip distances up to 2500 mi on a single hop.

During the day, the F layer often splits into two identifiable and distinct sublayers, designated the F1 and F2 layers. The F1 layer is found approximately 100 to 150 mi above the earth's surface, with the F2 layer above the F1, extending up to the 270- to 300-mi limit. Beginning at local sundown, however, the lower regions of the F1 layer begin to deionize as positive and negative ions recombine. Sometime after local sunset, the F1 and F2 layers effectively merge to become a single reduced layer beginning at about 175 mi.

The height and degree of ionization of the F2 layer varies with local sun time, with the season of the year, and with the 27-day and 11-year sunspot cycles. The F2 layer begins to form shortly after local sunrise and reaches maximum shortly after noon. During the afternoon, the F2-layer ionization begins to decay in an exponential manner until, for purposes of radio propagation, it disappears sometime after local sunset. During sunspot maxima, F-layer ionization does not completely disappear overnight, and the 20-m amateur band may be open worldwide 24/7.

Measures of Ionospheric Propagation

At any given time, several different characteristics of the ionosphere can be measured and used to make predictions of radio activity and long-distance propagation.

The *critical frequency* (f_c) and *maximum usable frequency* (MUF) are indices that tell us something of the state of ionization and communications ability. These frequencies increase rapidly after sunrise, enabling international communications at higher and higher frequencies as the MUF rises with the sun.

Critical Frequency (f_c)

The *critical frequency*, f_c, is the highest frequency that can be reflected when a signal strikes the ionosphere as a vertical (90 degrees with respect to the reflecting surface) incident wave. The critical frequency is determined from an *ionogram*, which is a cathode-ray tube (CRT) oscilloscope display of the height of the ionosphere as a function of frequency. The ionogram is made by firing a pulse vertically (Fig. 2.30) at the ionosphere from the transmitting station. The critical frequency, f_c, is then the highest frequency for which a reflected signal is received back at the transmitter site. Values of f_c can be as low as 3 MHz during the nighttime hours and as high as 10 to 15 MHz during the day.

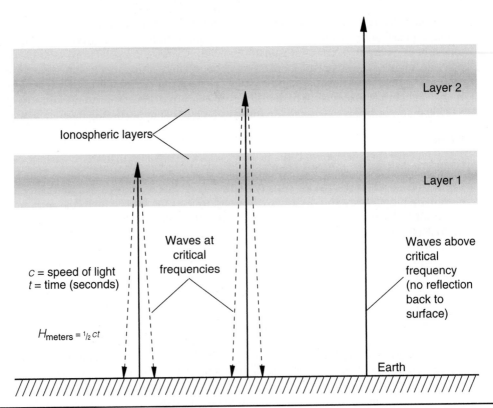

FIGURE 2.30   Finding the critical frequency of the ionosphere.

Virtual Height

Radio waves are refracted in the ionosphere, and those below the critical frequency are refracted so much that they return to earth. Such waves appear to have been reflected from an invisible radio "mirror." The height of this apparent mirror (point C′ in Fig. 2.27) is called the *virtual height* of the ionosphere. Virtual height is determined by measuring the time interval required for an *ionosonde* pulse (similar to that used to measure critical frequency) to travel between the transmitting station and a nearby receiving station (Fig. 2.31). By observing the time interval between the transmitted and the received pulses and applying a correction factor for the estimated speed of the wave through the atmosphere, the virtual height of the ionosphere can be calculated.

Maximum Useable Frequency (MUF)

The *maximum useable frequency* is the highest frequency at which communications can take place via the ionosphere over a given path. Normally the MUF for a given transmitter site is found to be at the farthest distance(s) the signal can graze the ionosphere in the direction of the sun. As a very rough rule of thumb, the MUF is approximately three times higher than the critical frequency.

Both the MUF and the critical frequency vary geographically, and they become higher at latitudes close to the equator. However, the critical frequency is determined by bouncing a signal off the ionosphere directly overhead; the measured MUF, on the

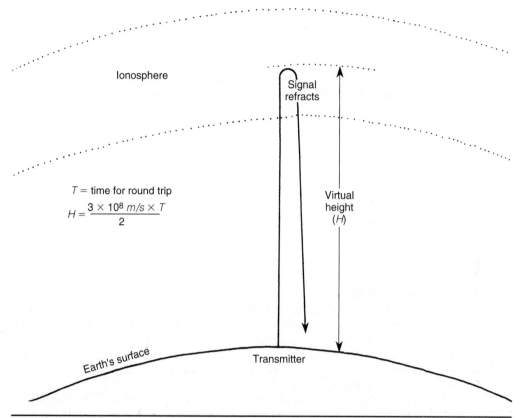

Ionosphere

Signal
refracts

T = time for round trip

$$H = \frac{3 \times 10^8 \; m/s \times T}{2}$$

Virtual
height
(H)

Earth's surface   Transmitter

FIGURE 2.31   Finding the virtual height of the ionosphere.

other hand, will depend on the state of the ionosphere 2500 mi away from the site where f_C is measured, so the two indices do not track perfectly.

On average, the best propagation occurs at frequencies just below the MUF. In fact, the *frequency of optimum traffic* (FOT) is approximately 85 percent of the MUF. Near the FOT, signal strengths are optimum because ionospheric absorption is at a minimum and signal-to-noise ratios (SNRs) are highest because atmospheric noise generally decreases with increasing frequency.

Lowest Useable Frequency (LUF)

Below a certain frequency, the combination of ionospheric absorption and atmospheric noise, static, etc., conspire to hinder and even prevent radio communications. The *lowest useable frequency* (LUF) is a measure intended to help identify the lower frequency limit for successful communications.

Unlike the MUF, the LUF is not dependent solely on atmospheric physics. The LUF of a system can be varied by controlling the signal-to-noise ratio of the transmitter-receiver path. Although many of the factors that contribute to SNRs are beyond our control, the *effective radiated power* (ERP) of the transmitter is one that often can be changed. For instance, a 2-MHz decrease in LUF is available for every 10-dB increase in the ERP of the transmitter. In practice, this can be accomplished through the use of higher power, lower transmission line losses, higher gain antennas at both ends of the circuit, narrower receiver filter bandwidth, or some combination of all four.

Ionospheric Variations and Disturbances

The ionosphere is an extremely dynamic region of the atmosphere, especially from a radio operator's point of view, because it significantly alters radio propagation on a minute-by-minute basis. For our purposes, it is convenient to divide the dynamics of the ionosphere into two categories: *regular variations* and *disturbances.*

Regular Ionospheric Variations

There are several different forms of variation seen on a regular basis in the ionosphere:

- *Diurnal* (daily)
- *27-day* (monthly)
- *Seasonal*
- *11-year and 22-year cycles*

Diurnal (Daily) Variation

The sun rises and sets on a 24-hour cycle; because it is the principal source of ionization of the upper atmosphere, you can thus expect diurnal variation. During daylight hours, the E and D levels exist, but these disappear at night. The height of the F2 layer increases until midday, and then it decreases until evening, when it disappears or merges with other layers. As a result of higher absorption in the E and D layers, lower frequencies are not as useful during daylight hours. At the same time, the F layers support reflection at higher frequencies during the day. As a *very* rough rule of thumb, for *ionospheric skip* communications purposes, the frequencies below 10 MHz can be thought of as primarily "nighttime" bands and the frequencies above 10 MHz are primarily "daylight" bands.

27-Day Cycle

Radiation from the sun also exhibits a 27-day cycle caused by the rotational period of the sun's surface. Sunspots form or come to the surface of the sun near its poles, then gradually drift toward the equatorial region during their visible lifetime. During this latitude migration they tend to rotate in synchronism with the surface of the sun as a whole, so they will face the earth only during a portion of each month. As new sunspots are formed, they are not visible on the earth until their region of the sun rotates earth-side. Sunspots are believed to be regions of swirling high-intensity magnetic fields and plasmas that somehow stimulate greater overall solar radiation. Paradoxically, as seen in Fig. 2.29B, sunspots are darker than the rest of the sun's surface, indicating that they are cooler. New aids, such as the STEREO satellite pair, coupled with widespread dissemination of monitored data, have vastly improved our ability to observe sunspot formation and movement as it happens.

Seasonal Variation

The earth's tilt varies the angle of exposure of our planet to the sun on a seasonal basis. In addition, the earth's annual orbit is elliptical, not circular. As a result, the intensity of the sun's energy that ionizes the upper atmosphere varies with the seasons of the year.

In general, all three layers are affected, but the F2 layer is only minimally involved. Ion density in the F2 layer tends to be highest in winter and less in summer. During the summer, the distinction between F1 and F2 layers is less obvious.

11-Year and 22-Year Cycles

The number of sunspots, statistically averaged or *smoothed*, varies on an approximately 11-year cycle. As a result, the ionospheric characteristics that affect radio propagation also vary on an 11-year cycle. Skip propagation on the HF and lower VHF bands is most prevalent when the smoothed sunspot number is at its highest. During the peak years (1957–61) of the strongest sunspot cycle of the radio era to date, true F-layer skip propagation was frequently enjoyed on 6 m (50–54 MHz) and higher, affecting even some VHF TV channels and possibly even the low end of the FM broadcast band! By comparison, during the latest minimum (2007–10), many populated areas of the midlatitudes went weeks or months with no evidence of skip activity above 20 MHz.

Sunspots are regions of high magnetization, and their polarity on either side of the sun's equator switches with every sunspot cycle. Thus, the true cycle of the sun's activity is a 22-year cycle, but the effect of sunspots on ionospheric propagation is unrelated to the magnetic polarity of the spots, so, for all practical purposes, we experience only the 11-year cycle.

Ionospheric Disturbances

Disturbances in the ionosphere can have a profound effect on radio communications—and most of them (but not all) are bad. This section briefly examines some of the more commonly encountered types.

Sporadic E Layer

A reflective cloud of ionization sometimes appears in the E layer of the ionosphere; this layer is sometimes called the *sporadic E, or E_s, layer*. Theories abound regarding the origins of E_s, and at this writing many experts believe the term E_s is a catchall for multiple unusual causes of E-layer ionization. In addition to the solar particle bombardment theory mentioned earlier, some believe E_s episodes originate in the effects of wind shear between masses of air moving in opposite directions. This action appears to redistribute ions into a thin layer that is radio-reflective.

Sporadic E propagation is normally thought of as a VHF phenomenon, with most activity between 30 and 100 MHz, and decreasing activity up to about 200 MHz. However, at times sporadic E propagation is observed on frequencies as low as 10 or 15 MHz. Reception over paths of 1400 to 2600 mi is possible in the 50-MHz region when sporadic E is present. In the northern hemisphere, the months of June and July are the most prevalent sporadic E months. When sporadic E is present, it typically lasts only a few hours. Hint: Amateur radio experimenters who specialize in E_s activity carefully track the progress of weather fronts in their part of the world.

Sudden Ionospheric Disturbances (SIDs)

The SID, or *Dellinger fade*, mechanism occurs suddenly and rarely gives any warning. The SID can last from a few minutes to many hours. SIDs often occur in conjunction with solar flares, or bright solar eruptions, that emit an immense amount of ultraviolet radiation over a short period. When this UV radiation reaches the earth, the SID causes a tremendous increase in D-layer ionization, and abnormally high levels of absorption result. The ionization is so intense that all MF/HF receivers on the earth's sunlit hemisphere experience profound loss of signal strength above about 1 MHz. Received signal strengths can drop by 60 to 90 dB in a few minutes at the onset of the SID, and it is not uncommon for amateurs, shortwave listeners, and others monitoring the HF bands to think the cause is a malfunction in their receivers or antennas! Many times during the record-setting sunspot maximum of 1957–61 the author was fooled into thinking his normally highly sensitive Hallicrafters receiver had "bit the dust".

SIDs are often accompanied by variations in terrestrial electrical currents and magnetism levels. Thus, there is good correlation between the occurrence of an SID and visual or radio *aurora borealis* (or *aurora australis* in the southern hemisphere) for a few hours or days thereafter. Because solar flare activity is far higher during periods of high sunspot activity, SIDs are most commonly observed during the peak years of the strongest sunspot cycles.

Ionospheric Storms

The *ionospheric storm*, which may last from several hours to a week or more, appears to be produced by an abnormally large rain of atomic particles in the upper atmosphere. These storms are often preceded by SIDs 18 to 24 hours earlier and by an unusually large collection of sunspots crossing the solar disk 48 to 72 hours earlier. The storms occur most frequently and with greatest severity in the higher latitudes, decreasing toward the equator. When such a storm commences, shortwave radio signals may begin to flutter rapidly and then drop out altogether. The upper ionosphere becomes chaotic; turbulence increases, and the normal stratification into layers, or zones, is disrupted.

Radio propagation may come and go over the course of the storm, but it is mostly "dead". The ionospheric storm, unlike the SID, which affects only the sunlit side of the earth, is global in its impact. Observers frequently note that both the MUF and f_c, the critical frequency, tend to drop rapidly as the storm commences.

An ionospheric disturbance observed on November 12, 1960, was preceded by about 30 minutes of extremely good, but abnormal, propagation. At 2000 hours Greenwich mean time (GMT), European stations were heard in the United States with S9+ signal strengths in the 7000- to 7100-kHz region of the spectrum—an extremely rare midafternoon occurrence when sunspot levels are high (as they were then). After about 30 minutes, the bottom dropped out and even AM broadcast band *skip* later that evening was nonexistent. At the time, the National Bureau of Standards radio station, WWV, was broadcasting a W2 propagation prediction (which is terrible; back then, "W" stood for "Warning!") each hour. It was difficult to hear even the 5-MHz WWV frequency on the east coast of North America in the early hours of the disturbance, and it disappeared altogether for the next 48 hours.

Of course, in retrospect it is obvious that the sequence observed on 40 m is explained by the MUF rapidly sliding through the band (the 30-minute period of abnormally good DX propagation would correspond to the MUF being in the 7.5- to 8-MHz range) and ultimately dropping to 1 MHz or below by evening, as evidenced by the total lack of broadcast band skip. It is interesting that this particular disturbance occurred on the weekend of the annual ARRL "Sweepstakes" contest, a competitive operating event that brings out thousands of amateurs in the United States and Canada. The ARRL's summary of results in the May 1961 issue of *QST* led off with the editorial comment, "Universal lament: [propagation] conditions that first weekend!"

Ionospheric Sky-Wave Propagation

Sky-wave propagation occurs because signals in the ionosphere are refracted sufficiently to be bent back toward the earth's surface. To observers on the ground, the signal appears to have been reflected from a radio "mirror" at the virtual height somewhere above the junction of the stratosphere with the lower boundary of the ionosphere. The *skip distance* is the shortest distance along the surface of the earth from the transmitter site (*A* in Fig. 2.32) to the nearest of all points on the earth (point *C* in Fig. 2.32) for which the ionosphere is capable of refracting the transmitted signal back to earth. At shorter distances, the higher-angle radiation from the transmitted signal either remains in the ionospheric layer or continues on out into outer space. The *ground-wave zone* is the distance from the transmitter site (*A* in Fig. 2.32) to the locus of points where the ground wave fades to an unusably low level (point *B* in Fig. 2.32). The *skip zone* is the distance from the outer edge of the ground-wave zone to the skip distance, or the distance from *B* to *C* in Fig. 2.32. Note carefully that only one of these distances, the *ground-wave zone*, is strongly dependent upon transmitter power level (or, more precisely, effective radiated power). The other distances are a function primarily of transmitter frequency and the state of the ionosphere at any given time.

At some frequencies and power levels the sky wave and the ground wave may interfere with each other. When this happens, the sky wave arrives at the receiver with a signal whose amplitude and phase relative to the ground-wave signal depend on the specifics of its path up to the ionosphere and back down. Since the ionosphere is in constant motion (think slow *undulations*, like a snake), the sky wave will arrive with randomly varying amplitude and phase relative to the ground wave. Thus, the sky wave can selectively strengthen or cancel the ground wave, giving rise to a class of fading mechanisms discussed further under "Ionospheric Fading".

Angle of Incidence and Radiation Angle

One of the factors determining skip distance is the angle at which the radio wave enters the ionosphere. If the transmit antenna is relatively omnidirectional, it "sprays" the

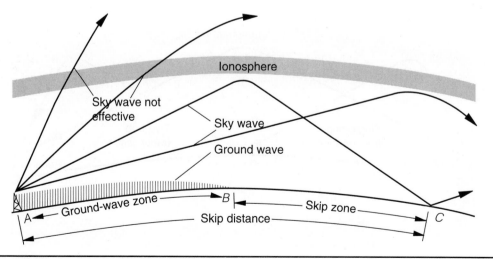

FIGURE 2.32 Ground-wave zone and skip zone.

lower ionospheric boundary with energy at *all* angles. As discussed earlier, the radiofrequency (RF) energy in angles nearly perpendicular to the boundary may or may not be refracted back to the earth, depending on the relationship between the transmitter frequency and f_C, the critical frequency. Nonetheless, for a low enough frequency, transmitted energy will, in fact, be refracted from the ionosphere at all transmit antenna takeoff angles from perfectly vertical to slightly below horizontal—that is, tangent to the earth's surface. (Exactly how far below horizontal depends on the height of the transmitting antenna above the spherical surface of the earth. For all practical purposes, however, horizontal is close enough.)

If the transmitted frequency is f_C or below, the transmitted energy for all takeoff angles will be returned to the earth, thus eliminating the *skip zone*. When that is the case, the ionosphere supports communication over a wide range of distances from the transmit antenna, potentially all the way out to where the refracted wave hits the earth some 2500 mi (on average) distant.

The practical effect of this is shown in Fig. 2.33. High-angle waves from the transmitting antenna are returned to the earth closer to the transmitter than low-angle waves are. When they hit the earth's surface (whether land or water) they are again refracted (or reflected) and once again head up to the ionosphere. Although Fig. 2.33 shows signal paths for only two specific takeoff angles, in fact there is an infinite number, all of which potentially bounce off the earth's surface and go back to the ionosphere. For extremely high angle waves, this process can occur tens of times before the wave ultimately reaches the receiving antenna far away.

In theory, the composite signal at the receiving antenna is the phased sum of all these different waves. In practice, each ground reflection introduces enough additional loss that waves encountering more than a handful of bounces (or *hops*) are far weaker at the receiving antenna than those that arrive after only a few hops.

Many of the propagation prediction programs readily available to the hobbyist calculate probable signal strengths at the receiving site for various propagation modes. The user can often bring up an auxiliary window that shows how many hops were

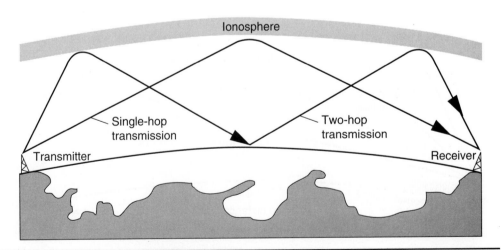

FIGURE 2.33 Single-hop and multihop skip communications.

employed and from which layers the reflections (or refractions, to be precise) probably occurred. Although it might seem obvious (especially for distances under 2500 mi) that fewer hops are better than more hops, in fact there are some tradeoffs to be considered. High-angle waves travel through less of the ionosphere during the process of refraction, so the path loss *per reflection* in the ionosphere is often less for a high-angle wave. Under normal conditions, the ground reflection loss is large enough to favor takeoff angles leading to fewer hops, but if the path is almost totally over saltwater, that may not be so true—especially if the ionosphere is disturbed (as indicated by higher A- and K-values).

Another factor in determining which takeoff angle will result in the strongest received signal is the relationship between the operating frequency and the MUF *at all points* along the path between the transmitter and the receiver. The emphasis is added because the MUF at any given instant varies with geographical location. For two sites located 2500 mi or less from each other, the simplest mode is one-hop. For the ionosphere to support single-hop propagation between those two sites, the MUF at the *midpoint* of the path, 1250 mi or less from the transmit site, must be greater than the operating frequency. The MUF at either end of the path is *not* the parameter of interest for single-hop propagation, since the only region where the radio wave is in contact with the ionosphere is near the midpoint of the path. As a general rule, the MUF is highest where average temperatures are highest—namely, in the equatorial regions—so higher-frequency communication is often possible on *transequatorial* paths when polar paths and midlatitude east-west paths are nonexistent.

For distances greater than 2500 mi, or for those occasions when the height of the F layer is lower than normal, multihop propagation will normally be required. For the two-hop path of Fig. 2.33, there are *two* MUFs that control whether the transmitted signal is heard at the receive site. *Both* MUFs must be greater than the operating frequency; if either one drops below it, the path will fail. If the sketch in Fig. 2.33 represents a transatlantic 20-m path in the northern hemisphere, most likely the MUF on the right side of the page (closer to Europe) will fail first, since the sun sets over the eastern Atlantic at least two or three hours before it does so on the western side.

As additional propagation modes, employing increasing numbers of hops, are analyzed, increasing numbers of MUF locations (often called *MUF control points*) must be examined. These additional modes cause the software to calculate predicted MUFs for a series of locations between the transmitter and the receiver. Hence, the previous statement that it is necessary to know the MUF "at all points along the path".

The specifics of the transmitting and receiving antenna installations also play a role in determining how many hops may provide the best signal over a given path. For ionospheric propagation paths (i.e., frequencies up through the lower VHF range), both the antenna itself and its height above local ground contribute to determining the relative amounts of energy radiated at each possible takeoff angle or received at each possible arrival angle. An antenna such as a ground-mounted vertical monopole has good low-angle radiation and minimal high-angle radiation. Although an infinite number of simultaneous propagation modes may exist between the two sites, the relatively strong low-angle radiation from the vertical will tend to make modes with the minimum number of hops the most probable paths to success.

In contrast, a horizontal dipole suspended $\lambda/8$ above ground will send most of its radiated energy nearly straight up, thus favoring multihop modes employing relatively high takeoff and arrival angles.

Two disadvantages of multihop modes to remember are that they are generally weaker and more subject to fading than a single-hop path.

Most important, keep in mind that there is no one "best" propagation mode or antenna radiation pattern. Some high-frequency bands (including the 11-m, or 27-MHz, citizens band) exhibit predominantly high-angle "short skip" during the summer and longer skip during the other months. Of course, when using higher angles of radiation at the higher frequencies, there will be a *critical angle* wave and *escape angle* waves that are not returned to the earth. These combinations are useless for terrestrial broadcasting or two-way communications.

So what is the ideal angle of radiation? The answer is: It depends! To date, the best answer to this question comes from the accumulated knowledge of an inquisitive band of dedicated DXers and contesters in the amateur fraternity. Using highly directive antennas at multiple heights in combination with sophisticated switching systems in their

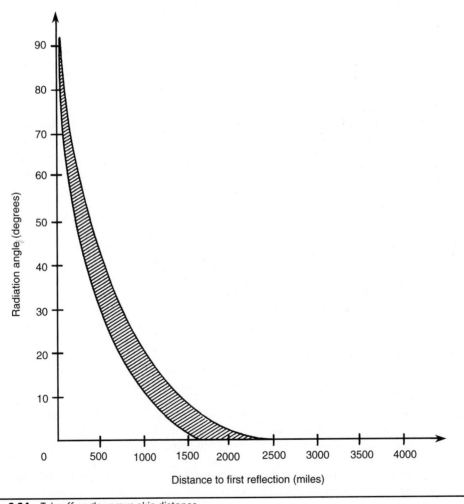

FIGURE 2.34 Takeoff angle versus skip distance.

radio rooms to allow rapid switching between antennas and combinations of antennas, they have compiled databases that make it clear no single angle of radiation is best. For instance, using long-boom Yagis on 14 MHz, they have learned that even on relatively short DX paths between, say, Europe and the east coast of North America, the optimum radiation angle varies throughout the day. Around sunrise and sunset it is not uncommon to hear stations from the other continent with very high antennas (on the order of 2 λ or higher) that clearly have superior low-angle performance. By midday, however, the best antennas often become the low ones (typically λ/2 to λ in height), with their superior pattern response at much higher takeoff angles. As we say elsewhere in this book, you can never have too many antennas!

Of course, if long-haul communication is not your objective, then different choices in antennas and/or their heights may result. For example, a resident of Virginia who wants to communicate with stations in the Carolinas or New York is well advised to select a higher angle of radiation with the help of Fig. 2.34 so that his or her signal has the best chance of being heard in those regions. On 80 m, for instance, that preference might cause the Virginian to eschew a vertical in favor of a low dipole.

Geography and the MUF

As Figs. 2.32 and 2.33 make clear, the distance covered on each hop by a transmitted signal is a function of the angle of radiation, often called the *takeoff angle*. Figure 2.34 plots the angle of radiation (relative to the horizon or ground) of the antenna versus distance to the first skip zone. As the angle of radiation is increased, shorter distances along the earth's surface are traversed with each hop. At an angle of about 30 degrees, for example, the typical distance per hop supported by the F layer is about 500 mi.

Although at first blush you might expect to see a single-line one-to-one relationship between the two variables on the graph, there is actually a zone shown (shaded). This occurs, of course, because the ionosphere is found at different heights throughout the 24-hour day/night cycle and throughout the course of a year. As a general guideline, however, in the absence of disturbed ionospheric conditions you can expect to attain between 1500 and 2500 statute miles per hop in the HF bands at low angles of radiation. Note, for example, that for a signal that is only a degree or two above the horizon the maximum skip distance is determined by the geometrical equations associated with the curvature of the earth's surface.

At distances greater than those shown in Fig. 2.34, the path requires the signal to make multiple hops. For instance, assuming an antenna with decent low-angle radiation and the ionosphere at an altitude capable of supporting one-hop skip communication out to 2500 mi, covering a distance of 7500 mi requires a minimum of three hops. However, depending on the instantaneous characteristics of the ionosphere at the distant reflection points and the relative gain of the transmitting and receiving antennas at different vertical angles, a greater number of shorter hops—involving a higher takeoff angle—may actually provide a stronger signal 7500 mi away than the three-hop path.

Thus, for communicating over short distances (a few hundred miles, say) at MF and HF, we are concerned with the characteristics of the ionosphere in the vicinity of the transmitting station. However, as the desired distances increase, the region(s) of the ionosphere that concern us are far enough away from the transmitting and receiving locations that we must be sure we know the characteristics of the ionosphere at any and all locations where the transmitted signal will come in contact with it. In particular, that means we are interested at a minimum in the MUF at the midpoint of each hop on the path between the transmitting and receiving sites.

For the 7500-mi path previously cited, we will normally want to be sure the MUF is above our chosen transmitting frequency in the vicinity of at least three places spaced 2500 mi apart from each other, each corresponding to the midpoint of one of the three hops minimally required to establish communications over that particular path.

Just as you can't determine the weather in New York City by sticking your head out the window in Los Angeles, you can't draw any conclusions about your ability to communicate with an amateur in China by the signal strength of an amateur in a neighboring state. As discussed earlier, the MUF at each specific place on this globe is a function of many variables. Here are two specific examples:

- MUFs are greatest near the equator and decline with increasing latitude, both north or south. As a general rule, on any given day noontime one-hop transequatorial communications having the midpoint of the hop close to the equator will be possible on higher frequencies than can be supported by paths anywhere else on the globe. Although multihop links are possible, the typically lower MUFs at the midpoints of the outer hops will force use of a lower frequency.

- MUFs usually "follow the sun", peaking shortly after noon. To maximize the likelihood of being able to communicate on a typical midlatitude east/west path, such as between Maine and Oregon or between California and Hawaii, choose a time when it's midafternoon at the eastern end of the circuit and mid- to late-morning at the western end. For best signal-to-noise ratios, pick the highest operating frequency that appears to be open that day.

Of course, these are guidelines for the daytime bands. For the nighttime bands, you typically want to select a time when the MUFs at the hop midpoints (called *control points*) are as low as possible, yet still above the desired operating frequency. This minimizes signal loss from D-layer absorption.

Great Circle Paths

As airplane pilots and seasoned air travelers well know, determining the correct direction in which to travel to get from one area of our globe to another is not always as simple as it appears. Using the traditional wall map found in schools, it would appear a traveler could board a jet plane one morning at Reagan International Airport outside Washington, D.C., and, by traveling due east, arrive in Lisbon, Portugal, in time for dinner. In truth, if you head due east from Washington, D.C., across the Atlantic, the first landfall will be West Africa, somewhere near Zaire or Angola. Why? Because the great circle bearing 90 degrees takes us far south of Portugal. Similarly, if we operate a Portuguese language shortwave broadcast station near Washington, D.C., and want to maximize the signal we deliver to listeners in Lisbon, our conventional wall map might cause us to beam our signal due east when the correct direction is northeast. The *geometry of spheres*, not flat surfaces, governs air travel and radio waves on our spherical world.

The problem of accurately representing the surface of a sphere on a flat two-dimensional surface such as a sheet of paper has confounded cartographers for centuries. Even today, despite the best efforts of the National Geographic Society and other experts, there is no one *projection* that is best for all possible uses.

For long-distance terrestrial radio communication, however, the single most important map projection of our earth is the *azimuthal equidistant projection*. That's quite a mouthful, but fortunately we can refer to it by its nickname: *great circle map*.

A *great circle* is any line drawn on the surface of a sphere that cuts the orb perfectly in half. Such a line must necessarily lie on a plane that passes through the center of the sphere. If we visualize passing an extremely large and rigid sheet of paper through our earth in such a way that two selected points on the sphere's surface and a point at the exact center of the sphere are all in contact with the paper, the intersection of the paper with the surface of the sphere forms a great circle connecting the two surface points. We will leave it to the mathematicians to prove that the great circle we just created is not only the largest circle that can be drawn on a sphere but also includes the shortest path for traveling between the two points while following the curvature of the surface.

Great circles are *circumferences* of a sphere. There can be an infinite number of them, but, except for one special situation, only *one* can be drawn between two specific points on the spherical surface. Generally, both a *short path* and a *long path* connect those two points. However, if the two points are exactly halfway around the globe from each other, they are called *antipodes* of each other, and the short path is identical in length to the long path. In fact, compass heading makes absolutely no difference when traveling between antipodal points—any road will take you there! Thus, an infinite number of great circles connects two points if they are antipodes.

To maximize the utility of the great circle approach to determining bearings, maps constructed according to *azimuthal equidistant projection* rules are employed. As implied by the name, distance along any radius line (azimuth) from the center of the map is precisely proportional to linear distance traveled over a spherical ground, and the same length regardless of compass bearing from the center of the map. (The same is definitely *not* true for *circumferential* distances on the map.) For example, if London and Los Angeles are equidistant from New York City, they will also be equidistant from the center of a great circle map centered on New York. This is *not* true of the Mercator projection; at higher latitudes (both positive and negative), an inch on the flat map corresponds to a much shorter distance on the earth than an inch near the equator does.

The other constraint for drawing an azimuthal equidistant projection is that each straight-line radius drawn from the center of the map (usually the user's location) to its periphery must correspond to a great circle on the earth. Thus, by tracing the single radius line that runs between the center of the map and any other location in the world, the user is assured of following the shortest possible path to that location.

Unlike a Mercator projection or other traditional wall map of the world, the shapes of land masses and bodies of water on a great circle map depend on the location chosen for the center of the map. When the rules for creating a great circle map are correctly followed, antipodal (or nearly so) continents or countries become "smeared" all around the periphery of the map. From the northeastern United States, for instance, Australia "morphs" to a rather strange shape.

For radio amateurs, the best centers for great circle maps are usually their own transmitting and receiving sites. But maps having their centers a short distance away are useable except for the most precise measurements and for very short distances from the center. Figure 2.35 is a great circle map drawn on Washington, D.C. For HF DXing, it is more than satisfactory for users located up and down virtually the entire east coast of the United States.

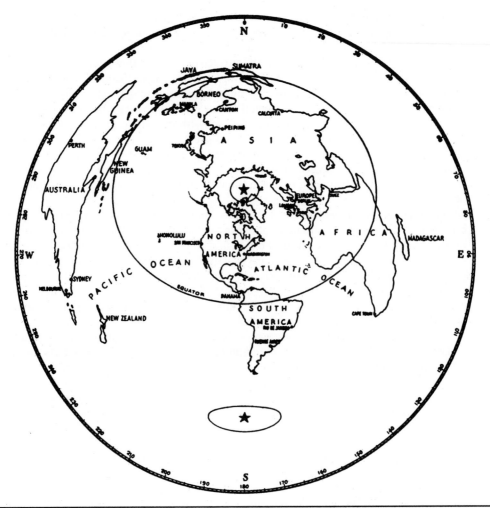

FIGURE 2.35 Azimuthal map centered on Washington, D.C. (*Courtesy of* The ARRL Antenna Book.)

For a while, antenna manuals published by the ARRL included a family of great circle maps centered on selected population centers around the world. Nowadays, plotting software is readily available for download from the Internet at no cost, either stand-alone or as part of a larger log-keeping and station control package. Whether using multiple hard-copy maps or recentering a software-generated map with a few keystrokes (to enter new latitude and longitude coordinates for the center of the map), seeing the world of HF radio propagation through the eyes of your fellow hobbyists on other continents is often highly informative.

Airplane crews and DXers both consult great circle maps to ascertain the compass heading that corresponds to the shortest distance between two points on (or just above) the surface of the earth. The navigator wants to get the airplane from here to there in the shortest possible time and with the smallest fuel expenditure; the radio operator wants to get a signal from here to there with the least path loss possible. As a general rule, the interests of both types of users are best served by selecting the *shorter* of the two great circle *segments* connecting the two points of interest, although that is not always the case for radio communications, as we shall see shortly.

Later in this book we take up *directional antennas*—antennas deliberately designed to maximize radiation in one direction or a small range of directions in exchange for accepting reduced radiation in other directions. With the exception of broadcast stations and government or commercial point-to-point stations, most directional antennas are intended to be rotated either electronically or mechanically. However, some highly directive MF and HF antennas for transmitting (the rhombic) or receiving (the Beverage) are too large to be rotated. Instead, one or more of each is constructed with its direction of maximum signal fixed on a desired region of our globe. An amateur low-band DXer in Maryland might choose to install a Beverage receiving antenna aimed at Europe. By referring to the great circle map of Fig. 2.35, he would conclude (correctly) that his Beverage should be oriented so as to provide maximum pickup of incoming signals centered on a *true compass bearing* of somewhere between 45 and 50 degrees. An amateur in Japan (roughly the same latitude as Washington, D.C., and Lisbon), on the other hand, would not aim due east or west to favor Europe on his receiving antenna; instead, he would first consult a great circle map centered on Tokyo and then aim his Beverage 25 to 30 degrees west of north!

Ionospheric Fading

It should be apparent from the preceding paragraphs that ionospheric communications are *statistical* in nature, rather than deterministic. One contributing phenomenon is *fading*—time-varying signal strength and phase at the receiver site.

Perhaps the most audibly dramatic form of fading—*selective fading*—is what we often hear when listening to analog voice transmissions on the MF and lower HF bands at night. Especially at night, at distances beyond the local ground-wave reception zone, the received signal will often consist of ground wave, sky wave, and multiple reflections of the sky wave from the ionosphere. Thanks primarily to the unceasing motion of the ionosphere, the received signal is an ever-changing mix of the transmitted signal arriving by uncountable numbers of paths, each having its own (time-varying) amplitude and phase relative to original signal. But the most distinctive aspect of this fading is caused by the fact that the ionosphere is *dispersive* and its reflection characteristics are different for even slightly different frequencies. Since an analog HF voice transmission consists of frequency components spread over a 3- to 20-kHz range around the carrier frequency depending on the service (amateur versus broadcast, for instance) and mode (amplitude-modulated double sideband with carrier versus single sideband suppressed carrier), the received signal is an eerie distortion of the original, similar to the *flanger* feature of modern-day sound mixers. It is such a distinctive sound that a pop music single released in 1959—"The Big Hurt" by Miss Toni Fisher—went all the way to number three on *Billboard* magazine's top 100 by replicating that sound.

Fading from any cause is a serious problem that can disrupt reliable communications and severely reduce intelligibility. Its effects can often be overcome by using one of several diversity reception systems. Three forms of diversity technique are popular: *frequency diversity, spatial diversity,* and *polarization diversity.*

In the frequency diversity system of Fig. 2.36, the transmitter delivers RF with identical modulating information to two or more frequencies simultaneously. Because the two frequencies will almost certainly fade differentially, one will always be stronger.

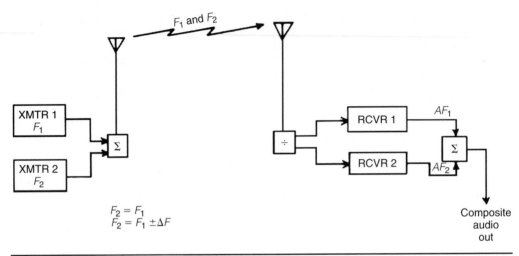

$F_2 = F_1$
$F_2 = F_1 \pm \Delta F$

FIGURE 2.36   Frequency diversity reduces fading.

The spatial diversity system of Fig. 2.37 employs a single transmitter frequency but two or more receiving antennas are used, typically spaced one half wavelength apart in the direction of the transmitter. The theory is that the signal will fade at one antenna while it grows stronger at the other. Diversity receivers utilize separate, but identical, *phase-locked* receivers tuned by the same master local oscillator and each connected to a separate antenna. Simple analog audio mixing or digital voting techniques based on the relative strengths of the signals keeps the audio output relatively constant while the RF signal at any one antenna fades in and out. In the past few years, use of diversity reception on the 160- and 80-m bands with this approach has become popular with serious weak-signal DXers, thanks to the Elecraft *K3* transceiver, which can be outfitted with a second, identical receiver that can be phase-locked to the main receiver. The audio outputs of the two receivers can be combined in an electronic mixer or kept separate to feed the left and right channels of stereo speakers or headphones. In the latter case, the operator's brain performs the final signal processing that results in enhanced ability to pull weak, fading signals out of the band noise.

Polarization diversity reception (Fig. 2.38) uses both vertically and horizontally polarized antennas to enhance reception. This form of diversity is based on the fact that our "always undulating" ionosphere slowly and randomly shifts the polarization of the transmitted signal as it refracts it back to the earth. As in the space diversity system, the outputs of the vertical and horizontal receivers are combined to produce a constant level output.

Using the Ionosphere

The refraction of MF, HF, and some lower VHF radio signals back to the earth via the ionosphere is the backbone of long-distance terrestrial MF/HF radio communications. As a direct consequence of the complex interplay between the three major ionospheric layers and their individual reactions to bombardment by the sun, in conjunction with path loss considerations, absorption, and the ever-changing position of the earth with respect to the sun, determining the best frequencies and times for communicating

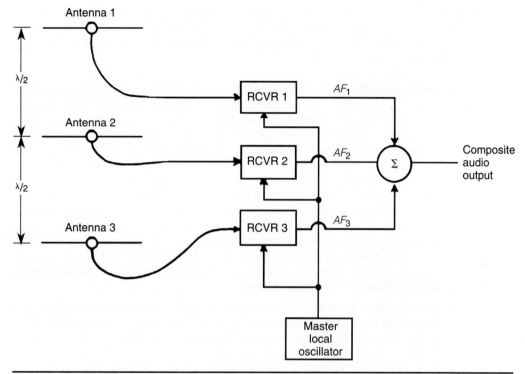

FIGURE 2.37   Spatial diversity.

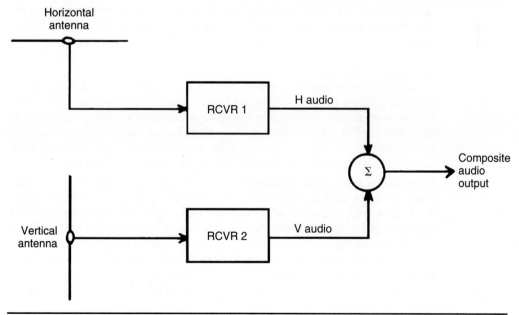

FIGURE 2.38   Polarization diversity.

with a fellow amateur or picking up a favorite shortwave broadcast station is not always easy. In this section we provide a handful of guidelines and "glittering generalities" intended to help readers more easily select a frequency or a time of day for best communications success.

- Because the MUF responds to the total level of radiation impinging on the ionosphere, it "follows" the sun. The higher the desired operating frequency, the greater the solar radiation levels needed to support ionospheric propagation. As a result, the 6-m amateur band at 50 to 54 MHz only rarely opens for F-layer DX propagation during the course of an average sunspot cycle. Similarly, the 10- and 12-m amateur bands may lie dormant for a few years of each sunspot minimum and go through periods of spotty openings on either side of the maximum. The same will be true to a lesser degree with the 15- and 17-m bands. Hence, all these bands are considered to be *daylight* bands, even though signals arriving by other propagation modes are sometimes heard on them during darkness hours. Because the critical frequency, f_c, is below many or all of these bands for much of a typical sunspot cycle, skip zones are almost always observed, although an outbreak of sporadic E propagation may occasionally make it appear that 20 m has no skip zone.

- Over the same cycle of sunspot activity, 20 m will most likely swing between being open for just a few hours each midday near sunspot minimum and bustling with activity around the clock near the maximum. Part of the popularity enjoyed by the 20-m band is that it is seldom completely "dead"; signals are almost always evident on the band, regardless of when activity is randomly checked.

- Frequencies below about 10 MHz are considered to be *nighttime* bands. High daytime absorption by the D layer is the primary reason they suffer a poor reputation when the sun is up, but this is largely a phenomenon of extended periods of extremely high sunspot counts such as was experienced in the 1957–61 time frame. Furthermore, moderate daytime absorption can be overcome or ameliorated by a combination of higher transmitter power and higher gain antennas at both ends of the circuit. Throughout most of an average sunspot cycle, both 80 and 160 m provide reasonable daytime propagation out to 1000 mi or so, and both expand their reach during the hours extending from shortly before sunset to shortly after sunrise to encompass worldwide DXing capability. During sunspot minima, f_c often drops below 3.5 MHz during the evening hours, with 80 m becoming useless for short skip contacts. Although f_c has been known to drop below 1.8 MHz and even into the AM broadcast band, that's a fairly rare happenstance.

- Forty meters is the great "transition" band. As such, it—more than any other band—exhibits multiple "personalities". During periods of high sunspot activity, f_c rises well above 7.5 MHz and the band serves as a great short- and medium-haul communications medium for much, if not all, of each 24-hour period. During these periods, there are apt to be many daylight hours each day when there will be no skip zone on 40. During sunspot minima, however, 40 is much more like 20 m—exhibiting a very pronounced skip zone (even at midday), providing surprising long-haul DX communications during daylight hours and shutting down completely at night as the MUF routinely drops below 7.0 MHz. These characteristics were especially evident during the 2008–09 minimum that signaled the end of solar cycle 23.

Because the characteristics of the MF and HF bands are so dependent on radiation from the sun, intelligently choosing a band to communicate with others requires some thought about the position of the sun with respect to the great circle path and the station at the other end. In general, paths that go through roughly equal segments of darkness and daylight are the most difficult to communicate over, especially during periods of low sunspot activity. In recent years, when sunspot activity has been generally low, attempts by North American amateurs to contact DXpeditions to the Himalayas, Afghanistan, and other remote central Asian regions have often led to disappointment. On one such recent occasion, by the time the sun had been shining in New York long enough for the MUF over the north Atlantic to support F-layer propagation at a compass bearing of 20 degrees, it was nearly midnight in Nepal, and 20 m had been "dead" for a few hours.

In contrast, when the sun is very active the MUF (even in the far northern or southern latitudes) may stay above 14 MHz long enough that 20 m may well be open to nearly all parts of the world simultaneously for at least a few hours out of each 24-hour period.

Skew Paths

Because the energy absorbed from the sun by the ionosphere is greatest near the equator and during the midday hours, MUF values at any given time vary substantially from point to point on the globe. In addition to causing the higher HF bands to be open only during daylight hours, sometimes equatorial MUFs are high enough to allow north/south propagation across the equator on a specific band while midlatitude or polar MUFs fail to support east/west or polar paths. If the 10-m band appears to be open between the United States and Chile or Argentina some noontime, communications from the United States to Europe or Asia will usually require dropping down a band or two to 12 m or 15 m, where, hopefully, the midlatitude and polar MUFs are high enough to support communications along those paths.

However, an alternative for trying to work east/west paths when the MUF is not high enough at midlatitudes or in the polar regions is to aim your antenna at a point on the globe that is open to you and that you have reason to believe is also open to the person or country you're trying to contact. Many times when the 10-m band is open only for north/south and transequatorial paths, it is still possible for stations in the United States and Europe to establish contact with each other by aiming their directional antennas at a region of the Atlantic Ocean near the equator and more or less midway between their respective longitudes. In this situation, the normal rules of aiming along the shorter great circle path do not apply, since the MUF in that direction is not high enough to support midlatitude paths. Instead, each station points its antenna at an angle to the other. This propagation mode is called *side scatter*.

One way to determine exactly where to aim your antenna is to find a clear frequency, run *full break-in* ("QSK") with one of today's transceivers, and transmit (call "CQ DX", for example) while rotating your antenna and listening for *backscatter* from your transmitted signal. With any luck, when you're aimed at the right area you'll be able to hear the last few milliseconds of your own transmissions (especially if it's international Morse code, generally referred to as *continuous wave*, or "CW") in your own receiver. By making small adjustments to the position of your antenna you'll be able to zero in on the best beam heading. Then you just have to hope that your faraway friends are doing the same thing!

If you were successful in working distant stations in this fashion, you would tell others that you worked Europe on 10 m, for instance, via a *skew path*. Over the years, skew paths have helped many DXers contact countries they might never reach otherwise. One of the most productive skew paths is obtained by aiming roughly due west from the northeastern United States and eastern Canadian provinces when trying to establish contact with stations in Japan, Korea, the Philippines, Southeast Asia, and the easternmost regions of Asiatic Russia—especially on 160 and 80 m. The reason for this can be seen by looking at the great circle map in Fig. 2.34. The great circle short path between the eastern part of the North American continent and the Far East goes right through the middle of the north auroral circle that surrounds the *magnetic north pole* (which is not a fixed location but which continually wanders around Alaska and northwestern Canada). Any perturbation of the earth's magnetic field increases signal absorption in the auroral regions and makes an otherwise difficult path almost impossible. In this example, the skew path is used to circumvent a region of high absorption rather than finding a way around an inadequate MUF.

Gray Line Propagation

The *gray line* is the twilight zone between the nighttime and daytime halves of the earth. This zone is also called the *planetary terminator* (Fig. 2.39). It tilts as much as 23 degrees on either side of the north-south longitudinal lines during the course of a year, depending on the season. During the vernal and autumnal equinoxes it runs directly north/south, and it reaches its most pronounced tilts on the longest and shortest days of the year.

The gray line is of interest because, for reasons not totally understood, signal levels between stations located on it are significantly enhanced—sometimes as much as 20 dB! Of course, the gray line is constantly in motion, moving in concert with the sun, so the effect is temporary but it can make the difference between a contact and no contact with

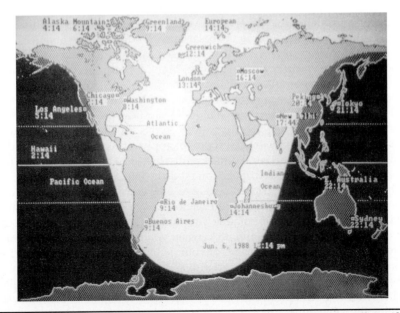

FIGURE 2.39 The planetary terminator ("gray line") provides some unusual propagation effects. (*Courtesy of MFJ Enterprises.*)

a rarely heard station in a remote part of the world. Because of the gray line, many serious DXers make sure they are carefully monitoring the HF bands during the twin twilight periods of their local dawn and dusk. It is not unusual for a gray line opening between two specific locations to be only 10 or 15 minutes in duration!

Because the tilt of the terminator changes over the course of a year, the paths that can benefit from it change as well. For instance, in Fig. 2.39, which is a gray line projection for early June, a station near Vancouver, British Columbia, Canada, might observe enhanced signal levels from stations in Bangladesh, Myanmar, or interior China. Three months earlier, in early March, that same Vancouver station might be more apt to hear surprisingly strong signals from Turkey, Saudi Arabia, and the horn of Africa.

Although our understanding of how gray line enhancement works is imperfect, some experts believe the effect is related to the D layer of the ionosphere. The D layer responds directly and rapidly to sunshine, disappearing very shortly after local sundown and returning shortly after the sun rises in the morning. Along the evening half of the gray line (Southeast Asia in Fig. 2.39), the D layer is rapidly decaying, but it has not yet built up along the morning half of the line (Vancouver). Thus, a fleeting period of less-than-normal absorption and enhanced signal levels may exist in a narrow "trough" along the terminator on any given day. Note that, for these long-distance paths, typically one station is near sunrise and the other near sunset.

Long Path

The great circle connecting any two points on the earth has two segments; for radio communications purposes we refer to them as the *long path* (major arc) and the *short path* (minor arc). If both paths are open, probably 99 percent of the time signal propagation between the two points is substantially more efficient along the short path—precisely because it is shorter! Not only is the incoming waveform at the receiving site greater when the transmitted power spreads out over a radiation envelope of smaller radius, but attenuation caused by extra hops and varying MUFs is less on the shorter path. However, despite greater path length, when the short path is not open at all there are many times when the long path can deliver an adequate signal to a specific location.

Figure 2.40 is a great circle map centered on New York, with night and day for 0200 Zulu (GMT) on April 1 shaded in. Suppose an amateur in the northeastern United States with a rotatable directional antenna had hoped to work a 5R8 station in Madagascar (off the east coast of Africa) on 15 m via short path (beam heading 80 degrees from New York) at that time. Assuming a solar flux level of about 100, most likely prolonged darkness over the Atlantic has caused the short-path MUFs between the two locations to fall below 21 MHz. But there is still a chance that the New York operator might be able to contact the 5R8 by turning his directional antenna 180 degrees to compass heading 260 degrees and testing the long path. Remember that the nearest important MUFs to the two stations on very long paths are about 1250 mi or so from each end of the path. So the first MUF of importance is about 1250 mi west of New York, in the southwestern United States, where the sun has just set. With any luck, the MUF there may still be above 21 MHz.

Similarly, at the other end of the link, the 5R8 has daylight an equal distance *east* of his station, directly opposite his short-path bearing to New York. Further, the entire path between his first MUF control point and the U.S. operator's control point is in daylight. Even though peak long-path signal strengths will likely be lower than short-path strengths (because the long path requires roughly twice as many hops between

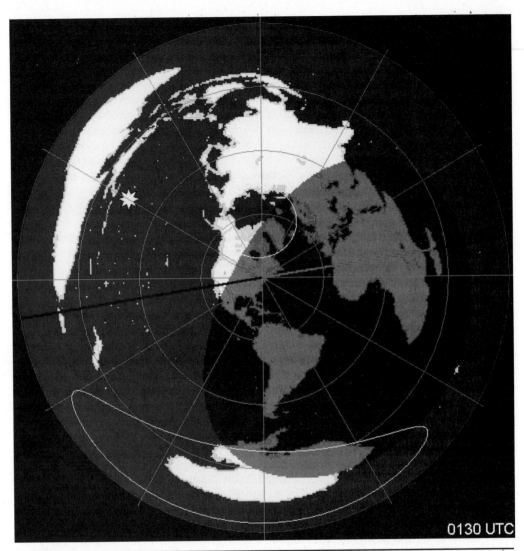

0130 UTC

FIGURE 2.40   0200Z April 1 great circle map showing long and short paths between northeastern United States and Madagascar.

New York and 5R8), there is still a moderately good chance that the two operators will be able to make contact the long way around. By switching to long path, we have put the path back into daylight where it counts—at the first MUF control point closest to each of us, and beyond.

A similar, but opposite, approach works well on 40 and 80 m between the northeastern United States and Japan during a winter afternoon on the U.S. east coast. In this case, the objective is to find MUF control points that are in darkness, even though either or both of the stations involved may be in daylight at the time of the contact. For a DXer in the northeastern part of the United States whose short-path propagation to the Asiatic Pacific region is often blocked by unsettled polar conditions, there are few HF operating experiences that compare with aiming the 40-m beam southeast on a winter afternoon and hearing an operator in the Far East answer a "CQ". It's *magic*.

Ionospheric Scatter Propagation

Ionospheric scatter propagation occurs when clouds of ions exist in the atmosphere. These clouds can exist in both the ionosphere and the troposphere, although the tropospheric model is more reliable for communications. Figure 2.41 shows the mechanism for scatter propagation. Radio signals from the transmitter are reflected from the cloud of ions to a receiver location that otherwise might not receive them.

There are at least three different modes of scatter from ionized clouds: *backscatter*, *side scatter*, and *forward scatter*. The backscatter mode is a bit like radar, in that the signal is returned to the transmitter site, or to regions close to the transmitter. Forward scatter occurs when the reflected signal continues in the same azimuthal direction (with respect to the transmitter), but is redirected toward the earth's surface. Side scatter is similar to forward scatter, but the azimuthal direction usually changes.

Unfortunately, there are often multiple reflections from the ionized cloud (shown as "multiple scatter" in Fig. 2.41). When these reflections are able to reach the receiving site, the result is a rapid, fluttery fading that can be quite deep.

Auroral Propagation

The visible *aurora* produces a luminescence in the upper atmosphere resulting from bursts of particles released from the sun 18 to 48 hours earlier. The light emitted is called the *northern lights (aurora borealis)* or the *southern lights (aurora australis)*. The ionized regions of the atmosphere that create the lights also form a reflective curtain at radio frequencies, especially at VHF and above, although auroral propagation is infrequently

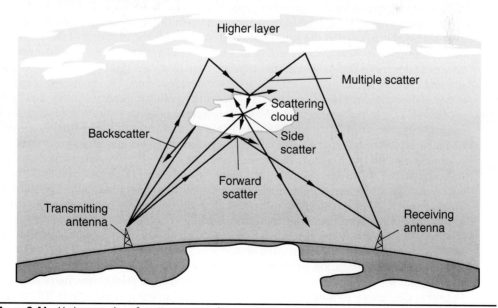

FIGURE 2.41   Various modes of scatter propagation.

observed on frequencies as low as 14 MHz. Auroral effects are normally seen at higher latitudes, although a few blockbuster auroras have been seen (and heard) as far south as Mexico! At those times, listeners in the southern tier of states in the United States are often treated to the reception of signals from the north being reflected from auroral clouds. Reflection off the aurora curtain is truly mirrorlike, and radio communication out to 1500 mi or so is possible. Numerous Internet sites, including some operated by U.S. government agencies, provide early aurora alerting services based on the expected time delay between visual observance of a flare on the sun's surface and the subsequent disruption of the earth's magnetosphere by arriving solar particles.

Meteor Scatter Propagation

Scatter propagation has been exploited mostly at VHF—not just with the ionosphere, but from meteor trails and man-made orbiting objects as well. When meteors enter the earth's atmosphere, they do more than simply burn up. The glowing meteor leaves a wide, but very short duration, transient cloud of ionized particles in its path. These ions act as a radio mirror that permits short bursts of reception—especially high-speed CW—between sites correctly situated. Meteor scatter reception is not terribly reliable. Its primary value for radio amateurs is to make communications possible over extended paths that are not normally available. However, at least two companies offer meteor scatter communications services for commercial users.

Propagation Predictions

Thanks to the relentless march of science and technology, propagation forecasting is gradually morphing from art to science. We now have a variety of solar sensors and monitors, both terrestrial and space-based, both simple and sophisticated, to help us better understand the ebb and flow of solar activity. Some propagation events—especially those triggered by the arrival of *particles* that can be seen leaving the sun hours before they arrive here—can be fairly well predicted in the short term, although we still may not be able to predict the timing of whatever caused the outburst of the particles themselves. We can reasonably consistently build a *causality trail* between the appearance of a sunspot or group of sunspots rotating back into view on the visible side of the sun and changes in the solar flux index over the next few days. We can follow weather fronts as they track across our continent and identify patterns likely to spawn clouds of sporadic E-skip in the next few hours. We can receive telephone alerts notifying us of the high likelihood we'll be able to see visible aurora borealis tonight or tomorrow night, based on the fact that a flare-up on the sun has sent a bundle of charged particles streaming toward the magnetic belts surround our globe. But we still cannot predict with any certainty or reliability what the instantaneous state of our ionospheric bands will be like at any given minute of any given hour on any given day. Our MF and HF ionospheric frequencies continue to surprise us in the details, even as we have an ever-improving ability to predict what we'll find overall.

In the decade following publication of the fourth edition of this book, distribution of propagation predictions has shifted almost totally from the print media to the Internet. Short-term propagation predictions that were a month old never were too great, anyway, but for looking at monthly averages, the family of prediction charts that appeared in *QST* every month (Fig. 2.42) was useful for seeing general trends. Today we are blessed to have not only near-instantaneous access to the forecasts of the best minds

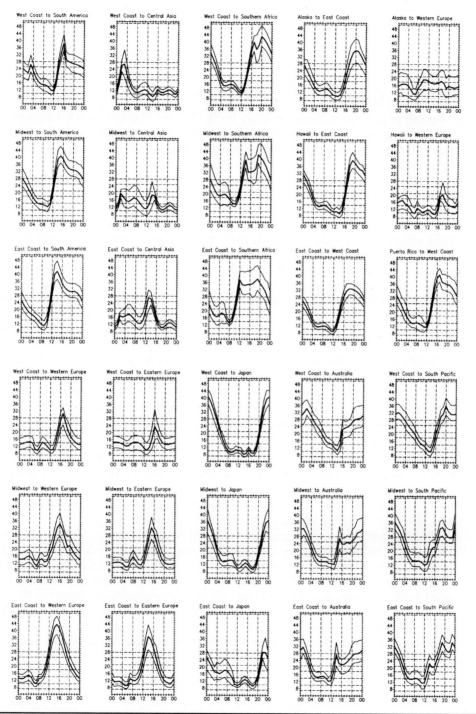

Figure 2.42 Propagation prediction charts from *QST* have been supplanted by freeware applications and the Internet. (*Courtesy of the American Radio Relay League.*)

in the business but also near-total access to as much of the raw data as we wish, should we choose to make our own predictions.

Appendix B includes an extensive list of Internet sites pertaining to HF propagation.

Propagation Software

A number of popular software packages for the PC and Mac have tackled ionospheric signal propagation from different directions. Although not as central to the topic of this book as the antenna modeling software described in Chap. 25, a few words are in order about some of them.

W6ELProp is freeware that can be found on the Internet. It is no longer maintained by its author, but it remains an easy-to-use Windows application with a user interface that is reminiscent of its DOS predecessor. *W6ELProp* focuses on conventional F-layer propagation; as such, it does not incorporate some of the "lesser" modes that may enhance signals, such as skew paths. It uses approximations that occasionally limit its accuracy, but it is a wonderful program for quickly getting the feel of propagation between any two points on the globe. In addition to the software itself, a number of other personal sites have published their own unofficial how-to guides for this chestnut.

HamCap is freeware offered by Alex Shovkoplyas, VE3NEA. It relies on a separate propagation prediction program called *VOACAP*, which must be downloaded and installed before HamCap can be used. *VOACAP* was originally developed by NTIA/ITS for the Voice of America.

VE3NEA also offers *IonoProbe*, which monitors and displays a number of valuable *space weather* parameters on a near-real-time basis.

Other software to check out on the Internet includes *CAPMan* (and its successor, *WinCAP*), *DXProp*, and *HF-Prop*. For VHF/UHF applications, look into *Radio Mobile Deluxe*, *Ground Wave Prediction System*, and *RFProp*.

Terrain Analysis

With few exceptions, antenna modeling packages for PCs or Macs assume a perfectly flat ground extending infinitely far in all directions from a point directly beneath the antenna in question. But few radio sites can lay claim to such a terrain, and the question naturally arises: What is the effect of the terrain on signals?

HFTA (High Frequency Terrain Assessment) is a program designed by R. Dean Straw, N6BV, to answer just that question. The user must supply details of his or her location to the software because some propagation statistics (arrival angles, for instance) vary with the user's region of the globe. The program can also download detailed public data files describing terrain contours in the immediate vicinity of the user. Alternatively, the user may be able to manually create a customized profile for his/her site by reading the necessary information off topographic maps of the area and manually entering it into the program. Once it has been provided with all the necessary data, HFTA assesses the terrain that may stand between the user's antenna site and a clear shot at the ionosphere at a variety of takeoff and arrival angles, examining both *reflection* and *diffraction* modes of propagation.

In operation, HFTA creates a profile for a single compass direction. If profiles for multiple headings are required, separate passes through the process must be completed. Typically, radio amateurs who wish to locate their antennas so as to maximize their ability to work DX find that a handful of profiles, one in the direction of each major population area, is sufficient. For example, a DXer or contest operator in the northeastern United States would profit most from profiles for 45 degrees (Europe, the Near East, and central Asia), 100 degrees (Africa), 170 degrees (Central and South America), 270 degrees (Australia and New Zealand), and 345 degrees (Western Australia, Japan, and the Far East). For this amateur, the 270-degree profile would also serve to identify terrain issues for most domestic U.S. beam headings, but amateurs in other parts of the country might need a sixth profile for that.

Of course, the greatest and best use of HFTA profiles is *before* buying a site intended to be a stellar radio location! But it also has great utility after the fact for identifying how high one's antennas for any given band must be to optimize the probabilities associated with the statistical nature of long-distance ionospheric propagation.

The Magic of Radio

Because so many of the known modes of propagation depend on external factors that are constantly varying over time, our ability to communicate—especially via the ionosphere—is a *statistical* one. Just because we were able to pick up the microphone and talk with another hobbyist halfway around the world on 20 m yesterday doesn't mean we'll be able to do it again today. The work we do to improve our antennas, the reading we do to understand the subtleties of propagation, the time we spend tuning across the bands, analyzing the noises we hear, listening for the weakest of signals—all are intended to tilt the odds more toward our favor. But ultimately we are at the mercy of Mother Nature, who is not above throwing us a curveball now and then.

Yet, for many of us, it is this statistical uncertainty that is at the very heart of the joy we find in communication by radio. For many, it is the unique sound of a weak signal fading in and out of the band noise that is the *magic* of radio. And, as in poker or bridge, where the number of different hands that can be dealt far exceeds the number that can be played in one's lifetime, so too is there a nearly infinite variety of possible band conditions awaiting us each time we sit down at our receiver and put on the headphones. The unpredictability of ionospheric HF propagation is without a doubt one of the lifelong draws of radio communications for tens of thousands, the author included. It's magic!

CHAPTER 3

Antenna Basics

An antenna is an example of a *transducer*—a device that converts one form of energy to another. Other examples include audio loudspeakers (electronic signals → sound waves), thermocouples (temperature changes → electrical signals), and woodstoves (stored chemical energy → heat).

An antenna converts time-varying electrical currents that are confined and guided within a circuit or transmission line to a radiated electromagnetic wave varying at the same rate and propagating outward through space—completely independent of the circuit that produced it. It is not unreasonable to visualize this process as "freeing" the electromagnetic waves created by the time-varying currents and "launching" them into space—much like a slingshot launches a projectile. As we shall see in future chapters, the *efficiency* with which an antenna performs this conversion is one of the key measures of its "goodness", and devices that accomplish this conversion efficiently share special physical and geometrical attributes.

In this chapter we shall summarize just enough basic electronics and electromagnetic theory to allow us to develop a feeling for the physics of antennas. That's about all we can do, because an in-depth knowledge of antenna theory requires a command of various advanced mathematics techniques, including calculus, coupled with completion of specialized antenna courses usually not taken until graduate school following the completion of a four- or five-year bachelor's degree in electrical engineering or physics.

Circuit Fundamentals

In this chapter and throughout this book, we will assume (without deriving or proving)—and frequently use—some fundamentals of physics and electronics engineering:

- *The superposition principle.* Unless otherwise noted, all components, circuits, systems, antennas, transmission lines, and propagation media discussed in this book are *linear* in their operation. That is to say, if an input voltage of 1 V applied to some device or system of devices results in an output of, say, 1 A (ampere) elsewhere in the system, an input of 5.32 V will result in an output of 5.32 A. In other words, output is *directly and exactly proportional* to input. Further, we assume the devices are *bilateral*. That is, if we switch things around and swap what we call "input" and "output", applying 5.32 A at the old output (now the new input) will result in a voltage of 1 V at the original input (now the new output). Finally, if we apply two different input signals to a device, the output from the device will be the sum, or *superposition*, of the two signals. Example: We apply a 1-V signal oscillating at a frequency of 3 MHz and a 4-V signal oscillating at 3.1 MHz to the original input of the aforementioned device. At the output we measure a 5.32-V 3-MHz signal and a 21.28-V 3.1-MHz signal.

- *Electric charge*. The smallest unit of electric charge is the charge of a single electron. By superposition, the total charge in a localized region is the sum of all the *net*, or *excess*, electrons in that region. For instance, if we deposit 1000 excess electrons on a glass rod by rubbing it with a cloth, the effect of that rod on a sensitive meter will be 1000 times the effect of a single electron. But note that like charges repel [each other]. If two electrons are forcibly brought near each other and then released, they will instantly try to separate and get as far from each other as the space they're in will allow. Thus, if we can add 1000 free electrons to a solid metal ball, for example, they will all move quickly to spread themselves around the surface of the ball, since this is a distribution pattern that represents the farthest they can get away from each other.

- *Current flow*. During his seminal experiments with electricity, Benjamin Franklin arbitrarily defined the "plus" and "minus" terminals of batteries, as well as "positive" and "negative" current flow in wires. Subsequently, other researchers determined that the primary conduction mechanism in metals consisted of electrons being broken away from their parent atoms and moving along the wire, from atom to atom, in response to an applied voltage. Unfortunately, Franklin's conventions turned out to be at odds with the electron model of the atom developed a century later, but they survived the conflict and continue in use to this day. As a result, even though we speak of positive current leaving the + terminal of a battery and flowing through a circuit and back into the – terminal of the battery, in truth electrons exit the – terminal of the battery, flow through the external circuit in the opposite direction to that of "conventional" current flow, and reenter the battery at its + terminal. Throughout this text we will often describe currents as if they comprise moving positive charges, but the reality is that it's the electrons that are actually in motion. This is a little easier to stomach if we consider the positive ion left behind by the mobile electron as having a "hole" where the electron used to be, and positive current flow is then the *apparent* motion of these holes along the length of wire, in a direction opposite to that of the electrons themselves.

- *Resistors*. A resistor is a two-terminal circuit element with the idealized characteristic

$$V = I \times R \ \ \text{or} \ \ R = \frac{V}{I} \tag{3.1}$$

which is equivalent to saying the *voltage V* (in volts) across the resistor is equal to the *current I* (in amperes) flowing through the resistor multiplied by the *resistance R* (in ohms) of that specific resistor. A useful analogy is a simple home plumbing system. The water pressure (analogous to voltage) causes the flow of water (current) through pipes (resistance). (The resistance of a pipe is directly proportional to its total length and inversely proportional to its cross-sectional area.) Similarly, the application of a constant voltage across the resistor terminals will result in a steady current flow through the resistor until the voltage is removed. The power dissipated in a resistor is given by

$$P = V \times I \text{ or, alternatively, } P = \frac{V^2}{R} \text{ or } P = I^2 \times R \tag{3.2}$$

Note that all three forms of Eq. (3.2) are equivalent and lead to the same answer.

The resistance of an ideal resistor (i.e., one with no capacitance or inductance) does not depend on frequency.

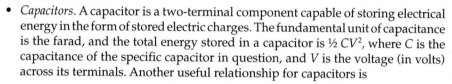

- *Capacitors.* A capacitor is a two-terminal component capable of storing electrical energy in the form of stored electric charges. The fundamental unit of capacitance is the farad, and the total energy stored in a capacitor is ½ CV^2, where C is the capacitance of the specific capacitor in question, and V is the voltage (in volts) across its terminals. Another useful relationship for capacitors is

$$I = C \times \frac{\Delta V}{\Delta t} \tag{3.3}$$

where I (in amperes) is the current through a capacitor of value C when the voltage across the capacitor terminals changes by some amount ΔV (in volts) over a short time interval Δt (in seconds). A simple capacitor you can make at home consists of two identical sheets of aluminum foil insulated from each other by a sheet of waxed paper of the same size.

NOTE *The only time current flows on the leads of a capacitor is when the voltage across its terminals is changing: I = C ΔV/Δt.*

A capacitor is said to exhibit *capacitive reactance*. Reactance works much like resistance to impede the flow of current in a circuit but the two are very different:

○ The reactance of an ideal capacitor does not dissipate any power.

○ The reactance of a capacitor is frequency dependent.

The magnitude of capacitive reactance is defined in Eq. (3.4):

$$X_C = -\frac{1}{2\pi f C} \tag{3.4}$$

where f is the frequency in hertz and C is the capacitance in farads. Alternatively, f can be in megahertz (MHz) and C in microfarads (μF). The minus sign is important in distinguishing the effect of capacitors from that of inductors.

- *Inductors.* An inductor is a two-terminal component capable of storing magnetic energy in the form of current, or moving electric charges. The fundamental unit of inductance is the henry (abbreviated H), and the total energy stored in an inductor is ½ LI^2, where L is the inductance of the device in question and I is the current (in amperes) going through it. Another useful relationship for inductors is

$$V = L \times \frac{\Delta I}{\Delta t} \qquad (3.5)$$

where V (in volts) is the voltage across the inductor of value L when the current through the inductor changes by some amount ΔI (in amperes) over a short time interval Δt (in seconds). A simple inductor you can make at home is formed from a length of insulated wire (such as the wire left over from the installation of your new garage door opener) wrapped around your hand a few times.

NOTE *The only time a voltage can exist across the two terminals of an inductor is when the current through the inductor is changing: $V = L \, \Delta I / \Delta t$.*

As discussed for capacitors, an ideal inductor does not dissipate power, nor are the current through, and the voltage across, the inductor in phase. The magnitude of inductive reactance is

$$X_L = 2\pi f L \qquad (3.6)$$

where f is in hertz and L is in henries (or megahertz and microhenries). Note that inductive reactance X_L is defined as positive, in contrast to capacitive reactance in Eq. (3.4).

- *Complex impedance.* Suppose we apply a sinusoidal wave of frequency f across two terminals or nodes in a circuit consisting of interconnected resistors, inductors, and capacitors. If we measure, with an oscilloscope or other appropriate piece of test equipment, the voltage and current relationship at those terminals, we will generally find that while the current also varies sinusoidally at frequency f, it is not exactly in phase with the voltage as it would be if the circuit consisted of only resistors. The capacitors and inductors in the circuit cause *phase shifts* of varying degrees between currents and voltages in different parts of the circuit.

  At any one frequency we can describe the relationship between the voltage and the current at the two measurement terminals in terms of a single value of resistance in series with a single value of reactance. In equation form, we say

$$Z_{IN} = R_{IN} + jX_{IN} \qquad (3.7)$$

or, more simply,

$$Z = R + jX \qquad (3.8)$$

This is known as a *complex impedance* because of the use of the *complex plane* mathematical operator j ($j = \sqrt{-1}$) to represent reactances. The magnitude of X may be either positive or negative, depending on whether the net reactance seen appears to be inductive or capacitive, respectively. For correctness, we speak of R as the *resistive part* and X as the *reactive part* of the impedance Z. Both R and X are in ohms or multiples thereof.

NOTE *Both R and X are functions of frequency, and their values depend in general on the values of all components in the circuit. Thus, Z_{IN} for an arbitrary circuit may be 50 + j47 Ω (ohms) at one frequency and 35 – j17 Ω at another frequency!*

- *Resonance.* If a simple series circuit contains a resistor, an inductor, and a capacitor all connected end to end, a meter across the two end terminals would "see" an input impedance at some frequency *f* of

$$Z_{IN} = R + jX_L + jX_C$$
$$= R + j(X_L + X_C) \tag{3.9}$$

Since X_C is always a negative number, this amounts to subtracting capacitive reactance from inductive reactance. In other words, looking into this simple network we can only tell what the *difference* between the inductive and capacitive reactance is. Another way of saying this is that we can see only the *net reactance*. It turns out that we can represent the two-terminal impedance of *any* configuration of *R*, *L*, and *C* at a single frequency with Eq. (3.9) as long as we understand that *R*, *L*, and *C* in the equation are functions of frequency and generally bear absolutely no relationship to *specific* resistors, inductors, or capacitors in the circuit.

A circuit is said to be *resonant* or *in resonance* at some frequency *f* when $X_C = -X_L$ so that the second (or "*j*") term in Eq. (3.9) goes to zero. When that occurs,

$$-\frac{1}{2\pi f C} = -2\pi f L \tag{3.10}$$

Dividing both sides by –1 and multiplying both sides by $2\pi f C$ leads to

$$1 = (2\pi f)^2 LC$$
$$= (2\pi)^2 f^2 LC \tag{3.11}$$

Swapping sides and rearranging a few factors produces

$$f^2 = \frac{1}{(2\pi)^2 LC} \tag{3.12}$$

or

$$f = \left(\frac{1}{2\pi\sqrt{LC}}\right) \tag{3.13}$$

Equation (3.13) is one of the most important equations of all radio and electronics design; we will have occasion to use this relationship at numerous points in this book, such as when we discuss the use of parallel-resonant lumped-component *traps* in shortened dipoles and Yagi beams.

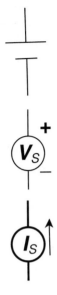

- *Sources.* Electrical and electronic circuits (including antennas and transmission lines) are made operational by being supplied with electrical charge. Devices designed for this specific purpose are called power supplies, batteries, or *sources*. A battery, for instance, is nothing but a supplier, or source, of readily accessible electric charge resulting from a carefully designed internal chemical reaction. Depending on the exact circuit or antenna we are examining, we may prefer to power it with a voltage source or a current source. These are idealized representations of practical batteries and power supplies. An ideal voltage source delivers a specified voltage to a pair of terminals on a circuit or antenna *regardless of the current through the source.* Similarly, an ideal current source passes a specified current between two terminals regardless of the resulting voltage across the source. In real life, no source is ideal but most are good enough that these approximations are valid for the range of practical currents and voltages we will deal with in this book. Unlike batteries and most power supplies, sources can supply alternating (ac) voltages and currents. This is especially important for radio frequency (RF) systems. Antennas, in particular, have value only when attached to a source of RF energy.

For virtually all the real-world antennas and RF circuits we will be considering in this book, the fundamental unit values (farads and henries, respectively) for capacitors and inductors are much too large. Thus, you will see reference to millihenries, micro-henries, and nanohenries, and to microfarads, nanofarads, and picofarads, where these prefixes have the following meanings:

milli   m 10^{-3} or 0.001 times

micro   μ 10^{-6} or 0.000 001 times

nano   n 10^{-9} or 0.000 000 001 times

pico   p 10^{-12} or 0.000 000 000 001 times

Similarly, often the fundamental unit of resistance (the ohm) is too small and we will find it more convenient to speak in terms of kilohms or megohms:

kilo   k 10^{3} or 1 000 times

mega   M 10^{6} or 1 000 000 times

Example 3.1   In commonly accepted terminology, a 0.001-μF capacitor is spoken of as a "double ought one". The "microfarad" portion is implied. A 0.001-μF capacitor is the same as a 1-nF (nanofarad) or 1000-pF (picofarad) capacitor (obtained by moving the decimal point either three or six places to the right, respectively).

◆

Fields

Originally conceived as a way of helping us understand and predict the observed effect of charges and magnets on other objects in the absence of any direct physical connection between the source and the affected objects, fields have acquired a life of their own. This "action at a distance" can be thought of as analogous to the way gravitational forces act.

If a voltage is applied between two points (in space, in a circuit, etc.), an *electric field* is said to exist between those points. Similarly, if a current flows in a wire or other conducting medium, a *magnetic field* is said to surround the conductor. A common high school physics experiment provides graphic proof of this: As shown in Fig. 3.1, currents flowing in the same direction in two parallel wires will cause the two wires to be *attracted* to each other, and unrestrained sections of the two wires will actually move toward each other! Reversing the direction of current flow in *one* of the wires will push the wires *apart*. The force between the wires is proportional to the product of the magnitudes of the two currents and inversely proportional to the distance between them. Note a very important aspect of magnetic fields: The force is at right angles to the direction of current flow!

Whether electric or magnetic, however, *all* fields originate with electric charges. A motionless electric charge creates a *static*, or non-time-varying, *electric field*. A *moving* electric charge traveling at a constant velocity creates a *static magnetic field* (which is at right angles to the direction of electron motion). Since a steady electric current in a wire is the result of many electric charges moving with a constant speed past any point in that wire, we can conclude that such a current will result in a steady magnetic field around the wire.

However, it is only when we rapidly *vary* the velocity of electric charges that electromagnetic radiation (and, hence, radio waves) becomes possible. In the same way that you cause your vehicle to go from being stopped at a traffic light to moving smoothly at 30 mph along a city street by *accelerating* it to the new speed after the light turns green, the only way to cause the velocity of electric charges to vary is by accelerating or decelerating those charges. In short, the *possibility* that radio waves can exist is the direct result of electric charges undergoing acceleration or deceleration. While there is more than one way to accelerate charged particles, the only method we need to concern ourselves with in this book is the application of rapidly varying voltages and currents to wires or other conducting structures.

Two other conditions are necessary for gener-

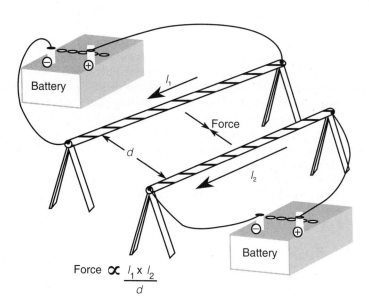

Force $\propto \dfrac{I_1 \times I_2}{d}$

FIGURE 3.1 The force between current-carrying parallel wires.

ating radio waves at a useful level. The first is that *linked* time-varying electric and magnetic fields must exist in the same region of space. (This is the thrust of Maxwell's equations.) The other requirement is that the *geometry* of the *radiator* must support and even "encourage" the tendency of these time-varying electromagnetic fields to break away from it and propagate into free space. Developing a familiarity with some of those geometries is one of the purposes of this book.

Voltage and the Electric Field

As implied by the aluminum foil suggestion earlier, one of the simplest capacitors consists of two parallel metal plates. The capacitance of this geometry is given by

$$C = \frac{\varepsilon A}{d} \qquad (3.14)$$

where C = capacitance in *farads*
    ε (epsilon)   = constant called the *permittivity* of the dielectric
    A = area of one of the plates
    d = spacing between plates in same system of units as A; very much
                 less than $\sqrt{A}$ (abbreviated as $<< \sqrt{A}$)

The magnitude of ε in a vacuum—called ε_0—is 8.85×10^{-9} F/m (farads per meter) or, more frequently, 8.85 pF/m. Other common dielectrics have larger permittivities that are usually expressed as a multiple of ε_0 by specifying a *relative permittivity*, ε_r, such that

$$\varepsilon = \varepsilon_r \varepsilon_o \qquad (3.15)$$

Examples of ε_r for some common materials include air (1.0006), plywood (2.1), waxed paper (3.7), formica (6), and distilled water (81).

When a battery is connected across the plates of a capacitor (Fig. 3.2), negative charges temporarily flow from the negative terminal of the battery to the lower plate, where they repel the negative charges on the upper plate. This causes the negative charges from the upper plate to travel along the upper wire into the battery's positive terminal, leaving the upper plate with a shortage of electrons—i.e., positively charged.

Because a voltage difference V_{BAT} now exists between the two parallel plates, an electric field E also exists between the plates. Its direction is (by definition) from the positive to the negative charges and at right angles to the conducting plates at the point of contact with the plates. For this geometry, its magnitude is constant everywhere between the plates (ignoring fringing around the edges) with a value $E = V/d$, where d is the distance (almost always measured in meters) between the plates. Following the very short initial charging interval, the E-field is static because the voltage on the capacitor plates is constant. Note that common convention for representing an electric field graphically not only shows the direction of the field at each point in space but uses the spacing, or *density*, of the drawn field lines as a way of showing variations in the approximate magnitude of the field throughout a region.

Current and the Magnetic Field

As shown in Fig. 3.3, a charge *moving* at a constant velocity along a straight wire constitutes a steady current that produces a *static* magnetic field that is everywhere at right

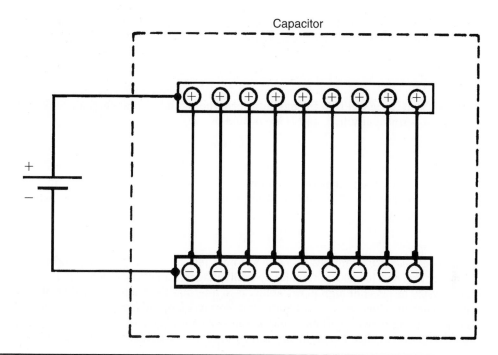

FIGURE 3.2   Charges on the plates of a capacitor.

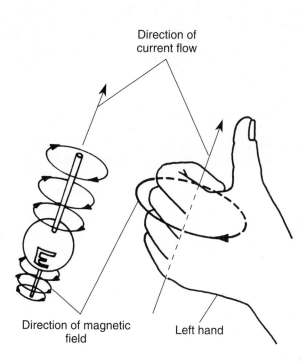

Direction of
current flow

Left hand

Direction of magnetic
field

angles to the current—namely, a circular field surrounding the conductor. Thus, magnetic fields and the currents that create them are always at right angles to each other. Similarly, the force on a moving electron (which constitutes a current, however small) from a magnetic field is proportional to the electron's velocity (i.e., to the current in a wire) and always at right angles to its direction of travel. Thus, a magnetic field can cause a moving electron to change direction, but it cannot cause it to speed up or slow down!

The strength and orientation of the magnetic field (usually represented by H) in a region of space is

FIGURE 3.3   Magnetic field around a half-wave antenna.

called the *flux density, B*. For an infinitely long, straight wire carrying a steady current I, the magnitude of the flux density at a point located a distance R from the wire is

$$B = \frac{\mu I}{2\pi R} \tag{3.16}$$

where μ is the *permeability* of the medium surrounding the current-carrying wire. In any region of permeability, μ, B, and H are related by

$$B = \mu H \tag{3.17}$$

With the exception of *ferromagnetic* materials, most permeabilities for common media are very close to that of a vacuum, $\mu_0 = 400\pi \times 10^{-9}$ H/m (henries per meter), or, more commonly, $\mu_0 = 400\pi$ nH/m. The permeabilities of other media are expressed relative to μ_0 by

$$\mu = \mu_r \mu_o \tag{3.18}$$

In *ferromagnetic* materials, μ_r ranges from a few hundred for cobalt and nickel to 5000 for iron and 100,000 or more for specialty magnetic materials such as *mu-metal* and *permalloy*. The effect of high permeability is to intensify the magnetic field created by a given current. Thus, power and audio transformers employ iron cores for highly efficient coupling between primary and secondary windings, and RF transformers often use ferrite cores called *toroids* for the same purpose, as we shall see in Chap. 24.

Displacement Current and Maxwell's Equations

If you have ever been running an appliance, such as a vacuum cleaner, in your home and accidentally unplugged the cord from the wall, you know the appliance immediately stopped. Its electrical circuit had been broken, and no current could flow to its motor. One of the earliest questions asked about capacitors was: How can (the temporary charging) current flow in the wires connecting the capacitor plates to the battery when the circuit is always broken in the space between the plates of the capacitor? A Scot, James Clerk Maxwell, answered this question in 1861, when he proposed the existence of a *displacement current* (in contrast to the *conduction current* in a wire) in the space between the plates. This current is the same $I = C\Delta V/\Delta t$ already mentioned in our definition of a capacitor.

Because the displacement current was unaffected by locating the capacitor plates in a vacuum, Maxwell concluded there must be some invisible yet material *medium* permeating all of space. He called this medium the *æther*, and it took scientists another half century to conclude there was no such thing! Nonetheless, the displacement current ultimately allowed Maxwell to show that a changing electric field creates a magnetic field and a changing magnetic field creates an electric field. This created symmetric cross-coupling terms to be added to then-existing equations that purported to describe the relationship between electric and magnetic fields, and allowed Maxwell to summarize all the important characteristics of electromagnetism in a family of interrelated equations. Once derived and written down on paper, the solutions to these equations were recognized by Maxwell and other scientists and mathematicians of the day as being of the same form as those describing the propagation of sound waves and other mechanical vibrations through media such as air and water.

Maxwell summarized all that he knew about electromagnetism in his *Treatise* of 1865, and it was there that he *predicted*, purely on the basis of his mathematical equations, the *possibility* of electromagnetic waves—presumably traveling through his mystical æther—as well as the speed at which these waves must travel through space. Maxwell then went a step further: Based on the most accurate measurements of the speed of light at the time, he concluded that visible light was itself comprised of electromagnetic waves!

Nearly a quarter century passed before the German scientist Heinrich Hertz demonstrated in 1887 that he could create, transmit, and receive what we now call radio waves. Hertz thus took the first baby steps in determining the physical geometry of conductors (antennas, in other words) that would allow the generation and detection of the EM radiation predicted by Maxwell. Of course, Hertz did not have any of the conveniences or inventions of the modern world; in particular, he did not have a tunable receiver or transmitter. Instead, he generated a spark (which contains a nearly infinite number of frequencies) across a gap between two curved conductors of a certain length with some laboratory equipment and attempted to detect the fields from that spark with a similar set of conductors a few feet away. The signal from his detector was then used to make a laboratory meter wiggle. And so it was that Heinrich Hertz was quite likely the first to experience the *magic* of radio!

Linked Electric and Magnetic Fields

Now let's return to the example of charging a capacitor from a battery. The flow of these charges constitutes a conduction current in the connecting wires that lasts for a very short time—only until the voltage across the plates of the capacitor is equal to that of the battery itself. That interval depends, of course, on the amplitude of the current, which initially is limited only by the combined resistance of the wires plus the internal resistance of the battery itself. (That's why the wire in automobile *jumper cables* is so fat!) The final value of the current is zero, just as the final voltage across the capacitor is the full battery voltage, V_{BAT}. As shown in Fig. 3.4, both voltage and current follow *exponential* curves during the transition period.

If R_s is very small, the voltage and current at the capacitor are in *quadrature* relationship to each other; we say they are 90 degrees out of phase, and that

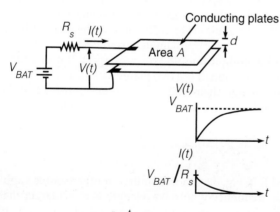

$$V_{CAP} = V(t) = V_{BAT}\left(1 - e^{-\frac{t}{R_s C}}\right)$$

$$I_{CHG} = I(t) = \frac{V_{BAT}}{R_s}\, e^{-\frac{t}{R_s C}}$$

R_s = the sum of all resistances in the circuit, including the resistance of the interconnecting wire and the internal resistance of the battery itself.

$$C = \frac{\epsilon A}{d}$$

Figure 3.4 Exponential voltage and current when charging a capacitor.

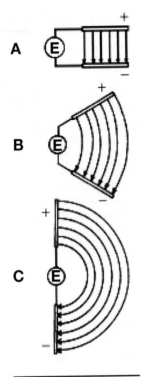

the current "leads" the voltage by that amount. Specifically, $I=C\Delta V/\Delta t$. In calculus, we would say the current flowing to the upper capacitor plate is the first derivative of the voltage across the capacitor. (Similarly, in an inductor the current "lags" the voltage by 90 degrees.)

Of course, once the capacitor plates are charged, a steady E-field exists between them. Since it is a *static* field, it does not create a magnetic field. However, during the very short instant when current *is* flowing in the wires connected to the capacitor, its amplitude is constantly changing as well, which means electrons are being accelerated or decelerated and a changing magnetic field encircles each of the wires. On each wire we have both a changing current and a changing voltage within the same region of space and thus the potential for radiation, depending on the length of the wires, the space between them, and other factors. (Have you ever turned on a lamp or appliance in your home and heard a "click" in an AM broadcast receiver? That's an example of radiation caused by the sudden change in current and voltage on the wire that runs between the switch and the appliance.)

Other geometries and orientations are possible for the two capacitor plates, and they will result in different E-field patterns in the space between the plates, as shown in Fig. 3.5. If the two plates of the capacitor are spread farther apart at one end than at the other, the electric field between them must curve to always meet the plates at right angles (Fig. 3.5B and C). The straight lines in A

Figure 3.5 Electric field between conducting plates at various angles.

become arcs in B, and approximately semicircles in C, where the plates are in a straight line, 180 degrees apart. But in addition to altering the orientation of the two plates with respect to each other, we can also change their shape. For instance, instead of the flat rectangular metal sheets we started with, we can make the two elements from cylindrical metal rods or wires.

The Hertzian Dipole

Now suppose the circle marked E in Fig. 3.6 is a transmitter or other source supplying sinusoidal RF energy at a single frequency whose wavelength is much larger than the length of the rods or wires in B. Electrons are simultaneously pushed onto one rod or wire and pulled from the other in direct response to the sinusoidal source voltage applied to the two rods. Of course, the amplitudes of the static and *quasi-static* E- and H-fields in the vicinity of the two rods are changing at the same sinusoidal rate as the applied voltage, but it turns out these fields exist only near the wires or rods and die out rapidly with increasing distance from the antenna.

Since the velocities of the electrons are constantly changing as they shuttle back and forth between the rods and the sinusoidal source, they are in a state of continuous acceleration and deceleration, thus setting up linked time-varying electric and magnetic fields in the space enclosing the rods. Above and beyond the conventional static and quasi-static fields, Maxwell's equations predict the existence of new fields resulting solely from the acceleration of electric charge in the region between the two rods. Spe-

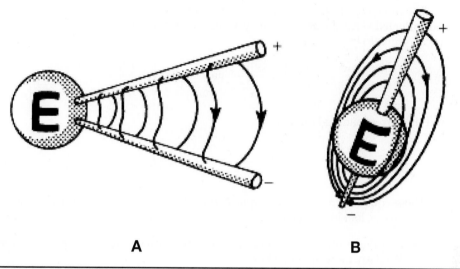

FIGURE 3.6   Electric field between wires at various angles.

cifically, for the thin rod geometry we've been using, accelerating charges along the rods create previously unpredicted magnetic fields and those "new" magnetic fields similarly create "new" electric fields, which in turn create . . . well, you get the picture.

Suppose we now attempt to observe the effects of these accelerating charges at a point in space far from the region around the rods. The acceleration of the electrons on the rods, coupled with the finite speed of light and *all* EM waves, causes us to detect an E-field that is at right angles to the radius line drawn between the rods and our observation point. Similarly, we detect a magnetic field that is also at right angles to the line between the rods and us, and at right angles to the E-field. In other words, both the E-field and the H-field are at right angles to the direction of travel from the antenna to the receiving point.

These fields are not the static and quasi-static fields of high school science classes, and they were totally unknown prior to Maxwell's treatise discovery of them! They can only be explained by the existence of a *transverse electromagnetic wave* originating from the region of the rods and traveling through space at the speed of light. It's as if the E- and H-fields associated with the accelerating electrons are detached from the rods and pushed out into space each time electrons slam against the end of either rod.

Further examination of the behavior of these fields at distant receiving points discloses the fact that these fields are an exact reproduction of the frequency and phase of the source voltage applied to the rods except for a delay that is proportional to the distance from the rods to the observation point. That delay is caused, of course, by the finite speed of light. Perhaps more surprising, however, is that if we attach a resistive load to our receiving antenna, we observe real power being dissipated! The E- and H-fields we are monitoring are two components of an electromagnetic wave propagating through space, and that wave is a *real* entity. That is, it carries energy with it and can do useful work wherever it goes, just as the sun's rays warm you by delivering energy in the form of heat when you lie out in them at the beach.

The received E-field at a point far from a very short (relative to wavelength) pair of rods oriented vertically and driven by a sine wave at frequency f_0 is described in spherical coordinates by

$$E_\theta = \frac{60\pi I_0 h}{\lambda r} \sin\theta \sin\left\{2\pi f_0\left(t - \frac{r}{c}\right)\right\} \tag{3.19}$$

and the corresponding H-field is

$$H_\phi = \frac{I_0 h}{2\lambda r} \sin\theta \sin\left\{2\pi f_0\left(t - \frac{r}{c}\right)\right\} \tag{3.20}$$

These equations are not as complicated as they may seem; let's break them down piece by piece, looking at each term:

- 60 is the result of combining a bunch of physical constants into a single number. We'll say more about it shortly.
- I_0 is the amplitude of the drive current at the center of the two rods; it simply tells us the obvious: Any field we measure at a distant point is going to be directly proportional to the drive current at the source.
- h is the total height, or length, of the antenna formed by the two rods, expressed as a fraction of the wavelength, λ. Implicit in the approximations allowing us to use this equation is the constraint that h is small with respect to λ. A good value might be $\lambda/20$ or less.
- r is the distance from the center of the two rods to our distant point.
- θ (pronounced "thay-ta") is the angle from the axis of the antenna. The $\sin\theta$ term tells us that the E-field drops to zero off the ends of the antenna and is maximum broadside to the antenna, where $\sin\theta = 1$. (See App. A for derivations and meaning of sine and cosine functions.)
- $\sin\{2\pi f_0(t - r/c)\}$ identifies the instantaneous amplitude of the E-field as a function of the drive signal frequency and the distance of the receiving point from the rods. Because this is an *argument* of a sine function, as it increases the resulting sine function simply goes through multiple cycles from –1 to +1 and back again.
- The subscript θ for the E-field means the only component of the E-field that has any amplitude far from the two rods is an E-field lying in the same direction as the axis of the rods. Although the axis of the rods is typically described as lying on the z axis of a cartesian coordinate system, the radiated wave is best analyzed in spherical coordinates. If not, we run the risk of E_θ becoming replaced by complicated equations for E_x, E_y, and E_z. An E-field comprised of solely E_θ is consistent with our earlier statement that the E- and H-fields are always at right angles to the direction of wave propagation.
- Similarly, the subscript ϕ for the H-field tells us that the only detectable component of the magnetic field at a distant point is in the ϕ direction as defined in

spherical coordinates; that is, it is a circular field whose axis is the same as that of the rods that form the source antenna.

For a given receiving antenna at a fixed, distant point we can combine all the geometrical and numerical factors of Eq. (3.19) into a single constant, k, that is valid for that particular spacing and orientation of the transmit and receive antennas. The resulting equation is

$$E_{RX} = k I_{TX(t-r/c)} \tag{3.21}$$

where $I_{TX(t-r/c)}$ is the drive current to the transmit antenna $t = r/c$ seconds prior to the time at which we measured E_{RX}.

As is true of our globe, a small surface area on a large sphere appears flat. So it is with radio waves: Although the waves break away from the source and propagate outward in space, forming an ever-expanding sphere, at any distant receiving site, the wave appears to be a *plane wave*.

If we divide Eq. (3.19) by Eq. (3.20), we find that all the terms cancel except for

$$\frac{E_\theta}{H_\phi} = \frac{60\pi}{1/2} = 120\pi \tag{3.22}$$

As it turns out,

$$\sqrt{\frac{\mu_0}{\varepsilon_0}} = \sqrt{\frac{4\pi \bullet 10^{-7} \text{ H/m}}{\frac{1}{36\pi} \bullet 10^{-9} \text{ F/m}}} = \sqrt{144\pi^2 \bullet 10^2} = 120\pi = 377 \ \Omega \tag{3.23}$$

μ_0 and ε_0 are, of course, the permeability and permittivity of a vacuum—free space, in other words. The square root of their ratio, or 377 Ω, is called the *impedance* of free space. Interestingly, the vacuum of free space has at least two characteristics (ε_0 and μ_0) and an impedance (measured in ohms). So the "nothing" of free space isn't exactly nothing, is it? Perhaps we should call it the *æther* . . . .

In the more general case, the E- and H-fields associated with a spherical or plane TEM wave propagating through an arbitrary medium characterized by μ and ε are related by

$$\frac{E_\theta}{H_\phi} = \sqrt{\frac{\mu}{\varepsilon}} \tag{3.24}$$

The antenna we have just described is known as a *hertzian dipole* because it is very short compared to the wavelength of the signal applied to it—exactly the situation with Heinrich Hertz's laboratory setup of 1887. The shape of this antenna's E-field pattern is totally described by the $\sin\theta$ term; it is doughnut-shaped in a three-dimensional view, and a figure eight when viewed in two dimensions with the axis of the dipole lying in the plane of observation. Compared to an isotropic radiator, which is totally fictitious but which—if it did exist—would radiate equally in *all* directions, the *directivity* of this antenna is 1.5, or 1.75 dBi (*decibels relative to isotropic*). From the standpoint of its pattern, this would not be a bad antenna, except for two "small" problems.

First, the input impedance of a hertzian dipole exhibits both resistance and capacitive reactance:

$$Z_{IN} = R_{RAD} + X_C \qquad (3.25)$$

where X_C is the quasi-static or low-frequency capacitance of the dipole's physical structure. For a very short dipole ($h \ll \lambda$),

$$R_{RAD} = 20 \left(\frac{2\pi h}{\lambda} \right)^2 = 800 \left(\frac{h}{\lambda} \right)^2 \qquad (3.26)$$

Thus, for a dipole with $h = \lambda/20$ (about the longest dipole this model is valid for), $R_{RAD} = 2\ \Omega$. As we will discuss in later chapters, antennas with very low feedpoint impedances are often not very efficient because the networks needed to match commonly available transmitter outputs to them are quite lossy themselves.

The other problem is the capacitive reactance term, which further complicates the problem of delivering power to the antenna in an efficient manner. In theory, a value of series inductance can be selected that will cancel out the capacitive reactance at f_0, but the $I^2 R$ losses in any practical inductor will dissipate most of the RF power delivered to the system.

As a result of these effects, the hertzian dipole is a marvelously impractical antenna, with typical efficiencies of well under 1 percent. However, it is a great reference device and serves as the basic mathematical building block for analyzing the vast majority of our real-life antennas.

The Half-Wave Dipole

Obtaining straightforward expressions for the far-field signal strength and input impedance of very short dipoles involved a lot of simplifying assumptions on the parts of many scientists and engineers over the past century or so. One of these assumptions is that the current in the short rods on the two sides of the signal source has the same magnitude and phase everywhere on the rods. This assumption is possible only because the longest dimensions of the rods are much, much shorter than a wavelength of the signal feeding the short dipole.

To obtain the corresponding equations for longer dipoles up to and including $\lambda/2$, a more complicated model, with even more simplifying assumptions, was called for. The mathematics needed is unbearably complicated, but the approach is reasonably straightforward. In particular, some of the key assumptions are:

- The longer dipole is modeled as a large number of very short dipoles (similar to the aforementioned one) laid end to end with the top (or right) end of one feeding the bottom (or left) end of the next.

- The amplitude and phase shift of the current in each small dipole element is assumed to be only slightly different from those of the elements on either side of it.

- At the ends of a longer dipole, the current in the dipole must go to zero even though the current at the feedpoint is still I_0.

The contribution to the received signal at a distant point from each short segment of the antenna is then summed at the distant location, taking into account the different propagation times to the receiving antenna from different segments of the source dipole.

In this model of a longer dipole, a current introduced at the feedpoint travels along the wire to or from each end. However, at the ends, where no current can flow, a *reflection* occurs and new currents start back toward the center. We will look at this process in more detail in the next section, but the net effect is that a *standing wave* of time-varying current (and a similar one for voltage) is created on the wire. This standing wave represents a vector combination of the outbound and returning currents at each point along the dipole. It is the amplitude and phase of this standing wave rather than the individual outbound and returning currents that we use for calculating the contribution of each tiny element of the model to the total radiation detected at a distant point from this dipole.

The most common models use either a straight-line decrease of current from the feedpoint to each end or a sinusoidal distribution (again based on reaching zero current at the ends) assuming a source frequency f_0.

The resulting E-field and H-field expressions for longer dipoles are identical to Eqs. (3.19) and (3.20) except for one difference: The $\sin\theta$ term of those equations is now replaced with

$$\frac{\cos\left(\dfrac{\pi h}{\lambda}\cos\theta\right) - \cos\dfrac{\pi h}{\lambda}}{\sin\theta} \tag{3.27}$$

For the special case of the half-wave dipole, Eq. (3.27) reduces to

$$\frac{\cos\left(\dfrac{\pi}{2}\cos\theta\right) - \cos\dfrac{\pi}{2}}{\sin\theta} \tag{3.28}$$

Of course, $\cos(\pi/2) = \cos 90^\circ = 0$, so the second term goes away.

Broadside to the axis of the dipole, $\theta = 90^\circ$, so the remaining term in the numerator becomes $\cos 0^\circ$ and the entire term has a value of 1.0. Figure 3.7 is a graphical representation of the geometrical variation given by Eq. 3.28—the well-known *doughnut pattern* of a half-wavelength dipole. When all the approximations are accounted for, the $\lambda/2$ dipole has 0.4 dB more gain in its main lobe than the hertzian dipole does, and the lobe is slightly narrower than that of the hertzian dipole—but these comparisons are meaningless to everyone except physics professors and their students since it is virtually impossible in the real world to deliver any significant transmitter power to an infinitesimally short dipole!

The *big* news with the $\lambda/2$ dipole is actually its feedpoint impedance, which in free space is 73 Ω resistive, with little or no reactive component. This high value of R_{RAD} makes the antenna easy to match with commonly available transmission lines, but, more important, makes it much easier to deliver most of the transmitter output power

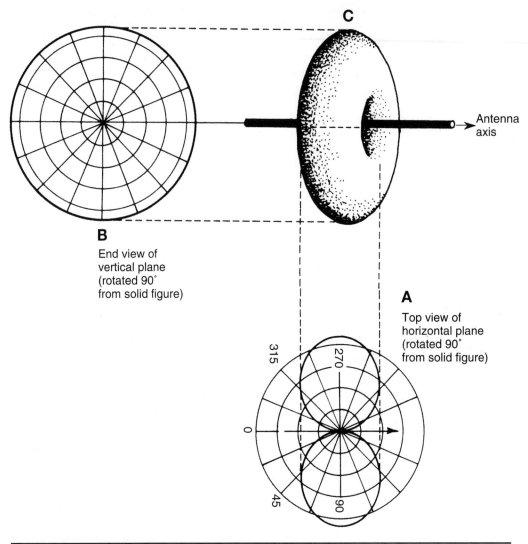

C

Antenna axis

B

End view of vertical plane (rotated 90° from solid figure)

A

Top view of horizontal plane (rotated 90° from solid figure)

315

270

0

45

90

FIGURE 3.7   Free-space radiation pattern of λ/2 dipole.

to the antenna rather than seeing it dissipate in the resistance of conductors and connectors.

Standing Waves

Assume that it is possible to have a wire conductor with one end extending infinitely, with a transmitter or other source of RF energy of single frequency f_0 connected to this wire. When the transmitter is turned on, an alternating current consisting of sine waves propagates along the wire. These waves are called *traveling waves*, and, although the resistance of the conductor gradually diminishes their amplitude, they continue to travel so long as the wire goes on "forever".

A real antenna, however, has finite length. Therefore, the traveling waves are interrupted when they reach the end of the conductor. To simplify our visualization of conditions on the line, assume that the transmitter is turned on just long enough to allow one cycle of RF (solid line) to drive the line (Fig. 3.8A). Because it is an alternating, or changing, current, it has a changing magnetic field associated with it. At the end of the conductor, the current path is broken, thus causing the magnetic field to collapse. That, in turn, creates an induced voltage at the end of the conductor that causes a new current to flow *back toward the source,* as in Fig. 3.8B. The current traveling from the transmitter toward the end is called the *incident,* or *forward,* current, and the returning current is called the *reflected* current. Similar terminology is used for the voltages on the line.

An observer standing at a point midway along one side of the dipole would thus see a single cycle of RF current and voltage followed sometime later by another cycle. The direction (or polarity) of the second current, the polarity of the second voltage compared to the original pulse of energy, and the time interval between them will depend on the length of the dipole with respect to the wavelength of f_0 and the observer's exact

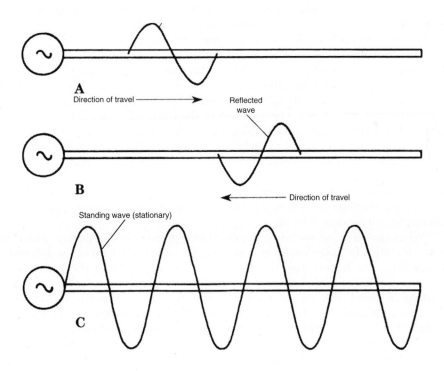

FIGURE 3.8 Standing wave formation on an antenna. (A) Incident (forward) traveling wave. (B) Reflected traveling wave. (C) Resultant standing wave.

position on the line. One example is shown in Fig. 3.8C, but it is only one of an infinite number of possible scenarios.

If now we let the transmitter send a continuous flow of sinusoidal energy of frequency f_0 to the dipole, a continuous flow of forward and reflected voltages and currents will result. (Of course, since the energy is RF energy, the instantaneous direction of the currents and the polarity of the voltages will be reversing many thousands or millions or billions of times each second.) Because they share a single conductor, the two waveforms must pass each other as they travel in opposite directions along the conductor. To our observer standing at the same point on the dipole, however, all that is evident to him at that point is a *single* voltage or current that is the vector sum of both voltages or both currents, respectively!

When, for example, the two voltages at that point are in phase with each other they reinforce, and the resultant seen by the observer is at its maximum value; when they are out of phase they cancel, and the resultant is at a minimum. The exact value of the maximum voltage at that point, and the phase of the resultant relative to the incident wave, depend on the exact length of the line in wavelengths or fractions thereof.

The truly amazing part, however, is that for a conductor of any finite length, such as our dipole antenna, the points at which the maxima and minima of the resultant voltage or current occur (Fig. 3.8C) are *stationary*! In other words, for a continuous drive signal of fixed f_0 the maxima and minima stand still even though both the incident and the reflected waves are constantly traveling out and back along the wire. Thus, the resultant is referred to as a *standing wave* of current or voltage.

The development of the standing wave on an antenna by actual addition of the traveling waves is illustrated in Fig. 3.9. In this sequence of "snapshots" of either currents or voltages taken at five different times on one side of a dipole of arbitrary length, the incident or forward waveform is represented by a thin solid line and the reflected waveform by a broken line. The resultant, or standing, wave along the antenna is represented by a heavy solid line. At the instant pictured in A, the forward and reflected waveforms perfectly coincide. (That is why the broken line is not visible.) The result is a standing wave having twice the amplitude of either traveling wave. In B, taken when the source voltage has progressed through one quarter (or 90 degrees) of a single cycle of f_0, the waveforms have traveled a little farther in opposite directions, and the amplitude of the resultant decreases, but the points of maximum and minimum standing current or voltage do not move.

When the traveling waves have moved to a position of 180 degrees phase difference, as shown in C, the resultant is zero along the entire length of the antenna. (The heavy black line representing the resultant is thus drawn along the centerline of the antenna itself.) The continuing movement of the traveling waves, shown in D another 90 degrees later, builds up a resultant in a direction opposite to that in A. Finally, the in-phase condition of the traveling waves results in a standing wave in E equal in amplitude, but 180 degrees out of phase with, the standing wave in A.

If a ruler or other straightedge is now laid over Fig. 3.9 from top to bottom so as to intersect all five curves at the same distance from the generator at the left-hand side of the drawing, as shown by the vertical dotted line in the figure, it will be clear that neither the maxima nor the minima of the resultant waveform move to the left or the right over the full 360 degrees of each cycle of RF. In other words, as the traveling waves move past each other, the standing wave changes only its amplitude and phase—never its position!

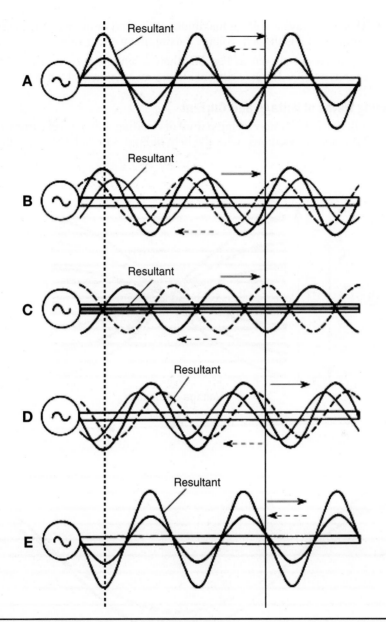

FIGURE 3.9 Development of standing wave from traveling wave.

The half-wave center-fed dipole is simply one specific length of the arbitrary-length antenna shown in Fig. 3.9. In fact, if we assume that the curves of Fig. 3.9 are describing voltage waveforms, the vertical dotted line in the figure can be thought of as corresponding to one end of a center-fed $\lambda/2$ dipole, and everything to the right of the dotted line can be discarded when discussing conditions on the dipole. The most important points to note for a $\lambda/2$ center-fed dipole are then:

- There is always a voltage maximum at the end of the wire on each side of the dipole, and it is the *only* voltage maximum on either side.

- Although not shown in Fig. 3.9, there is always a current maximum at the source, and it is the *only* current maximum anywhere along the antenna.

Standing Waves of Voltage and Current

Just to fully clarify the time-varying nature of standing voltage and current waveforms on a $\lambda/2$ dipole, let's examine, with the help of Figs. 3.10 and 3.11, what happens in a

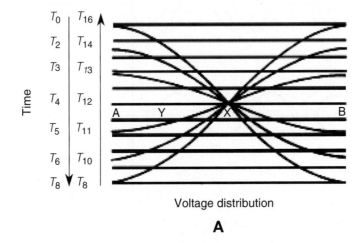

Voltage distribution

A

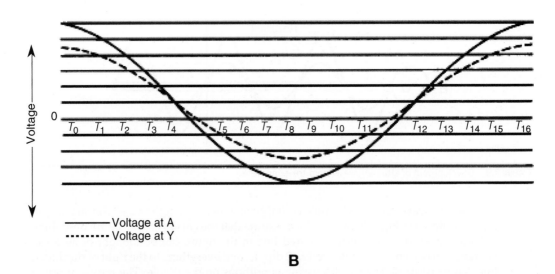

——— Voltage at A
········· Voltage at Y

B

Figure 3.10 Standing waves of voltage at two points on an antenna.

small region somewhere along either side of the dipole over the course of one complete cycle of RF energy from the transmitter.

In Fig. 3.10A, standing voltage waveforms occurring at equal time intervals over the course of one cycle of RF energy are brought together on one axis, *AB*, corresponding to the total length of a half-wave antenna. As before, we assume that RF energy to drive the antenna is being injected at the center (point X).

Figure 3.10B plots the standing voltage waveform as a function of time throughout one cycle of RF energy for two different points (A and Y) on the dipole. Note that the standing voltages seen at these two points are in phase—that is, their amplitudes go up and down simultaneously even though the peak amplitudes of the two points are, in general, different. This curve is valid for *any* pair of points or any number of points on the dipole.

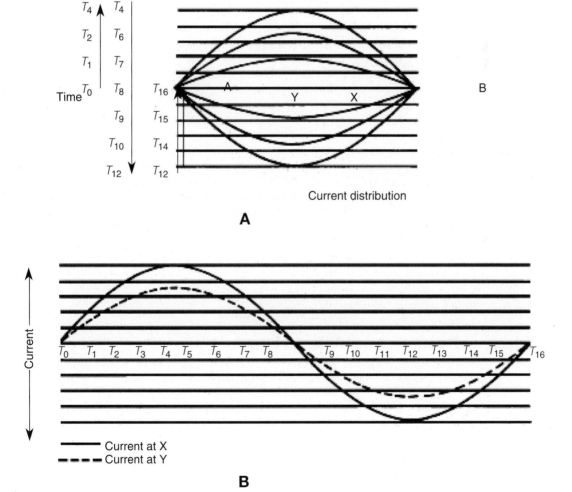

FIGURE 3.11 Standing waves of current at two points on an antenna.

Figure 3.11 presents the same information for the standing current waveforms on a $\lambda/2$ dipole.

Important facts to take away from Figs. 3.10 and 3.11 are:

- The standing voltage waveform is always minimum at the center of a $\lambda/2$ dipole.
- The amplitude of the standing voltage waveform is always maximum, but of opposing polarities, at the two ends of the dipole.
- The standing current waveform is always maximum at the center of the $\lambda/2$ dipole.
- The standing current waveform is always zero at both ends of the dipole (because of the boundary condition there—it's an open circuit!).

NOTE *If a radiator is longer than $\lambda/2$, the standing wave of current reverses direction every half-wavelength along the conductor. Starting with the boundary condition that the current at either end of a center-fed dipole must be zero, we arbitrarily choose a polarity for the current throughout the first half-wavelength from either end as we move toward the feedpoint. When we reach the next current node, $\lambda/2$ back from that end, the current reverses polarity for the next half-wavelength along the radiator (or as much of a half-wavelength as exists before reaching the feedpoint). In the center-fed dipole of Fig. 3.9, for instance, the length of each side is 3 λ, or six half-wavelength sections. Observe that the standing wave of current is positive on every other one and negative on the alternate three.*

To see why the current reverses, we can use the following argument: If a positive current in a $\lambda/2$ section corresponds to the left end of the section having a positive voltage and the right end having a negative voltage, a negative current in such a section must correspond to the right end having a positive voltage. Since the right end of one $\lambda/2$ section is the same point on the conductor as the left end of the next $\lambda/2$ section (except, of course, when we get to the end), having the same (positive or negative) voltage on the right end of one section and the left end of the next section must necessarily mean that the currents are of opposite polarity, or direction, in the two adjacent sections.

Remembering that the current reverses in each adjacent $\lambda/2$ section of a conductor will be key to understanding the patterns of loops, folded dipoles, collinear arrays, and other antenna types discussed in later chapters.

Velocity of Propagation and Antenna Length

In free space, electromagnetic waves travel at a constant velocity of 300,000 km (or approximately 186,000 mi) per second, according to the equation

$$c = \frac{1}{\sqrt{\mu_o \varepsilon_o}} \tag{3.29}$$

where μ_0, the *permeability* $= 4\pi \cdot 10^{-7}$ H/m

ε_0, the *permittivity* $= \dfrac{1}{36\pi} \cdot 10^{-9}$ F/m

RF energy on an antenna, however, moves at a velocity somewhat less than that of the radiated energy in free space because the antenna's *dielectric constant*, ε, is greater than the $ε_0$ of free space.

Because of the difference in velocity between the wave in free space and the wave on the antenna, the *physical* length of the antenna no longer corresponds to its *electrical* length. An antenna that is exactly a half-wavelength electrically will be somewhat shorter than this physically. This is also reflected in the formula for the velocity of electromagnetic waves,

$$v_P = f\lambda \tag{3.30}$$

where v_P is the velocity of propagation, f is the frequency, and λ is the wavelength. Since the frequency of the RF energy must remain constant, a decrease in the velocity results in a decrease in the wavelength. Therefore, the RF wave traveling in an antenna has a shorter wavelength than the same wave traveling in free space, and the physical length of a $\lambda/2$ dipole will be shorter.

The actual difference between the physical length and the electrical length of the antenna depends on several factors. A thin wire antenna, for example, has less effect on wave velocity than an antenna with a large cross section. As the circumference of the antenna increases, the wave velocity becomes progressively lower relative to its free-space velocity. The effect of antenna circumference on wave velocity is illustrated in the graph of Fig. 3.12.

Other factors can also lower wave velocity on the antenna. Stray capacitance, for example, increases the dielectric constant. In many installations this capacitance is dominated by the insulators used to give physical support to the antenna, as well as nearby objects made of metallic or high dielectric materials. The change in velocity resulting from stray capacitance is called *end effect* because the ends of the antenna act as though they are farther apart electrically than they are physically. End effect for a typical dipole of wire that is thin compared to its length is counteracted by making the physical length about 5 percent shorter than the electrical length, as expressed in the formula

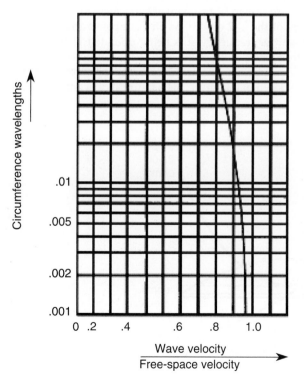

Figure 3.12 Effect of antenna circumference on wave velocity.

$$L = 0.95 \left(\frac{492}{f} \right) \tag{3.31}$$

$$= \frac{468}{f}$$

where L is the physical length in feet and f is the frequency in megahertz. This formula is reasonably accurate for determining the physical length of a half-wavelength antenna at the operating frequency, but every dipole installation is different, and "one size" definitely does *not* fit all.

The capacitive end effect also impacts the standing voltage and current waveforms. When the standing waves are measured, it is found that the nodes have some value and do not reach zero, because some current is necessary to charge the stray capacitance—especially that of the end insulators. One associated result is to lower the apparent impedance of the dipole at its ends; that which theoretically should be infinite is actually a few thousand ohms in practical dipoles, as suggested in Fig. 3.13.

Dipole Resonance

The antenna is a circuit element having distributed constants of inductance, capacitance, and resistance, which together form a resonant circuit. The half-wave antenna is the shortest resonant length of antenna, but antennas that are an integer multiple of $\lambda/2$ can also be resonant. Such antennas are said to be resonant at *harmonic* frequencies of f_0, the *fundamental* or design frequency of the $\lambda/2$ dipole. As an example, if an antenna is four half-wavelengths at the transmitter frequency, it is being operated at the fourth harmonic of its lowest resonant frequency, f_0. In other words, this antenna is a half-wavelength at one-quarter of the frequency of operation. The center-fed antenna of Fig. 3.9 is a 6λ antenna at the frequency shown in the figure. (The figure shows only one side of the antenna.)

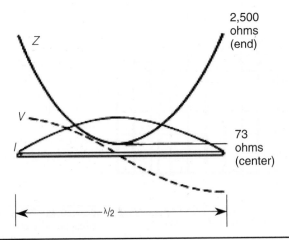

Figure 3.13   Impedance along half-wave antenna.

Antenna Resistance and Losses

A current flowing in the antenna encounters three kinds of resistance:

- Radiated power expended in the form of radiation can be thought of as an I^2R_R loss. R_R (also written R_{RAD}) is called the *radiation resistance*.

- Current flowing along the antenna conductor dissipates a certain amount of energy in the form of heat. In this I^2R_θ loss, R_θ is the *ohmic resistance*. The ohmic resistance is *not* the same as the wire's dc resistance as measured on a conventional ohmmeter or volt-ohmmeter (VOM); rather, it is somewhat higher because of the presence of *skin effect*, which limits the flow of RF energy to the surface of the conductor according to a complicated function of frequency and the specific parameters of the conducting material.

- There is also an I^2R_{SH} loss because of the *leakage*, or *shunt*, *resistance* of dielectric elements such as insulators. This effect of this resistance is often lumped in with the ohmic resistance.

The purpose of any antenna system is to convert as much source energy as possible to electromagnetic radiation. The energy dissipated by what we call the radiation resistance R_R is thus the useful part of the total power applied to the antenna system. Because the actual power from the source is split between R_{RAD} and the ohmic and shunt resistance terms, the latter two resistances should be kept as low as possible. The radiation resistance of a center-fed half-wave antenna in free space is 73 Ω; in a properly designed installation, the radiation resistance should be large compared to the loss resistances, and most of the available energy will be radiated as useful signal. The half-wave antenna is, therefore, a very efficient radiator under most circumstances. In theory, the very short dipole we used at the beginning of our analysis can have a radiated signal comparable to that of the λ/2 dipole (with the main lobe theoretically only 0.4 dB less), but its R_{RAD} is so low and its X_C so large that it is virtually impossible to avoid dissipating most of the transmitter output power in R_O and R_{SH} of the antenna as well as similar or larger loss resistances in the matching network for such a short antenna.

Half-Wave Dipole Feedpoint Impedance

A half-wave dipole can be fed anywhere along its length; it does not have to be fed at the center, even though there are many advantages to doing so. Because the voltage and current vary along the length of the half-wave dipole, and because the impedance at any point on the dipole is equal to the voltage at that point divided by the net current passing through that point, the impedance will vary along the length of the antenna. If V is divided by I at each point of the voltage and current curves of Fig. 3.13, the result is the impedance curve Z. The impedance is a minimum of about 73 Ω at the center point and rises to 2500 Ω or more at the ends. In general, however, the impedance is complex (a mixture of resistive and reactive terms) throughout most of the antenna length.

If the antenna is cut to a length corresponding to exact resonance the feedpoint impedance at the center will be purely resistive. However, if the antenna is longer or shorter than resonance, reactance reappears. When the antenna is shorter than its resonant length, the feedpoint impedance will exhibit capacitive reactance; conversely, when the antenna is made somewhat longer than the resonant length, inductive reactance will be seen.

Note: If the transmitted frequency is changed, the *electrical* length of the antenna also changes in accordance with Eq. (3.30). If the frequency is increased, the physical length of a given antenna stays the same but the electrical length becomes a greater fraction of a wavelength, with a corresponding increase in the amount of inductive reactance seen at the feedpoint. Conversely, if the frequency is lowered, the electrical length decreases, and the feedpoint impedance becomes more capacitive in nature.

No transmitter power is lost in a pure reactance. However, the mismatch caused by the reactive part of the feedpoint impedance *does* have the potential to cause reduced RF output from the transmitter, depending on the design of the transmitter. Further, for lengths other than an odd multiple of $\lambda/2$, most dipoles will exhibit sufficient feedpoint reactance to make the use of a matching network in the circuit somewhere between the transmitter and the feedpoint mandatory. Circulating currents in any wires or inductive components in the network will also contribute to resistive losses, lowering the efficiency of the antenna even more.

CHAPTER 4

Transmission Lines and Impedance Matching

Transmission lines and waveguides are conduits for transporting RF energy between elements of a radio system. For example, in a typical station with a transmitter (or *exciter*) and a power amplifier, one transmission line carries exciter output to the amplifier input and a second line delivers transmitter output energy to the antenna. A third line may carry the incoming RF energy from a transmit/receive switch or separate receiving antenna to the station receiver. Still other lines may switch filters in and out, or pass transmitter RF through signal monitors, power meters, and other station accessories.

A good analogy for visualizing how transmission lines do their job is a forced hot water heating system for your home or office. In such a system, copper pipes carry hot water from a central furnace or boiler to distant radiating units that extract a large percentage of the heat from the water and deliver it to the air or the floors in your building. In both radio and heating systems:

- The objective is to transfer as much energy as possible to the radiating unit(s) and minimize the unintentional loss of energy in the transmission line or copper pipes.

- A return path for the delivery medium back to the energy source must be provided.

- The longer the distance between the source and the radiating unit(s), the greater the potential for loss of energy in the transmission (distribution) system.

Because of the requirement for a return path, virtually all transmission lines below microwave frequencies (where waveguides are an important exception to this statement) have at least two conductors: At the same time RF current is flowing *toward* the radiating unit (antenna) in one conductor it is flowing *away* from the antenna and back to the source in the other.

Similarly, to minimize the unintentional loss of energy from the transmission line, virtually all lines in use today keep the conductors very close together for the entire distance between the source and the antenna. (This is where our analogy with the heating system breaks down because the return lines in many, if not most, heating systems travel completely different paths back to the furnace.)

To summarize: The geometry of almost all transmission lines has been designed to *minimize* the tendency of the lines to act as antennas and radiate on their own. In contrast, the geometry of an antenna has been designed to *maximize* its tendency to radiate!

If we know the geometry of the conductors in a transmission line, as well as dielectric properties of the material that occupies the space between them, we can use equations found later in this chapter to calculate the *characteristic impedance,* or Z_0, of that line. While it is possible to construct transmission lines of any desired Z_0 over a very wide range of impedances, certain favored impedances have evolved over the years and today represent almost the entirety of commercial transmission line production. Selection of a characteristic impedance, also known as the *surge impedance,* for a system is a very important decision that should be made early in the design of that system. Certain standard choices exist today: Most amateur, CB, and marine communications installations utilize a system impedance of 50 to 52 Ω (ohms). Most cable television systems (in the United States, at least) have standardized on 75 Ω. There are also historical and practical reasons for 300-, 450-, and 600-Ω lines, too. And a few transmission lines for specialized applications have more than two conductors.

Types of Transmission Lines

One of the earliest types of transmission lines was a single copper wire connected between the transmitter and some point on a horizontal antenna. Then, as now, the return path for this type of line was the earth or other ground path between the antenna and the transmitter. In almost all installations, this form of transmission line exhibits many undesirable characteristics, and we shall not dwell on it because it has, for the most part, been supplanted by far superior two-conductor lines with controlled geometries.

Parallel Conductor Lines

Next to appear were *parallel conductor lines,* whose general shape is shown in Figs. 4.1A through 4.1E. Figure 4.1A shows a cross-sectional, or end, view of a typical parallel conductor transmission line. Two conductors, both of diameter *d,* are separated by a dielectric (which might be air) at a spacing *S.* (These designations will be used in calculations later.) At first, all parallel lines were *open-wire line* (OWL), shown in Fig. 4.1B. Here, the wire conductors are separated by an air dielectric; spac-

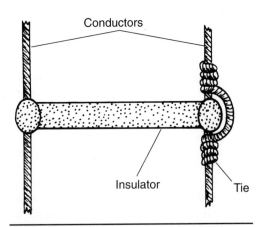

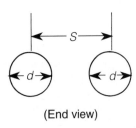

(End view)

FIGURE 4.1A Parallel line transmission line (end view).

FIGURE 4.1B Parallel open-wire line construction details.

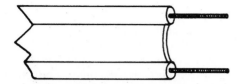

FIGURE 4.1C   Twin-lead transmission line.

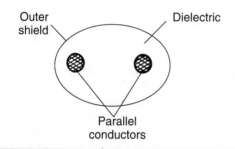

FIGURE 4.1D   Shielded twin-lead transmission line.

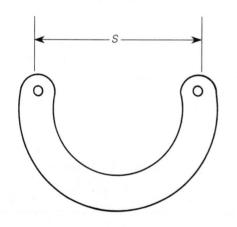

FIGURE 4.1E   Horseshoe parallel line spreader.

ing between the two wires is maintained at a fixed distance by rigid insulators. Early radio experimenters employed wood dowels soaked in paraffin for protection against the elements, but, as the market grew, ceramic spacers became available. More recently, molded plastic formulations have been employed to reduce both the cost and the weight of the total assembly.

Parallel lines have been used at VLF, MF, and HF for a century. Even antennas for the lower VHF bands are often fed with parallel lines. For years, the VHF, UHF, and microwave application of parallel lines was limited to educational laboratories, where they are well suited to performing experiments (to about 2 GHz) with simple, low-cost instrumentation. Today, however, printed circuit and hybrid semiconductor packaging have given parallel lines a new lease on life and a burgeoning market presence.

Figure 4.1C shows a type of parallel line called *twin-lead*. This is the venerable television antenna transmission line. It consists of two parallel conductors encased in, and separated by, a flexible plastic dielectric. The dimensions of TV-type twin-lead were established years ago in conjunction with the chosen dielectric to give the final product a *characteristic impedance* of 300 Ω to simplify impedance matching to the *folded dipole* driven element of the typical TV antenna. Other kinds of twin-lead are manufactured with other target impedances; some forms of transmitting twin-lead have an impedance of 450 Ω, for instance.

Figure 4.1D shows a form of parallel line called *shielded twin-lead*. This type of line uses the same structure as TV-type twin-lead, but adds a shield layer surrounding it. The shield may be braided, a thin layer of aluminum foil, or both. This feature can make it less susceptible to electrical noise and other problems.

Some users of open line prefer the spacer shown in Fig. 4.1E. Generally formed from plastic or ceramic, its U shape reduces losses, especially in wet weather, by providing increased leakage path length between the two conductors relative to their spacing, *S*.

Occasionally one will run across an installation using multiwire parallel line. Four-wire and six-wire transmission lines most often find use when the distance between transmitter and antenna is extremely long (i.e., a substantial portion of a mile).

Coaxial Lines

The second form of transmission line commonly used at sub-microwave frequencies is *coaxial cable* (Figs. 4.1F through 4.1L), often abbreviated as *coax* (pronounced "co-ax"). This form of line consists of two cylindrical conductors sharing a common axis (hence "coaxial"), and separated by a dielectric (Fig. 4.1F). Of necessity, the outer conducting cylinder is hollow, but the inner one is usually solid (although that is not a requirement). For low frequencies (in flexible cables) the dielectric may be polyethylene or polyethylene foam, but at higher frequencies *Teflon* and other specialized materials are common. In most of the inexpensive cables on the market, the dielectric completely fills the space between the two conductors over the entire length of the cable. But for extremely low loss requirements, circular discs are spaced along the inside of the line to keep the inner conductor centered with respect to the outer conductor, and the line is often filled with dry air or dry nitrogen and kept slightly pressurized.

Several additional variations in coaxial line construction and characteristics are available. Flexible coaxial cable, characterized by the RG-8, RG-58, RG-59, RG-213, et al., families, is perhaps the most common form. The outer conductor in such cable is made of either braid or foil (Fig. 4.1G). Cable and satellite TV system home installations are wired with coaxial cables, which should never be confused with audio cables. Another form of flexible or semiflexible coaxial line is *helical line* (Fig. 4.1H), in which the outer conductor is spiral wound. *Hardline* (Fig. 4.1I) is coaxial cable that uses a thin-wall aluminum tube as the outer conductor; it is ubiquitous in the cable television industry's outdoor distribution systems, often seen midway up the utility poles in our neighborhoods and as *drop cables* to our homes. Some hardline used at microwave frequencies has a rigid outer conductor and a solid dielectric.

Gas-filled line is a special case of hardline that is hollow (Fig. 4.1J); the center conductor is supported by a series of thin ceramic or Teflon insulators. The dielectric is usually anhydrous (i.e., dry) nitrogen or some other inert gas.

Some flexible microwave coaxial cable uses a solid "air-articulated" dielectric (Fig. 4.1K), in which the inner insulator is not continuous around the center conductor but, rather, is ridged. Reduced dielec-

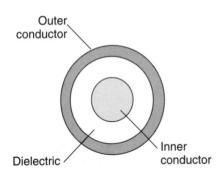

FIGURE 4.1F   Coaxial cable (end view).

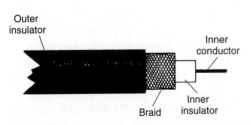

FIGURE 4.1G   Coaxial cable (side view).

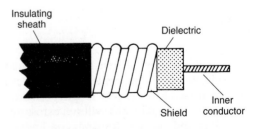

FIGURE 4.1H   Helical coaxial cable.

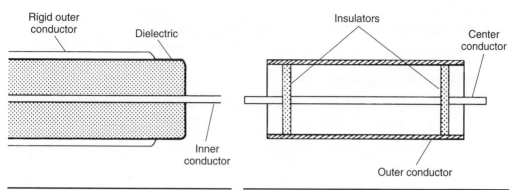

FIGURE 4.1I   Rigid coaxial line (*hardline*).

FIGURE 4.1J   Gas-filled hollow coaxial line.

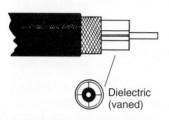

FIGURE 4.1K   Articulated coaxial line.

tric losses increase the usefulness of the cable at higher frequencies. Double-shielded coaxial cable (Fig. 4.1L) provides an extra measure of protection against radiation from the line, as well as from *electromagnetic interference* (EMI) from outside sources getting into the system.

Stripline, also called *microstripline* (Fig. 4.1M), is a form of transmission line used at high UHF and microwave frequencies. The stripline consists of a critically sized conductor over a ground-plane conductor, and separated from it by a dielectric. Some striplines are sandwiched between two ground planes and are separated from each by the dielectric board.

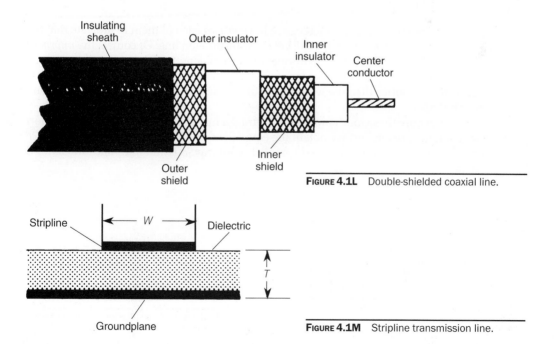

FIGURE 4.1L   Double-shielded coaxial line.

FIGURE 4.1M   Stripline transmission line.

Transmission Line Characteristic Impedance (Z_0)

If we take a very short length of two-conductor transmission line of any kind and measure its electrical characteristics at some operating frequency with simple test equipment, we will discover four parameters:

- Each of the two conductors exhibits a very small *series resistance* between its two ends, measured at the operating frequency. In a balanced line, the resistance of the two conductors is identical, but that is not necessarily the case for coaxial or other unbalanced lines. Regardless of the type of line, we shall define R' as the *sum* of the resistive losses in the two conductors.

- Similarly, each of the two conductors has a small *series inductance* between its two ends. All wires—even straight ones—have some inductance. Again, this inductance is the same for both conductors of perfectly balanced lines but may differ in unbalanced lines. We shall define L' as the *sum* of these inductances.

- At one end or the other we measure a small *shunt capacitance, C'*, between the two conductors. This is a measure of the capacitive coupling between two closely spaced wires; of course, it depends greatly on the dielectric material between them.

- At the same end of the short length of transmission line we measure a very high *shunt resistance* (usually alternatively stated as a very low *shunt conductance, G'*, where G' is the reciprocal of the shunt resistance) across the two conductors. This conductance is a characteristic of the dielectric material(s) filling the space between the two conductors. Sometimes the dominant part of it (lowest resistance, highest G') is unintended—as might result from unwanted moisture in the line!

In defining these four parameters, we have *primed* each of them to indicate that the value we are measuring is for a *unit length* of transmission line. Of course, the *unprimed* values of all four are directly proportional to the total length of the transmission line.

How short is "very short"? The length of transmission line used to obtain these data must be much shorter than the wavelength of the highest frequency we intend to send through the line.

If we now use these four parameters to draw a lumped element model of this short length of transmission line, we obtain the equivalent circuit of Fig. 4.2. This *single-ended* circuit is equally valid for balanced (parallel wire) transmission lines and unbalanced (coaxial) lines.

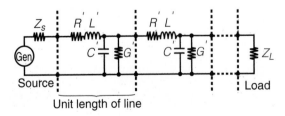

FIGURE 4.2 Transmission line equivalent circuit.

In other words, each short section of any transmission line can be represented by a (relatively) simple *RLC* network. If we then use network analysis and calculus to examine the current and voltage relationships at the input of a long transmission line made up of an infinite number of infinitesimally short sections daisy-chained together, we find the transmission line exhibits an impedance at its input terminals that is independent of the length of the line. This *characteristic impedance Z_0*, also sometimes called the line's *surge impedance*, is a function of the four parameters previously defined, which in turn are a function of the physical geometry of the line, including conductor size, shape, and spacing, and the dielectric constant of the insulating material between the conductors.

$$Z_0 = \sqrt{\frac{R' + j\omega L'}{G' + j\omega C'}} \tag{4.1}$$

where Z_0 = characteristic impedance of line, in ohms
R' = total series resistance per unit length of two conductors, in ohms
G' = shunt conductance between two conductors per unit length, in mhos
L' = total series inductance per unit length of two conductors, in henrys
C' = shunt capacitance per unit length between conductors, in farads
j = imaginary number $\sqrt{-1}$
ω = *angular frequency* in radians per second ($\omega = 2\pi f$)

In the general case, Z_0 is complex; that is, it has both a resistive component and a reactance component. Most real transmission lines fall in this category. The existence of a reactance component leads to attenuation and delay across each unit length that is often frequency dependent. When that is the case, signals applied to one end of the line are said to suffer *dispersion* or *frequency dispersion* as they propagate along the line. Thus, a complex waveform consisting of multiple-frequency components will become distorted by the time it reaches the far end of the line, since the different frequencies will have experienced differing amounts of attenuation over their common path.

For certain values of R', G', L', and C', the transmission is *lossless* and Z_0 is purely resistive or so nearly so that we can work with it as if it were. Three such cases are:

- $R' = G' = 0$. In an ideal cable, the series resistance is zero and the shunt resistance is infinite, so Eq. (4.1) reduces to the following simplified form for a *lossless cable*:

$$Z_0 = \sqrt{\frac{L'}{C'}} \tag{4.2}$$

- $\omega L' \gg R'$ and $\omega C' \gg G'$. Although not zero, the series resistance and shunt conductance are negligible with respect to the series inductance and shunt capacitance, respectively, at the frequencies of operation. Note, however, that R' is a function of frequency because the effective resistance of a wire at RF is modified by a frequency-dependent phenomenon called *skin effect*. It is not appropriate, for instance, to blindly use the dc or *ohmic* value of R at RF.

- $R'/L' = G'/C'$ and both R' and G' are small, but not necessarily negligible. If this condition is met, the line is *distortionless* or *dispersionless* even though it may not be *lossless*.

Because a unit length of transmission line contains capacitance and inductance, we should not be surprised to see that some of its characteristics depend on ε (permittivity) and μ (permeability). In particular, the velocity of propagation in a transmission line is

$$L'C' = \mu\varepsilon = \frac{1}{v^2} \tag{4.3}$$

or

$$v = \frac{1}{\sqrt{L'C'}} \tag{4.4}$$

In the usual transmission line, μ is essentially the same as for free space, but ε will be somewhat larger than ε_0, with the result that v in the line will be less than c, the speed of propagation in free space. For lossless lines, the velocity of propagation on the line is

$$v = \frac{c}{\sqrt{\varepsilon_r}} \tag{4.5}$$

and the wavelength of the wave in the line is

$$\lambda = \frac{\lambda_0}{\sqrt{\varepsilon_r}} \tag{4.6}$$

Example 4.1   A nearly lossless transmission line (R' and G' are very small) has an inductance of 3.75 nH per unit length and a capacitance of 1.5 pF per unit length. Find the characteristic impedance, Z_0.

Solution

$$Z_0 = \sqrt{\frac{L'}{C'}}$$

$$= \sqrt{\frac{3.75 \times 10^{-9} \text{ H}}{1.5 \times 10^{-12} \text{ F}}}$$

$$= \sqrt{2.5 \times 10^3} = 50 \ \Omega$$

◆

Following are equations for the characteristic impedance of various kinds of two-conductor transmission lines.
(a) Parallel line:

$$Z_0 = \frac{276}{\sqrt{\varepsilon}} \log\left(\frac{2S}{d}\right) \tag{4.7}$$

where Z_0 = characteristic impedance, in ohms
$\quad\varepsilon$ = dielectric constant of insulator between conductors
$\quad S$ = center-to-center spacing of conductors
$\quad d$ = diameter of (identical) conductors

(b) Coaxial line:

$$Z_0 = \frac{138}{\sqrt{\varepsilon}} \log\left(\frac{D}{d}\right) \tag{4.8}$$

where D = diameter of inside surface of outer conductor
$\quad d$ = diameter of inner conductor

(c) Shielded parallel line:

$$Z_0 = \frac{276}{\sqrt{\varepsilon}} \log\left(2A\frac{\left(1-B^2\right)}{\left(1+B^2\right)}\right) \tag{4.9}$$

where $A = s/d$
$\quad B = s/D$

(d) Stripline:

$$Z_0 = \frac{377}{\sqrt{\varepsilon_t}}\left(\frac{T}{W}\right) \tag{4.10}$$

where ε_t = relative dielectric constant of printed wiring board (PWB)
$\quad T$ = thickness of PWB
$\quad W$ = width of stripline conductor

The relative dielectric constant ε_t used here differs from the normal dielectric constant of the material used in the PWB. The relative and normal dielectric constants move closer together for larger values of the ratio W/T.

Example 4.2   A stripline transmission line is built on a 4-mm-thick printed wiring board that has a relative dielectric constant of 5.5. Calculate the characteristic impedance of the stripline if the width of the strip is 2 mm.

Solution

$$\begin{aligned}
Z_0 &= \frac{377}{\sqrt{\varepsilon_t}}\left(\frac{T}{W}\right) \\
&= \frac{377}{\sqrt{5.5}}\left(\frac{4}{2}\right) \\
&= \frac{377}{2.35}(2) \\
&= 321\ \Omega
\end{aligned}$$

◆

In practical situations, we usually don't need to calculate the characteristic impedance of a stripline but, rather, we need to design the line to fit a specific system impedance (e.g., 50 Ω). We can make some choices of printed circuit material (hence, dielectric
constant) and thickness, but even these are usually limited in practice by the availability
of standardized boards. Thus, stripline *width* is the variable parameter. Equation (4.10)
can be rearranged into the form

$$W = \frac{377\,T}{Z_0\sqrt{\varepsilon_t}} \tag{4.11}$$

An impedance of 50 Ω is accepted as standard for RF systems, except in the cable TV
industry and its allied fields. The reason for this dichotomy is that power-handling ability and low-loss operation don't occur at the same characteristic impedance. For example, the maximum power-handling ability for coaxial cables occurs at 30 Ω, while the
lowest loss occurs at 77 Ω; 50 Ω is therefore a reasonable tradeoff between the two
points. In the cable TV industry, however, the RF power levels are minuscule, but distribution lines are long. Cable TV uses 75 Ω as the standard system impedance in order
to take advantage of the reduced attenuation factor.

Other Transmission Line Characteristics

Velocity Factor

In the preceding section, we saw that the velocity of the wave (or signal) in the transmission line is less than the free-space velocity (i.e., less than the speed of light). Further,
we discovered in Eq. (4.5) that wave velocity on the line is related to the dielectric constant of the insulating material that separates the conductors in the transmission line.
Velocity factor v_F is usually specified as a decimal fraction of c, the speed of light (3×10^8
m/s). For example, if the velocity factor of a transmission line is rated at 0.66, then the
velocity of the wave is 0.66c, or (0.66) (3×10^8 m/s) = 1.98×10^8 m/s.

Velocity factor becomes important when designing *transmission line transformers* or
any other device in which the length of the line is important. In most cases, the transmission line length is specified in terms of *electrical length*, which can be either an angular measurement (e.g., 180 degrees or π radians), or a relative measure keyed to
wavelength (e.g., one half-wavelength, which is the same as 180 degrees). The *physical
length* of the line (as measured in wavelengths) is *always* shorter than its electrical
length.

Example 4.3   Find the physical length of a 1-GHz half-wavelength transmission line.

Solution   A wavelength (in meters) in free space is 0.30/f, where frequency f is expressed in gigahertz; therefore, a half-wavelength is 0.15/f. At 1 GHz, the line must be
0.15 m/1 GHz = 0.15 m long. If the velocity factor is 0.80, then the *physical length* of the
transmission line that will achieve the desired *electrical length* is [(0.15 m) (v)]/f =
[(0.15 m) (0.80)]/1 GHz = 0.12 m.

◆

Some practical considerations arise from the fact that the physical length of a real transmission line is shorter than its electrical length. For example, in certain types of phased-array antenna designs, radiating elements are spaced a half-wavelength apart and must be fed 180 degrees (half-wave) out of phase with each other. The simplest and least expensive connection between the two elements is a transmission line of a length that provides an electrical half-wavelength. Unfortunately, because of the velocity factor, the physical length for a one-half *electrical* wavelength cable is *shorter* than the free-space half-wave distance between elements. In other words, the cable will be too short to reach between the radiating elements!

Clearly, velocity factors must be known before transmission lines can be selected and cut to length for specific situations. Table 4.1 lists the velocity factors for several popular types of transmission lines. Because these are at best *nominal* values, the actual velocity factor for any given line should be measured, using techniques and instruments described in Chap. 27.

Type of Line	Z_0 (ohms)	Velocity Factor v
½-in TV parallel line (air dielectric)	300	0.95
1-in TV parallel line (air dielectric)	450	0.95
TV twin-lead	300	0.82
UHF TV twin-lead	300	0.80
Polyethylene coaxial cable	*	0.66
Polyethylene foam coaxial cable	*	0.79
Air-space polyethylene foam coaxial cable	*	0.86
Teflon coaxial cable	*	0.70
CATV hardline	75	0.8–0.9

* Various impedances depending upon cable type.

TABLE 4.1 Transmission Line Characteristics

Loss in Transmission Lines

As we saw earlier, loss in a two-conductor transmission line is modeled in the series R' and shunt G' terms of Eq. (4.1), repeated here as Eq. (4.12) for convenience. Up until this point we have discussed only lines with zero loss or negligible loss.

$$Z_0 = \sqrt{\frac{R' + j\omega L'}{G' + j\omega C'}} \tag{4.12}$$

As you can see, when R' and G' are zero, the expression under the square root sign simplifies to L'/C'. When either R' or G' or both are large enough to affect things, the math gets too messy for us to go through in detail, so much of the following will have to be taken on faith.

Earlier we modeled the series impedance of a short length of transmission line as $R' + j\omega L'$, and the shunt admittance as $G' + j\omega C'$. The *propagation constant* for a lossy transmission line is defined as

$$\gamma = \sqrt{(R' + j\omega L')(G' + j\omega C')} \tag{4.13}$$

The propagation constant can also be written in terms of the *line attenuation constant* α and *phase constant* β:

$$\gamma = \alpha + j\beta \tag{4.14}$$

Thus, when the line has loss, we can represent a sinusoidal (single-frequency) waveform with the following equation:

$$V(z,t) = V_0 e^{-\alpha z} \cos(2\pi f t - \beta z) \tag{4.15}$$

where $V(z,t)$ = notation that tells us the observed amplitude of the wave is a function of (i.e., depends upon) its location along the line (the z axis) and the time of observation

V_0 = amplitude of the input waveform at the peak of its sine wave or cycle

$e^{-\alpha z}$ = decaying exponential term representing the effect of line loss, called the *attenuation factor*; it is a function of wave frequency

$2\pi f t$ = portion of the cosine argument that mathematically accounts for the sinusoidal or cyclic variation in the amplitude of the wave with the passage of time

βz = portion of the cosine argument that mathematically accounts for the sinusoidal spatial variation in the amplitude of the wave at any instant; β is "shorthand" for $2\pi/\lambda$ where λ is the wavelength of the wave *in the transmission line*

The loss term is exponential because the loss mechanism in a transmission line is a linear function of distance along the line; that is, the peak amplitude of the wave a little farther along the line is, say, 99 percent of the peak wave amplitude at this point on the line. The peak amplitude of the wave at a second point farther along the line is then 99 percent of its amplitude at the previous point. And so on. In theory, the wave never quite goes to zero, no matter how long the line may be. α is called the *attenuation constant* of the transmission line, and it is a dimensionless number with units of *nepers per meter* (Np/m). It is related to *natural logarithms* the same way bels and dB are related to base-10 logarithms. (See App. A.)

Unfortunately, in the general case where we have no feeling for the relative importance of the R' and G' terms, α and β are unbearably complicated and Z_0 is a complex quantity—that is, it includes reactance terms. However, for the case of a line with just a *little* bit of attenuation, we can "cheat" by allowing the argument of the cosine term to be the same as for the lossless line and pretending the only significant effect of the loss is the addition of the exponential term in Eq. (4.15).

In that case, α can be inferred from the attenuation specification for the line, and β can be rewritten in terms of parameters we know. Specifically, we know the velocity of propagation, v, in the transmission line is given by $v = f\lambda$. Since $\beta = 2\pi/\lambda$, we can write

$$\beta = \frac{2\pi f}{cv_F} \qquad\qquad (4.16)$$

Thus we can predict the variation in the amplitude of a wave of frequency f simply from knowing the attenuation (usually expressed in dB/100 ft) and the velocity factor, v_F, of the transmission line.

Transmission Line Responses

When we employ a transmission line to deliver RF energy from a transmitter to an antenna we are normally interested in its performance at one frequency (as when we are on-off CW keying the transmitter) or for a very small range of frequencies (as in voice transmissions). For much of our analysis of the antenna and feedline system we need consider only the cable's *steady-state ac response*, especially when the load (antenna) impedance is perfectly matched to the *characteristic impedance* of the transmission line.

Often, however, the antenna and transmission line impedances are not exactly the same and we need to analyze what happens in such cases. Then it becomes extremely helpful to our understanding of how transmission lines function if we consider the line's *step-function response*. Step-function analysis involves the application of a single voltage transition at the input of the line: The input voltage snaps from zero (or an *initial* steady value) to a nonzero (or *final* steady) value "instantaneously", and is held there until all action along the entire transmission line has died out.

Through a mathematical technique known as *Fourier analysis*, a voltage step can be shown to consist of many individual sine waves of differing amplitudes and phases and extending over an extremely wide range of frequencies. Thus, by exciting the transmitter end of the transmission line with a single voltage step and examining the resulting waveforms at other points along the line and at the junction with the load, we can get an immediate feeling for the behavior of the line over a very wide range of frequencies, as well as a detailed understanding of what happens when there is a mismatch between the line and the load.

Step-Function Response of a Transmission Line

Before we get into the details of transmission line step response, let's look at a mechanical analogy that should be familiar to all of us. A taut rope (Fig. 4.3A) is tied to a rigid wall that does not absorb any of the energy in the pulse propagated down the rope. When the free end of the rope is given a vertical displacement (Fig. 4.3B) by yanking up on it once, a wave is propagated along the rope, toward the wall, at velocity v (Fig. 4.3C). When the pulse hits the wall (Fig. 4.3D), it is reflected (Fig. 4.3E) and propagates back *along* the rope toward the free end (Fig. 4.3F).

If a second pulse is propagated down the line before the first pulse dies out, then there will be two pulses on the line at the same time (Fig. 4.4). When the two pulses are both present on the rope, the resultant deflection of the rope will be the algebraic sum of the two pulses. If a continuous *train* of pulses is applied to the line at certain *resonant* repetition rates, the net deflection resulting from the combined incident and reflected pulse amplitudes at each point along the rope will create *standing waves*—one example of which is shown in Fig. 4.5—that do not propagate in either direction on the rope but, instead, go straight up and down at each point on the rope.

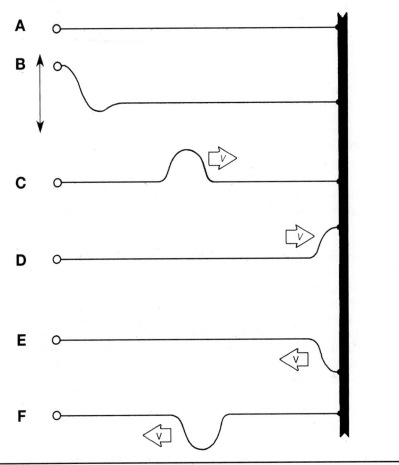

FIGURE 4.3   Rope analogy to transmission line.

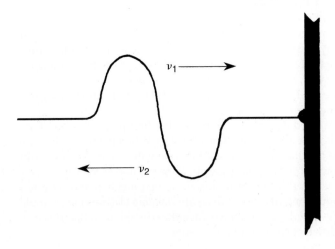

FIGURE 4.4   Interfering opposite waves.

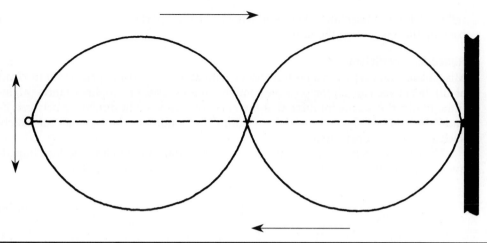

Figure 4.5 Standing waves.

Now let's look at a transmission line with characteristic impedance Z_0 connected to a load impedance Z_L, as shown in Fig. 4.2. The generator at the input of the line consists of a voltage source V with source impedance Z_S in series with a switch S_1. Assume for the present that both the source and the load impedances are pure resistances (i.e., $R + j0$) equal to Z_0, the characteristic impedance of the transmission line.

When the switch is closed at time T_0 (Fig. 4.6A), the voltage at the input of the line (V_{IN}) jumps to $V/2$. In Fig. 4.2, you may have noticed that the $L'C'$ circuit resembles a delay line. As might be expected, therefore, the voltage wavefront propagates along the line at a velocity v:

$$v = \frac{1}{\sqrt{L'C'}} \qquad (4.17)$$

where v = velocity, in meters per second
L' = inductance, in henrys per meter
C' = capacitance, in farads per meter

At time T_1 (Fig. 4.6B), the wavefront has propagated one-half the distance L, and by T_d it has propagated the entire length of the cable (Fig. 4.6C).

If the load is perfectly matched to the line (i.e., $Z_L = Z_0$), the load absorbs the wave and no component is

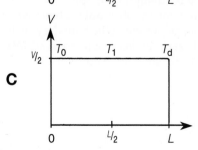

Figure 4.6 Step-function propagation along transmission line at three points.

reflected. But in a mismatched system (where $Z_L \neq Z_0$), a portion of the wave is reflected back up the line toward the source.

Reflection Coefficient

Mismatches can vary from a perfect short circuit at the load to an open circuit (no load at all). In between are all the possible combinations of resistance and reactance—an infinite number! It seems intuitive that the effect of a very slight difference between Z_L and Z_0 should have less effect on the resulting line conditions than a major mismatch, such as a short or open circuit.

The *reflection coefficient* Γ of a circuit containing a transmission line and load impedance is a measure of how well the load is matched to the transmission line:

$$\Gamma = \frac{V_{\text{REF}}}{V_{\text{FWD}}} \tag{4.18}$$

where V_{REF} = reflected voltage
V_{FWD} = *forward*, or incident, voltage

The absolute value of the reflection coefficient varies from -1 to $+1$, depending upon the nature of the reflection; $\Gamma = 0$ indicates a perfect match with no reflection, while -1 indicates a short-circuited load ($Z_L = 0$), and $+1$ indicates an open circuit ($Z_L = \infty$). To see how this comes about, consider the set of *boundary conditions* that must be true at the junction of the transmission line and the load (antenna).

- Just before arriving at the load, the advancing step function knows only that it's traveling on a transmission line of characteristic impedance Z_0. Therefore, at each and every point on the transmission line prior to the step function first reaching the load, the voltage is related to the current by $V_{\text{FWD}} = I_{\text{FWD}}Z_0$.

- The voltage across the load (usually an antenna) must always be related to the current through it by $V_L = I_L Z_L$.

- At the junction of the line and the load, the current must be continuous; that is, the current in the transmission line has no place else to go except into the load. Thus, if $I_L = I_{\text{FWD}}$, then $V_L = V_{\text{FWD}}$ only if $Z_0 = Z_L$. This is the matched condition, for which $\Gamma = 0$.

- Suppose, however, that we substitute a short circuit for the original load. Now $Z_L = 0$, so $V_L = I_L Z_L = 0$. The advancing wave is attempting to apply a voltage $V_{\text{FWD}} = I_{\text{FWD}}Z_0$ to the boundary, yet we measure a voltage of zero at the boundary. For this to be true, there must be a second voltage present at the boundary such that the sum of V_{FWD} and this new voltage is zero (by *linear superposition*). We call this new voltage V_{REF}, and it has to be exactly equal in magnitude to V_{FWD}, but of opposite polarity. In other words, $V_{\text{REF}} = -V_{\text{FWD}}$, hence

$$\Gamma = \frac{V_{\text{REF}}}{V_{\text{FWD}}} = \frac{-V_{\text{FWD}}}{V_{\text{FWD}}} = -1$$

- Similarly, suppose we disconnect the load completely, so that the transmission line is looking into an open circuit. In this case, $Z_L = \infty$ and $I_L = 0$. But if the current in the advancing wave is to be continuous, there must be an equal and opposite current flowing at the end of the transmission line to make the *net* load current zero. We call this new current I_{REF}, and it must be exactly equal in magnitude to I_{FWD} but of opposite polarity through the load. In other words, $I_{REF} = -I_{FWD}$. The minus sign simply means the reflected current has the same amplitude as the forward current, but it is directed in the opposite direction—i.e., back toward the source. If we could stand on the junction of the line and the load, we would see two waves, moving in opposite directions, but since the currents are 180 degrees out of phase with each other, we see no net current in either direction.

Because there are both forward and reflected waves at the junction of the line and the load, V and I there are the sum of the forward and reflected voltages and currents, respectively. Therefore:

$$Z_L = \frac{V_{TOTAL}}{I_{TOTAL}} \tag{4.19}$$

$$Z_L = \frac{V_{FWD} + V_{REF}}{I_{FWD} + I_{REF}} \tag{4.20}$$

where V_{FWD} = incident (i.e., forward) voltage
$\ V_{REF}$ = reflected voltage
$\ I_{FWD}$ = incident current
$\ I_{REF}$ = reflected current

Using Ohm's law and algebraic substitution to define the currents at the boundary in terms of the voltage components and the characteristic impedance of the line:

$$I_{FWD} = \frac{V_{FWD}}{Z_0} \tag{4.21}$$

and

$$I_{REF} = \frac{-V_{REF}}{Z_0} \tag{4.22}$$

(The minus sign in Eq. [4.22] indicates that I_{REF} is in the opposite direction.)

The two expressions for current (Eqs. [4.21] and [4.22]) may be substituted into Eqs. (4.19) and (4.20) to yield

$$Z_L = \frac{V_{FWD} + V_{REF}}{\dfrac{V_{FWD}}{Z_0} - \dfrac{V_{REF}}{Z_0}} \tag{4.23}$$

Repeating Eq. (4.18), which defined the reflection coefficient Γ as the ratio of reflected voltage to incident voltage:

$$\Gamma = \frac{V_{REF}}{V_{FWD}} \tag{4.24}$$

Rearranging Eq. (4.23) to combine all terms involving V_{REF} separately from those involving V_{FWD} and substituting into Eq. (4.24) gives

$$\Gamma = \frac{Z_L - Z_0}{Z_L + Z_0} \tag{4.25}$$

Example 4.4   A 50-Ω transmission line is connected to a 30-Ω resistive load. Calculate the reflection coefficient Γ.

Solution

$$\Gamma = \frac{Z_L - Z_0}{Z_L + Z_0}$$

$$= \frac{(50\,\Omega) - (30\,\Omega)}{(50\,\Omega) + (30\,\Omega)}$$

$$= \frac{20}{80}$$

$$= 0.25$$

Example 4.5   In Example 4.4, the incident (forward) voltage is 3 V rms. Calculate the reflected voltage.

Solution   If

$$\Gamma = \frac{V_{REF}}{V_{FWD}}$$

then

$$V_{REF} = \Gamma V_{FWD} \tag{4.26}$$

$$= (0.25)(3V)$$

$$= 0.75V$$

The phase of the reflected voltage can be deduced from the relationship between load impedance and transmission line characteristic impedance.

For resistive loads ($Z_L = R_L + j0$):

- If the ratio Z_L/Z_0 is 1.0, there is no reflection.

- If Z_L/Z_0 is less than 1.0, the reflected signal is 180 degrees out of phase with the incident voltage.

- If the ratio Z_L/Z_0 is greater than 1.0, the reflected signal is in phase with the incident voltage.

The step-function (or pulse) response of the transmission line leads to a powerful means of analyzing both the line and the load on an oscilloscope. Figure 4.7A shows (in schematic form) the test setup for *time domain reflectometry* (TDR) measurements. (As discussed in Chap. 27, the *vector network analyzer*, or VNA, is capable of providing TDR-like measurements at far lower cost to the hobbyist.) An oscilloscope and a pulse (or square-wave) generator are connected in parallel across the input end of the transmission line. Figure 4.7B shows a pulse test jig built by the author for testing lines at HF. The small shielded box contains a square-wave oscillator circuit built with a few inexpensive integrated circuits. Although a crystal oscillator can be used, an *RC*-timed circuit running at 1 MHz or less is sufficient. In Fig. 4.7B, the test pulse generator box is connected in parallel with the cable under test and the input of the oscilloscope. A

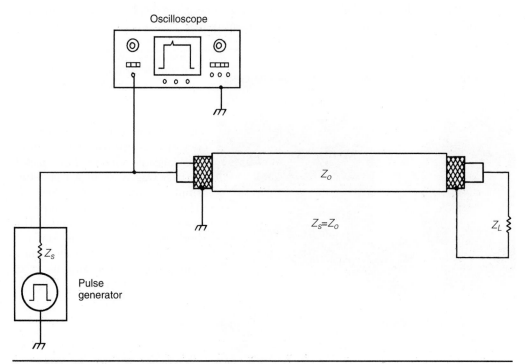

FIGURE 4.7A Time domain reflectometry setup.

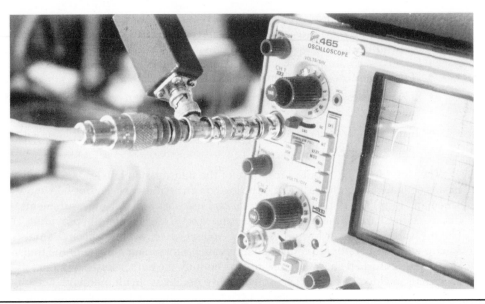

FIGURE 4.7B   Test setup for impromptu time domain reflectometry.

closer look is provided in Fig. 4.7C. A BNC "tee" connector and a double male BNC adapter are used to interconnect the box with the 'scope.

If a periodic waveform is supplied by the generator, the display on the oscilloscope will represent the sum of reflected and incident pulses. The duration of the pulse (i.e., pulse width), or one-half the period of the square wave, is adjusted so that the returning reflected pulse arrives approximately in the center of the incident pulse.

The images of Fig. 4.8 are TDR displays for several different values of Z_L. Approximately 30 m of coaxial cable with a velocity factor of 0.66 was used in a test setup similar to that shown in Fig. 4.7. The pulse width was approximately 0.9 µs (microseconds). The horizontal sweep time on the 'scope was adjusted to show only one pulse—which, in this case, represented one-half of a 550-kHz square wave (Fig. 4.8B).

FIGURE 4.7C   Close-up of RF connections.

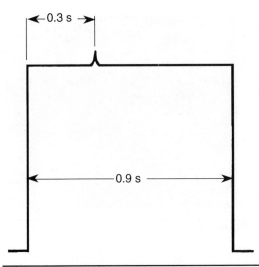

FIGURE 4.8A   Idealized TDR pulse. Small "pip" on top is reflected signal interfering with forward pulse.

FIGURE 4.8B   TDR pulse with tiny "blip". ($Z_L = Z_0$.)

The trace in Fig. 4.8B corresponds to a load matched to the line ($Z_L = Z_0$). A slight discontinuity, representing a small reflected wave, can be seen on the high side of the pulse. Even though the load and line are matched, any connector or other dimensional change may present a slight impedance discontinuity or *bump* that shows up on the 'scope. In general, any discontinuity in the line, any damage to the line, any too-sharp bend or other anomaly, can cause a slight impedance variation and, hence, a reflection.

Observe that the anomaly occurs approximately one-third of the 0.9-μs duration (or 0.3 μs) after the onset of the pulse. This tells us that the reflected wave arrives back at the source 0.3 μs after the incident wave leaves. Because this time period represents a round-trip, you can conclude that the wave required 0.3 μs/2, or 0.15 μs, to propagate the length of the line. Since we know the velocity factor, let's calculate the approximate length of the cable:

$$\text{Length} = cv_F T$$

$$= (3 \times 10^8 \,\text{m/s}) \times (0.66) \times (1.5 \times 10^{-7} \,\text{s})$$

$$= 29.7 \,\text{m}$$

(4.27)

which agrees "within experimental accuracy" with the 30 m actual length prepared for the experiment ahead of time. Thus, the simulated TDR setup (or an actual TDR instrument or VNA) can be used to measure the length of a transmission line. A general equation is

$$L \,(\text{meters}) = \frac{cv_F T_d}{2}$$

(4.28)

where L = length of transmission line, in meters
c = velocity of light (3×10^8 m/s)
v_F = velocity factor of transmission line
T_d = round-trip time between onset of pulse and first reflection

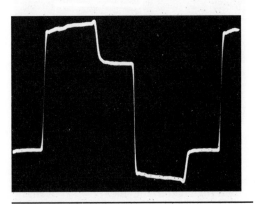

FIGURE 4.8C $Z_L < Z_0$.

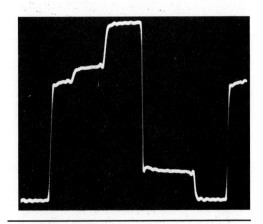

FIGURE 4.8D $Z_L > Z_0$.

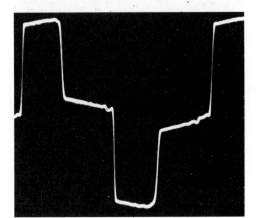

FIGURE 4.8E $Z_L = 0$.

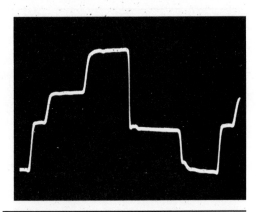

FIGURE 4.8F $Z_L = \infty$.

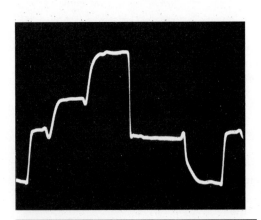

FIGURE 4.8G $Z_L = 50 - jX_C$.

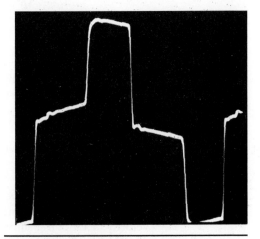

FIGURE 4.8 H $Z_L = 50 + jX_L$.

Figures 4.8C through 4.8H show the signal waveform at the input end of the transmission line resulting from a step-function input for various values of mismatched load impedance (i.e., Z_L not equal to Z_0). When the load impedance is less than the line impedance (in this case, $0.5Z_0$), the reflected wave is inverted as seen in Fig. 4.8C and sums with the incident wave along the top of the pulse. The reflection coefficient can be determined by examining the relative amplitudes of the two waves.

When $Z_L = 2Z_0$, the waveform of Fig. 4.8D results. In this case, the reflected wave is in phase with the incident wave, so it adds to the incident wave as shown. Waveforms for a short-circuited load and an open-circuited load are shown in Figs. 4.8E and 4.8F, respectively. Finally, examples of waveforms resulting from adding capacitive and inductive reactance to a 50-Ω load are displayed in Figs. 4.8G and 4.8H, respectively. The waveform in Fig. 4.8G resulted from a capacitance in series with a 50-Ω (matched) resistance; the waveform in Fig. 4.8H resulted from a 50-Ω resistance in series with an inductance.

Steady-State Response of the Transmission Line

When a CW RF signal is applied to a transmission line, the excitation is sinusoidal (Fig. 4.9), so investigation of the steady-state ac response of the line is useful. By *steady-state* we mean a sine-wave excitation of constant amplitude, phase, and frequency.

Our initial examination of the reflected wave that results when the load impedance does not match Z_0 of the transmission line used a step function, but the same reflection occurs when the transmission line is excited with a pure sinusoid at a single frequency. Here, too, when a transmission line is not matched to its load, some of the energy is absorbed by the load and some is reflected back up the line toward the source. The in-

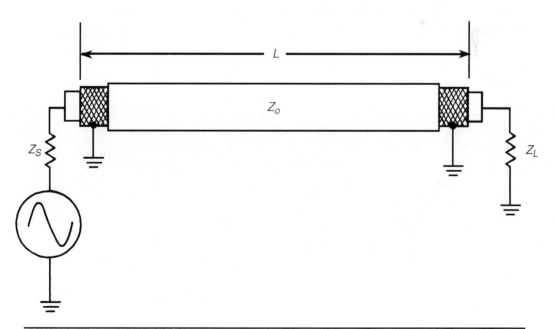

FIGURE 4.9 AC-excited transmission line.

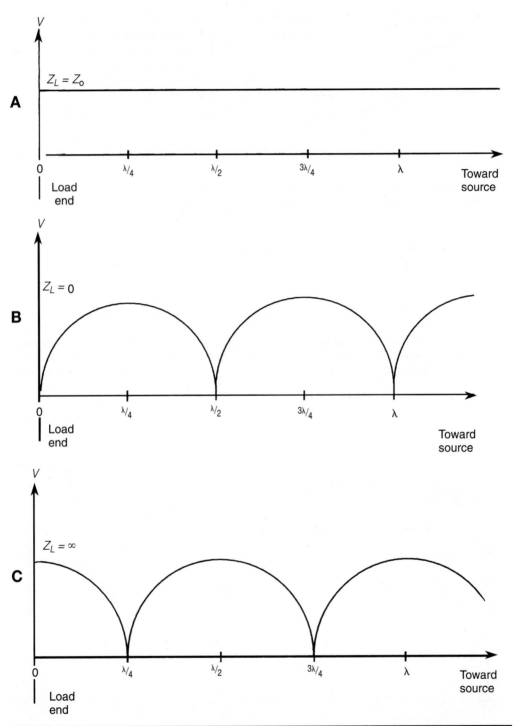

FIGURE 4.10   Voltage versus electrical length. (A) Matched impedances. (B) $Z_L = 0$, (C) Z_L = infinite.

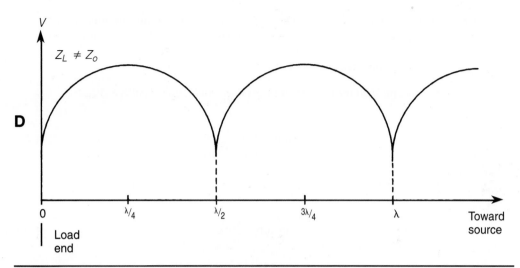

D

FIGURE 4.10   Continued: (D) Z_L not equal to Z_0.

terference of incident (or "forward") and reflected (or "reverse") waves creates *standing waves* on the transmission line.

If the voltage or current is measured along the line, it will vary, depending on the load, according to Fig. 4.10. In the absence of special instrumentation, such as a directional coupler, the voltage or current seen at any point will be the vector sum of the forward and reflected waves.

Figure 4.10A is a voltage-versus-length curve for a matched line—i.e., where $Z_L = Z_0$. The line is said to be "flat" because both voltage and current are constant all along the line. But now consider Fig. 4.10B and C.

Figure 4.10B shows the voltage distribution over the length of the line when the load end of the line is *shorted* (i.e., $Z_L = 0$). Of course, at the load end the voltage is zero, which results from the boundary condition of a short circuit. As noted earlier, the same conditions are repeated every half-wavelength along the line back toward the generator.

When the line is unterminated or open-circuited (i.e., $Z_L = \infty$), the pattern in Fig. 4.10C results. Note that it is the same shape as Fig. 4.10B for the shorted line, but the phase is shifted 90 degrees, or a quarter-wavelength, along the line. In both cases, the reflection is 100 percent, but the phase relationships in the reflected wave are exactly opposite each other.

Of course, if Z_L is not equal to Z_0 but is neither zero nor infinite, yet another set of conditions will prevail on the transmission line, as depicted in Fig. 4.10D. In this case, the nodes exhibit some finite voltage, V_{MIN}, instead of zero.

Standing Wave Ratio

One measure of the impact of load mismatch is the *standing wave ratio* (SWR) on a line.

If the current along the line is measured, the pattern will resemble the patterns of Fig. 4.10. The SWR is then called ISWR, to indicate the fact that it came from a current measurement. Similarly, if the SWR is derived from voltage measurements it is called VSWR. Perhaps because voltage is easier to measure, VSWR (or just plain SWR) is the term most commonly used in most radio work.

VSWR can be specified in any of several equivalent ways:

- *From incident (or forward) voltage (V_{FWD}) and reflected voltage (V_{REF}):*

$$\text{VSWR} = \frac{V_{FWD} + V_{REF}}{V_{FWD} - V_{REF}} \qquad (4.29)$$

- *From transmission line voltage measurements* (Fig. 4.10D):

$$\text{VSWR} = \frac{V_{MAX}}{V_{MIN}} \qquad (4.30)$$

- *From load and line characteristic impedances:*

$$(Z_L > Z_0)\text{VSWR} = Z_L / Z_0 \qquad (4.31)$$

$$(Z_L < Z_0)\text{VSWR} = Z_0 / Z_L \qquad (4.32)$$

- *From forward (P_{FWD}) and reflected (P_{REF}) power:*

$$\text{VSWR} = \frac{1 + \sqrt{P_{REF} / P_{FWD}}}{1 - \sqrt{P_{REF} / P_{FWD}}} \qquad (4.33)$$

- *From reflection coefficient (Γ):*

$$\text{VSWR} = \frac{1 + \Gamma}{1 - \Gamma} \qquad (4.34)$$

Of course, it's also possible to determine the reflection coefficient Γ from a knowledge of VSWR:

$$\Gamma = \frac{\text{VSWR} - 1}{\text{VSWR} + 1} \qquad (4.35)$$

VSWR is expressed as a *ratio.* For example, when Z_L is 100 Ω and Z_0 is 50 Ω, the VSWR is $Z_L/Z_0 = 100\ \Omega/50\ \Omega = 2$, which is usually expressed as VSWR = 2.0:1. However, VSWR can also be expressed in decibel form:

$$\text{VSWR} = 20 \log (\text{VSWR}) \qquad (4.36)$$

Example 4.6   A transmission line is connected to a mismatched load. Calculate both the VSWR and the VSWR decibel equivalent if the reflection coefficient Γ is 0.25.

Solution

$$(a)\ \text{VSWR} = \frac{1+\Gamma}{1-\Gamma}$$

$$= \frac{1+0.25}{1-0.25}$$

$$= \frac{1.25}{0.75}$$

$$= 1.67 : 1$$

$$(b)\ \text{VSWR}_{dB} = 20 \log (\text{VSWR})$$

$$= (20)\,(\log 1.67)$$

$$= (20)(0.22)$$

$$= 4.4\ \text{dB}$$

◆

SWR on a transmission line is important for several reasons.

- The higher the SWR, the higher the maximum voltages on a transmission line for a given power applied or delivered to the load. High SWR can cause unexpected destruction of the line and any electronics equipment attached to it.

- Many of today's solid-state transmitters, transceivers, and amplifiers reduce RF output or shut down completely in the presence of high SWR.

- The reflected wave on a transmission line represents energy "rejected" by the load. For a specified power delivered to the antenna, all components must be "supersized" for a line with a high SWR.

- A transmission line operating with high SWR will exhibit greater loss per unit length than its specified loss for a matched load impedance.

Mismatch (VSWR) Losses

The power reflected from a mismatched load represents a potential loss and can have profound implications, depending on the installation. For example, one result might be a slight loss of signal strength at a distant point from an antenna. A more serious problem can result in the destruction of components in the output stage of a transmitter. The latter problem so plagued early solid-state transmitters that designers opted to include shutdown circuitry to sense high VSWR and turn down output power proportionately—a practice that remains to this day.

In microwave measurements, VSWR on the transmission lines (that interconnect devices under test, instruments, and signal sources) can cause erroneous readings—and invalid measurements.

Determination of VSWR losses must take into account *two* VSWR situations. Figure 4.9 shows a transmission line of impedance Z_0 interconnecting a load impedance Z_L to a source with output impedance Z_S. There is a potential for impedance mismatch at *both ends* of the line.

When one end of the line is matched (either Z_S or Z_L), the *mismatch loss* (ML) caused by SWR at the mismatched end is

$$ML = -10 \log \left[1 - \left(\frac{SWR - 1}{SWR + 1} \right)^2 \right] \tag{4.37}$$

which from Eq. (4.35) is

$$ML = -10 \log (1 - \Gamma^2) \tag{4.38}$$

Example 4.7   A coaxial transmission line with a characteristic impedance of 50 Ω is connected to the 50-Ω output (Z_0) of a signal generator, and also to a 20W load impedance Z_L. Calculate the mismatch loss.

Solution
(a) First find the VSWR:

$$VSWR = Z_0 / Z_L$$

$$= (50\,\Omega) / (20\,\Omega)$$

$$= 2.5:1$$

(b) Mismatch loss:

$$ML = -10 \log \left[1 - \left(\frac{SWR - 1}{SWR + 1} \right)^2 \right]$$

$$= -10 \log \left[1 - \left(\frac{2.5 - 1}{2.5 + 1} \right)^2 \right]$$

$$= -10 \log \left[1 - \left(\frac{1.5}{3.5} \right)^2 \right]$$

$$= -10 \log \left[1 - (0.43)^2 \right]$$

$$= -10 \log [1 - 0.185]$$

$$= -10 \log [0.815]$$

$$= (-10)(-0.089)$$

$$= 0.89$$

◆

When both ends of the line are mismatched, a different equation is required:

$$\mathrm{ML} = 20 \log\left[1 \pm (\Gamma_1 \times \Gamma_2)\right] \qquad (4.39)$$

where Γ_1 = reflection coefficient at source end of line $(\mathrm{VSWR}_1 - 1)/(\mathrm{VSWR}_1 + 1)$
Γ_2 = reflection coefficient at load end of line $(\mathrm{VSWR}_2 - 1)/(\mathrm{VSWR}_2 + 1)$

Note that the solution to Eq. (4.39) has two values: $[1 + (\Gamma_1\Gamma_2)]$ and $[1 - (\Gamma_1\Gamma_2)]$.

The equations reflect the mismatch loss solution for low-loss or "lossless" transmission lines. This is a close approximation in many situations; however, even cables having minimal loss at HF exhibit substantially higher losses at microwave frequencies. Interference between forward and reflected waves produces increased current at certain antinodes—which increases ohmic losses—and increased voltage at certain antinodes—which increases dielectric losses. It is the latter that increases with frequency. Equation (4.40) relates line losses to the reflection coefficient to determine total loss on a line with a given VSWR.

$$\mathrm{Loss} = 10 \log\left(\frac{n^2 - \Gamma^2}{n - n\Gamma^2}\right) \qquad (4.40)$$

where loss = total line loss in decibels
Γ = reflection coefficient
n = quantity $10^{(A/10)}$
A = total attenuation presented by line, in decibels, when line is properly matched $(Z_L = Z_0)$

Example 4.8 A 50-Ω transmission line is terminated in a 30-Ω resistive impedance. The line is rated at a loss of 3 dB/100 ft at 1 GHz. Calculate (*a*) loss in 5 ft of line, (*b*) reflection coefficient, and (*c*) total loss in a 5-ft line mismatched per above.

Solution

$$(a) \quad A = \frac{3\,\mathrm{dB}}{100\,\mathrm{ft}} \times 5\,\mathrm{ft}$$

$$= 0.15\,\mathrm{dB}$$

$$(b) \quad \Gamma = \frac{Z_L - Z_0}{Z_L + Z_0}$$

$$= \frac{50 - 30}{50 + 30}$$

$$= 20/80$$

$$= 0.25$$

$$(c) \quad n = 10^{(A/10)}$$

$$= 10^{(0.15/10)}$$

$$= 10^{(0.015)}$$

$$= 1.04$$

$$\text{Loss} = 10 \log \left(\frac{n^2 - \Gamma^2}{n - n\Gamma^2} \right)$$

$$= 10 \log \left[\frac{(1.04)^2 - (0.25)^2}{1.04 - (1.04)(0.25)^2} \right]$$

$$= 10 \log \left[\frac{1.082 - 0.063}{1.04 - (1.04)(0.063)} \right]$$

$$= 10 \log \left(\frac{1.019}{1.04 - 0.066} \right)$$

$$= 10 \log \left(\frac{1.019}{0.974} \right)$$

$$= 10 \log (1.046)$$

$$= (10)(0.02)$$

$$= 0.2 \text{ dB}$$

Compare the matched line loss ($A = 0.15$ dB) with the total loss (Loss $= 0.2$ dB), which includes mismatch loss and line loss. The difference (i.e., Loss $- A$) is only 0.05 dB. If the VSWR or the total line length were considerably larger, however, the loss would rise.

Impedance Matching in Antenna Systems

Although antennas are *reciprocal* devices, a receiving antenna can do an excellent job even in the absence of perfect matches between the antenna and the feedline or between the feedline and the receiver. This is especially true in the MF and lower HF regions, where reception is almost always limited by atmospheric noise.

In contrast, proper matching of the transmitter to the feedline and the feedline to the antenna is essential for maximizing our radiated signal. That is because maximum power transfer between a source and a load always occurs when the system imped-ances are matched.

Of course, the trivial situation is when all three sections of our system—transmitter, feedline, and antenna—have the same impedance. The most obvious example would be an antenna (such as a half-wave dipole in free space) with a 75-Ω resistive feedpoint impedance fed from garden variety 75-Ω coaxial cable or hardline that is connected to a transmitter with a 75-Ω output impedance.

However, in real life the feedpoint impedance is rarely what the books say it should be. Even our ubiquitous $\lambda/2$ dipole seldom appears to be exactly 75 Ω or purely resis-tive to its feedline. Then there are our wire arrays, our Yagis and our loops—all with feedpoint impedances substantially higher or lower and/or more reactive than the ideal λ/2 dipole. So we can be reasonably certain that a matching network at the junction of the feedline and the antenna is likely to be needed.

For most of us, the transmitters, transceivers, and amplifiers we are apt to buy or build are designed for maximum power transfer to a 50-Ω resistive load. In many cases we can simplify our task by choosing to use 52-Ω or 75-Ω transmission line from the transmitter to a matching network closer to the antenna. Most of the following techniques for obtaining a match assume just such a configuration.

In general there are two broad categories of matching techniques:

- Distributed matching systems (stubs, transmission line sections, etc.)
- Lumped element tuners (*antenna tuning units*, or ATUs)

The second category is covered in Chap. 24.

Distributed Matching Networks

Networks that depend on the propagation of the signal along a carefully chosen length of conductor for their matching capabilities are called *distributed* networks. The popular examples described in the following sections are typically employed either as a series element placed between the antenna and feedline or as a shunt element directly across the antenna feedpoint or a short distance along the feedline from it.

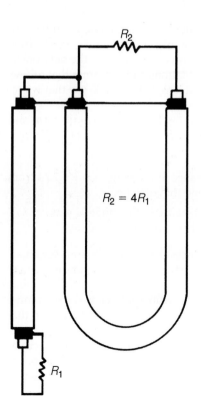

Keep in mind that since the dimensions of most distributed matching networks are a function of the operating frequency (or wavelength), a specific implementation will work as advertised only over a limited frequency span centered on the design frequency. Most of the networks described here are most useful, therefore, as matching devices for single-band antennas and arrays.

Coaxial Cable Baluns

Figure 4.11 describes a balun that transforms impedances at a 4:1 ratio, with $R_2 = 4 \times R_1$. The length of the balun (or U-shaped) section of coaxial cable is

$$L = \frac{492 v_F}{F} \qquad (4.41)$$

where L = measured physical length of U-shaped section of cable, in feet

   v_F = velocity factor of coaxial cable (typically 0.66–0.85)

   F = operating frequency, in megahertz

FIGURE 4.11   Coaxial balun transformer.

In general, a balun can be used in either a *step-up* or a *step-down* configuration. For instance, this balun is frequently used as shown to transform the 300-Ω feedpoint impedance of a folded dipole down to 75 Ω for a coaxial transmission line. The radiation (or feedpoint) resistance of the dipole corresponds to R_2 in the figure, and R_1 represents the transmitter output impedance. However, the owner of a multielement Yagi or a short ($< \lambda/4$) vertical monopole might employ the balun in a reverse configuration; now R_1 represents the low feedpoint impedance of the antenna, while R_2 corresponds to the Z_0 of a higher-impedance transmission line.

Matching with Stubs

The input impedance of a random length of lossless (or nearly so) transmission line is given by:

$$Z_{\text{IN}} = Z_0 \left[\frac{Z_L + jZ_0 \tan(\beta l)}{Z_0 + jZ_L \tan(\beta l)} \right] \tag{4.42}$$

where Z_{IN} = input impedance of line looking into end opposite load
$\quad\quad Z_L$ = load impedance
$\quad\quad Z_0$ = characteristic impedance of transmission line
$\quad\quad \beta$ = $2\pi/\lambda$
$\quad\quad l$ = length of transmission line, stated in terms of λ

The term (βl) in the argument of the tangent functions in Eq. (4.42) includes λ in both the numerator and the denominator, so it does not matter whether we use the free-space wavelength or the wavelength in the transmission line, as long as the *same* one is used for both β and l. However, in some of the manipulations and outcomes that follow, λ refers to the wavelength *in the transmission line,* so it is wise to always use that wavelength when working with stubs.

Most of the time, Z_{IN} will be complex—that is, having both a resistive (R) term and a reactive (jX) term—even for a purely resistive load, except when either of the following situations is true:

- The input end of the line is one specific distance in the range $0 \le l \le \lambda/4$ from the load, or an integer multiple of $\lambda/2$ beyond that distance.

- The input end of the line is at one specific distance in the range $\lambda/4 \le l \le \lambda/2$ from the load, or an integer multiple of $\lambda/2$ beyond that distance.

Since there is an infinite number of possible impedances that Z_{IN} can assume for an arbitrary line length, the odds of finding a purely resistive value of Z_{IN} are extremely small—unless, of course, $Z_L = Z_0$.

A very important characteristic of transmission lines comes into view if we examine Eq. (4.42) a little further. Suppose we are free to adjust the length (l) of the line so that the input end of the line is at a value of l that causes the tangent functions in Eq. (4.42) to be zero. Since that occurs when the argument (βl) of both tangent functions is zero or at phase angles that are integer multiples of π radians, l must be an integer multiple of $\lambda/2$. In fact, when $l = n\lambda/2$ (for integer values of n), $Z_{\text{IN}} = Z_L$, regardless of whether Z_L is resistive, reactive, or complex.

In general, the simplest ways to determine Z_{IN} as a function of feedline length for a specific load are to use the graphical Smith chart techniques of Chap. 26 or one of the many online calculators available. However, when the load on a nearly lossless length of line is either a short circuit or an open circuit, the math required to find Z_{IN} for a given length of line is quite a bit simpler and yields interesting and useful results.

For a short circuit at the far end of the line, $Z_L = 0$, and Eq. (4.42) reduces to

$$Z_{\text{IN-SHORT}} = Z_0 \left[\frac{0 + jZ_0 \tan(\beta l)}{Z_0 + j0} \right] = jZ_0 \tan(\beta l) \tag{4.43}$$

Since the only term in the solution is imaginary, Z_{IN} is purely reactive—alternating between capacitive and inductive reactance as the length of the short-circuited line is varied and $\tan(\beta l)$ swings negative or positive, respectively.

Similarly, for an open circuit at the far end of the line, $Z_L = \infty$, and Eq. (4.42) reduces to

$$Z_{\text{IN-OPEN}} = Z_0 \left[\frac{1}{j \tan(\beta l)} \right] = -jZ_0 \cot(\beta l) \tag{4.44}$$

NOTE: *If you have a network analyzer, VNA, or antenna analyzer that provides R and X for an unknown impedance attached to its terminals, you can determine Z_0 and the velocity factor (v_F) for a mystery piece of transmission line. First obtain $Z_{\text{IN-OPEN}}$ and $Z_{\text{IN-SHORT}}$ for the line. Now multiply Eq. (4.43) by Eq. (4.44) to obtain Eq. (4.45):*

$$\left(Z_{\text{IN-SHORT}} \right)\left(Z_{\text{IN-OPEN}} \right) = \left[jZ_0 \tan(\beta l) \right]\left[-Z_0 \cot(\beta l) \right] \tag{4.45}$$

But tan x = 1/cot x, so Eq. (4.45) reduces to

$$\left(Z_{\text{IN-SHORT}} \right)\left(Z_{\text{IN-OPEN}} \right) = -j^2 Z_0^2 = Z_0^2 \tag{4.46}$$

In other words, we can determine the characteristic impedance of our unlabeled transmission line by taking the square root of the product of the short-circuit and open-circuit impedance measurements we made.

If now we obtain the ratio of Eqs. (4.43) and (4.44), we have

$$\frac{Z_{\text{IN-SHORT}}}{Z_{\text{IN-OPEN}}} = \frac{jZ_0 \tan(\beta l)}{-jZ_0 \cot(\beta l)} = -\tan^2(\beta l) \tag{4.47}$$

Multiplying both sides by −1 and taking the positive square root, we obtain

$$\tan(\beta l) = \sqrt{\frac{-Z_{\text{IN-SHORT}}}{Z_{\text{IN-OPEN}}}} \tag{4.48}$$

Now we know l and the tangent of βl, so we can find β, the phase constant of the line from trig tables or our scientific calculator.

More likely, we would like to know v_F, the velocity factor of the line. Rearranging Eq. (4.16), we obtain

$$v_F = \frac{2\pi f}{c\beta} \qquad (4.49)$$

Because a shorted or open length of transmission line always exhibits a pure reactance at the other end, such lines are often used to match an antenna to its transmission line. A specific reactance, either capacitive or inductive, can be obtained simply by adjusting the length of the line, or *stub*.

In most cases, however, not only do we need to cancel out the reactive part of the feedpoint impedance, we also want to change the resistive part to match the Z_0 of the feedline. Simply adjusting the length of the stub cannot compensate for *both* the resistive and the reactive mismatches. A second adjustable variable is needed. That turns out to be the distance of the stub back toward the transmitter from the antenna feedpoint.

For a variety of reasons, both electrical and mechanical, stubs are more often found connected in parallel, across the feedline, than in series with one side of the feedline. In such cases, it is far easier to work with the reciprocal of the impedances involved. These *admittances*, denoted by Y, consist of a shunt *conductance* G (the purely resistive part of the admittance) in parallel with a shunt *susceptance* B, which is the purely reactive part. Mathematically, if $Z = R + jX$, then

$$Y = \frac{1}{Z} = \frac{1}{R + jX} = G + jB \qquad (4.50)$$

To put $\dfrac{1}{R + jX}$ into the right form, we multiply both top and bottom by $R - jX$, the

complex conjugate of $R + jX$. This gives us

$$\left[\frac{1}{R + jX}\right]\left[\frac{R - jX}{R - jX}\right] = \frac{R - jX}{R^2 + X^2} \qquad (4.51)$$

Thus,

$$G = \frac{R}{R^2 + X^2} \qquad (4.52)$$

and

$$B = \frac{-jX}{R^2 + X^2} \qquad (4.53)$$

In a small percentage of stub-matching cases, it turns out to be possible to locate the stub right at the antenna feed terminals. Most often, however, a stub must be located at some specified distance from the antenna feedpoint.

Without the use of paper or online Smith charts, stub matching is far more difficult than it needs to be. Chapter 26 includes a detailed example of stub matching using a Smith chart.

Special Cases

If we use an impedance bridge or other instrument to determine the input impedance at one end of an infinitely long transmission line, we will obtain the value Z_0—the *characteristic* or *surge impedance* for that particular type of line. If we attach a load impedance equal to Z_0 at the far end of a much shorter length of transmission line, we will again measure Z_0 at the input. But if we attach a load impedance $Z_L \neq Z_0$, we have no idea what input impedance we will see unless we know Z_L, Z_0 and the exact length of the line in electrical wavelengths. As we have seen in the preceding paragraphs, that is because the impedance "looking into" a transmission line at the input is the result of the line transforming the load impedance in a manner determined by the length of the line and the relationship between the line and load impedances. Here are some specific combinations that are important to know.

Matched Line and Load Impedances

Assuming a reasonably low loss transmission line, a purely resistive load impedance $Z_L = Z_0$ appears as Z_0 at the line input for *any* length of line. Any other load impedance, whether a pure resistance $\neq Z_0$ or a resistive component R_0 accompanied by a reactive component, will transform to some other $R \pm jX$ value at the line input; the value (relative to Z_L) will be solely determined by the length of the line (in wavelengths).

Half-Wavelength Section of Line

A half-wavelength section of transmission line is "magical" because it repeats *any* load impedance, whether matched or not. Thus, for a single frequency or band of frequencies, a transmission line that is any integer multiple of $\lambda/2$ long can be used to remotely measure and monitor the feedpoint impedance of an antenna. The two caveats are:

- An accurate knowledge of the line's velocity factor is needed before cutting the line to length.

- A given length of line is $\lambda/2$ only over a very narrow range of frequencies.

Quarter-Wave Matching Section

A quarter-wavelength line exhibits a transforming property, converting low-resistance loads to higher ones, and capacitive reactances to inductive. The best way to understand and use this feature is with the *Smith chart* (explained in detail in Chap. 26) or a similar tool. One important application of this property is to transform a load impedance to a higher (or lower) main transmission line impedance. For instance, suppose an antenna for 10 MHz has a feedpoint impedance of 100 Ω resistive. It's not a very good match for 52-Ω coaxial cable, but a $\lambda/4$ section of 75-Ω RG-59 or RG-11 coaxial cable for television systems will transform the antenna feedpoint impedance from 100 to 52 Ω.

Impedances for the $\lambda/4$ transformer are calculated (for resistive loads) as follows:

$$Z_L Z_S = Z_0^2 \tag{4.54}$$

where Z_L = load or antenna feedpoint impedance
Z_0 = characteristic impedance of $\lambda/4$ section
Z_S = characteristic impedance of system transmission line

The $\lambda/4$ transmission line transformer, shown in Fig. 4.12, is sometimes called a *Q-section matching network*.

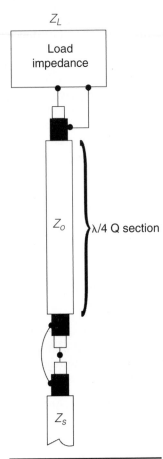

Z_L

Load
impedance

Z_O } λ/4 Q section

Z_S

FIGURE 4.12   Quarter-wavelength
Q section.

Example 4.9   A 50-Ω source must be matched to a load impedance of 36 Ω. Find the characteristic impedance required of a
Q-section matching network.

Solution

$$Z_0 = \sqrt{Z_S Z_L}$$

$$= \sqrt{(50\,\Omega)\,(36\,\Omega)}$$

$$= \sqrt{1800\,\Omega^2}$$

$$= 42\,\Omega$$

◆

Clearly, the key to success is to have available a piece of
transmission line having the required characteristic impedance. Unfortunately, the number of different coaxial cable impedances available for the Q-section is limited, and so its
utility for matching an arbitrary pair of impedances is also
limited.

On open-wire transmission line systems, on the other
hand, it is quite easy to achieve the correct impedance for the
matching section. As before, we use the preceding equation to
find a value for Z_0, and then calculate the conductor spacing
and diameter needed for a short length of a custom transmission line! Armed with information about the diameter of common wire sizes, once we know the feedpoint impedance, we
can use Eq. (4.55) to calculate the appropriate conductor spacing:

$$S = d \times 10^{(Z_0/276)} \tag{4.55}$$

where S = spacing between conductors of parallel line
       d = conductor diameter expressed in same units as S
       Z_0 = desired surge impedance for Q-section

Assuming true open-wire line (with air dielectric) is used, the length of the
quarter-wave section in feet can be calculated using the familiar $246/F$, where F is in
megahertz.

A *quarter-wavelength shorted stub* is a special case of the stub concept that finds particular application in microwave circuits. *Waveguides* (Chap. 20) are based on the properties of the quarter-wavelength shorted stub. Figure 4.13 shows a quarter-wave stub
and its current distribution. The current is maximum across the short, but wave cancellation forces it to zero at the terminals. Because $Z = V/I$, when I goes to zero, the impedance becomes infinite. Thus, a quarter-wavelength stub has an infinite impedance at its
resonant frequency, and acts as an insulator. This concept may be hard to swallow, but
the stub is a "metal insulator".

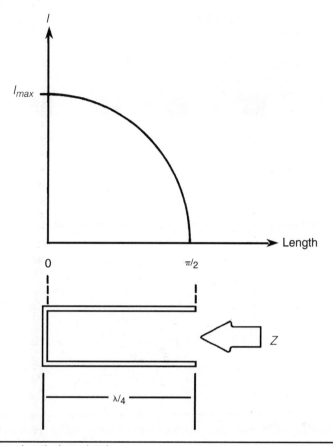

FIGURE 4.13   Quarter-wavelength shorted stub.

Series Matching Section

The *series matching section*—also called the *series section transformer*—is a generalization of the Q-section that permits us to build an impedance transformer that overcomes the Q-section's limitations. According to *The ARRL Antenna Book*, with appropriate choices of transmission lines this form of transformer is capable of matching any load resistance between about 5 and 1200 Ω. In addition, the transformer section does not have to be located at the antenna feedpoint.

Figure 4.14 shows the basic layout of the series matching section. Three lengths of coaxial cable, L_1, L_2, and L_3, serve to form the entire feedline connecting the antenna to the transmitter. Lengths L_2 and the line to the transmitter (L_3, which is any convenient length and doesn't appear in the equations) have the same characteristic impedance, usually 75 Ω, while section L_1 has a different impedance. Note that only standard, easily obtainable values of impedance are used here.

The design procedure for this transformer consists of finding the correct lengths for L_1 and L_2. You must know the characteristic impedance of the two lines (75 Ω and 50 Ω, respectively, in this example), along with the complex feedpoint impedance of the antenna, $Z_L = R_L + jX_L$.

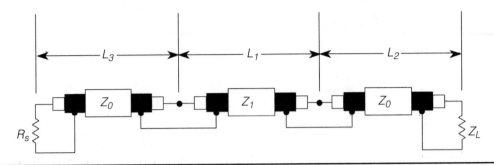

Figure 4.14   Series matching section.

The first task in designing the transformer is to define *normalized* impedances:

$$N = \frac{Z_1}{Z_0} \tag{4.56}$$

$$R = \frac{R_L}{Z_0} \tag{4.57}$$

$$X = \frac{X_L}{Z_0} \tag{4.58}$$

Adopting ARRL notation and defining $A = \tan L_1$ and $B = \tan L_2$, the following equations can be written:

$$\tan L_2 = B = \pm \sqrt{\frac{(R-1)^2 + X^2}{R\{N - (1/N)^2\} - (R-1)^2 - X^2}} \tag{4.59}$$

$$\tan L_1 = A = \frac{\{N - (R/N)\} \times B + X}{R + (XNB) - 1} \tag{4.60}$$

The lengths of L_1 and L_2 determined from Eqs. (4.59) and (4.60) are in electrical degrees. To find their physical length in feet:

$$L_1(\text{feet}) = \frac{L_1(\text{degrees})\lambda}{360} \tag{4.61}$$

$$L_2(\text{feet}) = \frac{L_2(\text{degrees})\lambda}{360} \tag{4.62}$$

where

$$\lambda = \frac{984 \times v_F}{f}$$ (4.63)

where v_F = velocity factor of each specific cable
   f = frequency in megahertz
   λ = expressed in feet

Note that λ is not the free-space wavelength but the (shorter) wavelength of the wave in the transmission line. Make sure to use the velocity factor for the correct cable in each of the two calculations!

Although the sign of B can be selected as either − or +, the use of + is preferred because a shorter matching section results. If the term inside the square root sign is negative, B is an imaginary number and the section cannot be constructed with the two line impedances selected. Typically that will happen if the two impedances are too similar and the required transformation is too great. The sign of A cannot be chosen by the user, but if it turns out to be negative, add 180 degrees to the result.

Series-section transformers can also be designed graphically with the help of the Smith charts of Chap. 26.

CHAPTER **5**

Antenna Arrays and Array Gain

Multiple antennas or antenna elements fed from a common source of radiofrequency (RF) energy are collectively called an *array*. (Similarly, multiple antennas feeding a single receiver input also form an array.) The received signal at any distant point from an array of transmitting antennas is a combination of the signals from *all* the fed elements. Because each signal arriving at the receiving antenna has traveled a unique distance and its drive signal may have been phase-shifted with respect to others, the received signal is the vectorial combination of the contributions from all the elements of the array.

Arrays are of interest because they make it possible, for a given transmitter power, to increase received signal strengths relative to a single radiating element. They are also important to understand because the effect of nearby earth or other conductors on the performance of antennas can often best be explained through array analysis. In this chapter we will explore the basics of arrays, using as our starting point *all-driven arrays*.

All-Driven Arrays

In Chap. 3 we saw that the strength of the received electromagnetic field far from a $\lambda/2$ dipole or $\lambda/4$ monopole transmitting (TX) antenna was proportional to the magnitude of the RF current at the feedpoint of the TX antenna. (See Fig. 5.1A.) The received signal will be a weak replica of the transmitted signal, delayed with respect to the original by an amount that will depend on the exact distance between the receiving and transmitting antennas. In simplified form, and ignoring the time it takes for the signal to reach the receiving point, the magnitude of the received E-field strength at a distant point is

$$E_{\text{RXorig}} = kI_{\text{TXorig}} \tag{5.1}$$

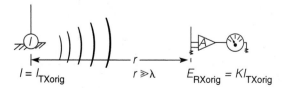

$I = I_{\text{TXorig}}$ 　　　　 $r \gg \lambda$ 　　 $E_{\text{RXorig}} = KI_{\text{TXorig}}$

FIGURE 5.1A Received signal at a distant point from a single transmitting antenna.

where E_{RXorig} = field strength at distant receiving antenna

I_{TXorig} = feedpoint current of transmitting antenna

k = constant that collects in a single number the total proportionality between E and I resulting from the specific characteristics of antennas, the orientation of the two antennas with respect to each other, and the attenuation along the path between them.

Suppose we now excite a *second* transmitting antenna with feedpoint current I_{TXorig} identical in all respects to the current in the first antenna; that is, a current having the same magnitude and phase (at the second feedpoint) as the first current. Further, let's suppose we locate this second TX antenna at $\lambda/2$ from the first but place it at exactly the same distance (Fig. 5.1B) from the receiving point so that the peak amplitudes of the two signals reach a maximum at the receive point at the same exact time. The principles of *superposition* and *linearity* tell us that the net, or total, received signal will now be *twice* that of the original because the receive antenna is intercepting two electromagnetic fields (that happen to be in phase) instead of one. Twice the current or voltage at the receiver input corresponds to a signal increase of 6 dB. (See App. A for an explanation of the use of logarithms and decibels to describe ratios of power, voltage, and current.)

Now let's cut the power to each of the two TX antennas in half, thus keeping the *total* power to the two antennas the same as the original power to the first TX antenna. At each TX antenna, then, $P_{\text{TXnew}} = \frac{1}{2} P_{\text{TXorig}}$. But since $P = I^2 R$,

$$I_{\text{TXnew}} = \frac{\sqrt{P_{\text{TXnew}}}}{\sqrt{R}} \tag{5.2}$$

where R is the feedpoint resistance of each antenna and I_{TXnew} is the feedpoint current resulting from the new input power, P_{TXnew}, to each element.

Putting the old and the new equations for power in a ratio, we obtain for the new feedpoint current in *each* TX antenna

$$I_{\text{TXnew}} = \frac{\sqrt{P_{\text{TXorig}}/2}}{\sqrt{R}} = \frac{\sqrt{I_{\text{TXorig}}^2 R/2}}{\sqrt{R}} = I_{\text{TXorig}}/\sqrt{2} = 0.707 I_{\text{TXorig}} \tag{5.3}$$

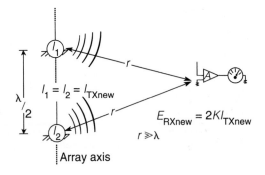

$l_1 = l_2 = l_{\text{TXnew}}$

$E_{\text{RXnew}} = 2K l_{\text{TXnew}}$

$r \gg \lambda$

Array axis

FIGURE 5.1B   Received signal at a distant point from two identical transmitting antennas.

In other words, by cutting the power to the original TX antenna in half, we have reduced the received E-field strength from that TX antenna to 71 percent of its original value. This is a consequence of power in a resistive load being proportional to the *square* of the current.

But there are *two* currents at the receiving site, each resulting from intercepting the radiated field from one of the two TX antennas, and their magnitudes are in phase, so thanks to the principle of superposition we can add the two:

$$E_{\text{RXnew}} = 2(0.707) k I_{\text{TXorig}} = 1.414 k I_{\text{TXorig}} \tag{5.4}$$

Thus, the new current at the receiver resulting from the same total power at the transmitting site is 1.414 times the original current—or 3 dB larger!

What we have just created is an *antenna array*—more specifically, a *two-element all-driven phased array* operated in its *broadside* mode. (In broadside mode, the element feed currents are in phase and the direction(s) of maximum radiation are on a line at right angles (hence, *broadside*) to an imaginary line—called the *array axis,* and indicated by the vertical dotted line in Fig. 5.1B—connecting the elements.) In a sense, we have created "something from nothing". Through the simple expedient of erecting a second antenna and splitting our allowed transmitter power equally between the two antennas, we have accomplished an increase in received signal strength at a distant point—an increase that would have required us to double our transmitter power (from, say, 100 to 200 W) if we had continued to use just the original antenna.

This seems magical and it is, in a way, but "there's no free lunch", as the saying goes. In this case, as with all the arrays we will discuss, we have increased our radiated field in *some* directions at the expense of the field or signal strength in *others*. It may be helpful to visualize the radiation field of the original antenna as a spherical balloon, and a given transmit power corresponds to blowing the balloon up to a certain size. Now wrap your two hands around the middle of the balloon and squeeze. Pushing in on part of it causes the skin of another part of the balloon to extend, even though the total air within the balloon remains constant. Similarly, an array of antenna elements modifies the radiation field of the original antenna, making it stronger in certain directions while weakening it in others. In fact, if we were to move our receiving antenna to a distant point anywhere on the array axis we would find the received E-field to be zero or nearly so.

Let's see why that is. The two transmitting verticals of Fig. 5.1B are spaced $\lambda/2$ apart, but their feed currents are in phase. When the field from antenna A reaches antenna B, it will have traveled a half-wavelength, and, hence, its waveform will be 180 degrees out of phase from its starting point at A and also 180 degrees out of phase with the waveform of antenna B's field (since we said at the beginning that the two TX feedpoint currents were in phase at the base of each antenna). Because A and B are so close together (compared to the distance to the receiving antenna), the magnitudes of the two fields from the two TX antennas are virtually identical at the receiving antenna. As the fields from antennas A and B propagate outward along an extension of the line connecting A and B, they are essentially the same strength at every point on the line, but 180 degrees out of phase. That is to say, they almost completely cancel at all receiving points on a line connecting the centers of the two antennas. The received signal strength from this two-element phased array is, for all intents and purposes, zero anywhere along the axis of the array. The array has "stolen" power from certain directions to increase the

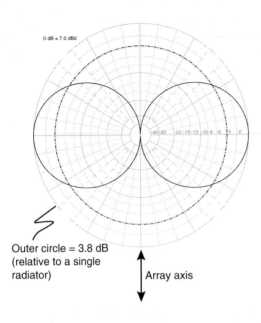

Outer circle = 3.8 dB
(relative to a single
radiator)

Array axis

FIGURE 5.1C Array pattern for the two-element array of (B) in broadside (fed in phase) mode.

radiated power in other directions—specifically, those broadside to the elements of the array.

If we move our receiving antenna around the compass at a constant distance from the center of the array, we find that the received signal varies from zero on the array axis to a maximum broadside to the array, as shown in Fig. 5.1C.

Any time the two elements of an array are odd multiples of $\lambda/2$ apart and the currents at the two elements are in phase, radiation along the axis of the array will be zero and radiation broadside to the line connecting the two elements will be a maximum. In our first example we calculated a gain of 3 dB over a single element, but that is only a rough approximation because we ignored possible interactions between the two elements. In truth, if the two elements are $\lambda/2$ apart, on-axis radiation is everywhere zero, but interaction between the two elements alters their feedpoint impedances and the resulting gain in the desired (broadside) direction is not exactly 3 dB greater than that of a single radiating element. In fact, for $\lambda/4$ verticals spaced $\lambda/2$ apart, it's a bit higher than 3 dB; as shown in Fig. 5.1C, it's 3.8 dB. For other spacings, such as $3\lambda/2$, $5\lambda/2$, etc., and for other antenna orientations (dipoles end to end, for instance) the mutual impedances of the elements can cause the array gain to be either greater than or less than 3 dB for a fixed input power to the array. (Also, for other than $\lambda/2$ spacing, additional *side lobes* form.) As we shall see in Chap. 12, it is these same interactions between closely spaced conductors that allow us to obtain array gains from multielement antennas having only one element directly connected to a feedline!

If now the RF signal being applied to one of the two radiating verticals in Fig. 5.1B is shifted 180 degrees so that the two antennas are being driven out of phase, the pattern of Fig. 5.1D results. Note that the pattern is not just rotated 90 degrees but is somewhat "squashed" compared

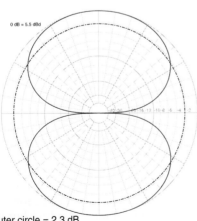

Outer circle = 2.3 dB
(relative to a single
radiator)

FIGURE 5.1D Array pattern for the two-element array of (B) in end-fire (fed 180 degrees out of phase) mode.

to the broadside pattern. The peak gain of this array's *end-fire* mode compared to a single vertical is 2.3 dB—about 1.5 dB less than that of the broadside mode.

Array Factor

One nice feature about arrays is that we can build them out of just about any kind of antenna element. From a purely mathematical viewpoint, the easiest arrays to analyze are those comprised of *isotropic radiators* because the individual elements have a very simple pattern—they radiate equally well in all directions. As a result, when two or more isotropic radiators are combined to form an array, the resulting pattern is the *array pattern* itself. The array pattern (also called the *array factor*) is an equation obtained from evaluation of the geometry of the element positions, as well as the relative amplitudes and phases of the individual element feedpoint currents. For the two-element broadside array discussed here and shown in Fig. 5.1B, the shape of the azimuthal array pattern is given by the equation

$$AP = \cos\left(\frac{\pi}{2}\cos\theta\right) \tag{5.5}$$

where θ is the angle between the array axis and a line drawn from the center of the array to a distant receiving antenna.

But isotropic radiators, as we saw in Chap. 3, are totally fictitious, so we are forced to form arrays out of "real" elements: dipoles, verticals, loops, and other basic antenna types. When we do that, the resulting radiation pattern developed by the array is the array pattern multiplied by the radiation pattern of an individual element. For example, if we replace the λ/4 monopoles of Fig. 5.1B with λ/2 dipoles, we still have a two-element broadside driven array with λ/2 element spacing. So the resulting radiation pattern will be the array pattern of Fig. 5.1C multiplied by the inherent radiation pattern of a conventional dipole (Fig. 3.7). Clearly, the exact orientation of the dipole elements relative to the orientation of the array itself will make a big difference in the overall pattern. An excellent rule of thumb is to try to use elements in such a way that their natural direction of maximum radiation coincides with at least one of the desired directions of maximum radiation for the array as a whole. Failure to do so will usually lead to arrays that are "temperamental"; that is, they will tend to have patterns that are sharp and/or multilobed (often in the wrong directions) and feedpoint impedances that are unstable and/or difficult to match.

Consider the broadside array of Fig. 5.2A, where we have replaced the verticals with λ/2 horizontal dipoles laid end to end with centers λ/2 apart. The pattern for this array is graphed in Fig. 5.2B. Note that the *directivity* of the main lobe is "sharper" or narrower than the main lobe of Fig. 5.1C as a result of the added directionality of the λ/2 dipoles. However, the peak *gain* of the main lobe relative to a single λ/2 dipole is only 1.5 dB because the mutual impedances between the two dipoles cause the feedpoint current in each dipole to be less than it would be if the

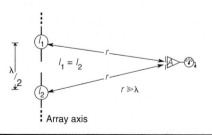

FIGURE 5.2A Two-element broadside array of λ/2 dipoles spaced λ/2 apart on their axis.

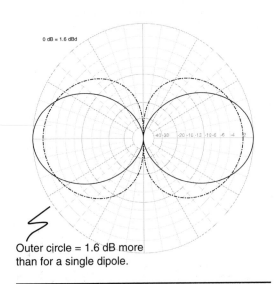

0 dB = 1.6 dBd

Outer circle = 1.6 dB more
than for a single dipole.

FIGURE 5.2B   Overall pattern for two-element broadside
array in (A).

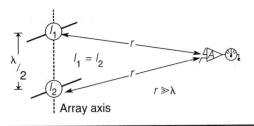

I_1

$I_1 = I_2$

$\lambda / 2$

I_2

r

r

$r \gg \lambda$

Array axis

FIGURE 5.2C   Two-element broadside array of dipoles
spaced $\lambda/2$ at right angles to the array axis.

two dipoles were far enough apart to have negligible interaction.

But suppose we place one of the dipoles $\lambda/2$ *above* the other, as depicted in Fig. 5.2C. Each dipole "sees" the other dipole differently, and their mutual impedances are different from those of Fig. 5.2A as a result. As reported in Fig. 5.2D, the overall gain of this two-element broadside array is 3.8 dB more than that of a single dipole. Clearly, if one is going to use two driven dipoles to maximize forward gain in a broadside array, the configuration of Fig. 5.2C is preferable to that of Fig. 5.2A. Note that in either case, however, the dipoles are oriented so that the direction(s) of maximum innate radiation for the basic dipole element coincides with the desired direction of maximum radiation for the array as a whole when operated in its broadside mode—that is, with in-phase currents at the feedpoints of the dipoles. Along the axis of the array (i.e., the extension of the line drawn between the two antennas of the array), then, not only is the array pattern trying to bring the radiated field strength to zero, so too is the inherent pattern of the dipoles themselves. The overall pattern of Fig. 5.2D will be of particular interest to us later when we discuss the effect of ground on a dipole's pattern—especially when we vary the height of the dipole.

Another popular form of array operation is the *end-fire* mode. Here the objective is to maximize radiation along the axis connecting the elements. One popular way to do this is by spacing the elements $\lambda/2$ apart and feeding them 180 degrees out of phase. By the time the out-of-phase field from element A travels along the axis to element B it will have shifted another 180 degrees and be back in phase with the field from B, thus effectively doubling the radiated field along the axis, relative to a single element. For two identical radiators spaced $\lambda/2$ apart and fed with out-of-phase excitations of equal amplitude, the shape of the array pattern is given by

$$AP = \sin\left(\frac{\pi}{2}\cos\theta\right) \qquad (5.6)$$

From our earlier discussion about attempting to align the inherent pattern of the elements of the array with the main lobe of the array pattern, we conclude that horizontal

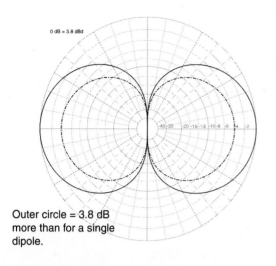

0 dB = 3.8 dBd

Outer circle = 3.8 dB
more than for a single
dipole.

FIGURE 5.2D Overall pattern for the two-element
broadside array in (C).

dipoles oriented parallel to the array axis are not the best choice for elements in this end-fire array. For instance, if we feed one dipole in the array of Fig. 5.3A with a drive current that is out of phase with the current in the other element, we obtain the azimuthal pattern of Fig. 5.3B—not totally useless, but probably not what we expected, either. Along the array axis, the dipole patterns radiate very little horizontally polarized energy to the far field, and at right angles to the array axis the element spacing and drive current phase relationships conspire to create a similar null. But the radiated energy has to go *somewhere*, and it is "easiest" for it to go off at 45 degrees to either axis, looking like a perfect four-leaf clover!

Instead, the horizontal dipoles should be rotated on the page 90 degrees, as shown in Fig. 5.3C. The resulting overall pattern of Fig. 5.3D is much more in line with our intentions and our expectations. (As noted before, the overall gain relative to a single antenna is a bit less than 3 dB because of the effect of the mutual impedances between the elements for this spacing and relative element orientation.)

If the user wishes to switch a single fixed physical array between broadside and end-fire patterns, the ideal element may well be a vertical dipole or monopole, since those antenna types enjoy omnidirectional azimuthal radiation patterns and can be equally at home in either broadside or end-fire applications without having to be physically rotated or relocated

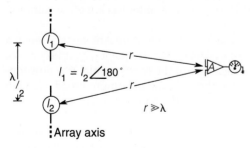

FIGURE 5.3A Two-element end-fire array of λ/2 dipoles spaced λ/2 apart on the array axis.

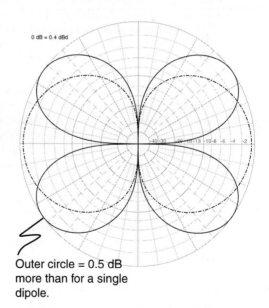

0 dB = 0.4 dBd

Outer circle = 0.5 dB
more than for a single
dipole.

FIGURE 5.3B Overall pattern for the two-element end-fire array in (A).

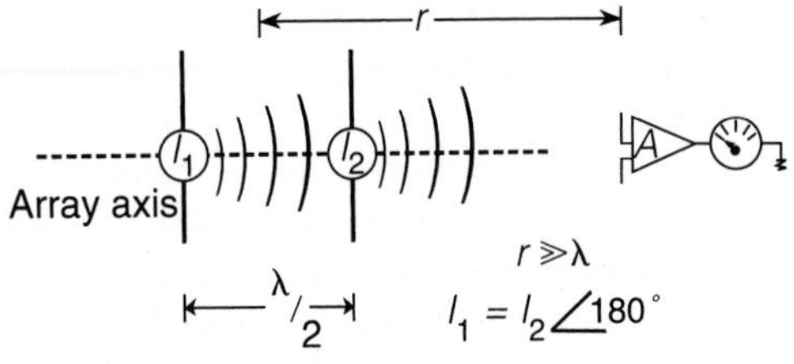

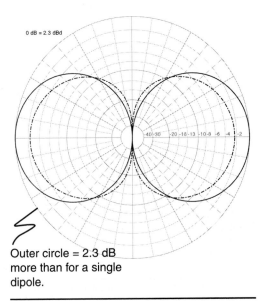

FIGURE 5.3C Two-element end-fire array of dipoles spaced λ/2 on the array axis.

Outer circle = 2.3 dB more than for a single dipole.

FIGURE 5.3D Overall pattern for the two-element end-fire array in (C).

when changing the preferred direction of maximum gain.

Nor is it necessary that the feedpoint currents be equal in all elements. Over the years, many AM broadcast stations have employed multielement arrays with unequal feed currents. One common configuration uses a *binomial* current distribution in the vertical elements. (A five-element array might have relative element currents of 1-4-6-4-1, for instance.) Combined with very close element spacings, these arrays are capable of producing extremely high gain patterns and deep nulls or *side lobes*. (The latter are important in the AM broadcast band to prevent *co-channel* and *adjacent channel* interference to "protected" stations on the same frequency in other parts of the country.)

Another form of array is the *collinear*. In this configuration, multiple "copies" of an element type are laid (or stacked, if vertical) end to end and electrically connected "heel to toe" through *phasing sections* designed to force the currents in all elements to be in phase even though only one element is actually driven by a signal from the transmitter. Many VHF and UHF verticals and mobile whips are collinears; the short lengths of antennas for those frequencies make stacking very practical from a mechanical standpoint. For a vertical, collinear stacking does not alter the shape of the azimuthal pattern at all; instead, the usual objective is to "sharpen" the elevation pattern so that more gain in all azimuthal directions is available at low elevation angles at the expense of (usually useless) high-angle radiation. At VHF and UHF, collinear antennas are often commercially sold as a complete array in a single assembly.

As we shall discuss in Chap. 8 ("Multiband and Tunable Wire Antennas"), when a half-wave dipole is operated substantially above its design frequency, f_C, it becomes a pair of collinear arrays because the wire on each side of the center insulator is now longer than λ/4. As the operating frequency rises above f_C, the maximum broadside gain of the antenna grows relative to its value at f_C because the additional wire length in the radiating element possesses a greater *length x current* product. Eventually, however, current reversal in adjacent half-wave segments begins to reduce the maximum broadside gain of the array, which occurs when the total length of the wire is 5/4 λ, corresponding to an operating frequency near 9 MHz for an 80-m dipole. Above that frequency, the maximum broadside gain starts to decrease with frequency, and eventually the array pattern breaks up into multiple smaller lobes. Thus, the broadside gain of an 80-m $\lambda/2$ dipole operated on 40 or 30 m is greater than the 80-m gain; on 40 m the increase is on the order of 1.5 dB. The increased gain is accomplished through a narrowing (or "sharpening") of the main radiation lobe. On 30 m, the gain of the main lobe is still about 0.5 dB greater than on 80, but it is accompanied by two additional lobes exhibiting gain equal to the main lobe but somewhat broader at their 3-dB points. A horizontal wire antenna $5/4\ \lambda$ long on any band is known as an *extended double Zepp* (EDZ) (or, alternatively, *double extended Zepp*) for that band. See Chap. 6 ("Dipoles and Doublets") for further discussion of the EDZ.

Of course, if one side of the dipole antenna is thrown away and the remaining side oriented vertically and fed against a ground plane as a vertical monopole, the length for maximum broadside gain is one half of $5/4\ \lambda$, or $5/8\ \lambda$; hence, the popularity of VHF and UHF whips of that length!

An entire class of antenna—the *loop*, which we'll discuss in later chapters—is actually a form of array. To see this, consider the *bent dipole*. This is nothing more than an ordinary $\lambda/2$ dipole with the outer half of each side bent straight down. If a second such bent dipole is turned upside down, attached at its ends to the corresponding ends of the first dipole, and one of the two feedpoints shorted out, we have what everyone calls a loop antenna. (See Fig. 5.4.) Because of the "automatic" phase reversal in adjacent $\lambda/2$ sections of a longwire, the currents at the centers of the two dipoles are always in phase and we have a two-element broadside array, with the elements spaced $\lambda/2$ apart. The loop can be suspended so that it lies in a horizontal plane, a vertical plane, or some plane partway in between. If suspended vertically and only a single high support is available, the loop takes a triangular form and is called a *delta loop*. If multiple loops are attached to a common boom, the resulting "array of arrays" is called a *quad*. Regardless of its name, however, each loop is essentially two half-waves in phase, and the direction of maximum radiation for that array is broadside to the plane formed by its component dipoles.

Further, there is nothing magic about feeding the loop at the center of the bottom dipole. As long as the entire wire path around the loop is one wavelength, the loop can

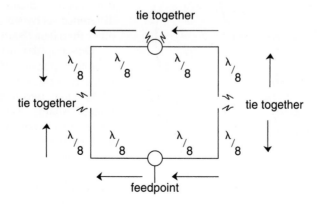

FIGURE 5.4   The full-wave loop array of two bent dipoles. Arrows show direction of current everywhere at a given instant.

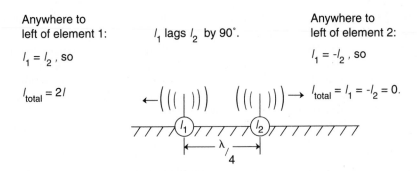

Anywhere to left of element 1:

$I_1 = I_2$, so

$I_{total} = 2I$

I_1 lags I_2 by 90°.

Anywhere to left of element 2:

$I_1 = -I_2$, so

$I_{total} = I_1 = -I_2 = 0$.

FIGURE 5.5A   Two-element quadrature or cardioid array.

be broken and fed anywhere on that path. If a vertically suspended loop is fed in the middle of a side, it becomes predominantly vertically polarized and may well exhibit better ability to work or hear long-haul DX. If a horizontally suspended loop is fed from the "next" side, the azimuthal orientation of maximum performance will rotate 90 degrees.

Unfortunately for the HF loop, low dipoles and arrays of low dipoles are (usually adversely) affected by the ground beneath them. The effect of the ground, in fact, is to create yet another array with its own *array factor*, so proper analysis of the operation of a loop antenna or array must take that into account. (Array factors multiply each other.) Ground reflection effects are discussed later in this chapter. Chapter 13 goes into loops and quads in much more detail.

Thus far we have limited the discussion of arrays to those with feedpoint currents that are either *in phase* or *out of phase*. Actually, *any* phase relationship between feedpoint currents can be used, but as a rule the math gets more complicated, the patterns less intuitive, and the physical implementations more difficult when the phase difference between fed elements is other than 0 or 180 degrees.

One popular array that employs *quadrature* phasing of element currents is the two-element *cardioid array* (Fig. 5.5A). Here two identical omnidirectional radiators (vertical monopoles, for instance) spaced λ/4 apart are fed with currents of identical magnitude phased 90 degrees apart. Visualizing how the array pattern is formed is not that much more difficult than for the broadside and in-line arrays: Suppose the current in element 1 lags the current in element 2 by 90 degrees. By the

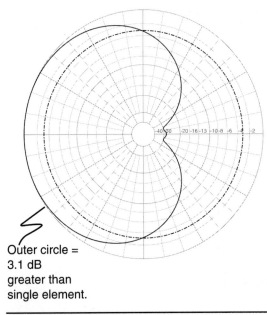

Outer circle = 3.1 dB greater than single element.

FIGURE 5.5B   Cardioid pattern using omnidirectional radiating elements.

time the radiated field from element 1 reaches 2, it has accumulated an additional lag of 90 degrees due to the distance it had to cover. Thus, radiation along the axis heading in the direction of 1 to 2 is the algebraic sum (from 2 outward) of two equal fields that are 180 degrees out of phase, and the radiation field is canceled in that direction.

In the opposite direction along the axis of the two elements, heading from 2 to 1 and beyond, the radiation from the feedpoint current in 2 (which *leads* the feedpoint current in A by 90 degrees) exactly loses the 90-degree lead as it covers the distance from 2 to 1. Thus, from 1 outward, radiation fields from 1 and 2 are perfectly in phase, and reinforced.

Between these two extremes, the amplitude of the radiation pattern at a distant receiving site is a smoothly changing value from maximum to minimum signal strength, symmetrically about each half-circle. The resulting equation for far-field signal strength as a function of azimuth (compass heading) is proportional to 1 + cosθ and has the shape shown in Fig. 5.5B. For a pair of λ/4 vertical monopoles over perfect ground, the array has a maximum gain of 3.1 dB relative to a single identical monopole.

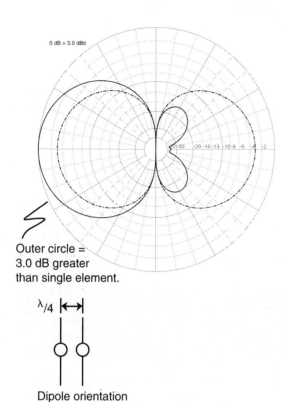

Outer circle = 3.0 dB greater than single element.

λ/4

Dipole orientation

FIGURE 5.5C Cardioid pattern from two λ/2 horizontal dipoles.

We say this two-element array produces a *unidirectional cardioid* pattern. (AM clear-channel broadcast station WBZ, 1030 kHz, has used this pattern from their two-tower site overlooking the Atlantic Ocean at Hull, Masssachusetts, to blanket the eastern United States with a commanding signal for decades!) For that reason, the best antenna elements to use for forming a cardioid are omnidirectional radiators in the azimuthal plane—verticals, in other words—but horizontal λ/2 dipoles at identical heights above ground and oriented so that their maximum radiation is broadside to the array pattern maximum are a possible alternative if additional side nulls and rear *fishtail* lobes (Fig. 5.5C) in the pattern are acceptable.

Feeding All-Driven Arrays

Thus far we have not discussed the practical issues associated with feeding all-driven arrays. In general, feed systems for the various members of the loop and collinear families are the easiest because the second element is driven by the first element, so there is little or nothing for the user to do to make sure currents in the two elements are equal or nearly so. In theory, a λ/2 dipole in free space has a feedpoint impedance of 73 Ω resistive. Most of us are not so lucky, however, as to be able to hang our dipoles in free space, so virtually all MF and HF dipoles are something other than 73 Ω, and they often have a reactive component to their impedance, as well. Even in free space, the second dipole in a loop may well cause the entire structure to have a feedpoint impedance other than 73 Ω. (The exact change depends on the distance between the centers of the two dipoles.) But the methods for canceling out the reactive part and transforming the real part to 50 Ω are the same as for a simple dipole. As a general rule, open-wire line, used in conjunction with an *antenna tuning unit* (ATU) at the transmitter end of the line or at ground level directly below the loop, is an excellent choice and usually represents less of a downward drag on the antenna. But 50-Ω coaxial cable may actually be a closer match to the loop's natural impedance.

Next in ease of implementation are the bidirectional line arrays with equal element currents. Whether in phase or 180 degrees out of phase, the principle of *symmetry* assures us that the element feedpoint impedances are identical, and transmitter power can usually be equally divided among them simply by feeding multiple identical transmission lines in parallel. Often it will be useful to incorporate one or more impedance-transforming sections of appropriate transmission line to convert the typically low junction impedance to something higher and more appropriate for the main run back to the transmitter.

When we move to element phase angles other than 0 and 180 degrees, feeding the all-driven array becomes much more difficult because the impedances of the individual elements differ from each other and, hence, have a tendency to take differing amounts of power from the source or passive component divider networks unless special techniques are employed for ensuring equal power or equal current at all the element feedpoints.

As we saw in Chap. 4, all transmission lines exhibit a velocity of propagation that is somewhat less than the speed of light in free space or air. As a result, a length of transmission line used to create a 90-degree delay at some operating frequency will be shorter than the physical distance between two array elements spaced λ/4 apart, as measured in air or free space. If the layout of the array is such that the physical distance of each such *delay line* from a central combining point to each element is substantially shorter than the distances between the elements themselves, it may be possible to phase each element with a minimum transmission line length. However, frequently it is necessary to make the phasing lines longer, such as using a 3λ/4 line to provide a 90-degree delay.

Some important points when constructing or procuring phasing lines for arrays with separately fed elements:

- Make all phasing lines out of the same transmission line. If the lines are coaxial, be sure to use cable from a single reputable vendor. Ideally, all phasing lines to individual elements of a given array should be cut from the same spool.

- Be sure you know the velocity of propagation (as defined by the cable's *propagation factor*) for the cable you are using. Don't guess, and don't rely on published specifications—measure it! Use a *time domain reflectometer* (TDR), a *vector network analyzer* (VNA), or other suitable instrument capable of allowing you to accurately calculate the propagation factor or equivalent line length of your chosen cable material. (See Chap. 27 for some instruments and techniques.) This is not quite as important for broadside arrays where all the elements are fed in phase through identical lengths of cable originating at a common junction, but it is a good habit to form in any case. Even in this situation, all cables between the common junction and individual elements should be obtained from the same spool or production lot.

- Measure your cable lengths accurately. Even on the MF and lower HF bands, you should strive to keep your lengths of "identical" phasing lines within a few inches of each other.

- Make sure any matching networks between the central distribution point and each element of the array are identical and not contributing unexpected phase shifts or delays in the signals being sent to the elements.

Although the discussion of arrays so far has assumed they are being used for transmitting, that isn't always so. Some of the most exciting work in HF arrays currently under way deals with receive-only arrays for the MF and lower HF bands, where the cost and structural demands of transmitting arrays can be largely avoided by using shortened elements, minimal RF ground systems, low-power components, and preamplifiers to compensate for very low signal levels, low feedpoint impedances, and the signal loss of very long cable runs back to the receiver. Nonetheless, all the points noted here about the importance of precise matching of phasing lines and networks are just as valid for receive-only arrays.

Parasitic Arrays

If a piece of metal or other conducting material with dimensions on the order of $\lambda/4$ or greater is placed in the vicinity of an antenna, it can cause "distortions" in the electromagnetic fields radiated or received by the antenna. By deliberately positioning one or more conductors near a simple radiating element such as dipole, we can increase the field strength in some directions at the expense of others. The resulting configuration is known as a *parasitic array,* and it is capable of producing patterns and array gains similar to the ones created in the *all-driven arrays* previously discussed.

The principal value of a parasitic array is its potential for simplifying the feed system. In general, parasitic arrays use a combination of element-to-element spacings and element dimensions to create desired array patterns without the use of phasing lines and multiple matching networks. Because of this, they are mechanically simpler to construct, and at 14 MHz and above, they are today the antenna of choice because antennas with array gains of 6 dB or more are easily rotated, allowing them to be aimed in any direction desired.

Parasitic arrays can be vertically or horizontally polarized; they can be made of $\lambda/4$ grounded elements or $\lambda/2$ symmetrical elements. They can be constructed from copper wire (typically on 160, 80, and 40 m), aluminum tubing (80 through 6 m), or solid aluminum rod (6 m and up). Vertical arrays for 160 through 40 are often constructed with triangular lattice tower sections—i.e., they provide their own support mechanism!

A mathematical explanation of how parasitic arrays work is beyond the scope of this book. Instead, a short qualitative description is provided as a guide to helping the reader develop an intuitive feel for their operation.

Picture a horizontal $\lambda/2$ dipole up in the air somewhere, fed with RF energy at its resonant frequency. With nothing around it, its radiation pattern will be the familiar doughnut shape of Fig. 3.7. Now suppose we place another thin metallic object, of approximately the same dimensions, near the dipole (typically within a wavelength or less) and in the dipole's broadside radiation lobe. Electromagnetic waves from the dipole will impinge upon this neighboring object, *inducing* currents and voltages of the same frequency along the object's surfaces. The new electromagnetic fields resulting from those currents also radiate into space in all directions, very much like the fields from the dipole itself. Thus, a distant receiving antenna will pick up a signal that is the linear combination (*superposition*) of the original signal from the dipole and the induced signal from the metallic object. But the latter signal arrives at a distant receiving point with amplitude and phase that are, in general, different from those of the dipole. Depending on the different paths traveled by the two waves and the degree to which the radiated fields from the added conductor differ in phase from those of the dipole, the signals may add or subtract at the receiving antenna.

The superposition at a distant receiving point of waves from the two radiating conductors is identical to the process we saw previously with all-driven arrays, so we can conclude that we have—even if only by accident—created an array despite the fact that only one of its elements is driven. We call that element the *driven element*, of course, and the other conductor is called a *parasitic element.*

With a driven array, the pattern and array gain are controlled by judicious choice of element spacing and by individually setting or forcing the feedpoint current amplitude and phase for each element of the array. With a parasitic array, however, the only control we have is the ability to adjust element dimensions and interelement spacings.

As you might suspect, if the parasitic element is identical to the dipole—that is, if it is resonant at the operating frequency—it is likely to be "receptive" to the generation of induced currents and voltages from the dipole's fields. However, reradiation by a perfectly self-resonant parasitic element does not generally lead to array patterns that are of much use.

But now suppose we slightly lengthen the parasitic element. This does not materially reduce its ability to pick up and reradiate a portion of the passing field from the dipole, but it does substantially alter the phase of the currents induced in it. For typical interelement spacings, the resulting field from the parasitic element opposes or partially cancels the dipole field in directions *beyond* the added element, such that the overall field seen by a distant receiving antenna is significantly reduced.

Similarly, if we now make the same element somewhat *shorter* than its self-resonant length at the operating frequency, the reradiated fields are again of a different phase from the dipole's fields, but this time the effect is usually to *add* to the dipole's fields at a distant receiving point beyond the added element.

In the first case we call the parasitic element a *reflector*, and in the second we call it a *director.* Depending on the specific design objectives for a parasitic array, there can be multiple reflectors and/or multiple directors. Compared to all-driven arrays, interelement spacings in parasitic arrays are often smaller; 0.1λ to 0.3λ is typical.

More than 85 years after its invention, arguably the most famous parasitic array is the *Yagi-Uda* beam antenna—often called *Yagi*, for short. It has attained such a level of popularity that an entire chapter (Chap. 12) of this book is devoted to it. But other kinds have existed for at least as long, including many wire arrays and combinations of grounded verticals. We will explore these in more detail in later chapters specifically dedicated to the different families of arrays.

Ground Effects

In the real world, very few antennas are found in "free space". Only at VHF and above can we hoist an antenna to heights that approximate free space. Even many satellite antennas have a conducting body under them or near them, so while their signals are truly traveling through space, they still have to be designed and their operation analyzed on the assumption that their radiation patterns might well be affected in some way by nearby conductors.

Thus, when discussing antenna operation and performance, we almost always have to incorporate the effects of nearby conductors on the antenna and the waves it radiates or receives. When the dimensions of the nearby conductor are large enough to materially alter the radiation pattern of the antenna at the chosen operating frequency, we call it a *ground plane* or, more simply, *ground*. An *ideal ground* is a perfectly conducting conductor stretching infinitely far in all directions. For many scenarios, especially those involving frequencies below VHF, that's as rare as "free space"! But for mobile VHF and UHF whips mounted atop a vehicle, a vehicular metal roof is often a "good enough" approximation to an ideal ground.

The performance of an antenna located above ground is affected in three ways:

- Like any other conductor in the vicinity of a basic antenna, a ground plane functions as a parasitic element by reradiating the EM fields originating at the antenna. The resulting signal strength far from the antenna is thus a combination of signals received directly from the antenna as well as via the ground plane.

- Some fraction of the reradiated energy from the ground plane returns to the antenna. Generally, its effect—through *superposition* at the feedpoint—is to alter the current and voltage relationships there, thus changing the input impedance of the antenna relative to its free-space value. Depending on the spacing between the antenna and the ground plane, the radiated field strength for a given RF power delivered to the antenna from the transmitter may be greater or smaller than in the free-space case.

- If the ground is not perfectly conducting, a portion of the RF energy that impinges upon it will be dissipated (and lost forever) in the resistive component of the ground characteristics. As we discuss in Chap. 9, ground-mounted vertical monopoles are more susceptible to this effect than horizontal dipoles because the return current for the vertical comes back to its feedpoint through lossy earth in the region immediately surrounding the vertical.

When we talk about the first two of these effects by nearby ground on an antenna we say RF energy from the antenna is "reflected" from the ground plane in the same way that we say light is reflected from a mirror. But what do we mean?

The electromagnetic principle that is at work here is this: A perfectly conducting material cannot support a voltage differential (and thus a nonzero E-field) between any two points on the material's surface (or inside the bulk of the material, for that matter). *Perfectly conducting* means the resistance (R) between any two points on the conductor is zero, so any voltage differential across those points would imply the existence of an infinitely large current flowing between them. (Anyone who has ever dropped a wrench across the two terminals of a car battery has a good grasp of this concept!) But a voltage differential is the result of an E-field between the two points, so there must be no

E-field, either. As we did in Chap. 4 ("Transmission Lines and Impedance Matching"), we can use the principle of *superposition* to see that a zero voltage between any pair of points in space or on the earth's surface can be the result of two separate waves having equal amplitude and opposite polarities in the region between the two points.

In other words, the *boundary conditions* at a perfectly conducting surface *require* us to conclude that any horizontal E-fields from the original antenna must be matched by identical fields of equal amplitude and opposite polarity from a second source. But it turns out (without any proof given here) the only magical entity that can do that for *all* locations on the ground plane beneath and near a specific antenna is something we call an *image antenna*, looking for all the world like the original antenna viewed in a mirror: It is located the same distance *below* the ground plane as the original antenna is above it, and the horizontal components of its currents, voltages, and fields are equal in amplitude but opposite in direction (or phase) to those of the original antenna.

Why do we bother with a fictitious *image antenna*? Because doing so allows us to predict the effect of a ground plane on the radiation pattern of a nearby antenna simply by removing the ground plane and replacing it with (fields from) the image antenna! In other words, we can determine the effect of a nearby ground plane on an antenna's pattern by removing the ground plane and treating the antenna and its image as a two-element array. Since we've already seen how to deal with two-element arrays whose elements are fed out of phase and in phase, we now can use the same visualizations to see what happens to far-field patterns from an antenna over perfectly conducting ground.

Example 5.1   What is the broadside elevation response for a horizontal $\lambda/2$-dipole of Fig. 5.6A erected to a height of $\lambda/4$ above ground?

Solution   The dipole is $\lambda/4$ above ground, so the image antenna must be $\lambda/4$ *below* the surface of the ground. Figure 5.6B shows the signal paths r_1 and r_2 from the two dipoles to a distant receiver. Because the original dipole is horizontal (with respect to the ground plane), the image dipole is also horizontal but is considered to be fed 180 degrees out of phase with it. In your imagination, remove the ground and observe that the total spacing between the original dipole and the image antenna is twice $\lambda/4$, or $\lambda/2$. Because the antennas are perfectly out of phase, the combined signal strength at any far-field receiving point that is equidistant from them (which corresponds to all points on the surface of the ground) is zero. Further, because the wave from the image antenna takes a half-cycle to travel the half-wavelength straight up toward the original dipole, it arrives in perfect

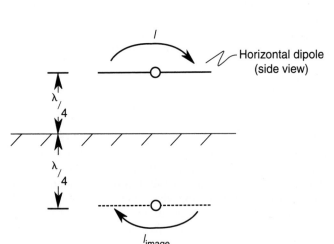

FIGURE 5.6A   A horizontal dipole and its image antenna.

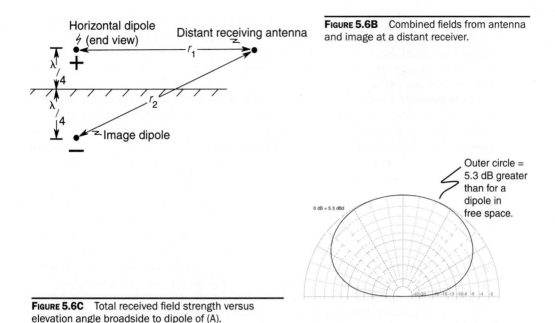

FIGURE 5.6B Combined fields from antenna and image at a distant receiver.

FIGURE 5.6C Total received field strength versus elevation angle broadside to dipole of (A).

phase at the other antenna, and the line of maximum field strength from this particular array is directly above the two dipoles. At elevation angles between straight up (90 degrees) and the horizon (0 degrees), the faraway field strength takes on intermediate values, as shown in Fig. 5.6C.

◆

Because the maximum radiated signal strength for a horizontal dipole at heights of $\lambda/4$ or less above a ground plane is straight up in the air, the antenna is often called a *cloud burner,* and it is ideal for *near-vertical incidence skywave* (NVIS) communications. However, the broadside pattern is actually quite robust until the vertical elevation (or takeoff) angle drops under 30 degrees (the 3-dB point). There are far, far worse antennas for all-around use on 80 and 40 m than a $\lambda/2$ dipole at a height of roughly $\lambda/4$ (35 ft)!

Above 30 degrees, at those elevation angles that result in the radiated fields from the original horizontal antenna and its image arriving at a distant point in phase, or nearly so, the resulting signal at the receiver is *twice* the amplitude of the signal that would have come from the original antenna in free space. The image antenna—which is really a way of looking at the ground reflection effect—has increased the received signal strength of our original antenna by nearly 6 dB relative to the free-space case!

In general, r_1 and r_2 (the distances from the dipole and its image to a remote receiving point) are not equal, so the phase shift between the two radiated signals is something other than zero. The degree of reinforcement or cancellation at any given elevation angle above an arbitrary horizontally polarized antenna depends on the height of that antenna above ground. As the height of the antenna above ground is increased, eventually the phase shift of the wave from the image antenna on its way to distant receiving points increases to where the maximum radiated field is no longer straight up in the air. With increasing height above ground, the elevation angle corresponding to maximum radiated field strength begins to drop. However, no matter how high the antenna is raised, as long as it is above a ground plane, the horizontally polarized radiation at 0 degrees elevation angle (i.e., the *horizon* when flat land is involved) will always be zero because the image antenna and the real antenna are always exactly out of phase on these equidistant paths.

Now consider a vertically polarized dipole not in contact with the earth, as shown in Fig. 5.7. Further consistent with the mirror analogy, boundary conditions at the ground plane require the image antenna's vertically polarized (with respect to the ground plane) E-field to be *in* phase with the corresponding component of the original antenna's radiated field. Thus, the ground beneath a vertical $\lambda/2$ dipole, for instance, is replaced with an identical dipole having the same vertical orientation and a standing current that is exactly in phase with the drive current of the aboveground dipole. Figure 5.10 at the end of the chapter consists of a series of graphs depicting the ground reflection factor for vertically polarized antennas as a function of their height above ground.

For tilted antennas, or antennas otherwise having a mix of both horizontal and vertical polarization, visualizing the two cases separately may be the easiest way to understand the overall effect of nearby ground on the radiated signal. Of course, if antenna modeling software is available, all the hard work is done for you by the computer!

For the specific case of a ground-mounted monopole, the image antenna is best visualized as an identical but upside-down vertical with its upper end just "touching" the base of the real antenna. If the current flow at some instant is downward in the real monopole, it is also downward in the image monopole. Because the vertical component of the two currents is in phase, the current distribution in the two verticals appears continuous across the ground plane; consequently, the effect of the ground-plane boundary conditions is to cause the image antenna to act like the other half of a vertical dipole. Like any other dipole, the standing current is maximum at the center of the resulting antenna structure and in phase across the entire combined length of the two pieces, as suggested by Fig. 5.8.

Because the radiated E-fields from the real and image vertical antennas are in phase and at right angles to the ground plane, there is no automatic cancellation of the radiated field at 0 degrees elevation angle. In the presence of a perfectly conducting ground plane, a vertical enjoys the same 6-dB "boost" in received signal strength at certain wave angles as do horizontal dipoles but—because its E-field is at right angles to the ground plane—this boost occurs at 0 degrees. The ground-mounted vertical is, therefore, unsurpassed in its ability to produce a strong signal at the horizon and very low elevation angles. As discussed in Chap. 2 ("Radio Wave Propagation"), however, the losses inherent in real earth cause the distant radiated field from any vertical to be zero near the horizon. Despite that, numerous DXpeditions have demonstrated that ground-mounted monopoles at the edge of ocean salt-water have a distinct advantage over virtually any other antenna for intercontinental communications on the MF and HF bands.

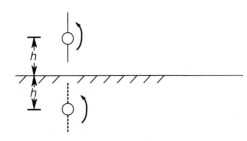

Figure 5.7 Image antenna for vertical dipole not in contact with the ground.

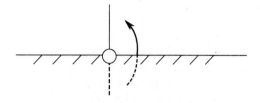

FIGURE 5.8 Image antenna for ground-mounted vertical monopole.

Lossy Ground

The discussion of ground reflection effects thus far assumes perfectly conducting ground; for average or poor grounds, the effect is to partially fill in some of the nulls in the elevation pattern (so they seldom get deeper than perhaps 15 dB down) and to reduce the maximum amplitude of the array factor in its favored directions—especially as the frequency increases. Lobes above the lowest angle lobe are reduced even further.

NOTE *Especially at the lower elevation angles, the ground characteristics that are important are not the ones for the ground directly beneath the antenna; rather, it is the ground many wavelengths from the antenna that establishes the efficiency of the reflection that can add as much as 6 dB to the signal compared to the same antenna in free space. For instance, for a horizontal dipole or Yagi at height λ, a 10-degree takeoff angle is reinforced by the ground reflection 5.7 λ away; on 14 MHz that's about 400 ft! At 1 degree (the "sweet spot" in the minds of many DXers), the first bounce is nearly a mile away!*

Also, in real life, with real earth beneath an antenna, the *effective ground* may well lie some distance beneath the physical surface. How far beneath will depend on the operating frequency and the actual characteristics of the earth at that frequency, but distances of a few inches to a few feet are common.

Figures 5.9 and 5.10 at the end of this chapter display the ground reflection factors for horizontally polarized and vertically polarized antennas, respectively, at selected heights above ground. (The charts of Fig. 5.9 are for an azimuth broadside to the axis of the dipole.) Each individual chart shows, for a given height, both the theoretical elevation pattern over perfectly conducting ground (broken line) and the equivalent pattern (solid line) over *average* ground ($\sigma = 0.005$ S/m and $\varepsilon_r = 13$). In all of these charts, the outer ring of the plot corresponds to 7.0 dBd—i.e., to a maximum signal strength that is 7.0 dB greater than would be obtained from an identical dipole in free space.

Clearly, the ability to determine the optimal height(s) for maximizing an antenna's utility over the intended range of communications distances is a very powerful tool. Especially for individuals planning to erect MF or HF antennas, Figs. 5.9 and 5.10 are potentially among the most useful charts of the entire book.

NOTE *The charts of Figs. 5.9 and 5.10 assume flat ground under and near the antenna. The effect of ground that gradually slopes down as one gets farther and farther from the antenna is to enhance the very low angle signal strength in the same manner that hoisting the antenna higher can. Conversely, ground that slopes up tends to mimic lower antenna heights. As discussed in Chap. 2, terrain analysis programs can help identify or predict the effect of a specific terrain contour on the performance of a specific antenna installation.*

Example 5.2   An amateur radio operator's favorite band for casual operating is 20 m, but she also spends time DXing on 15 and 10 m when those bands are open. She is planning to purchase a triband Yagi antenna but is trying to decide whether to put up a 35-ft house-bracketed tower or a guyed 70-ft tower to support the antenna. From what we know about the characteristics of those bands throughout most years of a typical sunspot cycle, we don't expect them to be open for short skip very often, so she should try to select a height that favors elevation angles below 50 degrees on 20 m, somewhat lower angles on 15, and even lower angles on 10. A height of $\lambda/2$ on any band provides a very broad lobe centered on 25 degrees—right in the middle of her desired range of angles. On 20 m, $\lambda/2$ corresponds to 35 ft, which is just about the practical minimum if the antenna is to be above nearby obstructions. On 10 m, however, a triband beam at 35 ft is a full λ high and the peak amplitude of the main lobe has moved lower in takeoff angle even as a second lobe, slightly weaker than the main lobe, appears in the elevation pattern, as seen in Fig. 5.9H. Of course, average ground will act more like a dielectric and less like a perfect mirror at 28 MHz, so the nulls will partially fill in. Except at times of extremely high sunspot activity or sporadic E-skip, it's very unlikely that the ionosphere will support high-angle F-layer paths at 28 MHz, so the most useful elevation angles on 10 m are almost always lower than those of 15 or 20.

If our operator is interested primarily in working very long haul DX on 20 m, she might want to consider heights closer to λ (70 ft) at the risk of inserting some nulls at useful angles in her 15- and 10-m elevation patterns. By raising her triband beam from 35 to 70 ft over average ground, she gains not only another decibel of forward gain at the peak of the main lobe, but the elevation angle of that peak drops from 25 degrees to 14 degrees. At the low elevation angles that are so useful for long-haul DXing, the resultant overall improvement from doubling the height of her beam from $\lambda/2$ to λ is about 5 dB—equivalent to tripling her transmitter output power! In return, a null appears at 30 degrees (reducing her signal for some short-haul elevation angles on 20 m) and the elevation patterns on 15 and 10 break up into multiple lobes, albeit with lower peak elevation angles. In short, a height of 70 ft or as close to it as she can get will put most of her output power in the right range of takeoff angles on the band where she will need it the most—20 m—at a cost of some higher-angle nulls on all three bands.

◆

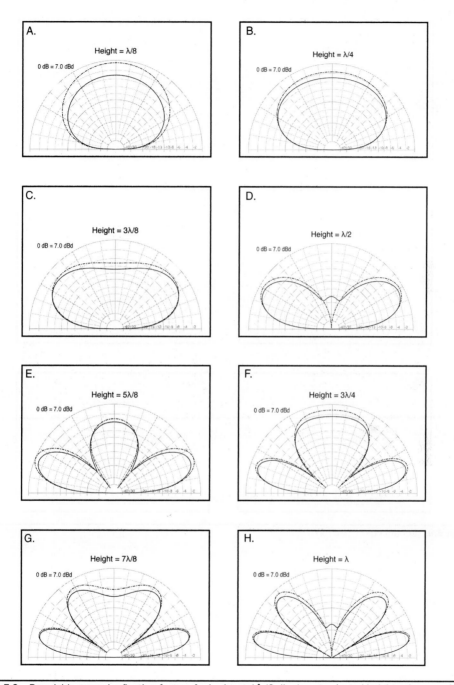

FIGURE 5.9 Broadside ground reflection factors for horizontal $\lambda/2$ dipoles at selected heights above perfect (broken line) and average (solid line) ground.

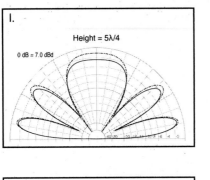

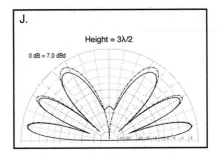

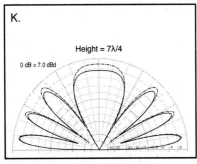

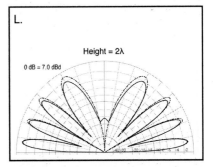

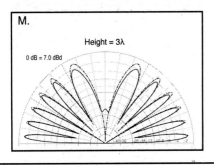

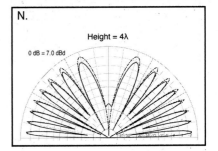

FIGURE 5.9 Continued.

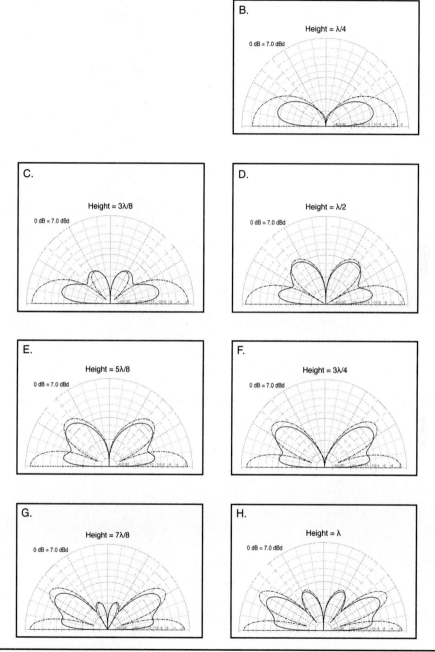

FIGURE 5.10 Ground reflection factors for vertical λ/2 dipoles at selected heights above perfect (broken line) and average (solid line) ground.

High-Frequency Building-Block Antennas

CHAPTER 6

Dipoles and Doublets

A common misconception about radio communication is that costly antennas and supports are required in order to hear or "get out" well. To the contrary, many of the world's strongest signals and most competitive amateur radio enthusiasts employ simple but effective antennas that can be erected by inexperienced people without any special tools or equipment. The key to success in these cases is not money—it is knowledge of a few key fundamental concepts, coupled with intelligent planning and *siting* of the antenna, all based on the user's objectives.

Leading all other antenna types in performance versus simplicity is the *half-wave dipole*. The half-wave dipole—consisting of nothing more than a length of wire—is an amazing antenna in its own right. For that reason alone, a solid understanding of how a $\lambda/2$ dipole functions is perhaps the single most important piece of knowledge to take away from this book. But it's *so* good, in fact, that it forms the basic building block for countless other types of antennas that we know by other names: the *Yagi*, the *quad*, the *collinear*, the *loop* . . . and so on. In short, whether used individually or in concert with one or more additional dipole-like elements, the half-wave dipole excels at efficiency of radiation and ease of matching to common types of feedlines.

The half-wave dipole is one member of a class of antennas named *Hertz*, or *hertzian*, for radio pioneer Heinrich Hertz. Hertz used a short dipole in his 1887 experiments that first demonstrated the existence of radio waves predicted by James Clerk Maxwell nearly a quarter-century earlier. In general, a hertzian antenna is one that does not rely on the presence of a connection to ground and is usually constructed and erected such that its two sides maintain a symmetrical relationship to ground. The most common example of an antenna with hertzian origins is the ubiquitous TV antenna. In contrast, *Marconi* antennas as a class typically are unbalanced with respect to ground and depend upon fields induced in either the natural ground or an artificial ground system beneath them for their proper operation. Commonly seen examples of a Marconi antenna are AM broadcast towers and citizens band ground-plane antennas.

The original *hertzian dipole* was not a *half-wave dipole*; Heinrich Hertz's experimental apparatus had dipole arms that were much shorter than the fundamental wavelength produced by his spark generator. However, over the years, usage of the term *hertzian* has been narrowed to refer primarily to antennas whose total length is approximately a half-wavelength. Today, the terms *dipole* and *half-wave dipole* are frequently used interchangeably.

Occasionally you may hear also the half-wave dipole called a *half-wave doublet*. Historically, the terms *dipole* and *doublet* referred to two elements of charge, one positive and one negative, driven to opposite ends of a fictitious infinitesimally short hertzian dipole by an equally fictitious ideal signal source. As noted in Chap. 3, many, if not most, graduate electrical engineering texts on antennas use the elementary doublet as their starting point for mathematical derivation of the characteristics of all dipole structures, including half-wave dipoles. For clarity, the authors of this book eschew the use of *doublet* in describing any "real" antennas and use the term *elementary doublet* only when referring to the infinitesimally short hertzian dipole that gives antenna theorists such joy.

The half-wavelength dipole is an electrically "balanced" antenna consisting of a single conductor one half-wavelength long. At HF and below, the antenna is usually installed horizontally with respect to the earth's surface or other local ground plane. The resulting electric fields at distances greater than a few wavelengths from the antenna lie parallel to the wire. Thus, by definition, the horizontal dipole produces a *horizontally polarized* signal. At VHF and above, dipoles are commonly found in either orientation.

Although the transmitter output can be injected anywhere along the length of the antenna, most commonly the wire is cut in the middle and a signal source injected there, as shown in Fig. 6.1. This configuration is the ubiquitous *center-fed halfwave dipole*.

Because the velocity of propagation of an electromagnetic wave along a wire is slightly less than in free space, the *physical* length of a half-wavelength on the antenna is about 5 percent shorter than a half-wavelength in free space. A free-space half-wavelength is found from

$$L\ (\text{feet}) = \frac{492}{F(\text{MHz})} \tag{6.1}$$

or, in metric terms,

$$L\ (\text{meters}) = \frac{150}{F(\text{MHz})} \tag{6.2}$$

where L = length of half-wavelength radiator, in feet
F = operating frequency, in megahertz

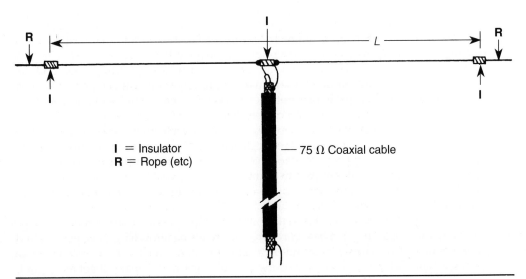

I = Insulator
R = Rope (etc)

— 75 Ω Coaxial cable

FIGURE 6.1   Simple half-wave dipole antenna.

In wires and other conductors, the length calculated here is too long. The physical length of a conductor having a diameter or cross-sectional side much, much smaller than its length is shortened by about 5 percent because of the velocity factor of the wire, any insulation on it, and capacitive effects of the end insulators. A more nearly correct approximation for such a "slender" half-wave dipole is

$$L \text{ (feet)} = \frac{468}{F(\text{MHz})} \qquad (6.3)$$

Example 6.1 Calculate the approximate physical length for a half-wavelength dipole operating on a frequency of 7.15 MHz, the middle of the U.S. 40-m band.

Solution

$$L = \frac{468}{F}$$
$$= \frac{468}{7.15}$$
$$= 65.45 \text{ ft}$$

or, stated in feet and inches:

$$L = 65 \text{ ft } 5\tfrac{1}{2} \text{ inches}$$

In practice, it would be smart to spool off about 70 ft of wire, fold it in half, and cut it at the midpoint. This leaves a foot or more of wire at each end of the two halves for threading through end and center insulators and wrapping back on itself.

Despite the precision in the preceding calculation of *L*, there's no real good reason for worrying about the exact length of this dipole. Several local factors—nearby objects, height above ground, and the length/diameter ratio of the conductor to name a few—might make it necessary to either add or trim a short length of wire on each side to achieve exact resonance. But to what end? The primary reason to aim for approximately a half-wavelength is to make the antenna wire *resonant*, or nearly so, at the frequency of operation so that we can minimize the reactive part of the feedpoint impedance that will have to be tuned out by the transmitter, an *antenna tuning unit* (ATU), or other matching network. Unfortunately, a lot of people accept Eq. (6.3) as a universal truth, a kind of immutable Law of the Universe. Perhaps abetted by books and articles on antennas that fail to reveal the full story, too many people spend too much time worrying about the wrong things when constructing and installing dipoles.

Whether the choice of design frequency in the preceding equations is critical or not depends on the percentage bandwidth of the band in question. For instance, for a half-wave dipole cut to the center (14.175 MHz) of the U.S. 20-m band, the antenna might be used at frequencies ranging between plus and minus 1.2 percent of the center frequency, and there will be a generally insignificant change in the feedpoint impedance over the entire band. An 80-m dipole, on the other hand, could conceivably be called upon to span a range of frequencies (in the United States, at least) up to plus and minus nearly 7 percent of the band center (3.750 MHz). For most homebrew and commercially available transmitters, the feedpoint characteristics will change too much for proper matching and full power transfer to the antenna over the entire range without inclusion of an adjustable matching network somewhere between the transmitter and the antenna, or without some method of *broadbanding* the antenna itself.

The Dipole Feedpoint

Figure 6.2 shows the voltage (V) and current (I) distributions along a half-wave dipole. If center-fed, the feedpoint is at a voltage minimum (*node*) and a current maximum (*loop*).

At resonance, the impedance of the feedpoint is purely resistive: $R_{ANT} = V/I$. There are two contributors to R_{ANT}:

- Ohmic losses R_Ω that generate nothing but heat when signal is applied. These ohmic losses originate in the small but finite resistance of real conductors and connections at wire junctions, and in any shunt leakage paths. In a properly constructed dipole having a length on the order of $\lambda/2$ and located in the clear, these losses are usually negligible. Because of *skin effect*, which is a complex function of frequency and conductor parameters, R_Ω is *not* the same as the dc resistance of the wire as measured with a conventional ohmmeter.

- The *radiation resistance* R_{RAD} of the antenna. This is the value of a fictitious lumped-component resistance that would dissipate the same amount of power that actually radiates from the antenna.

Example 6.2 Suppose we use as an antenna a large-diameter conductor with negligible ohmic losses at the transmitter frequency. If 1000 W of RF power is applied to the feedpoint, and a current of 3.7 A RMS is measured, what is the radiation resistance?

Solution

$$R_{RAD}(\text{ohms}) = \frac{P}{I^2} \tag{6.4}$$

$$= \frac{1000}{(3.7)^2}$$

$$= 73\,\Omega$$

◆

In this example, matching the antenna feedpoint impedance would appear to be simplicity itself because the free-space feedpoint impedance of a simple dipole is about 73 Ω, seemingly a good match to 75-Ω coaxial cable or hardline. Unfortunately, the 73-Ω feedpoint impedance is almost a myth except in free space, far from the earth or any other conducting surface.

Figure 6.3 shows a plot of approximate radiation resistance (R_{RAD}) versus height above ground (as measured in wavelengths); note that the radiation resistance varies from less than 10 Ω to around 100 Ω as a function of height. As we saw in Chap. 5, this

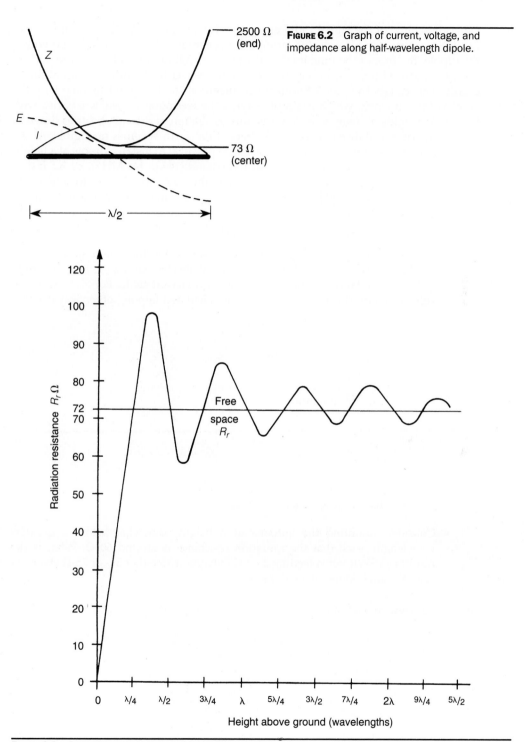

FIGURE 6.2 Graph of current, voltage, and impedance along half-wavelength dipole.

— 2500 Ω (end)

— 73 Ω (center)

Z

E

I

λ/2

Radiation resistance R_r, Ω

Free space R_r

Height above ground (wavelengths)

FIGURE 6.3 Radiation resistance versus height above ground.

is because some of the signal radiated by the dipole strikes the ground directly beneath the antenna and is reflected straight up, where it induces a second field within the dipole. Because of the time it takes the radiated field to reach the ground and return, the induced field is shifted in phase with respect to the original signal in the dipole. This causes the ratio of voltage to current at the feedpoint to vary with the height of the antenna, which is simply saying the feedpoint impedance varies with height. At heights of many wavelengths, this oscillation of the curve settles down to the free-space impedance (73 Ω) as the effect of the earth becomes negligible. At the higher frequencies (VHF and above), it becomes practical to install a dipole at a height of many wavelengths. In the 2-m amateur radio band (144 to 148 MHz), $\lambda \approx 6.5$ ft (i.e., 2 m $\times$ 3.28 ft/m), so a height of "many wavelengths" is relatively easy to achieve. In the 80-m band (3.5 to 4.0 MHz), however, $\lambda \approx 262$ ft, so "many wavelengths" is a virtual impossibility for most.

Any of a number of alternative responses can be chosen:

- Ignore the problem altogether. In many installations, the height above ground will be such that the radiation resistance will be close enough to represent only a slight impedance mismatch to a standard coaxial cable. For $Z_0 > R_{RAD}$, the voltage standing wave ratio (VSWR) is calculated (among other ways) as the ratio

$$\text{VSWR} = \frac{Z_0}{R_{RAD}} \qquad (6.5)$$

For $Z_0 < R_{RAD}$, it is the ratio

$$\text{VSWR} = \frac{R_{RAD}}{Z_0} \qquad (6.6)$$

where Z_0 = coaxial cable characteristic impedance
R_{RAD} = radiation resistance of antenna

- Consider mounting the antenna at a height somewhat below a quarter-wavelength, such that the radiation resistance is around 60 Ω. What is the resulting VSWR when feeding a 60-Ω antenna with either 52- or 75-Ω standard coaxial cable? Some calculations reveal:

For 75-Ω coaxial cable:

$$\text{VSWR} = \frac{Z_0}{R_{RAD}}$$

$$= \frac{75}{60}$$

$$= 1.25 : 1$$

For 52-Ω coaxial cable:

$$\text{VSWR} = \frac{R_{RAD}}{Z_0}$$

$$= \frac{60}{52}$$

$$= 1.15:1$$

In neither case is the VSWR created by the mismatch too terribly upsetting. Note that the same low VSWR results for an antenna fed by 52-Ω coaxial cable when it is slightly less than $\lambda/4$ above ground and $R_{RAD} = 45\ \Omega$. An 80-m dipole strung at a height of 50 ft is a very competent antenna and typical of what the author has used for many years.

- Mount the antenna at whatever height is convenient, and use an impedance-matching scheme at or near the feedpoint to reduce the VSWR. Chapter 4 contains information on various suitable broadband impedance-matching methods including Q-sections, coaxial impedance transformers, and broadband RF transformers. Homebrew and commercially available transformers are available to cover most impedance transformation tasks.

- Increase R_{RAD} by lengthening the two sides of the dipole equally, and use a *stub* or other matching method to tune out the reactance introduced by the longer elements.

- Use open-wire line between the antenna and an ATU. The ATU converts the complex impedance seen at the transmitter end of the feedline to 50 Ω (usually the preferred impedance to present to the transmitter). There are many reasons to do this:

  o For a given length, open-wire line generally exhibits lower loss than coaxial cable or other solid-dielectric transmission lines—especially in the presence of high VSWR on the line.

  o True open-wire line (as contrasted to receiving type television *twin-lead* or similar lines) has a higher breakdown voltage specification than most of the commonly available coaxial cables. One of the penalties of a high VSWR on a transmission line is that for a given transmitter power the maximum voltage on the line increases with increasing VSWR.

  o Feeding a balanced antenna (the horizontal center-fed dipole) with balanced transmission line helps maintain balance in the antenna, thus minimizing pattern skew and unwanted radiation from (or signal pickup by) the feedline.

  o Open-wire line is usually lighter than coaxial cable, and the center of a dipole supported only at its ends will sag less.

  o An ATU located at the transmitter end of the feedline is usually more easily accessed if adjustment for a good match is necessary.

The Dipole Radiation Pattern

In Chap. 3 we developed the pattern of a very short (relative to a wavelength) dipole in free space and compared it to that of a totally fictitious reference antenna called the *isotropic radiator*. Compared to the latter, which is truly *omnidirectional* ("all directions"), the short dipole generates a greater radiated signal broadside to the axis of the wire and has little or no radiation off its ends. We describe the short dipole in part by saying it has 1.5-dB *gain* relative to an isotropic radiator in its favored direction, but we simultaneously understand that the gain we're referring to came at the expense of radiated signal in other directions. Thus, gain in an antenna is a direct consequence of its *directivity*.

In fact, much of the time what we call *gain* (G) is actually *directivity* (D). Specifically,

$$G = \xi D \tag{6.7}$$

where ξ is the *efficiency* of the antenna. For dipoles and similar devices having lengths of $\lambda/4$ or longer, efficiency is usually close to 100 percent ($\xi = 1.0$), and we can be a little careless with our use of the word *gain*. But for some antenna types that we cover later in this book, ξ is much lower. A Beverage wire, for instance, has very low efficiency and terrible gain, yet can have excellent directivity.

Always keep in mind that *directivity* and *gain* are specified in *three dimensions*. Too many times, people oversimplify the topic by publishing only the *azimuthal* (compass heading) view of the radiation pattern. In other words, the reader is given a pattern as viewed from directly above the antenna that shows the directivity in the horizontal plane. But a signal does not propagate away from an antenna in an infinitely thin sheet, as such presentations seem to imply; rather, it radiates to varying degrees at *all* elevation levels above ground. Thus, proper evaluation of an antenna takes into consideration both horizontal and vertical plane patterns.

Figure 6.4 shows the radiation pattern of a dipole antenna in free space as seen from different perspectives. In the horizontal plane (Fig. 6.4A), when viewed from above, the pattern is a figure eight that exhibits bidirectional gain broadside to the axis of the antenna. What appear to be two main "lobes" contain the bulk of the radiated RF energy, with little or no power off the ends of the antenna axis. This pattern is the classical dipole pattern that is published in most antenna books.

Also shown, however, is the vertical plane pattern for a dipole antenna in free space (Fig. 6.4B). Note that when sliced this way the radiation pattern appears circular. When the two patterns are combined in the round, you can see the three-dimensional doughnut-shaped pattern (Fig. 6.4C) that helps us visualize the true pattern of an unobstructed dipole in free space.

When a dipole is installed "close" to the earth's surface or other ground beneath it, as is the case for most HF antennas, the antenna pattern is distorted from that of Fig. 6.4. Two effects must be taken into consideration:

- Radiated energy from the antenna that strikes the ground directly below the antenna and is reflected back up to the antenna induces additional currents and voltages in the antenna. *The phase of these relative to the original transmitted wave is a function of the height of the antenna above the reflecting plane.* This changes the

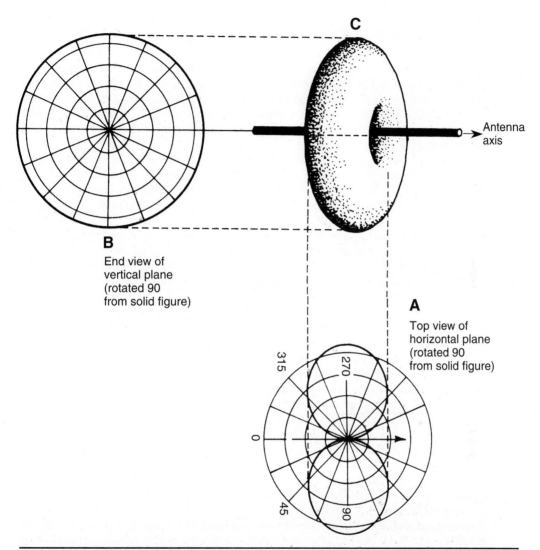

C

Antenna axis

B
End view of
vertical plane
(rotated 90
from solid figure)

A
Top view of
horizontal plane
(rotated 90
from solid figure)

315

270

0

45

90

FIGURE 6.4   Idealized dipole radiation pattern.

feedpoint impedance of the antenna relative to its free space value and causes small changes in the gain of the antenna as its height is varied.

- Energy radiated downward from the antenna at all angles below horizontal is reflected from the earth's surface back into the atmosphere. As discussed in Chap. 5, the ground acts like an *image antenna* and the resulting *array effect* alters the radiation pattern of the dipole *in the vertical plane* as measured at distant receiving points. Figures 5.9 and 5.10 at the end of the previous chapter plot typical radiation pattern versus elevation for horizontal and vertical dipole antennas installed within a few wavelengths of the earth's surface.

Tuning the Dipole Antenna

There are generally two tasks to address when erecting and adjusting a dipole antenna: *achieving resonance* and *impedance matching* the feedline to the antenna. Although frequently treated in the literature as the same issue, they are not. Resonance is a characteristic of the antenna itself; it has nothing to do with the feedline used to supply power to the antenna. (But if the user does not take proper care in locating the feedline relative to the antenna, the feedline may *become* an unintended part of the antenna!) We will take these topics up separately.

As we saw earlier, as the height of a center-fed dipole is raised or lowered with respect to the ground plane beneath it, its feedpoint impedance at resonance varies over about a 10:1 range. Therefore, simply achieving resonance for the dipole we just erected does not necessarily mean we can state with any certainty what its final feedpoint impedance is.

There is only one proper way to tune a dipole antenna to resonance: *Adjust the length of the antenna elements,* not the transmission line! It was in order to make these adjustments that we recommended you initially cut the wire for the dipole a bit longer than the equations would suggest. (It's a lot easier to *remove* a little wire than to *add* it back on!)

Resonance

We start by defining the *resonant frequency* of a dipole antenna as the lowest frequency at which the dipole feedpoint impedance is purely resistive—that is, it is the frequency at which the dipole exhibits neither inductive nor capacitive reactance at its feedpoint.

Resonance is indicated by a minimum in the VSWR curve. Figure 6.5 shows a graph of VSWR versus frequency for several different cases. Curve A represents a disaster: a high VSWR all across the band. The actual value of VSWR can be anything from about 3.5:1 to 40:1 or more, but the cause is nonetheless the same: The antenna is either open or shorted, or it is so far off resonance as to appear to be open or shorted to an SWR meter.

Curves B and C represent antennas that are resonant within the band of interest. Curve B represents a broadbanded antenna that is relatively flat all across the band and does not exhibit excessive VSWR until the frequency is outside of the band. Curve C is also resonant within the band, but this antenna has a much higher Q than curve B. We might naively think the broadbanded antenna is always better, but that's true only if broadness is not a symptom of lower efficiency. In particular, resistive losses tend to broaden an antenna's response curve (the best response curve of all is obtained from a dummy load!), but of course those same losses also reduce its efficiency, ξ. In most cases where the dipole is being used for transmitting purposes, giving up efficiency to obtain a wider SWR bandwidth is a poor trade.

Curves D and E are resonant outside the band of interest. In D, the antenna is resonant below the band, implying that the dipole is too long. In this case, shortening the antenna a bit should raise the resonant point to the desired center frequency inside the band. Similarly, in E the antenna is resonant above the upper limit of the band, so this antenna is too short and must be lengthened. (Hence, our earlier suggestion to leave a little extra wire folded back at each end insulator.)

How much to cut? That depends on two factors: how far from the desired frequency the resonant point is found, and which band is being used. The latter requirement

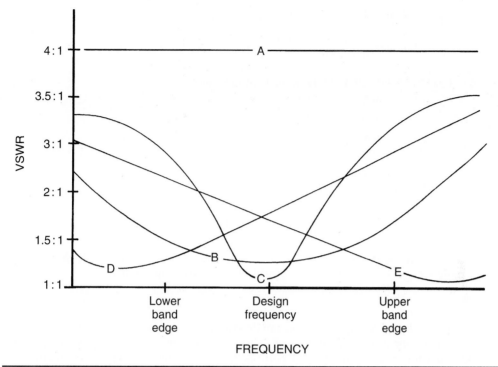

FIGURE 6.5 VSWR versus frequency for several cases.

comes from the fact that the *frequncy per unit length* varies from one band to another. Here is a simple procedure for calculating this figure:

1. Calculate the length required for the upper band edge.
2. Calculate the length required for the lower band edge.
3. Subtract the first length from the second.
4. Determine the width of the band in kilohertz by subtracting the lower band edge (in kilohertz) from the upper band edge.
5. Divide the bandwidth obtained in Step 4 by the difference in length obtained in Step 3; the result is the variation in kilohertz per unit length.

Example 6.3 Calculate the frequency change per unit of length for 80 m and for 15 m.

Solution For 80 m (3.5 to 4.0 MHz in the United States):

1. $L_{UPPER} = \dfrac{468}{4} = 117 \text{ ft}$

2. $L_{LOWER} = \dfrac{468}{3.5} = 133.7 \text{ ft}$

3. Difference in length: 133.7 ft – 117 ft = 16.7 ft

4. Frequency difference: 4000 kHz – 3500 kHz = 500 kHz

5. 5. Calculate $\dfrac{\text{frequency}}{\text{unit length}}$: $\dfrac{500\text{kHz}}{16.7\text{ft}} = 30\text{kHz/ft or }2.5\text{kHz/in}$

For 15 m (21.0 – 21.45 MHz in the United States):

1. $L_{\text{UPPER}} = \dfrac{468}{21.45} = 21.82$ ft

2. $L_{\text{LOWER}} = \dfrac{468}{21} = 22.29$ ft

3. Difference in length: 22.29 ft – 21.82 ft = 0.47 ft

4. Frequency difference: 21,450 kHz – 21,000 kHz = 450 kHz

5. 6. Calculate $\dfrac{\text{frequency}}{\text{unit length}}$: $\dfrac{450}{0.47} = 960$ kHz/ft or 80 kHz/in

At 80 m the frequency change per foot is small, but at 15 m small changes in a dipole's length can result in very large shifts in its resonant frequency.

Using the preceding steps, you can determine approximately how much to add (or subtract) from an antenna under test. If, for example, you design an antenna for 21,390 kHz but find the actual resonant point is 21,150 kHz, the frequency shift required is 21,390 – 21,150, or 240 kHz. To determine how much to add or subtract (as a first guess):

1. The factor for 15 m is 80 kHz/in, which is the same as saying 1 in/80 kHz.

2. The required frequency shift is 240 kHz.

3. Therefore:

$$\text{Change in length} = 240\text{kHz} \times \frac{1 \text{ in}}{80\text{kHz}}$$

$$= 3 \text{ in}$$

Each side of the antenna must be changed by half of the length calculated, or 1.5 in. Because you wish to *raise* the resonant frequency of the dipole, each side should be *shortened* 1.5 in. Once the length is correct, as determined by new VSWR measurements, the connections at the center insulator are soldered and made permanent, and the antenna is hoisted back up to its original height.

NOTE *This is an excellent application for an SWR analyzer or antenna analyzer (see Chap. 27), especially if the original point of minimum SWR seems to be outside the frequency range you are authorized to transmit in.*

Impedance Matching

While the procedures detailed here may help you place the frequency of minimum SWR where you want it, they do not directly address the elimination of the reactive part of the antenna feedpoint impedance. The difference between resonance and impedance matching is seen in the value of the VSWR minimum. While the minimum indicates the resonant point, its value is a measure of the relationship between the feedpoint impedance of the antenna and the characteristic impedance of the transmission line.

In general, your work is done if you can keep the VSWR below 2.0:1 across the transmitting frequency range of interest. (It's unlikely that VSWRs modestly greater than 2.0:1 are of any consequence when using the antenna for receiving purposes.) With rare exception, today's transceivers and amplifiers are capable of delivering close to rated output power to loads with a VSWR of 2.0:1 or less without incurring any damage.

If these procedures do not result in a suitable SWR for your specific installation, there are two options open to you:

- Add an ATU between the antenna and the transmitter.
- Add some form of stub matching or lumped components to the antenna feedpoint.

As a general rule, the second option will be a trial-and-error exercise in frustration without a suitable antenna analyzer or vector network analyzer (VNA; Chap. 27) to provide knowledge of the resistive and reactive parts of the antenna impedance before and after each adjustment to the antenna matching network.

Other Dipoles

Thus far, the dipoles covered in this chapter have been classical half-wave dipoles, in which a "thin" half-wavelength conductor is "broken" at its center and each half is connected to one side of a two-conductor transmission line. This antenna is typically installed horizontally at a half-wavelength or more above the earth's surface. If the antenna is made of wire, it is usually suspended under tension from a high support at each end. If the dipole is made from rigid metal tubing or rod stock—most commonly, aluminum—it is often supported at its center, as we shall see in Chap. 12 ("The Yagi-Uda Beam Antenna"). But on the lower HF bands, where available space and suitably tall supports are often in short supply, ideal dipole installations are often hard to implement. In such cases, the antennas described here have found acceptance by thousands of users the world over. Some of these antennas are in every way the equal of the horizontal dipole, while others are basically compromise configurations to be used when a conventional dipole is not practical. All share a very important attribute, however: They are all derivatives of the basic half-wave dipole, so a good understanding of conventional $\lambda/2$-dipole radiation patterns, standing current distributions, and feedpoint characteristics will make qualitative analysis and pattern visualization for these antennas much easier!

Inverted-V Dipole

Arguably the most popular substitute for a dipole on the lower HF bands is the *inverted-V* (or *inverted-vee*) *dipole*. This is simply a half-wave dipole whose ends are significantly lower than its center. Like a conventional dipole, it is usually fed in the center with a balanced or unbalanced transmission line, and it is often the antenna of choice when only a single high support is available. Because the inverted-vee is supported at its center, heavy feedlines, such as those made of RG-8 or RG-213 coaxial cable, can be suspended from the same center support, eliminating any downward pull on the center of the antenna. Another mechanical advantage of the inverted-vee is that there is no need for substantial tension in the antenna wires themselves, since they are not being called upon to support the weight of the feedline or the center insulator. Hence, there is no reason an inverted-vee can't be constructed from #18 or smaller gauge wire.

Figure 6.6 shows a typical inverted-vee installation. The operating frequency and the height of the center support will determine the approximate height above ground of the ends, but there is no reason the ends can't come almost to the ground if provision for preventing people and animals from touching the antenna is included. Angle α can be almost anything convenient, provided that $\alpha > 90$ degrees; typically, α is between 90 and 120 degrees. On 80 m, the minimum practical height for the center support is about 35 ft for $\alpha = 120$ degrees and the overall length of the antenna's *footprint* is about 110 ft. Conversely, if backyard space is at a premium, reducing the included angle to 90 degrees leads to a minimum center support height of nearly 50 ft, but the footprint is only slightly more than 90 ft. An inverted-vee for 40 m could have minimum dimensions—both height and footprint—of approximately 50 percent to 60 percent of these figures.

Sloping the antenna elements down from the horizontal to an angle (as shown in Fig. 6.6) lowers the resonant frequency for a given length of wire. Thus, the wire lengths may need to be cut slightly shorter for any given frequency than for a true horizontal dipole cut for the same frequency. There is no precise equation for calculation of the overall length of the antenna elements; the concept of "absolute" length holds even less for the inverted-vee than it does for regular dipoles. There is, however, a rule of thumb that can be followed for a starting point: Make the antenna about 6 percent shorter than a dipole for the same frequency. The initial cut of the antenna element lengths (each quarter-wavelength) is then

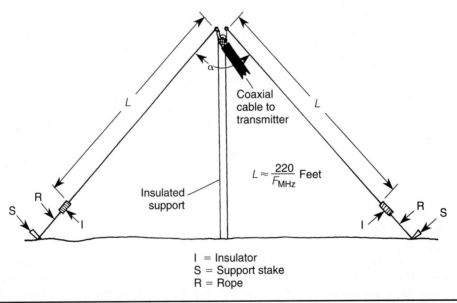

$$L \approx \frac{220}{F_{MHz}} \text{ Feet}$$

I = Insulator
S = Support stake
R = Rope

FIGURE 6.6   Inverted-vee dipole.

$$L \text{ (feet)} = \frac{220}{F(\text{MHz})} \tag{6.8}$$

After the length L is calculated, the actual length is found from the same cut-and-try method used to tune the dipole in the previous section.

Bending the elements downward also changes the feedpoint impedance of the antenna and narrows its bandwidth. Thus, some adjustment in these departments is in order. You might want to use an impedance-matching scheme at the feedpoint or an antenna tuner at the transmitter.

As we shall note numerous times in this book, there is no "free lunch". The inverted-vee is definitely a compromise antenna. In effect, we gain some mechanical advantages (previously described) while incurring the following electromagnetic disadvantages:

- Part of the radiation from each leg of the inverted-vee is horizontally polarized and part is vertically polarized. For $\alpha = 90$ degrees, these two components are roughly equal; simple trigonometry then tells us each is about 70 percent of the magnitude of the original horizontally polarized E-field from a horizontal dipole.

- Remembering that the radiation field far from the antenna is the sum of the radiation from all the many very small segments of wire making up the antenna, we note that the average height of the high-current portions of the antenna is less than that of a comparable dipole whose entire length is at the same height as the inverted-vee's center. Thus, the horizontally polarized radiation from the inverted-vee tends to favor higher takeoff angles, compared to a horizontal dipole at the same height as the center of the inverted-vee.

"Aha!" you say. "Since some of the radiation is now vertically polarized, doesn't that mean that there will be an increased amount of low takeoff angle radiation for working distant stations?"

The answer is, "Not where you're expecting it!", as explained here:

- Just as the currents on opposite sides of a balanced two-wire transmission line feeding a dipole are 180 degrees out of phase with each other (thus minimizing radiation from the feedline at distant points by cancellation of the fields from the two sides), the *vertical* component of the inverted-vee's radiation on one side of the center insulator is out of phase with the vertical component on the other side. Since the average spacing of the high-current portions of the two sides of the inverted-vee is much wider than that of a parallel-wire transmission line—perhaps $\lambda/8$—there is incomplete cancellation and some vertically oriented energy is, in fact, radiated. However, the direction of maximum radiation is *not* broadside to the wire, as it is for the horizontal component. Rather, the vertically polarized pattern is typically maximum along the wire axis, at right angles to the optimum direction of horizontal radiation, and over average ground about 4 dB lower in amplitude at a 20 degree elevation angle than a simple $\lambda/4$ vertical would be.

Figure 6.7 shows the modeled patterns for the horizontally and vertically polarized E-field components of an 80-m inverted-vee with a 90-degree included angle and a center support height of 50 ft over average ground. Note the 90-degree difference in the compass headings of the horizontal and vertical E-field lobes.

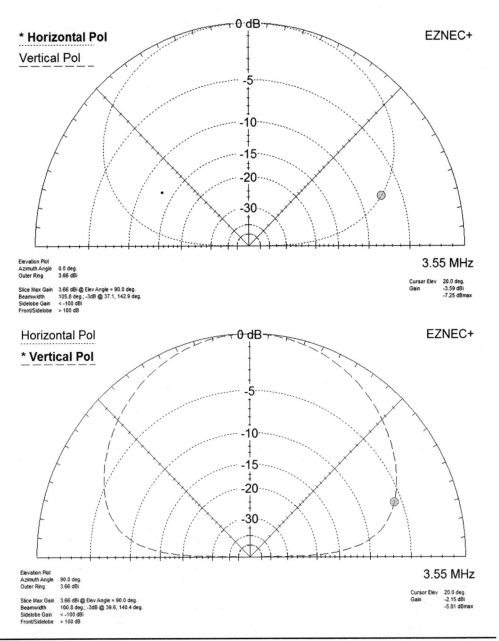

FIGURE 6.7   Inverted-vee radiation patterns.

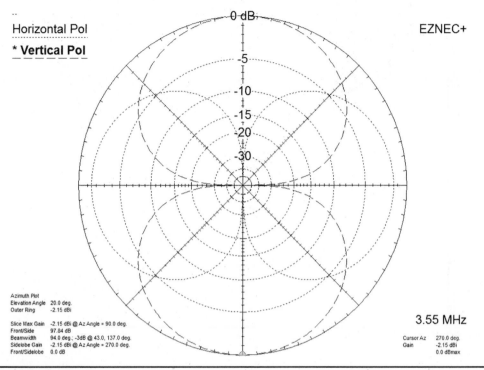

Horizontal Pol

* **Vertical Pol**

EZNEC+

3.55 MHz

Azimuth Plot
Elevation Angle 20.0 deg.
Outer Ring   -2.15 dBi

Slice Max Gain -2.15 dBi @ Az Angle = 90.0 deg.
Front/Side   97.84 dB
Beamwidth   94.0 deg.; -3dB @ 43.0, 137.0 deg.
Sidelobe Gain   -2.15 dBi @ Az Angle = 270.0 deg.
Front/Sidelobe   0.0 dB

Cursor Az   270.0 deg.
Gain   -2.15 dBi
            0.0 dBmax

FIGURE 6.7 Continued.

Sloping Dipole ("Sloper")

The *sloping dipole* (Fig. 6.8) is popular with those operators who need a low angle of radiation and have at least one tall support. This antenna, informally referred to as a *sloper,* is a half-wavelength dipole that is erected with one end at the top of a support and the other end close to the ground so as to make a 45-degree (more or less) angle with the support. Just like a horizontal dipole, it is usually fed in the center by coaxial cable or two-wire balanced transmission line. When fed with a coaxial (unbalanced) line, common practice is to connect the outer conductor or shield of the line to the *lower* half of the antenna, but there is little provable basis for doing so. In fact, the antenna is so unbalanced with respect to nearby ground that how the coaxial cable is connected is essentially immaterial.

Some of the comments made regarding the inverted-vee antenna also apply to the sloping dipole, so please see that section also. Like the inverted-vee, a sloping dipole sacrifices the high-current portions of the antenna to lower heights, but unlike the inverted-vee, the vertically polarized contributions to the radiated field from the two halves of the antenna are in phase because the two halves of the antenna are not folded back toward each other. The vertical components of the E-field are, however, offset spatially, with the result that the antenna exhibits a small amount of gain in the direction of the lower end. That "forward" gain is accompanied by a modest reduction at the rear (i.e., in the direction of the higher end), so the antenna exhibits a small but noticeable front-to-back ratio, as shown in Fig. 6.9.

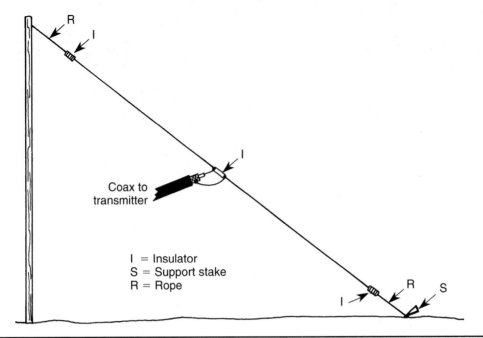

FIGURE 6.8   Sloping dipole ("sloper").

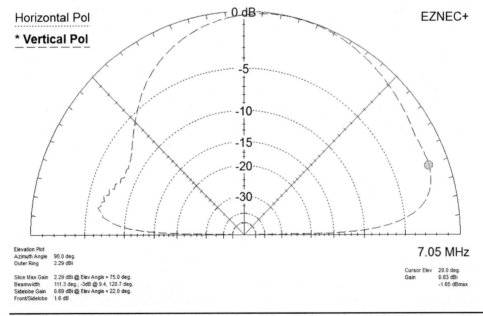

FIGURE 6.9   Sloping dipole radiation patterns.

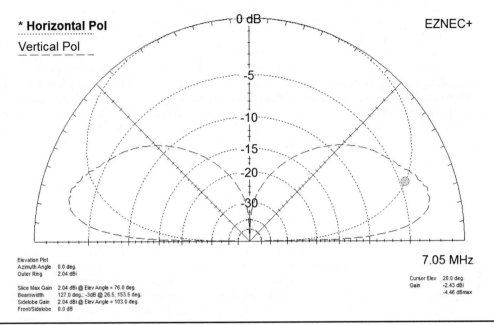

* Horizontal Pol

Vertical Pol

EZNEC+

7.05 MHz

Elevation Plot
Azimuth Angle   0.0 deg.
Outer Ring   2.04 dBi

Slice Max Gain   2.04 dBi @ Elev Angle = 76.0 deg.
Beamwidth   127.0 deg.; -3dB @ 26.5, 153.5 deg.
Sidelobe Gain   2.04 dBi @ Elev Angle = 103.0 deg.
Front/Sidelobe   0.0 dB

Cursor Elev   20.0 deg.
Gain   -2.43 dBi
              -4.46 dBmax

FIGURE **6.9**   Continued.

Space permitting, some operators like to arrange four sloping dipoles from the same mast such that they point in different directions around the compass (Fig. 6.10). A single four-position coaxial switch midway up the center support facilitates switching a directional beam around the compass to favor four different compass headings. If the coaxial cables to the four-position switch are of identical length that is chosen based on whether the remote switch short-circuits the unused cables or leaves them open, additional front-to-back rejection (in the order of 10 dB) from the three idle slopers is possible.

Broadbanded Dipoles

One of the rarely discussed aspects of antenna construction is that the length/diameter ratio of the conductor used for the antenna element is a factor in determining the bandwidth of the antenna. In general, a larger cross-sectional area makes the antenna more broadbanded. When mechanical considerations permit, this suggests the use of aluminum tubing instead of copper wire for the antenna radiator. In fact, on 14 MHz and above, that is almost always the

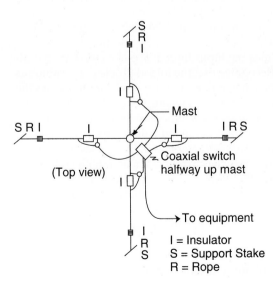

(Top view)

Mast

Coaxial switch
halfway up mast

To equipment

I = Insulator
S = Support Stake
R = Rope

FIGURE **6.10**   Directional antenna made of four slopers (top view).

preferred solution. At those frequencies, dipoles constructed from aluminum tubing typically require support only at their centers, so a dipole made of aluminum tubing may be relatively inexpensive when compared to the cost of wire *plus* supporting materials such as rope and pulleys. As frequency decreases, the weight becomes greater because the tubing is both longer and (for structural strength) of greater diameter. On 40 m and below, dipoles of aluminum tubing require special supporting mechanisms and are generally not cost effective. Yet coverage of the entire 80-m band (in the United States, at least) is a significant problem, especially for newer transmitting equipment, because the band is 500 kHz wide, which corresponds to a bandwidth of ±7 percent if the antenna is tuned for 3.75 MHz. Most solid-state exciters and transceivers today include *fold-back* circuitry that shuts down the amplifier stages or greatly reduces their output power levels in the presence of high VSWR. High-power amplifiers, whether employing vacuum tubes or solid-state devices, include VSWR-driven shutdown protection circuits that often require the operator to resolve the problem and manually reset the amplifier controls before they can be used again.

Of course, an ATU can be inserted between the output of the transmitter and the feedline. But such units either need to be manually tuned when changing frequency or, if they are self-tuning, are very expensive. What other alternatives are there?

Here are four approaches to widening the inherent bandwidth of a center-fed dipole antenna: *folded dipole, bowtie dipole, cage dipole,* and *fan dipole*.

Folded Dipole

Figure 6.11A shows the *folded dipole* antenna. This antenna consists of two or more half-wavelength conductors of identical diameter shorted together at both ends and fed in the middle of *one* conductor. Except at the ends, the two conductors are held a few inches apart (at HF, at least) by insulated spacers. The two-wire folded dipole is often made with 300-Ω television antenna twin-lead transmission line. Because the free-space feedpoint impedance is nearly 300 Ω, or four times that of a conventional dipole, the antenna is a reasonably good match (depending on the height of the antenna above ground) for the same type of twin-lead used as the transmission line. Such a dipole will exhibit a 50 percent improvement in 2:1 VWSR bandwidth compared to a single-wire dipole. On 80 m, for instance, a folded dipole resonant at 3.750 MHz will cover from 3.6 to 3.9 MHz with an SWR of 2:1 or lower.

Why is the input impedance *four* times the input impedance of a conventional dipole when one of the two wires is fed? Recognize that the folded dipole is a continuous 1-λ loop. At each end of the fed wire, the current does a U-turn and continues into the unfed wire. If this current were of the same phase everywhere, it would thus be flowing in a spatial direction opposite to the current near the feedpoint. But it's also true that at or near the antenna design frequency the current in the fed half-wave element goes to zero at each end. Since the current in a wire that is longer than λ/2 reverses direction in adjacent λ/2 sections (i.e., at each current *node*), the current reverses at both ends of the folded dipole. When viewed from inside the wire (if that were possible), the current flowing in the unfed wire is thus 180 degrees out of phase *electrically* with the drive current, but its nominal *spatial* direction is the opposite of the current in the driven wire. The two effects (wire path and phase reversal) combine to put the unfed wire's current in phase with the drive current, as viewed from outside the wires. The current in the unfed wire is virtually identical in amplitude to that in the fed wire, suffering only a very slight reduction in amplitude from ohmic and radiation losses.

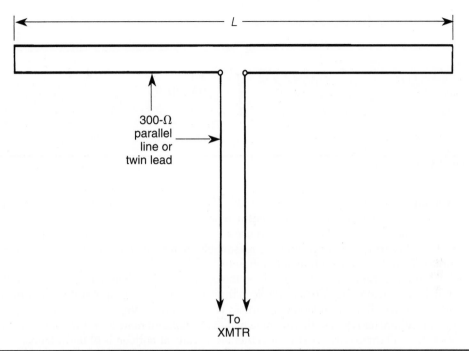

FIGURE 6.11A   Folded dipole fed with 300-Ω line.

Assuming we can match the 300-Ω line to the transmitter, we have the same power going into the folded dipole as we did into the single-wire dipole. For equal power, if one feedpoint current is half the other, its corresponding feedpoint voltage must be twice the other (because $P = VI$ = constant). Thus, in free space the transmission line "sees" a feedpoint impedance of

$$
\begin{aligned}
Z_{\text{FOLDED}} &= \frac{2V_{\text{DIPOLE}}}{\dfrac{I_{\text{DIPOLE}}}{2}} \\[2mm]
&= 4\frac{V_{\text{DIPOLE}}}{I_{\text{DIPOLE}}} \\[2mm]
&= 4Z_{\text{DIPOLE}} \\[2mm]
&\approx 290\ \Omega
\end{aligned}
\tag{6.9}
$$

The *impedance step-up ratio* for a two-wire folded dipole is 4:1 if the two conductors are of equal diameter. When they're not, the relationship is more complicated but the impedance step-up is generally proportional to the ratio of the unfed wire diameter to the fed wire diameter. Thus, a folded dipole can be designed to provide a specific feedpoint impedance to the transmitter and transmission line, within limits, by making one of the two wires larger than the other.

In many installations, the best feedline for the folded dipole will be 300-Ω twin-lead or, better yet, open-wire line connected to a balanced wire ATU at the transmitter end.

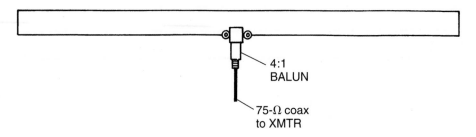

Figure 6.11B   Folded dipole fed with coaxial cable.

Although coaxial cable can be connected to the antenna feedpoint directly, it will experience a roughly 6:1 VSWR over its entire length.

Another popular alternative is to use a 4:1 balun transformer at the feedpoint (Fig. 6.11B). This arrangement provides a reasonable match between the folded dipole and either 52- or 75-Ω coaxial transmission line.

Adding a third wire, also unfed, on the opposite side of the fed wire from the first unfed wire, creates the *three-wire dipole.* A three-wire center-fed dipole offers *twice* the 2:1 SWR bandwidth of a conventional dipole. The impedance step-up ratio for conductors of equal diameter is 9:1, so this antenna makes an excellent mate to 600-Ω open-wire line. As before, all three wires are tied together electrically at both ends of the antenna.

For both the folded dipole and the three-wire dipole, the fundamental tradeoffs when compared to a single-wire dipole are increased bandwidth on the design band versus increased weight and loss of multiband capability.

Bow-Tie Dipole

Another method for broadbanding a half-wave dipole is to use two identical dipoles fed from the same transmission line and arranged to form a bow-tie shape, as shown in Fig. 6.12. The use of two identical dipole elements on each side of the transmission line increases the apparent conductor cross-sectional area so that the antenna has a slightly improved length/diameter ratio.

The *bow-tie dipole* was popular in the 1930s and 1940s, and became the basis for the earliest television receiver antennas. (Each analog TV channel is 6 MHz wide, so a broadband antenna is required to cover each group of adjacent channels: low VHF, high VHF, and UHF.) The bow-tie also attained some measure of popularity in amateur cir-

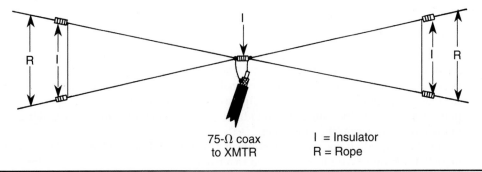

75-Ω coax
to XMTR

I = Insulator
R = Rope

Figure 6.12   Bowtie dipole.

cles during the 1950s as the so-called Wonder Bar antenna for 10 m. It still finds use, but its popularity has faded in the intervening period.

Cage Dipole

The *cage dipole* (Fig. 6.13) is similar in concept, if not construction, to both the bow-tie and the three-wire dipole. ("If three wires are good, five or six must be better, right?") Again, the idea is to connect several parallel dipoles to the same transmission line in an effort to increase the apparent cross-sectional area. Insulated spreaders to keep the wires separated are typically made from plexiglass, lucite, or ceramic. They can also be made from materials such as wood, if the wood is properly treated with varnish or polyurethane, or from any other lightweight insulating material that has been rendered impervious to moisture. The spreader disks are held in place with wire jumpers (see inset to Fig. 6.13) that are soldered to—or tightly wrapped around—the main element wires.

Fan Dipole

A tactic used by some 80-m and 10-m amateurs is to parallel two or more dipoles cut for different parts of the same band, fanning the individual wires on each side of the common center insulator out with distance from the center. Unlike the broadbanding schemes previously described, the wires at each end are not connected to each other. This "stagger tuning" method sends most of the RF to the shorter dipole at the upper end of the band and to the longer dipole at the lower end of the band. The overall result is to somewhat flatten variations in VSWR across the entire band. Modeling this configuration beforehand, to get a handle on the desired difference in lengths for optimum coverage of the desired frequency, is highly recommended. If three or four separate half-wavelength elements are employed, it should be possible to overlap even narrower sections of the band in order to create an even flatter VSWR characteristic. Remember: The ends of the wires in fan dipoles should *not* be tied together!

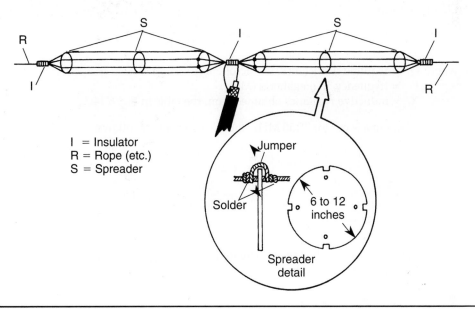

I = Insulator
R = Rope (etc.)
S = Spreader

Jumper

Solder

6 to 12 inches

Spreader detail

FIGURE 6.13   Cage dipole.

Shortened Coil-Loaded Dipoles

At lower frequencies, the half-wavelength dipole is often too long for installations where real estate is at a premium. The solution for many operators is to use a coil-loaded shortened dipole such as one of those shown in Fig. 6.14. In the absence of any loading, a shortened dipole (i.e., one that is appreciably less than a half-wavelength) exhibits capacitive reactance at its feedpoint. This reactance can be canceled with an inductance placed at almost any point along the radiator; in Fig. 6.14A and B, coils are placed at 0 percent (i.e., at the feedpoint) and 50 percent of the element length, respectively.

Figure 6.14C is a table of inductive reactances (in ohms) as a function of the shortened radiator's length, expressed as a percentage of a half-wavelength. It is likely that the maximum allowable percentage will be dictated by your specific installation, but the general rule is to pick the largest percentage that will fit within the available space.

For overall antenna efficiency, coils in the middle of both sides are preferable to coils at the feedpoint, for two reasons:

- Coils at the feedpoint have more current going through them, hence greater ohmic (resistive) losses.

- Coils at the feedpoint are replacing the highest current portions of the antenna with lumped components that don't radiate anywhere near as well.

Example 6.4   Suppose you have about 40 ft of backyard available for a 40-m antenna that normally needs about 65 ft for a half-wavelength. What value of inductor do you need?

Solution   Because 39 ft is 60 percent of 65 ft, you could use this value as the design point for this antenna. From the table, a shortened dipole that is 60 percent of a full $\lambda/2$ dipole's length requires an additional inductive reactance of 950 Ω with the loading coils at the midpoint of each radiator element to allow the feedline to "see" a purely resistive feedpoint impedance. Rearrange the standard inductive reactance equation ($X_L = 6.28$ FL) to the form

$$L \ (\mu H) = \frac{X_L}{6.28 F(\text{MHz})} \tag{6.10}$$

where L = required inductance, in microhenrys
F = frequency, in megahertz (MHz)
X_L = inductive reactance obtained from the table in Fig. 6.14C.

If the antenna is cut for 7.150 MHz, the calculation is as follows:

$$L(\mu H) = \frac{X_L}{6.28 F(\text{MHz})}$$

$$= \frac{(950)}{(6.28)(7.15)}$$

$$= 20.7 \mu H$$

Keep in mind that the inductance calculated here is approximate; it might have to be altered by cut-and-try methods.

◆

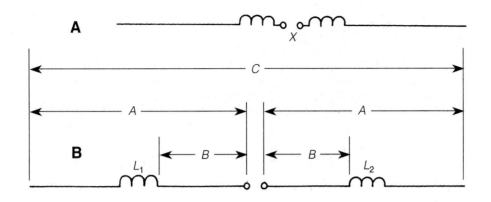

$L_1 = L_2 = L$

$C = 2A$

$A = \frac{1}{2}C$

$$C = \frac{468M}{F_{MHz}}$$

$O \leq M \leq 1$

Percent of half-wavelength	Coils at feedpoint (Ω)	Coils at middle of radiators (Ω)
20	1800	2800
30	950	1800
40	700	1800
50	500	1300
60	360	950
70	260	700
80	160	500
90	75	160
95	38	80
98	15	30

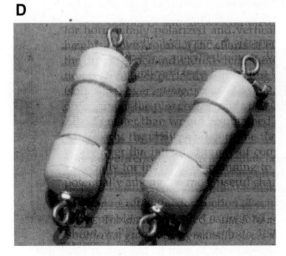

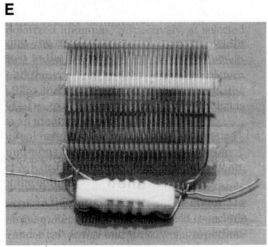

FIGURE 6.14 Shortened or loaded dipole. (A) Inductors at feedpoint. (B) Inductors midway along elements. (C) Chart of reactances for coils. (D) Commercially available coils. (E) Homemade coil based on commercial coil stock.

The shorter the dipole is, relative to a half-wavelength, the greater the amount of loading required and the narrower the bandwidth of the combined assembly. The loaded dipole is a very sharply tuned antenna. Because of this, you must either confine operation to one segment of the band or provide an antenna tuner to compensate for the sharpness of the bandwidth characteristic. However, efficiency drops markedly far from resonance even with a transmission line tuner.

Figure 6.14D and E shows two methods for constructing a coil-loaded dipole antenna. Figure 6.14D shows a pair of commercially available loading coils especially designed for this purpose. Unless you have a way to measure their inductance, they must be used with the wire segment lengths specified by the manufacturer. The ones shown are for 40 m, but other models are also available. The inductor shown in Fig. 6.14E is a section of commercial coil stock connected to a conventional end insulator or center insulator. No structural stress is absorbed by the coil—all forces are applied to the insulator, which is designed for normal dipole tensions.

Inductance values for other lengths of antennas can be approximated from the graph in Fig. 6.15. This graph contains three curves for coil-loaded short dipoles that are 10, 50, and 90 percent of the normal half-wavelength size. Find the proposed location of the coil, as a percentage of the wire element length, along the horizontal axis. The intersection of a vertical line from that point and one of the three curves yields the required inductive reactance (see along vertical axis). Inductances for other overall lengths can be rough-guessed by interpolating between the three available curves, and then validated by cut and try.

Two points should be made with respect to Fig. 6.15:

- Loading coils are most effective (in terms of how much inductance is necessary to "replace" a specified length of the dipole) at the center (or feedpoint) of the dipole.

- Once again, there's no "free lunch". The shorter the total wire length of the dipole is, the less the overall efficiency of the antenna structure as a whole will be.

The Bent Dipole

In Chap. 3 we pointed out that the bulk of the radiated field from a $\lambda/2$ dipole originates in the center half of its total length. That is, the outer $\lambda/8$ section on each end of the dipole is important for establishing resonance but it contributes little to the total signal received at a distant point. The *bent* dipole allows the use of a full $\lambda/2$ dipole in limited space by bending the outer portions of the dipole. The simplest approach is to let them dangle (perhaps with a slight weight or pull-down cord at each end to keep them from swinging in strong winds), but the ends can also be bent horizontally, instead. In fact, the ends can be bent just about any way you want! Just keep them away from the horizontal center portion of the dipole and avoid doubling back at any points along the wires.

Figure 6.16 shows a bent dipole with dangling ends. The main lobe of the horizontally polarized field is down about 0.5 dB in free space and 0.6 dB when the center portion is 0.3 λ above typical ground. The feedpoint impedance will be lower than for the conventional (totally horizontal) $\lambda/2$ dipole but still potentially a decent match to 52-Ω coaxial cable. Of course, open-wire line is a great way to feed it. The author had great success on the 80- and 40-m bands for a number of years with a low 80-m bent dipole like that of Fig. 6.16; the antenna was looped over the branches of a series of maple trees separating his apartment from the cemetery out back!

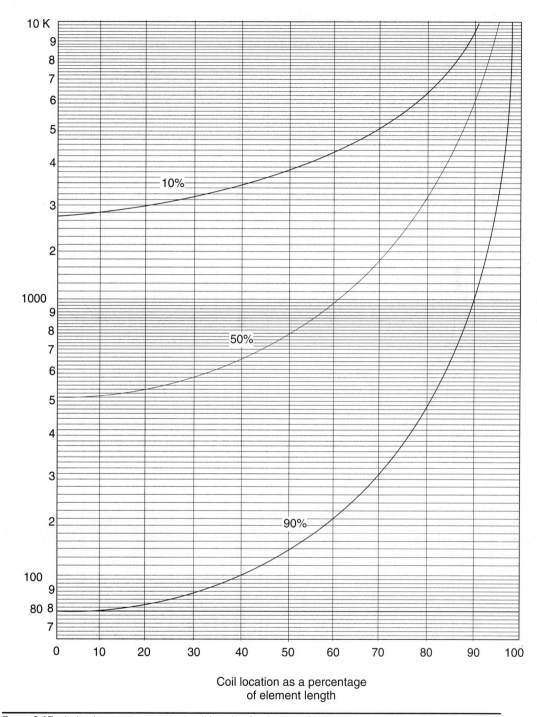

Figure 6.15 Inductive reactance versus coil location for shortened inductance-loaded dipole.

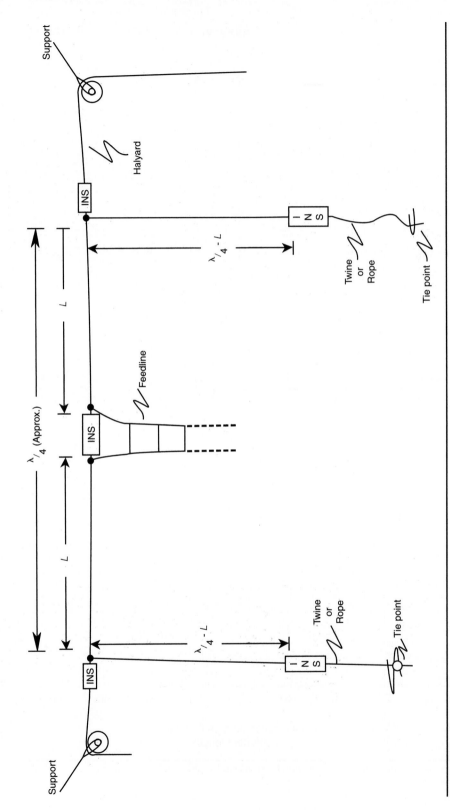

FIGURE 6.16 Bent dipole.

Extended Double Zepp Antenna

As the length of each side of a half-wave dipole is lengthened beyond $\lambda/4$, the gain (or directivity) of the main lobe increases. This continues until each side is $5\lambda/8$ long, at which point current reversal in adjacent $\lambda/2$ sections begins to progressively subtract from the main lobe. Thus, the maximum gain of a standard dipole "doughnut" or figure eight pattern occurs when the total length of the antenna is $2 \times 5/8\ \lambda$, or $5/4\ \lambda$. This configuration (Fig. 6.17) is known as the *extended double Zepp* (EDZ) antenna, and it provides a gain of about 2 dB over a $\lambda/2$ dipole at right angles to the antenna wire plane. It consists of two sections of wire, each one of a length

$$L_1(\text{feet}) = \frac{600}{F(\text{MHz})} \tag{6.11}$$

Typical lengths for L_1 (or half the total wire length) are 20.7 ft on 10 m, 28 ft on 15 m, 42 ft on 20 m, 84 ft on 40 m, and 168 ft on 80 m.

Because the feedpoint of the extended double Zepp is midway between a current node and a voltage node, the input impedance is also midway between high and low, and contains reactance as well as resistance. The antenna can be fed directly with 450-Ω open-wire line, especially if a balanced antenna tuner is available at the equipment end of the line. Alternatively, it can be fed from a quarter-wavelength matching section (made of 450-Ω twin-lead or equivalent open air parallel line) L_2, as shown, or a balun if coaxial cable is preferred. The length of the matching section should be

$$L_2(\text{feet}) = \frac{103}{F(\text{MHz})} \tag{6.12}$$

The extended double Zepp will work on several different bands. For example, a 20-m EDZ will work as an EDZ on the design band, as a slightly overlength $\lambda/2$ dipole on 40 m, as a slightly short $\lambda/2$ dipole on 80 m, and as a four-lobe cloverleaf-patterned antenna on frequencies above the design band. If multiband use is intended, the best feed system is open-wire line back to an ATU.

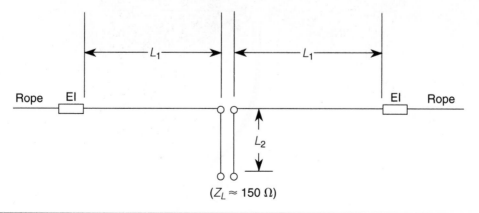

FIGURE 6.17 Extended double Zepp (EDZ) antenna.

Multiband Fan Dipole

The basic half-wavelength dipole antenna is a very good performer, especially when cost is a factor. The dipole yields consistently good performance for practically no investment. A bonus is that a standard half-wavelength dipole (but not any form of *folded* dipole) exhibits a current maximum at its feedpoint not only on its fundamental frequency (where it is λ/2 long) but also on its third harmonic (where it is 3/2 λ in length). Thus, a 40-m half-wavelength dipole fed with 52-Ω coaxial cable on 7 MHz can also be used on 21 MHz with reasonably comparable VSWR. However, its pattern changes from the classic bidirectional figure eight pattern on its fundamental frequency to a four-lobe cloverleaf pattern at its third harmonic.

Dipoles for multiple bands can be hung in parallel and fed from the same feedline and center insulator. Because a dipole operated far from its resonant frequency exhibits a much higher impedance, most of the transmitter power will go into the dipole that is resonant on the band being used. Figure 6.18 shows three dipoles (A1–A2, B1–B2, and C1–C2) cut for different bands, operating from a common feedline and balun transformer. Each of these antennas is a half-wavelength on a different band.

When building this antenna, try to keep the dipoles separated from each other a bit. Note that it will not be necessary to include a separate dipole for any band having a wavelength one-third that of a lower-frequency band that already has a dipole. For example, if you cut A1–A2 for 7 MHz, don't bother including a separate dipole for 21 MHz. If you do, the feedpoint impedance will not be what was expected.

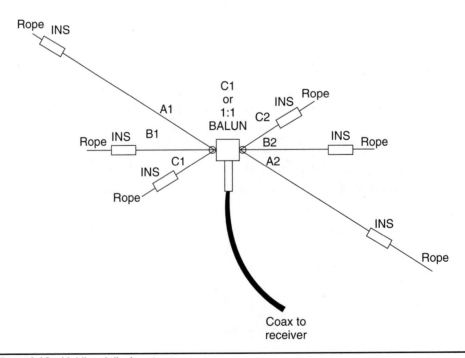

FIGURE 6.18   Multiband dipole antenna.

Summary

From HF to the UHF region, the $\lambda/2$ dipole is the cornerstone of many, if not most, antenna installations in the world. It and all its derivatives that we shall meet in subsequent chapters provide easy and inexpensive access to the airwaves for everyone. For every variation on the basic dipole presented in this chapter, there are countless others in use somewhere in the world as you read this. The author's first antenna was—believe it or not—an 80-m three-wire folded dipole erected at the mind-blowing height of 30 ft! Today, many decades later, the author enjoys *two* conventional 80-m dipoles, strung at right angles to each other and fed independently through separate E. F. Johnson *Matchboxes* manufactured 50 years ago—a reminder that the more things change, the more they stay the same.

CHAPTER 7

Large Wire Loop Antennas

Loop antennas are characterized as small or large depending on the dimensions of the loop relative to the wavelength of the operating frequency. These two types of loops have different characteristics, work according to different principles, and serve different purposes. *Small loops* are those in which the current flowing in the wire has approximately the same phase and amplitude at every point in the loop (which fact implies a very short wire length, i.e., less than about 0.2λ). Such loops respond to the magnetic field component of the electromagnetic radio wave. A *large loop* antenna has a wire length greater than 0.2λ, with most being $\lambda/2$, 1λ, or 2λ. The current in a large loop varies along the length of the wire in a manner similar to other wire antennas whose length is comparable to $\lambda/2$. In particular, the direction of current flow reverses in each $\lambda/2$ section— a fact that is especially useful in the simplified analyses of loops we will use in the following sections.

$\lambda/2$ Large Loops

The radiation pattern of a large wire loop antenna is strongly dependent on its size. Figure 7.1 shows a *half-wavelength loop* (i.e., one in which each of the four sides is $\lambda/8$ long). Assuming S_1 is in its closed position, the real part of the feedpoint $(X_1 - X_2)$ impedance is on the order of 3 kΩ because it occurs at a voltage maximum. Unfortunately, the imaginary, or reactive, part of the impedance is closer to 20,000 Ω! This is an antenna that few antenna couplers can match, and one that has little to recommend it compared to a bent dipole in the same physical configuration with a 6-in insulator inserted in the midpoint of the wire segment opposite the feedpoint (corresponding to S_1 being open).

The fundamental difficulty with the $\lambda/2$ loop is that the two ends of the wire appearing at the terminals of S_1, each $\lambda/4$ from the feedpoint, carry high voltages that are *out of phase* with each other in normal operation of a half-wave dipole. Tying the ends together (by closing S_1) forces the wire into an "unnatural" mode, resulting in abnormally high resistive and reactive components of feedpoint impedance.

Nonetheless, a simple trick can "tame" the difficult input impedance of the $\lambda/2$ loop. In Fig. 7.2, an inductor (L_1 or L_2) is inserted into the circuit at the midpoint of each wire segment adjacent to the driven segment. These inductors should have an inductive reactance X_L of about 370 Ω at the center of the chosen operating band. The inductance of each coil is

$$L_1 = L_2 = \frac{3.6 \times 10^8}{2\pi F} \tag{7.1}$$

where L = coil inductance, in microhenrys (µH)
F = midband frequency, in hertz (Hz)

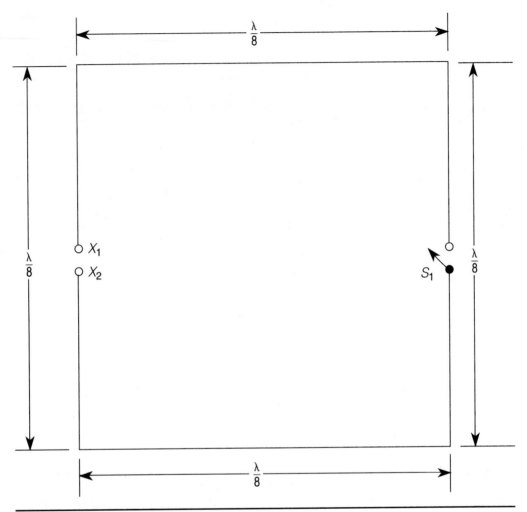

Figure 7.1 Half-wave square loop antenna.

Example 7.1   Find the inductance for each coil in a loaded half-wavelength closed-loop antenna that must operate in a band centered on 10.125 MHz.

Solution   Use Eq. (7.1).
(*Note:* 10.125 MHz = 10,125,000 Hz.)

$$L = \frac{3.6 \times 10^8}{2\pi(10,125,000)} = 5.7\mu\text{H}$$

◆

The coils add lumped circuit inductive loading to the loop, giving it a greater effective length in a small space. The currents flowing in the antenna can be quite high, so,

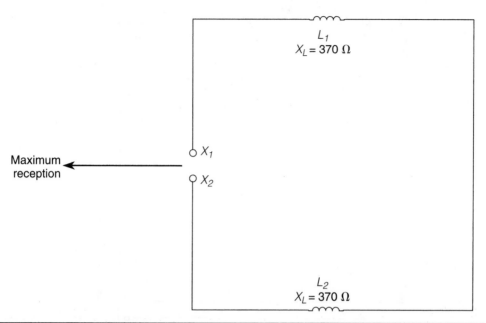

Figure 7.2   Inductive loading improves λ/2 loop.

when making the coils, be sure to use a size that is sufficient for the power and current levels anticipated. The 2- to 3-in-diameter B&W Air-Dux style coils are sufficient for most amateur radio use. Smaller coils are available on the market, but their use should be limited to low-power situations.

1λ Large Loops

If space constraints are not forcing you to a λ/2 loop, then a 1λ loop might be just the ticket. Such a loop has many desirable features, including a manageable feedpoint impedance, more gain than a dipole in favored directions, and ease of analysis.

The simplest way to analyze the 1λ loop of Fig. 7.3 is to treat it as two horizontal half-wave "bent" dipoles whose outer halves are bent toward each other and connected together electrically. As we have seen in an earlier chapter, the center half of a λ/2 dipole is responsible for most of the radiated field strength; the ends establish resonance, minimize feedpoint reactance, and help raise the input or feedpoint impedance to a reasonable value. Thus, bending the outer halves of the dipoles 90 degrees has limited effect on the operation of the elements.

Only one of the two dipoles is fed directly from the transmission line, and a short circuit is placed across the feedpoint of the other. Remember that current reverses direction in adjacent half-wave sections of a collinear array of dipoles. Thus, after an even number of half-wave segments, the drive current naturally wants to be in phase with the current coming around the full periphery of the loop. At the resonant operating frequency, the current goes to zero at what would be each end of the driven dipole and then reverses direction in the second dipole. This results in the currents in the horizontal sections of the two dipoles being in phase *spatially*—that is, they both point in the same direction (left or right in Fig. 7.3) at all times. In free space the 1λ loop is a *two-*

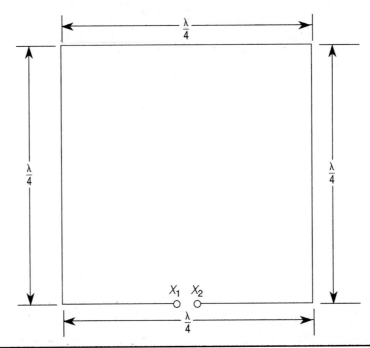

FIGURE 7.3   One-wavelength square loop (single-element quad).

element driven array, whereby the λ/2 section opposite the dipole connected to the feed-line is driven at its high-impedance ends, rather than at its center, from voltages set up by the fed dipole.

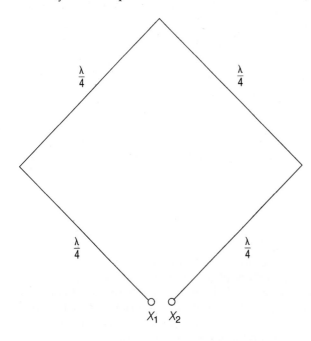

The 1λ loop of Fig. 7.3 produces a bidirectional gain of about +2 dB over a dipole in directions perpendicular to the plane of the loop—that is, into and out of the page. The elevation pattern formed by this loop is somewhat "squashed" because the top and bottom horizontal sections are in quadrature phasing straight up and down.

The loop of Figure 7.4 also contains one wavelength of wire, but the square has been put into a *diamond* orientation so that only one tall support is needed. In effect, we have replaced two bent dipoles with two inverted-vees facing each other, one above the other. Note that the con-

FIGURE 7.4   Bottom-fed diamond loop.

nection to the feedline is no longer in the center of a leg but at a vertex joining two legs. Broadside to the loop (in and out of the page), the horizontal components of their radiation are in phase, but since the effective spacing of the two antennas is somewhat less than $\lambda/2$, they interact a bit differently from the square loop of Fig. 7.3, and the net gain compared to a single inverted-vee is compromised a bit. The vertically polarized component of radiation cancels out completely broadside to the loop and creates a weak cloverleaf in the plane of the loop (to the left and right on the page).

Broadside patterns of both the square and diamond loops look quite good in free space, showing 2- to 3-dB gain over a single dipole or inverted-vee, respectively. However, the presence of real ground beneath them compromises their array gain, especially on the lower HF bands, where few users will be able to raise the bottom of the loop much higher than $\lambda/8$ above ground, and array gains of 1 to 2 dB are more likely.

Delta Loop

The *delta loop* antenna, like the Greek uppercase letter "delta" (Δ) from which it draws its name, is triangular (Fig. 7.5). Although delta loops for VHF frequencies are often oriented point-down (as shown in the figure) to put the horizontally polarized radiating portion of the loop as high as possible, HF delta loops are far more often erected point-up so as to require only one high support. The wire required to form a delta loop is a full wavelength, with elements approximately 2 percent longer than the natural wavelength (like the quad). The actual length will be a function of the proximity and nature of the underlying ground, so some experimentation is necessary. For the isosceles triangle of Fig. 7.5, the approximate preadjustment lengths of the sides are found from:

$$L_1(\text{feet}) = \frac{437}{F(\text{MHz})} \tag{7.2}$$

$$L_2 = L_3 = \frac{296}{F\,(\text{MHz})}\ (\text{in feet}) \tag{7.3}$$

There is no reason, however, why your delta loop can't be an equilateral triangle, in which case the starting point for each side is

$$L_1 = L_2 = L_3 = \frac{343}{F(\text{MHz})}\ (\text{in feet}) \tag{7.4}$$

Delta loops for 14 Mhz and above can be constructed from either copper wire or aluminum tubing, and are usually mounted on rotatable masts. Delta loops for 7 MHz and below are too large to be easily rotated, but fixed wire versions are popular with many amateurs.

The sharpness of the included angles on the delta loop is not particularly desirable and results in the antenna having less broadside gain than the square (quad) loop, but the triangular shape allows it to be suspended from a single high support. In all cases, maximum gain is in the two directions perpendicular to the plane of the loop.

The overall length of wire needed to build any of the preceding three 1λ loop antennas is

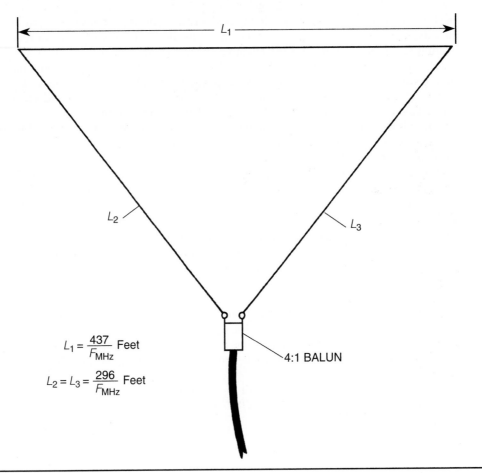

$$L_1 = \frac{437}{F_{MHz}} \text{ Feet}$$

$$L_2 = L_3 = \frac{296}{F_{MHz}} \text{ Feet}$$

4:1 BALUN

Figure 7.5 Delta loop antenna.

$$L \text{ (feet)} = \frac{1005}{F \text{ (MHz)}} \qquad (7.5)$$

The polarization of the loop antennas fed as shown in Figs. 7.3 through 7.5 is horizontal because of the location of the feedpoint. That is because the highest-current sections of a 1λ loop are the two $\lambda/8$ sections that straddle the feedpoint, and the corresponding continuous $\lambda/4$ section halfway around the loop from the feedline. But there is nothing magic about that particular feedpoint location. On the square loop, moving the feedpoint to the middle of either vertical side will change the polarization to vertical. Similarly, on the diamond loop, vertical polarization is realized by moving the feedpoint to either of the two side vertices. On the delta loop, placing the feedpoint at either of the two other vertices produces a diagonal polarization that offers approximately equal vertical and horizontal polarization components, but vertical polarization is easily attained by moving the feedpoint one-fourth of the way up (or down, as the case may be) either side.

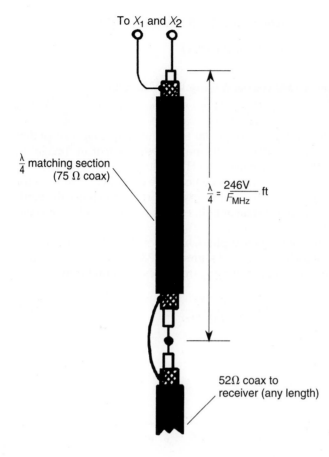

To X_1 and X_2

$\frac{\lambda}{4}$ matching section (75 Ω coax)

$\frac{\lambda}{4} = \frac{246V}{F_{\text{MHz}}}$ ft

52Ω coax to receiver (any length)

FIGURE 7.6 Quarter-wavelength coaxial matching section.

The feedpoint impedance of a 1λ loop is the vector combination of three components:

• Baseline feedpoint impedance of a simple $\lambda/2$ "bent" dipole or inverted-vee in free space

• Effect of radiated fields from the second $\lambda/2$ section on the fed section

• Ground reflection effects, if too close to ground to use free-space assumptions

The resulting input impedance for the array is around 110 Ω for any reasonable height above ground, so it provides a slight mismatch to 75-Ω coax and a 2:1 mismatch to 52-Ω coax. A very good matching to 52-Ω coax can be produced using the scheme of Fig. 7.6. Here, a quarter-wavelength coaxial cable matching section (see Chap. 4) is made of 75-Ω coaxial cable. The length of this cable should be

$$L = \frac{246 v_F}{F} \tag{7.6}$$

where L = length, in feet (ft)
   v_F = velocity factor of coaxial cable
   F = frequency, in megahertz (MHz)

The impedance Z_0 of the cable used for the matching section should be

$$Z_0 = \sqrt{Z_L Z_S} \tag{7.7}$$

where Z_0 = characteristic impedance of coax used in matching section, in ohms
   Z_L = feedpoint impedance of antenna, in ohms
   Z_S = source impedance (i.e., 52-Ω characteristic impedance of line to receiver in standard systems)

When Eq. (7.4) is applied to a system having $Z_s = 52\ \Omega$ and $Z_L = 110\ \Omega$:

$$Z_0 = \sqrt{(110)(52)} = 76\Omega \qquad (7.8)$$

This is a very good application of 75-Ω coaxial cable or CATV hardline.

Half-Delta Sloper (HDS)

The *half-delta sloper* (HDS) antenna (Fig. 7.7) is similar to the full delta loop, except that (like the quarter-wavelength vertical) half of the antenna is in the form of an "image" in the ground. Gains of 1.5 to 2 dB are achievable. The HDS antenna consists of two elements: a $\lambda/3$-wavelength sloping wire and a $\lambda/6$ vertical wire on an insulated mast (or a $\lambda/6$ metal mast). Because the ground currents are very important, much like the vertical antenna, either an extensive radial system at both ends is needed or a base ground return wire must be provided.

The HDS will work on its design frequency, plus harmonics of the design frequency. For a fundamental frequency of 5 MHz, a vertical segment of 33 ft and a sloping section of 66 ft are needed. The approximate lengths for any frequency are found from

$$d_1(\text{feet}) = \frac{\lambda}{3} = \frac{328}{F(\text{MHz})} \qquad (7.9)$$

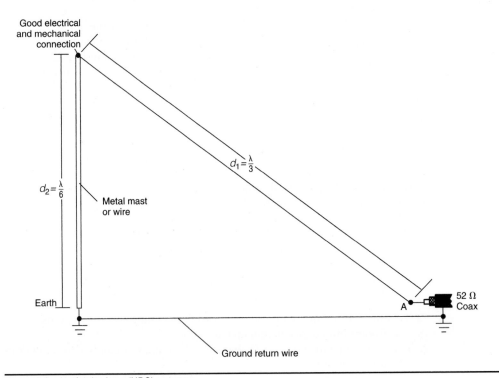

FIGURE 7.7 Half-delta loop (HDS) antenna.

and

$$d_2(\text{feet}) = \frac{\lambda}{6} = \frac{164}{F\,(\text{MHz})} \tag{7.10}$$

The HDS is fed at one corner, close to the ground. If only the fundamental frequency is desired, then you can feed it with 52-Ω coaxial cable. But at harmonics, the feedpoint impedance changes to as high as 1000 Ω. If harmonic operation is intended, then an antenna tuning unit (ATU) is needed at point A to match these impedances.

As with the delta loop, the "sharp" interior angle (i.e., <90 degrees) of the HDS is not ideal because components of the radiated fields from the different sections of the antenna will cancel in directions where other components are reinforcing.

2λ Bisquare Loop Antenna

When the size of the basic large loop is again doubled—this time to 2λ—loop analysis and patterns are no longer quite so clean and simple. As a general guideline, loops larger than 1λ have most gain in the main lobe (relative to a simple dipole) and "reasonable" feedpoint impedances only when the number and arrangement of $\lambda/2$ sections are such that the fields from multiple sections consistently reinforce each other in a few specific directions. One way to ensure this is to make sure the physical configuration of the loop results in currents on opposing sides of the loop being in phase with each other. Unfortunately, this becomes much more complicated for 2λ loops.

As an example, let's return to the 1λ loop of Fig. 7.3. When fed at the center of the bottom horizontal $\lambda/4$ element, the horizontally polarized radiation of the top element reinforces that of the bottom element, and the primary gain is broadside to the plane of the loop, and horizontally polarized. Radiation from the upper half of each *side* element is canceled by out-of-phase radiation from the bottom half. So the predominant mode of operation for this antenna with the feedpoint located as shown is as a two-element *horizontally* polarized array.

If we now increase the length of each side to $\lambda/2$, we can consider this larger loop as comprised of four $\lambda/2$ dipoles, one on each side of the square, connected end to end and fed, as before, at the center of the bottom element. Noting that the current reverses in adjacent dipoles, we see that the current in the top element is now exactly out of phase *spatially* with the current in the bottom element. The same situation is true for the vertical currents in the two side elements. Because the top and bottom are $\lambda/2$ apart, as are the two sides, maximum horizontally polarized radiation is straight up and down (a *cloud burner*, in other words) and maximum vertically polarized radiation is to the left and right on the page. Thus, this loop antenna radiates best as a *vertically* polarized two-element array in the plane of the loop! Of course, in free space there is no such thing as "vertical" or "horizontal" and it doesn't matter which side of the loop is fed—no matter what, it radiates *end fire*, not broadside. Unfortunately, the peak gain of the main vertically polarized lobe is only about 0.6 dB greater than for a single $\lambda/2$ vertical dipole. In the presence of a nearby ground, some small horizontally polarized lobes appear, but over any kind of real ground up to a height of 1.25λ for the top wire, the antenna does best as an end-fire vertically polarized array. If its overall length is adjusted for minimum reactance at the feedpoint, a 4:1 balun will make it a decent match to either 52-Ω or 75-Ω transmission line.

The 2λ *bisquare antenna*, shown in Fig. 7.8, is built like the diamond loop shown earlier (i.e., it is a large square loop fed at a vertex located at the bottom of the assembly)

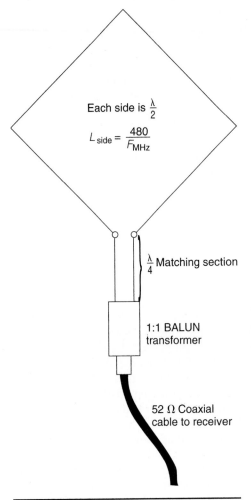

Each side is $\frac{\lambda}{2}$

$L_{\text{side}} = \dfrac{480}{F_{\text{MHz}}}$

$\frac{\lambda}{4}$ Matching section

1:1 BALUN transformer

52 Ω Coaxial cable to receiver

FIGURE 7.8   Bisquare 2λ square loop antenna.

except that each side is now λ/2 in length. If the reactance is canceled out through slight shortening or lengthening of the total loop, the feedpoint impedance is a good match for either 52-Ω or 75-Ω cable directly. The difference in feedpoint impedances between the 2λ diamond loop and the 2λ quad loop is a direct result of changing the location of the feedpoint from the midpoint of a side to the corner between two sides.

When operated at its design frequency and fed as shown in Fig. 7.8, the bisquare antenna offers maximum horizontally polarized radiation in free space at a 45-degree elevation angle broadside to the plane of the antenna (42 degrees when the feedpoint is λ/4 above ground). The peaks of the vertically polarized radiation pattern are as before—that is, in the plane of the wire (to the left and right on the page), but the gain is 2 dB or so *less* than that of a single vertical λ/2 dipole, thus proving that "more" is not always "better"!

In general, there is little to recommend the added mechanical complexity of a 2λ loop unless the phase relationship between opposing sections is "corrected" with the use of phasing lines or inductors.

Near-Vertical Incidence Skywave (NVIS) Antenna

The preceding descriptions of large loop operation in the presence of ground are based on hanging each loop from one or two high supports. However, these loop antennas can instead be suspended so that the plane of the loop of wire is parallel to the ground beneath it. When that is done, at practical heights above ground these loops will send most of their radiation straight up in the air, since the innate array pattern of the 1λ loops generates maximum radiation at right angles to the plane of the loop.

Given the high-angle horizontal polarization of the bisquare loop, which favors short-haul communications links, comparable results can be obtained by laying the same amount of wire "on its side", supported at a more practical height. A 2λ loop of wire arranged in a diamond of uniform height λ/8 or more can provide nearly the same gain as the bisquare, albeit in a more omnidirectional pattern. Actually, the pattern of the 2λ NVIS antenna consists of two separate four-lobed cloverleafs: one with horizontally polarized lobes at right angles to the four wires, and the other with vertically polarized lobes midway in between.

CHAPTER 8

Multiband and Tunable Wire Antennas

M̲ost communications operators require more than one band, and that makes the antenna problem exactly that—a problem to be solved. Amateur, commercial, and military operators are especially likely to need either multiple antennas for different bands or a multiband antenna that operates on any number of different bands. This situation is especially likely on the high-frequency (HF) bands from 3.5 to 29.7 MHz.

Another problem regards the tunability of an antenna. Some amateur bands are very wide (several hundred kilohertz), and that causes the feedpoint impedance of most antennas to be highly variable from one end of the band to the other. It is typical for amateurs to design an antenna for the portion of the band that they use most often and then tolerate a high *voltage standing wave ratio* (VSWR) at the other frequencies. Unfortunately, when you see an antenna that seems to offer a low VSWR over such a wide range, it is almost certain that this broad response is the result of resistive losses that are reducing the antenna efficiency. However, it is possible to tune an antenna over a wide bandwidth using an adjustable *antenna tuning unit* (ATU) similar to those described in Chap. 24.

Multiband Antennas

For users whose interest (or license) is limited to narrow frequency ranges at a number of different places in the frequency spectrum, *multiband antennas* provide a reasonable approach to putting a good signal out on each of the desired bands. The sections that follow provide examples of some time-tested ways to do this.

Trap Dipoles

Perhaps the most common form of multi-band wire antenna is the *trap dipole* shown in Fig. 8.1A. In this antenna category, paral-

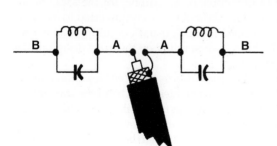

A=10-meter segment
A+B=15-meter segment

Figure 8.1A   Trap dipole for multiband operation. (*Courtesy of Hands-On Electronics and Popular Electronics*)

lel resonant traps are used in combination with shortened segments of wire or aluminum tubing to provide the equivalent of quarter-wavelength half-elements on each band of interest. The technique can be applied to monopoles or to both sides of a dipole or other balanced antenna. Typically, each trap is parallel resonant at one of the desired operating frequency ranges. The high impedance associated with a parallel-resonant circuit allows very little radiofrequency (RF) energy at the trap frequency to pass from one side of the trap to the other. At frequencies other than the trap design frequency, the RF excitation of the element passes through the trap relatively unattenuated. Below the trap design frequency, the trap appears as a net inductive component and above the design frequency it appears as a net capacitive component.

Let's look at how one pair of traps can provide two-band operation with low SWR for a simple dipole. In the example of Fig. 8.1A, the trap on each side of the center insulator is parallel resonant on 10 m. Because of the high impedance of the trap on that band, very little 10-m energy gets beyond the traps, and only the sections of wire labeled "A" have any appreciable RF in them. If the length of each section A is approximately one quarter-wavelength (or about 8 ft long on the 10-m band), the antenna will function as a resonant 10-m dipole on that band.

If a transmitted signal on a *lower* frequency (say, 15 m) is applied across the center of this antenna, the reactance of the trap capacitor will increase but the reactance of the inductor will decrease. Since the capacitor and the inductor are in parallel, the capacitor will not have a major impact on the operation of the antenna on 15 m, but the inductor will provide a modest amount of lumped-element "loading". As a result, the sum of the lengths of A and B will be less than a full $\lambda/4$ (the natural nontrap length) on 15 m.

In general, trap dipoles are shorter than nontrap dipoles cut for the same band. The actual amount of shortening depends upon the values of the components in the traps, so consult the manufacturer's data for each model of trap purchased.

Some trap antennas employ multiple traps on each side of the center insulator to cover three or more bands with a single feedline. The most popular combinations are probably 20-15-10 and 80-40-20, employing two separate parallel-resonant traps on each side of the insulator. The principle of operation is as previously described, except that a third wire segment is usually found between the two sets of traps on a side. Where more than one pair of traps is used in the antenna, make sure they are of the same brand and are intended to work together.

On all bands except the highest one, the radiation efficiency of the trap antenna will be somewhat less than that of a full $\lambda/4$ monopole or $\lambda/2$ dipole for the same band. That is because a small portion of the radiating element has been converted to a nonradiating lumped inductance. However, the effect is small, usually resulting in a net reduction in radiated field strength of 0.5 dB or so, depending on the exact design of the traps and their distance from the feedpoint.

A disadvantage of trap antennas is that they provide less harmonic rejection than an antenna designed for a single band. The antenna has no idea what band the transmitter "thinks" it's transmitting on, so any harmonic energy that falls within any of the design bands of the trap antenna will be radiated with the same efficiency. As a result, users of multiband antennas need to take every reasonable precaution to be sure that harmonics and other out-of-band spurious emissions from their transmitters or transceivers and associated amplifiers are as low as possible.

Multiple Dipoles

Another approach to multiband operation of an antenna consists of two or more $\lambda/2$ dipoles fed from a single transmission line, as shown in Fig. 8.1B. There is no theoretical limit to how many dipoles can be accommodated, although there is certainly a practical limit based on the total weight of multiple dipoles and their spacers. One trick is to remember that bands related to each other by a 3:1 ratio of frequencies can probably be covered by a single dipole cut for the lower-frequency band. Such is the case, for instance, with 40 and 15 m and possibly even with 80 and 30 m.

Assuming the dipoles are at least $\lambda/2$ above ground at the lowest frequency, a reasonably good match to the feedpoint impedance is provided by either 50-Ω or 75-Ω coaxial cable. The two sides of the coax (center conductor and shield) can be connected to the center of the *multi-dipole* directly or through a 1:1 balun transformer, as shown in Fig. 8.1B. Each antenna (A-A, B-B, or C-C) is cut to $\lambda/2$ at its design frequency, so the approximate overall length of each dipole can be found from the standard expression for dipole length.

Overall length (A + A, B + B, or C + C):

$$L \text{ (feet)} = \frac{468}{F \text{ (MHz)}} \tag{8.1}$$

or, for each side of the dipole (*A*, *B*, or *C*):

$$L \text{ (feet)} = \frac{234}{F \text{ (MHz)}} \tag{8.2}$$

As always, close to the earth's surface these equations are approximations and are not to be taken too literally. Some experimentation will probably be necessary to optimize resonance on each band. Also, be aware that the drooping dipoles (B and C in this case) may act more like an inverted-vee antenna (see Chap. 6) than a straight dipole, so the equation length will be just a few percent too short. In any event, a little "spritzing" with this antenna will yield acceptable results.

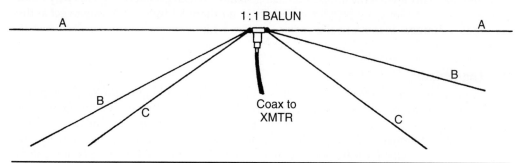

FIGURE 8.1B   Multiband dipole consists of several dipoles fed from a common feedline.

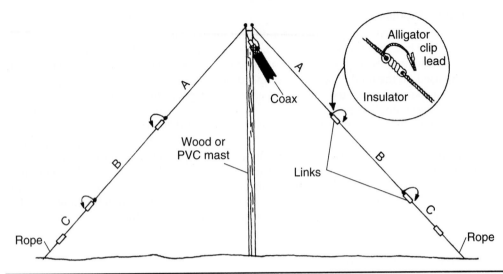

FIGURE 8.1C   Multiband inverted-vee uses shorting links to change bands.

Some amateurs build the multiple-band dipole from four- or five-wire flat TV rotator cable. Starting with the highest frequency, cut each wire to the required length and strip off any unused portions.

Another possibility is the "jumper-tuned" inverted-vee shown in Fig. 8.1C. In this situation, a single conductor, broken into segments A, B, and C (or more, if desired), is used for each leg of the antenna. Adjacent segments are separated by inline insulators. (Standard end insulators are suitable.) Segment A is a quarter-wavelength on the highest-frequency band of operation, A + B is a quarter-wavelength on the next highest band of operation, and A + B + C is a quarter-wavelength on the lowest-frequency band of operation.

The antenna is "tuned" to a specific band by either connecting or disconnecting a wire (see inset) jumper across the insulator between the segments. Either a switch or an alligator clip jumper will short out the insulator to effectively lengthen the antenna for a lower band.

A big disadvantage to this type of multiband antenna is that you must go outside and manually switch the jumpers to change bands, which probably explains why other antennas are a lot more popular, especially in northern latitudes. One way to get at the higher jumpers is to make the center support a tilt-over mechanism (as described in Chap. 28).

Log-Periodic

From a distance, the *log-periodic* (LP) antenna resembles a long boom Yagi (Chap. 12) parasitic array, but it is not. Rather, it is an all-driven array that derives its name from the fact that each element length and the spacing from that element to the next one is a constant percentage of the previous element's length and spacing, respectively. An open-wire parallel transmission line that runs the length of the boom feeds the centers of all elements, which are split and insulated at their midpoints, but the transmission line is flipped as it passes from element to element so that each element is fed 180 degrees out of phase with respect to the ones on either side of it. A commonly seen example of log-periodics is the modern multichannel VHF TV antenna, but much larger HF "LPs" can be found at many military bases around the world.

With proper design, a log-periodic antenna can attain over a 2:1 or greater frequency span the same performance characteristics (forward gain, front-to-back ratio, etc.) that a typical three-element Yagi exhibits over a very narrow band of frequencies. To accomplish this, however, the LP requires a boom length that is on the order of a wavelength or greater at the lowest operating frequency. LPs can make sense for certain wideband applications (broadcast television reception, military frequency-hopping, to name a couple) but they are mechanically quite cumbersome compared to the available alternatives for amateurs and others with widely separated operating bands. Compared to a trap triband Yagi for 20-15-10 m, for instance, an LP of equivalent electrical performance requires a substantially stronger supporting tower and rotator, and consumes a far larger turning radius.

Tuned Feeder Antennas

One of the most popular approaches over the years to coaxing multiband operation from a single antenna is found in Fig. 8.2—the *tuned feeder* type of antenna. When fed this way, an 80-m dipole can be used from 160 through 10 m, but it requires a superior ATU and a length of balanced (parallel wire) transmission line.

What is a "superior" ATU? One that can match a very wide range of impedances—both resistive and reactive—and that has heavy-duty components to minimize internal losses caused by high circulating currents in the coupler components and the possibility of arc-over caused by high voltages. Even at power levels of 150 W or less, an ATU designed for 1500 W into a matched load is a smart investment. Additionally, a superior

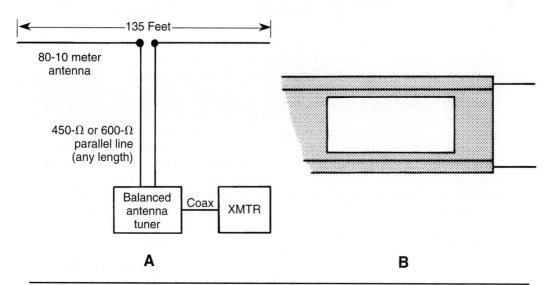

FIGURE 8.2 Tuned feeder antenna can be used on several bands. (*Courtesy of Hands-On Electronics and Popular Electronics*)

ATU will allow wide-range impedance matching of both balanced and unbalanced loads (antenna + feedline).

High-frequency 1000W or greater ATUs on the market at this writing include units from Nye Viking, LDG, Ameritron, MFJ, Palstar, Vectronics, and others. Two excellent units from "yesteryear" that can still be found at hamfests, flea markets, and on the various Internet used equipment sites are the E. F. Johnson *Kilowatt Matchbox* (which can match a distinctly wider range of impedances than its little brother, the 275-watt *Matchbox*) and the Dentron *Super Super Tuner*. Unfortunately, the Johnson *Matchboxes* are not of much use below 3.5 MHz. Be aware, when searching for the proper ATU, that "low loss" and wide matching range translate into a large enclosure. In today's world, a low-loss, legal-limit ATU is often the largest "box" in the radio shack, dwarfing transceivers and power amplifiers alike!

The antenna of Fig. 8.2 is a center-fed dipole that exhibits the familiar figure eight or doughnut-shaped radiation pattern at or near its fundamental frequency—i.e., when each side of the antenna is approximately $\lambda/4$ in length. As discussed in Chap. 6, as the operating frequency is increased, the dipole legs become longer and longer in terms of wavelength. The effect of this is to cause the peak amplitude of the main radiation lobe first to grow even larger than it is for a $\lambda/2$ dipole and then to decline and break into additional lobes that begin to pop up at other angles relative to the axis of the antenna. Fig. 8.3 shows the radiation patterns for an 80-m dipole operated in free space at a few selected higher frequencies. When used closer to earth, however, the nulls in the radiation pattern are nowhere near as sharp and as deep as shown here. As a result, this antenna will actually be useable regardless of the direction of the signal you're attempting to hear or work. At higher frequencies, a dipole in free space that is resonant on 3.6 MHz will exhibit resistive input impedances ranging from 40 or 50 Ω up to 5K Ω or so (the second harmonic is often the worst), and reactances up to perhaps 2000 Ω (both positive and negative). Adding a 40-m dipole to the same feedpoint can substantially lessen the range of impedances that must be matched on the higher bands.

There is no requirement that a dipole be fed in the center; that's simply a convenience to simplify matching of transmitters to feedlines and feedlines to antenna feedpoint impedances. Nor does the pattern of a $\lambda/2$ wire change as a result of where the feedpoint is located. Feeding a $\lambda/2$ dipole at one end means the feedpoint impedance will be very high (a few thousand ohms), but a good ATU should be able to handle this. Figure 8.4 shows the once-popular *end-fed Zepp* antenna. This antenna uses a half-wavelength radiator but it is fed at a voltage node rather than a current node (i.e., the end of the antenna rather than the center). Typically, 450- or 600-Ω parallel-conductor air-dielectric open-wire transmission line is used to feed the Zepp because of the high voltages on the line as a result from the extreme impedance mismatch between the line and the antenna. In theory, the line can be any length, but the task of the ATU is simplified if the length is an odd number of quarter-wavelengths for those bands where the antenna length is a multiple of $\lambda/2$ at the operating frequency. When that condition is met, the transmission line transforms the high feedpoint impedance to a much lower value that is more apt to fall within the ATU's range. For example, a $\lambda/4$ section of 600-Ω open-wire line will transform a 3000-Ω feedpoint impedance at one end of a $\lambda/2$ dipole down to 120 Ω—usually an easy match for an ATU!

As the operating frequency is raised above the point where the wire is $\lambda/2$ in length, the radiation pattern begins to depend on the location of the feedpoint. At the fundamental operating frequency (80 m in our example), there is no difference in radiation

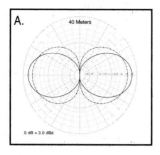

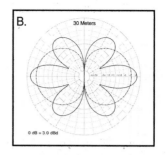

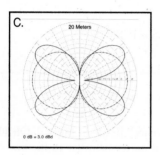

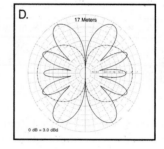

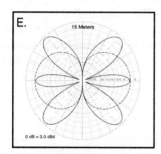

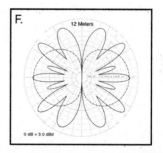

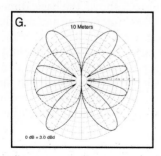

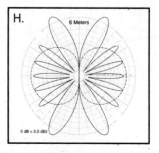

FIGURE 8.3 Patterns for an 80-m center-fed dipole on higher frequencies (solid line) versus its 80-m pattern (broken line).

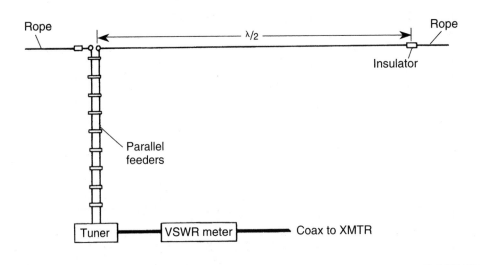

FIGURE 8.4   End-fed Zepp antenna.

pattern between the end-fed Zepp and the center-fed dipole of Fig. 8.2. On 40 m, however, the center-fed dipole still has a figure eight pattern similar to the one on 80, but sporting about 2 dB more gain in the main lobe, which is slightly narrower as well. The end-fed Zepp's pattern, however, resembles a four-leaf clover (actually, two figure eights at right angles to each other) with the pattern peaks at 45 degrees to the axis of the wire. In fact, the end-fed Zepp's 40-m pattern is remarkably similar to the center-fed dipole's 20-m pattern! If we were to continue to double the operating frequency a number of times and compare patterns between the two antennas, we would observe that a given amount of main lobe splitting occurs at a frequency for the center-fed antenna that is twice as high as the Zepp's frequency.

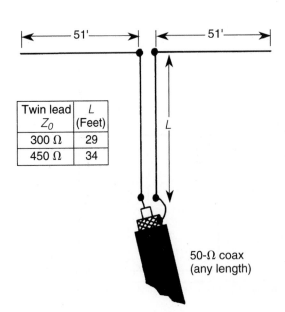

Twin lead Z_0	L (Feet)
300 Ω	29
450 Ω	34

G5RV Multiband Dipole

Figure 8.5 shows the basic dimensions of the *G5RV* antenna, an antenna that has enjoyed much popularity over the years. It has been erected as a horizontal dipole, a sloper, or an inverted-vee antenna. In its standard configuration, each side of the dipole is 51 ft long, and it is fed in the center with a matching section of either 29 ft

FIGURE 8.5   G5RV antenna. (*Courtesy of Hands-On Electronics and Popular Electronics*)

of 300-Ω line or 34 ft of 450-Ω line. Thus, the antenna itself is shorter than a $\lambda/2$ 80-m dipole but longer than a 40-m $\lambda/2$ dipole. In short, the dipole itself is not a particularly good match on 80, 40, 20, or 15 m, but the effect of the short matching section is to bring the feedpoint impedance of the combination closer to a reasonable match for either 50-Ω or 75-Ω coaxial cable on those bands. It is not hard to see that the VSWR of the assembly varies significantly across the HF spectrum, and the addition of amateur bands at 12, 17, and 30 m has blunted the utility of the G5RV noticeably.

Of course, with a good ATU, the antenna can be matched throughout much, if not all, of the HF spectrum but then there is no need for a specific length of 300-Ω or 450-Ω line.

Longwire Antenna

If, instead of using a feedline, a single wire is brought from the end-fed antenna directly into the radio room or to the ATU (wherever it may be located), the antenna is simply a *longwire* antenna. In this case, it can be thought of as having a single-wire feedline, but in truth the feedline is part of the antenna radiating system and should be analyzed or modeled as such. Such an antenna configuration can work (the author used one about 30 ft high to workstations around the world with his 10W transmitter for the first four years of his HF amateur radio activities), but the user must realize the most important thing he or she can do to improve the antenna is to make sure there is an excellent RF ground system (see Chap. 30) attached to the chassis of the ATU or transmitter where the wire connects. Keep in mind that bringing the radiating portion of the antenna into the radio room, coupled with an inadequate RF ground, may have some unintended consequences: greater exposure to RF fields, greater likelihood of interference to audio equipment and telephones, RF burns to the fingers caused by touching the metal chassis of the transmitter, etc.

By longstanding convention, a "true" longwire is a full wavelength or longer at the lowest frequency of operation. In Fig. 8.6 we see a longwire, or "random-length", antenna fed from a tuning unit. As the operating frequency is varied, the feedpoint impedance of the long wire will also vary significantly, with a reactive component that can be quite substantial. Again, use of a superior ATU is important for matching the generally unpredictable feedpoint impedance of the antenna to the typical transmitter output impedance of 50 Ω.

Off-Center-Fed Dipole

The radiated field from a half-wave dipole operated at its fundamental resonant frequency is the result of a standing wave of current and voltage along the length of the dipole. As we saw in an earlier chapter, the center of such a dipole is a point of maximum current and minimum voltage. Thus, if we feed the $\lambda/2$ dipole at its center, the feedpoint impedance is 73 Ω in free space, and it oscillates between a few ohms and 100 Ω as the the antenna is brought closer to ground. If we feed the same dipole at one end instead of in the center, the feedpoint impedance is quite high, perhaps between 3000 Ω and 6000 Ω, depending on the effects of insulators and other objects near the ends of the wire. By feeding the dipole at an intermediate location between the center and one end it is often possible to find a more attractive match to a specific feedline's characteristic impedance.

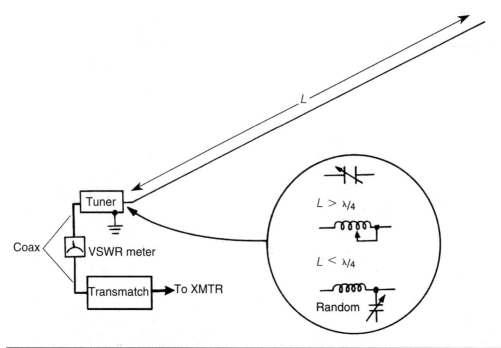

FIGURE 8.6   Random-length longwire antenna.

However, off-center feedlines suffer from an inherent lack of balance between the two sides of the feedline because there is no electrical or physical symmetry at the point where they attach to the antenna. As a result, it is much harder to keep the feedline from radiating and becoming an unintended part of the antenna.

Windom

The Windom antenna (Fig. 8.7) has been popular since the 1920s. Although Loren Windom is credited with the design, there were actually multiple contributors. Coworkers at the University of Illinois with Windom who should be cocredited were John Byrne, E. F. Brooke, and W. L. Everett. The designation of Windom as the inventor was probably due to the publication of the idea (credited to Windom) in the July 1926 issue of *QST* magazine. Additional (later) contributions were rendered by G2BI and GM1IAA.

The Windom is a roughly half-wavelength antenna that will also work on even harmonics of the fundamental frequency. Like the off-center-fed dipole, the basic premise is that a dipole's feedpoint resistance varies from about 50 Ω at the center to about 5000 Ω at either end, depending upon the location of the feedpoint. In the Windom antenna of Fig. 8.7A, the feedpoint is placed about one-third the way from one end, presumably where the impedance is about 600 Ω.

The Windom antenna works "moderately" well—but with some caveats. It is important to again recognize that the return path for a single-conductor feedline is the ground system underneath the antenna and feedline. In distinct contrast to similar horizontal dipoles that are center fed, the extent and quality of the ground beneath the antenna is a major factor in the overall radiation efficiency of the Windom. Further, this

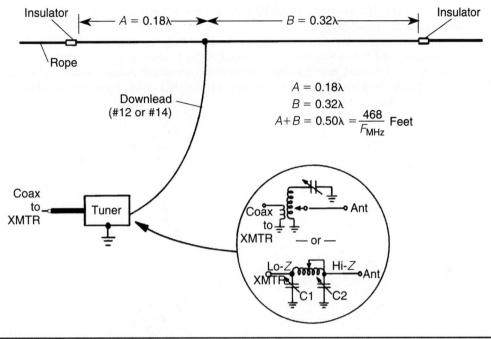

FIGURE 8.7A Windom antenna.

is an inherently unbalanced radiating system with all the concomitant issues of "RF in the shack". One could just as easily view the Windom as a lopsided "T" antenna or as an inverted-L with a secondary section of top loading attached; in either of those cases, the feedline itself is the primary radiator.

The choice of tuning unit for the Windom will depend on the frequencies it is to be used on, but, as mentioned earlier, it is likely that a very good tuner capable of remov-

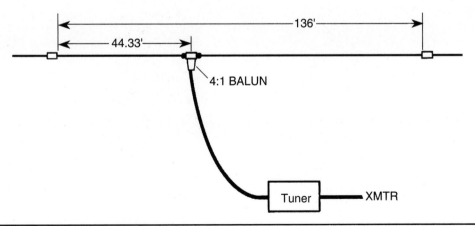

FIGURE 8.7B Coaxial-fed Windom.

ing large amounts of reactance will be required for at least some of the HF amateur bands available nowadays.

An alternative to the single-conductor feedline Windom is shown in Fig. 8.7B. In this antenna a 4:1 balun transformer is placed at the feedpoint, and this in turn is connected to 75-Ω coaxial transmission line to the transmitter. A transmatch, or similar antenna tuner, may also be needed, presumably located at the transmitter end of the transmission line.

CHAPTER 9

Vertically Polarized Antennas

In previous chapters, we defined the polarization of an antenna as being the same as the orientation of the electrical (E) field for the antenna. The direction of the electric field for a specific antenna design is a function of the geometry of the radiating element(s). For a complex structure with different portions creating E-fields having different orientations and phasing, the "effective" E-field is the vector sum of all the individual E-fields and may not even be stationary. However, for the simple case of an antenna (or radiating element) consisting of a single wire or thin rod oriented vertically, the E-field points in the same direction as the long dimension of the antenna, so the antenna polarization is also vertical, by definition.

But what do we mean by "vertical"? In free space, just as an astronaut freely floating outside the space shuttle has no particular up or down or sideways reference, a wire in free space is neither vertical nor horizontal nor anything in between. In fact, the words "vertical" and "horizontal" have meaning for antennas only in respect to some reference plane possessing some amount of electrical conductivity. For most of us, there are only a few such reference planes that we will ever be concerned with:

- The earth's surface (for most land-based or fixed-station antennas)

- The roof or trunk of an automobile (for most mobile antennas)

- The skin of a satellite (for most satellite-mounted antennas)

- The roof of a tall building (for a city dweller's antennas)

We define the polarization of the antenna primarily because many antenna characteristics (radiation pattern and input impedance, to name just two) are greatly affected by any metallic or conducting bodies in close proximity to it. All of the surfaces listed here are conductors to one degree or another, so the orientation and proximity of the antenna relative to any of these surfaces are potentially critical to the proper functioning of the antenna.

Throughout this book, we will assume—unless otherwise stated—that we are always talking about antenna orientations relative to the earth below, with the assumption that our globe is perfectly spherical. In other words, "up", "down", and "vertical" all coincide with the direction of gravity's pull directly beneath the antenna. "Horizontal" will be, by definition, 90 degrees away, parallel to the surface of the earth.

Vertical Dipole

To develop our understanding of vertical antennas, we start with a half-wave dipole, oriented with one end directly "above" the other relative to earth, as in Fig. 9.1. If we

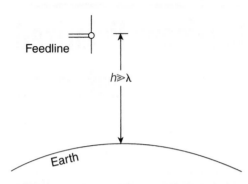

place this antenna many, many wavelengths above the earth's surface, we can guess that its pattern will be very close to its free-space pattern, yet we have preserved a sense of what is "vertical" and what is "horizontal" to make our discussion of this antenna easier to follow.

The vertical dipole is constructed in exactly the same manner as the horizontal dipole but is mounted or supported in the vertical plane. In general practice, the radiating element that is closer to the ground is connected to the shield or outer conductor of any unbalanced feedline, such as coaxial cable or CATV hard line, but there is no factual basis for doing so in a properly engineered installation.

Figure 9.1 λ/2 vertical dipole far from earth's surface.

Like the horizontal dipole, the approximate length of the vertical dipole is calculated from

$$L \text{ (feet)} = \frac{468}{F \text{ (MHz)}} \tag{9.1}$$

where L = *total* length of dipole, in feet
       F = operating frequency, in megahertz

Example 9.1   Calculate the length of a half-wavelength vertical dipole for operation on a frequency of 14.250 MHz in the 20-m amateur radio band.

Solution

$$L = \frac{468}{F}$$

$$L = \frac{468}{14.250} = 32.8 \text{ ft}$$

Note: The 0.8-ft part of this calculated length can be converted to inches by multiplying by 12:

$$0.8 \times 12 = 9.6 \text{ in}$$

Each leg of the vertical dipole in this example is one half of the calculated length, or

$$\frac{32.8}{2} = 16.4 \text{ ft}$$

◆

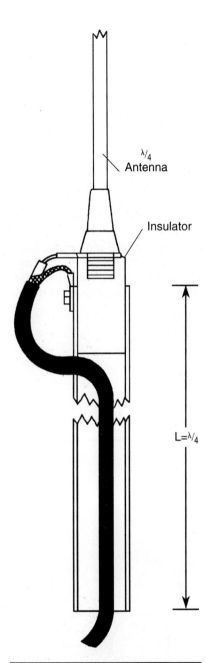

$\lambda/4$
Antenna

Insulator

$L=\lambda/4$

FIGURE 9.2   Coaxial vertical dipole.

The vertical dipole antenna is used in many locations where it is impossible to properly mount a horizontal dipole or where a roof- or mast-mounted antenna is impossible to install because of logistics, a hostile landlord, or a homeowners' association. Some row house and town house dwellers, for example, have been successful with the vertical dipole.

As a general rule, it is wise to try to dress the feedline away from a dipole at right angles for as much of the feedline length as possible. The primary purpose of this is to avoid unbalancing the pattern of the dipole as well as the currents in the two sides of the transmission line. Unfortunately, any dipole for the HF bands is likely close enough to earth ground or other structural grounds that it already has been compromised.

A *coaxial vertical* is similar to the vertical dipole (and, in fact, it can be argued that it is a form of vertical dipole) in that it uses a pair of vertical radiator elements. Such an antenna is an extreme case of imbalance because the feedline comes away from the dipole completely in line with it! In this antenna the radiating element that is closer to the ground is coaxial with the transmission line as shown in Fig. 9.2. The use of a transmission line choke immediately beneath the coaxial sleeve will minimize the tendency of the feedline to act as if it were part of the antenna. One form of choke consists of an appropriate number of turns of the coaxial cable looped through one or more ferrite cores of the right material for the frequency involved.

The coaxial vertical antenna was once popular with CB operators. In some cases, you can find hardware from these antennas at hamfests or on the used equipment Web sites, and the pieces can be modified for amateur radio use. For use on 10 m, it is a simple matter to cut the 11-m CB antenna for operation on a slightly higher frequency. Conversion for bands below 27 MHz is a little more difficult; most likely only the insulator and mounting assembly are salvageable. Keep in mind, however, that adjacent sizes of 0.058-in-thick aluminum tubing are designed such that the inside diameter (ID) of the larger piece is a slip-fit for the outside diameter (OD) of the smaller piece. You can, therefore, connect adjacent sizes of aluminum tubing together without the need for special couplers, etc. With that in mind, salvaged insulator assemblies can be cut off with just 6 to 10 in of the former radiator and sleeve, and new radiators can be created from *telescoped* tubing sections.

Grounded Vertical Monopole

The vertical dipole is a perfectly fine antenna, and it has formed the basis for many VHF and UHF broadcast antennas for decades. The reader can build it by using exactly the same dimensions as required for the horizontal dipole of Chap. 6. Its "real life" pattern is as shown in Fig. 9.3 as long as it is kept far (many wavelengths) from conducting objects with dimensions comparable to, or larger than, itself. That turns out to be fairly easy (for fixed stations, at least) at VHF and UHF frequencies, and not so easy at MF, HF, and microwave frequencies.

It is, in fact, the *mechanical* difficulty of keeping vertical dipoles far from conducting surfaces that makes the grounded quarter-wave vertical so popular. To suspend a vertical dipole for 80 m, for instance, above the earth requires a support more than 150 ft high! And just bringing the transmission line away from the center of the dipole at right angles would present a completely separate challenge.

We saw in Chap. 5 ("Antenna Arrays and Array Gain") that we can use the earth beneath us as a substitute for one side of a dipole. So, as the saying goes, "If you can't beat 'em, join 'em!" Over the range of frequencies where we would have to incur tremendous cost and effort to lift a vertical dipole into the air high enough to avoid unbal-

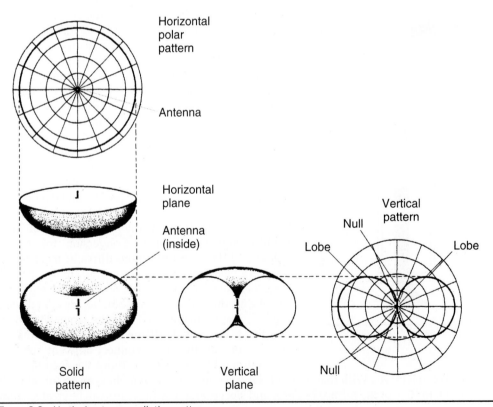

FIGURE 9.3   Vertical antenna radiation pattern.

anced currents in the two sides, we typically choose to use a quarter-wave monopole operated against the ground beneath.

For the MF and lower HF bands (up through perhaps 5 MHz or so), the simplest approach is to mount the vertical on the earth's surface. For still higher frequencies, lifting the vertical into the air (to get above any nearby obstructions) works well as long as we also lift an artificial ground plane—consisting of a finite number of (usually equally spaced) radials—with it.

The equivalence of grounded $\lambda/4$ verticals to $\lambda/2$ vertical dipoles is easily seen if you think of the bottom half of the original vertical dipole as being made of stranded wire. Starting with the dipole, unwrap the strands of the lower wire all the way back to the center insulator and spread them out horizontally, equally spaced around a circle. You have created a ground-plane antenna. Next, take each of the strands and with a very sharp (imaginary) knife, slice it into hundreds of thinner strands. Now spread all of those out equally, as well. Ultimately, you will have a perfectly conducting ground underneath your $\lambda/4$ vertical.

Figure 9.4A shows the basic geometry of the vertical monopole antenna. Here a source of RF is applied at the base of a radiator of length L. Although most commonly encountered verticals are a quarter-wavelength ($L = \lambda/4$) long, that length is not the only permissible length. For now, however, we will limit our discussion to $\lambda/4$ verticals.

Figure 9.4B shows the current and voltage distribution in a quarter-wavelength vertical. Like the half-wave dipole, the $\lambda/4$ vertical is fed at a current node, so the feedpoint impedance is at a minimum—typically less than 37 Ω, depending upon nearby objects, diameter of the radiating element, and other factors.

Figures 9.4C and 9.4D show the two configurations previously discussed for a $\lambda/4$ HF vertical antenna. In Fig. 9.4C the radiator element is mounted at ground level but electrically insulated from ground and fed with 52-Ω coaxial cable. The inner conductor of the coaxial cable is connected to the radiator element, while the cable shield is connected to ground at the base of the vertical. For a $\lambda/4$ radiator, the feedpoint impedance will be lower than 52 Ω, but in most cases, the resulting voltage standing wave ratio (VSWR) is an acceptable tradeoff for simplicity and allows elimination of a matching network or ATU. If the antenna has a feedpoint impedance of 37 Ω, the VSWR will be 52 Ω/37 Ω, or 1.4:1 at the design frequency.

An elevated ground-plane vertical is shown in Fig. 9.4D. This antenna is equally as popular as the ground-mounted quarter-wave vertical, especially on 40 m and above. Amateurs and CB operators find it easy to construct this form of HF antenna because the lightweight vertical can be mounted at a reasonable height (15 to 60 ft) using a television antenna slip-up telescoping mast that is reasonably low in cost. As discussed earlier, this antenna replaces the lower half of a vertical dipole with an artificial ground comprised of quarter-wavelength radials.

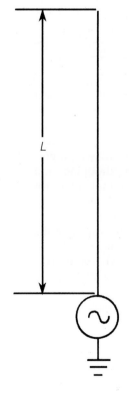

FIGURE 9.4A   Basic vertical monopole.

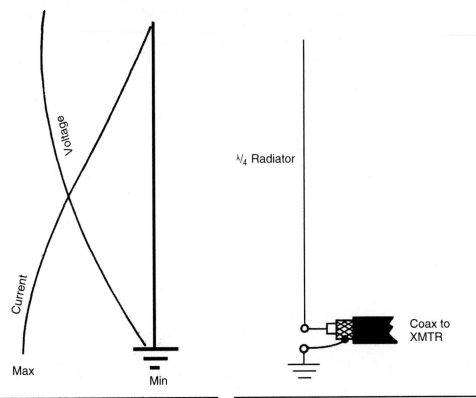

FIGURE 9.4B   Current and voltage distribution along λ/4 vertical.

FIGURE 9.4C   Simple coaxial-fed vertical antenna.

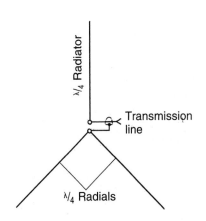

The radials of a ground-plane vertical antenna are typically installed at some vertical angle between horizontal (0 degrees) and, say, 45 degrees below horizontal. Figure 9.4D is an example of the latter case, showing a ground-plane antenna made with "drooping" radials. Similarly, Fig. 9.4E is an example of a ground-plane vertical having perfectly horizontal radials. Sometimes the feedpoint VSWR of the antenna can be improved by adjusting the degree of "droop" in the vertical's radials.

Unlike ground-mounted verticals, the "ground plane" formed by the radials of Fig. 9.4E does not preclude the existence of a far E-field *below* the horizontal plane of the radials, as suggested by the (solid line) pattern in the figure.

FIGURE 9.4D   Mast-mounted vertical uses radials as a counterpoise ground.

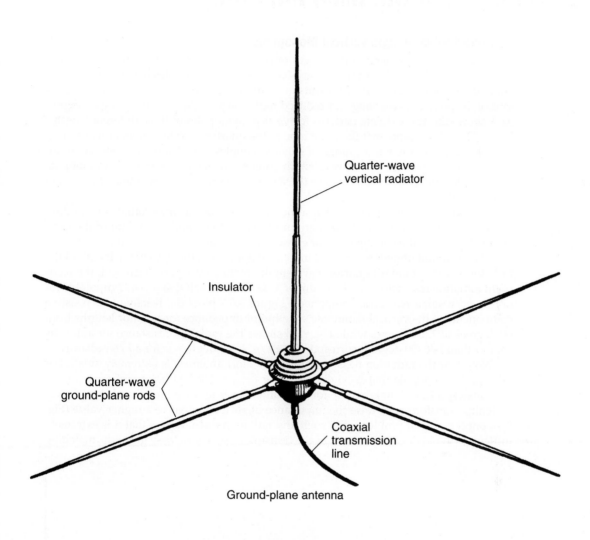

Quarter-wave
vertical radiator

Insulator

Quarter-wave
ground-plane rods

Coaxial
transmission
line

Ground-plane antenna

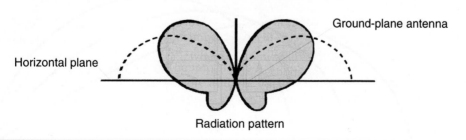

Ground-plane antenna

Horizontal plane

Radiation pattern

FIGURE 9.4E   Ground-plane vertical antenna.

Non-Quarter-Wavelength Vertical Monopoles

If now we gradually increase the length of our grounded monopole, we observe the same flattening of the doughnut shape as we saw with the dipoles of Chap. 6. Lengthening the monopole beyond $\lambda/4$ results in additional RF energy in the low-angle main lobe of the pattern, in exchange for reduced radiation at higher elevation angles. Figure 9.5A shows the approximate patterns for vertical monopoles of three different lengths: $\lambda/4$, $\lambda/2$, and $5/8\lambda$. Note that the main lobe of the quarter-wavelength antenna has reasonable gain over the widest range of elevation angles, but also the lowest maximum gain of the three cases. The $5/8$-wavelength antenna, which is the grounded monopole equivalent of the *extended double Zepp* (EDZ), enjoys both the lowest angle of radiation and the highest maximum gain.

The patterns shown in Fig. 9.5A assume a perfectly conducting ground surrounding the antenna. The effect of real-earth ground losses is to eliminate any hint of the radiated field at distant receiving sites for elevation angles near the horizon (Fig. 9.5B).

The feedpoint impedance of a grounded monopole is a function of the length of the radiator. For the standard quarter-wavelength antenna over perfect ground, the feedpoint radiation resistance is a maximum of 37 Ω (i.e., one half that of a $\lambda/2$ dipole), with only a very small reactance component. Figure 9.6A plots the reactive and resistive components of the vertical monopole's feedpoint impedance for radiator lengths from 60 degrees to 120 degrees, while Fig. 9.6B shows the radiation resistance for antennas shorter than $\lambda/6$ (90 degrees corresponds to $\lambda/4$, so 60 degrees is for a $\lambda/6$ radiator).

Note that the radiation resistance for such short antennas is extremely small. For example, a monopole that is 30 degrees long ($30/360 = 0.083\ \lambda$) has a resistance of approximately 3 Ω. General practice for such antennas is to use a broadband impedance-matching transformer to raise the impedance of such antennas to a higher value (Fig. 9.7), but the biggest problem with such low-radiation resistances is that it is extremely difficult to avoid dissipating most of the transmitter power in lossy grounds, matching

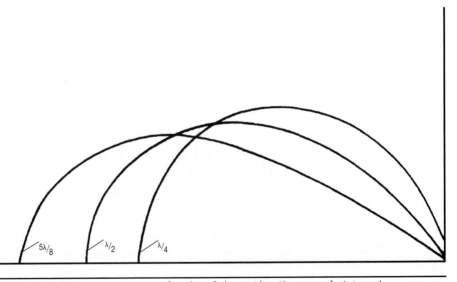

FIGURE 9.5A   Vertical radiation pattern versus a function of element length over perfect ground.

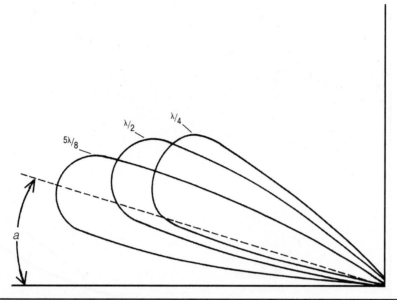

Figure 9.5B   Accounting for ground losses over real earth.

network components, and the connections themselves—just ask any serious user of HF mobile equipment!

Ground Systems for the Grounded Monopole

Earlier in this chapter, we discussed how the vertical antenna radiates equally well in all (azimuthal) directions. Over the years, some wags have said the vertical radiates equally *poorly* in all directions. Which of these opinions is correct?

In fact, *either* can be true. The secret of success is this: The grounded vertical monopole antenna works well only when placed over a *good* ground system. (This comment does not apply to a vertical dipole in free space, many wavelengths away from the nearest conductors.) As we discussed earlier, in a ground-mounted vertical the ground system provides an electrical "return" for antenna currents; it is the replacement for the "other" half of the dipole. To the extent that the ground return is "lossy" or incomplete in its coverage of the radiation field of the vertical, a significant ground loss resistance is added to the total resistance seen by the RF energy delivered to the antenna terminals by the transmitter and feedline. If drawn schematically, ground losses and antenna radiation resistance are in series and, hence, the transmitter output power is divided between the radiation resistance of the antenna and the loss resistance of the ground return system. With radiation resistances in the 2- to 37-Ω range, it does not take much loss resistance to steal half or more of your output power!

Ground return loss is one of the primary reasons mobile installations generally do not get out as well as home station antennas. Because mobile whips (usually operated as grounded verticals) are so much shorter than a quarter-wavelength, their input impedance is *very* low—usually just a few ohms, at best.

The usual way to provide a good ground for a *ground-mounted* vertical is to use a system of radials such as those shown in Fig. 9.8. (We will take up elevated ground-

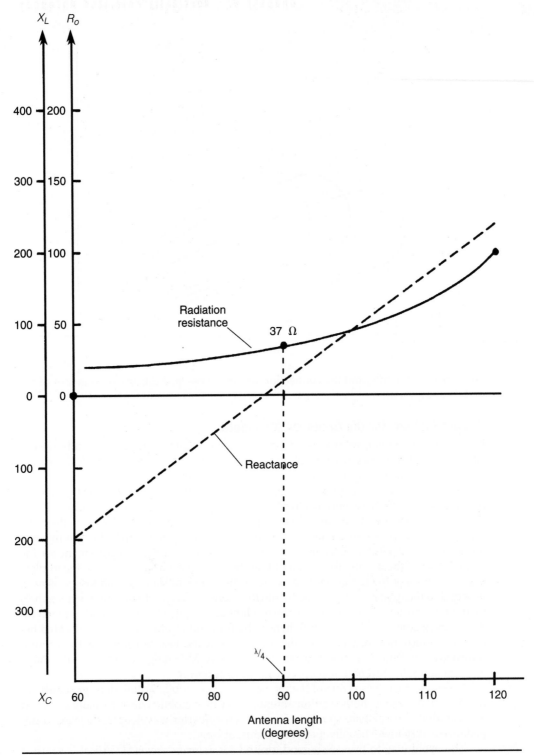

FIGURE 9.6A Antenna impedance as a function of antenna length.

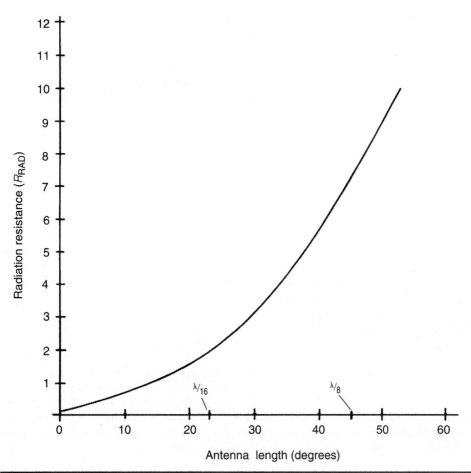

FIGURE 9.6B   Radiation resistance as a function of antenna length.

plane antennas shortly.) There we see (by looking down from directly above) 16 radials equally spaced to cover the full circle around the antenna. At one time it was commonly understood that each radial was to be a quarter-wavelength, where L (feet) = 246/F (MHz). In truth, radials in direct contact with the ground are sufficiently detuned from their free-space characteristics that an exact length is not at all important. Instead, what *is* important is that enough radials are installed that their far ends are typically no more than 0.05 λ apart. A little thought will lead to the counterintuitive realization that the *shorter* your radials are, the *fewer* of them you need!

And this is, in fact, true. However, it is also true that the shorter your radials are, the less efficient your ground system will be and the higher your ground resistance losses will be. The basic point being made is this: If you have a fixed amount of wire to apply to your system of radials, a few long radials are not as helpful as a greater number of shorter ones. Stated yet another way, it's better to have longer radials than shorter ones, but if you can only have shorter ones, you don't need as many before you will have maxed out the performance gains you can accomplish by adding more radials.

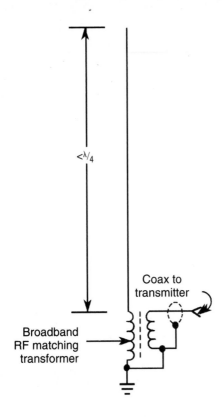

FIGURE 9.7 Impedance matching through broadband transformer.

A little math will show that designing your radial system such that radial tips are no more than 0.05 λ apart leads to the following relationship:

$$N \text{ (number of radials)} = L \text{ (in wavelengths)} \times 128/\lambda$$

Thus, for quarter-wavelength radials, N is about 32. There is nothing magic or abrupt, of course, about the 0.05 λ figure; increasing the spacing from 0.05 λ to 0.10 λ, for instance, simply means a slight increase in ground system losses and a corresponding slight decrease in signal strength at the distant receiving station in return for needing to install half as many radials.

There is also nothing magic about making all the radials equal in length. If you are boxed in by property boundaries, a city lot, or structures that get in the way, simply do the best you can. Radials are important to the efficient operation of verticals, but many different compromise geometries can lead to equally excellent results. Over the past two decades, one of the authors has had a commanding signal on 160 m from two different homes with highly lopsided radial fields of between 16 and 24 radials.

It is also true that the taller the vertical, the longer the radials should be for optimum antenna system efficiency. That is because a taller vertical tends to put return currents into the earth at a greater distance from its base than a shorter vertical does. Again, as the radials are lengthened, the number of them should increase. A "safe" rule of thumb is that your radials should be approximately the same length as your vertical's physical height.

Probably the most important thing to keep in mind is that transmitter output power headed for the antenna actually gets split, or "used up", in three ways: feedline losses, antenna and ground system losses, and antenna radiation resistance. Of those, only power dissipated in the antenna radiation resistance contributes to the received signal strength far away. How much power is lost in either of the other two categories is determined with Ohm's law calculations on a simple resistive divider. The lower the radiation resistance of the antenna, the more attention that should be paid to feedline losses and losses in the ground system. Thus, a radial field used in conjunction with a short transmitting vertical (less than λ/4 tall) requires much more wire than one for a full-size λ/4 or taller monopole if the radiation efficiencies of the two antennas are to be comparable.

Prior to World War II, scientists and engineers performed a series of tests on radial fields for AM broadcast stations. They concluded that anything more extensive than a

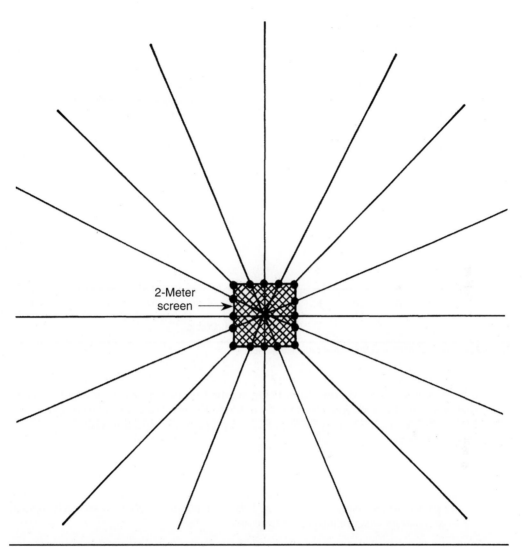

FIGURE 9.8 Comprehensive ground system for vertical antenna.

radial field consisting of 120 quarter-wave radials provided insignificant improvements to radiation efficiency and signal strength. Figure 9.9 is typical of the kind of information that these experiments provided. These test results have formed the basis of much folklore over the years and only recently have new experiments and new reports clarified matters. Today we know that a system of 50 to 60 radials is more than plenty for a quarter-wave ground-mounted vertical in virtually any application, and that the exact length of radials on or in the ground is immaterial.

One of the least important characteristics of radials is wire size. Antenna return currents are split among the many radials, and the conductivity of even tiny wire sizes is far superior to that of all known soils. Bigger issues are survivability, mechanical strength, and visibility. Wires laid on the ground or on grass are ultimately subjected to

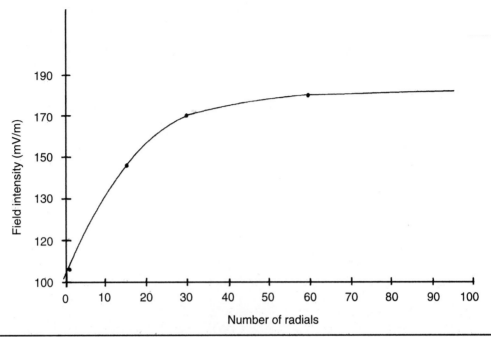

Figure 9.9 Method for connecting many radials together.

a variety of stresses, whether from being stepped on or tripped over, or after easing their way into the sod. Some terrains do not lend themselves to burial techniques, and highly visible insulation may be important to prevent unsuspecting visitors from tripping over them.

Caution *If you decide to use an aboveground radial system, be sure to include measures to prevent people from tripping over it or (if it is an elevated system) running into it. Depending on the law where you live, there can be liability implications when a visitor gets injured, even if that person is an intruder or trespasser.*

The return currents in radials result from displacement currents flowing in the capacitor formed by the vertical element and the ground system. Displacement currents (see Chap. 2) are RF alternating currents that do not require a conducting medium for their existence. Consequently, radials can be either insulated or bare, although bare metals used outdoors will develop an oxide insulating layer on them sooner than insulated conductors. Radials can be copper or aluminum. Many amateurs have formed their radial fields out of scrap wiring from multiple sources, while others have managed to create additional wire by separating multiconductor cable into individual strands. If buying wire new, it's hard to beat the price of 500- or 1000-ft spools of interior house wiring at the large discount home supply stores.

Because of the orientation of the electromagnetic fields around the vertical, only the ground resistance along radial lines from the base of the vertical is important. That is to say, there is little merit to connecting the tips of the radials together or adding other nonradial wire interconnections to the basic radial system. The same is true of ground rods, which, in and of themselves, are virtually useless in improving the performance of grounded verticals.

Typically, all radials are connected together at the base of the antenna, and the ground side of the transmission line is connected to this system. When installing radials, try not to put them too far below the surface of your soil; the farther beneath the surface they are, the less effective they become.

While a radial system can be designed to serve as part of your lightning protection and grounding system (Chap. 30), it does not have to be. Although conventional solder connections are eventually destroyed outdoors and will vaporize in the path of a lightning strike, as long as it's understood they're not "forever", there's nothing wrong with soldering your copper radials to a copper bussbar or flexible plumber's tubing circling the base of your vertical. (See Fig. 9.10.)

FIGURE 9.10   Field strength versus number of radials.

Extremely important, however, is to avoid putting copper in direct contact with the legs or cross-braces of your galvanized tower if that is what you are using for your vertical. Appropriate clamps can often be obtained at electrical supply houses. In any event, sliding a sheet of thin stainless steel between any copper and the tower leg will eliminate the problem.

Some amateurs prefer to place a copper wire screen at the center of the radial system. The minimum size of this screen for it to be useful is a function of the vertical's operating frequency range and the height of the vertical element. A possible concern is the creation of *rectifying junctions* after extended exposure to the elements at each location where two wires in the screen cross.

Feeding the Ground-Mounted Vertical

The feedpoint resistance of a vertical monopole over perfect ground is one-half that of a dipole in free space. Thus, the input impedance of a $\lambda/4$ vertical radiator fed against ground is about 37 Ω. For many applications this is close enough to 52 Ω that no further matching is required, and 52-Ω coaxial cable or hard line can be connected directly to the antenna. If minimizing the standing wave ratio (SWR) on the transmission line is important, the matching network can be a commercially available antenna coupler (protected from the weather by some form of housing) or a simple L-network. In either case, the purpose of the coupler or network is to raise the resistive part of the impedance seen by the transmission line to 50 or 75 Ω while simultaneously canceling out the reactive portion of the antenna impedance. In general, the coupler settings or L-network fixed component values will be suitable for only a narrow range of frequencies; typically, the values or settings used for, say, 40 m will not be useful on any other amateur bands. Of course, direct feed implies that the base of the vertical is not in electrical contact with the ground system beneath it. This is often called *series feed*.

Alternatively, the base of the vertical radiator can be tied to ground and the resulting antenna fed through a tap point on the vertical section some distance above the base. Known as *shunt feed*, many forms of this have been devised and are discussed in detail elsewhere in this book. In the shunt-feed arrangement, the connection partway up the vertical element functions much like an autotransformer. This is best visualized by observing that since the tip of a quarter-wave vertical is a high-impedance point (no current, high voltage) and the base is a low-impedance point (high current, low voltage), the impedance observed along the vertical element gradually increases as one moves the observation point from its base up toward its tip. One advantage of this approach is that the shunt-feed network generally allows for close matching of the feedpoint impedance to an arbitrary transmission line impedance.

There are three methods of shunt-feeding a grounded vertical antenna in common use today: *delta, gamma,* and *omega*. In general, shunt feed is useful when the resistive component of the antenna impedance on the centerline of a balanced element or the junction of the vertical element and ground is significantly lower than the characteristic impedance of the transmission line feeding it.

The delta feed system is shown in Fig. 9.11A. In this case, a taut feed wire is connected between a point on the antenna, which represents a specific impedance on the antenna, and an antenna tuner. At one time, this method of feed was common on AM broadcast antennas. Although you would think that the sloping feed wire would distort the pattern, that is not the case; any such distortion is minimal.

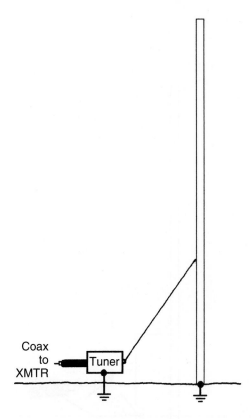

FIGURE 9.11A Delta-fed grounded vertical.

The gamma feed system is shown in Fig. 9.11B. This method is commonly used by amateurs to feed Yagi beam antennas, so it is quite familiar to many in the amateur radio world. In effect, the gamma match and its cousins use *distributed components* to accomplish the same impedance-matching functions a coupler or L-network does in the series-feed case. The two critical dimensions for a gamma match are the distance from the base or centerline of the antenna to the tap point on the antenna (i.e., the active length of the *gamma rod*) and the spacing of the gamma rod away from the antenna element. The lower end of the gamma rod is not grounded; rather, it is connected to the hot side of the transmission line. It is important that the rod not be anywhere near a quarter-wavelength, or it would become a vertical antenna in its own right, and in fact would resemble the *J-pole antenna*.

One advantage of the gamma match and other shunt-feed techniques is that the vertical antenna element retains a direct connection to ground, providing superior static discharge during and ahead of approaching thunderstorms. This is especially useful on the lower amateur frequencies, where verticals are often built using triangular guyed tower sections and it is less expensive to construct the tower with a direct electrical connection (through the tower legs) to ground.

The biggest disadvantage of the gamma matched vertical is that it is much more difficult to initially adjust because the gamma rod tap point may be well beyond the user's reach—especially on the lower bands. A secondary disadvantage is that some methods of feed for multielement phased arrays of verticals use current-forcing techniques that are easier to control with the series feed.

Unless the user has the ability to adjust not only the gamma rod spacing and length but also the overall length of the vertical element, the gamma match may require a simple matching network between the base of the antenna and the transmission line. In the easiest cases, a single series capacitor or inductor between the base of the gamma rod and the hot side of the feedline may be all that is needed to obtain a satisfactory match and SWR. More generally, an *omega match* (Fig. 9.11C) can be inserted, although the user may need to experiment with which side of the series capacitor (C_S) the hot end of the gamma capacitor (C_G) should attach to.

Chapter 18 ("Antennas for 160 Meters") includes details of a gamma match for a top-loaded tower on 160 m. In some difficult cases, when the tower length or available loading is not ideal, an omega match feed system (shown in Fig. 9.11C) may be necessary. The omega match adds a shunt capacitor to the basic gamma match.

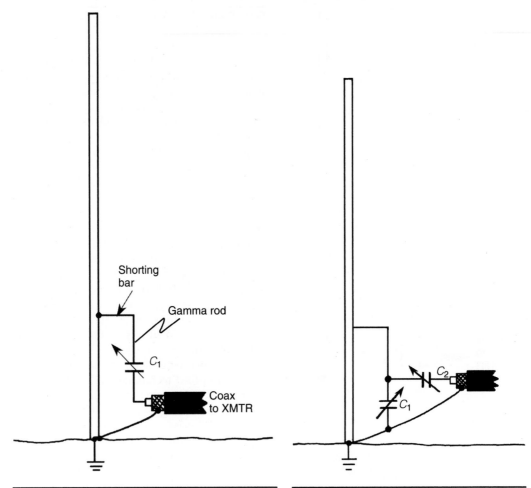

FIGURE 9.11B Gamma-fed grounded vertical.   **FIGURE 9.11C** Omega-fed grounded vertical.

As the radiating element of a vertical monopole increases beyond λ/4, both the resistance and the reactance seen at the series feedpoint increase, as well. Thus, for any significant increase in the length of the vertical radiator, direct matching to commonly available transmission lines will result in increased SWR on the transmission line, so some form of matching network or ATU should be employed. Again, an alternative is to directly ground the base of the radiator and use some form of shunt-feed network to accomplish the proper match.

The base impedance of a ⅝-wavelength antenna is about 1600 Ω—certainly not an acceptable match for anything but a custom-designed transmission line. So some form of impedance matching is needed.

For a single-band ⅝λ antenna, one option is to use a form of *stub matching* employing unbalanced coaxial cable, such as shown in Fig. 9.12. The lengths are given by

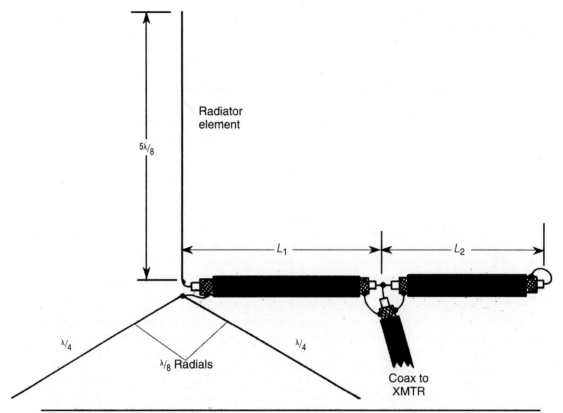

FIGURE 9.12   *Q*-section for match feedpoint impedance to line.

$$L_1 \text{ (feet)} = \frac{122}{F(\text{MHz})} \tag{9.2}$$

or

$$L_1 \text{ (meters)} = \frac{38}{F(\text{MHz})} \tag{9.3}$$

and

$$L_2 \text{ (feet)} = \frac{30}{F(\text{MHz})} \tag{9.4}$$

or

$$L_2 \text{ (meters)} = \frac{9}{F(\text{MHz})} \tag{9.5}$$

Elevated Ground-Plane Antenna

There are three key differences between a ground-mounted vertical and an elevated ground-plane vertical:

- Radials located some distance above ground must be treated as resonant elements.

- Fewer radials are needed to ensure that most of the transmitter power is delivered to the radiating element instead of being dissipated in lossy earth.

- An elevated GP vertical may have fewer nearby obstacles (e.g., trees and buildings) and thus enjoy a cleaner shot at low elevation angles.

Once the base of an elevated ground-plane vertical is raised above a certain height (many consider $\lambda/8$ a reasonable minimum), its radials act less like long, skinny capacitors and more like the other half of a vertical dipole. As such, it is important for maximum radiation plus ease of matching that their lengths to be approximately $\lambda/4$ at the operating frequency, although the exact length will vary somewhat with radial "droop"—the vertical angle at which the radials come away from the base of the monopole.

However, the farther away the lossy earth is from these elevated radials, the less effect it has and the easier it is to capture the bulk of the return currents of the vertical radiating element in a small number of radials. Recent experiments based on detailed modeling using NEC-4 (see Chap. 25) indicate that somewhere between two and eight radials per band is quite adequate for elevated antennas, although unless at least three equispaced radials are used, there will be some variation in field strength with azimuth (compass heading). The improvement in radiated field characteristics for more than three or four radials is quite subtle, however, as long as the GP antenna is kept at least $\lambda/8$ off the ground. Models of a $\lambda/4$ antenna show little difference between horizontal and drooping radials other than to raise the feedpoint impedance from 25 to 40 Ω.

Keep in mind that these discussions of height above earth ground refer to the height of the *electrical* ground, not the sod. Depending upon ground conductivity and groundwater content, the effective height of earth ground may lie some distance beneath the surface. The actual depth is best found from experimentation and may, unfortunately, vary with precipitation and with the season—especially if the ground freezes and/or the local water table changes greatly.

Vertical versus Horizontal Polarization

A question frequently asked is: "Would I be better off putting up a horizontal dipole or a vertical?" As usual, the answer is: "It depends." Here are some factors often found to be helpful in arriving at an answer:

- Is there a "convention" regarding antenna orientation (polarization)? Usually because of historical patterns of usage that have evolved, certain groups have standardized on specific polarizations. The 11-m citizens band, for instance, uses vertical polarization because of the high percentage of vehicular mobile users. As a result, almost all house- or tower-mounted antennas—even multielement Yagis and quads—for 27 MHz are vertically polarized. The same is true for most FM mobile and repeater operation on 2 m and above, whereas weak signal SSB or CW DXing on those bands is accomplished with long-boom, high-gain Yagis employing horizontal polarization. Almost all broadcast television receivers were originally located in residential dwellings, so that service began with horizontal polarization as the standard, today's video-equipped RVs and limousines notwithstanding. Today some FM broadcasting is done with the transmitted signal split between horizontal and vertical polarization, in recognition of the fact that substantial numbers of antennas in both "flavors" are in use.

- What frequencies do you wish to use? What is your objective for a particular antenna? On the HF bands there is generally a 6-dB ground reflection advantage at each end of the path at higher elevation angles for horizontal antennas, but below about 10 MHz verticals often outperform all but the highest dipoles and beams at low radiation angles. Otherwise, the source polarization is immaterial for skip propagation because the ionosphere tends to randomly rotate the polarization of waves passing through it regardless of what was delivered to it. If you are an inveterate DXer, you will most likely enjoy more success on the MF and lower HF bands with a vertical than with a dipole, but if your principal operating activity is a regional or local network of friends, a low horizontal dipole or loop may be your best choice.

- Are you bothered by local noise sources, and, if so, do they favor one polarization over the other? If you can effect an improvement in received *signal-to-noise ratio* (SNR) through the proper choice of polarization, that may be the single most important factor for your specific installation.

- What space or support limitations do you have? A treeless residential backyard may be better suited for an 80-m ground-mounted $\lambda/4$ vertical than for a $\lambda/2$ dipole, while some college dormitory residents have had great success with a simple dipole or end-fed longwire stretched from rooftop to rooftop across the courtyard below.

Perhaps the best answer to this question is: "Put up at least one of each!" Some years ago the author had both a dipole and an elevated ground-plane vertical for 40 m behind his house. In actual use, there were "dipole nights" and "vertical nights" on 40, with received signal strength differences on transatlantic paths shifting back and forth between the two antennas from night to night by as much as 20 dB! No one antenna can do it all; if you have the space and the time to put up two or more antennas.

Supporting the Vertical Antenna

Verticals for frequencies below about 5 MHz are substantial structures; their support requirements are akin to those of 60-ft or greater towers. Therefore, information on supporting verticals for the MF and lower HF bands is found in Chap. 29 ("Towers").

Above 5 MHz, supporting a $\lambda/4$ vertical is a much simpler task. For these antennas, the techniques of Chap. 28 ("Supports for Wires and Verticals") are typical.

Directional High-Frequency Antenna Arrays

CHAPTER 10

Wire Arrays

When a single wire gets substantially longer than a half-wavelength, or when multiple wires are employed, we enter the realm of wire arrays. In general, the purpose of using an array instead of a simpler antenna is to get higher forward gain, better rejection off the back and/or sides, or a combination of both. Few antenna projects bring as much reward at so low a cost as wires, and that's especially true of wire arrays.

Longwire Antennas

The *end-fed longwire* (or, more simply, *longwire*) is a name given to any of several types of resonant and nonresonant antennas. The longwire antenna is capable of providing gain over a dipole and (if it is long enough) a low angle of radiation (which is great for DX operators!). But these advantages become substantial only when the antenna is several wavelengths long. Nonetheless, longwire antennas have been popular for decades—often because the layout of a particular property lends itself far more readily to locating one end of an HF wire close to the radio room, if not actually in it!

Any given longwire antenna may be either resonant or nonresonant, depending upon the operating frequencies used. In the old days, most longwires for use on the amateur MF and HF bands could be cut to be reasonably close to resonant on all the HF bands because those bands were harmonically related to each other. But with the addition of the 5-, 10-, 18-, and 24-MHz band segments, that's not possible anymore.

Figure 10.1 shows the classic random length, nonresonant longwire antenna. It consists of a wire radiator that is at least a quarter-wavelength long but is most often much longer. The specific length is not critical, but by convention it is usually assumed to be a wavelength or more at the lowest frequency of operation anticipated. Nonetheless, a 70-ft wire fed at one end against a good RF ground will acquit itself well on all HF bands above 3.5 MHz.

The longwire is end-fed. Therefore, when it is an even number of quarter-wavelengths it will exhibit a very high feedpoint impedance (a few thousand ohms), while for lengths that are odd multiples of $\lambda/4$ it will present a much lower impedance (perhaps 25 to 100 Ω). Thus, it is imperative that an antenna tuner (ATU) be used, since even those transceivers with built-in antenna tuners typically cannot match the very high feedpoint impedances of end-fed wires that are approximately a multiple of $\lambda/2$ in length.

Often, the most important factor affecting the performance of an end-fed longwire is not the antenna—it's the *ground* system it's fed against! Although we may be aware

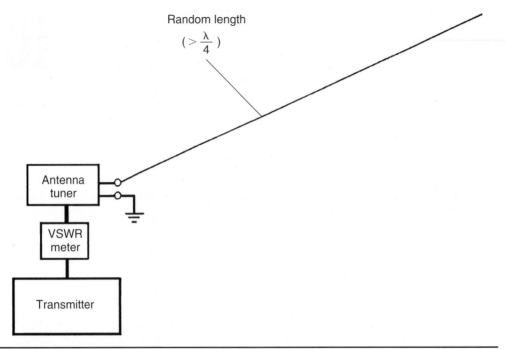

Random length

$$(> \frac{\lambda}{4})$$

FIGURE 10.1 Random-length antenna.

that a $\lambda/4$ monopole radiates well only with the help of the ground system beneath it, *any* antenna with a single-wire feedline depends on the nearby ground to pass the *return currents* back to the transmitter circuits. If the transmitter is trying to push electrons into a longwire on each half-cycle of RF, it must necessarily be receiving or pulling an equal number of electrons from a return circuit of some kind. In the worst case, where no attention at all is paid to the ground return paths, the transmitter typically gets its return path from a mix of anything and everything that's connected to the chassis and the ground of the power amplifier: ac power line, mic cable shield, audio cable shields, etc. Lack of attention to an appropriate ground return path is a major reason why equipment in a radio transmitting environment becomes "hot" whenever the transmitter is operated. The issue is especially relevant to the installation of a longwire antenna, since, in many cases, the fed end of the antenna actually comes right into the radio room.

Failure to include an adequate ground return for a longwire raises the apparent feedpoint impedance and causes a large portion of the transmitter output power to be dissipated in the resistive losses of other ground return paths rather than in the radiation resistance of the antenna. Ideally, many radials of lengths greater than or equal to $\lambda/4$ at the lowest frequency of operation, as shown in Fig. 10.2, should be included as part of any longwire antenna project.

The kind of ground we're talking about is not at all like the power company's ground, which is a safety ground. The longwire requires an RF ground, which is discussed in detail in Chap. 30. At the very least, a $\lambda/4$ radial can be connected to the

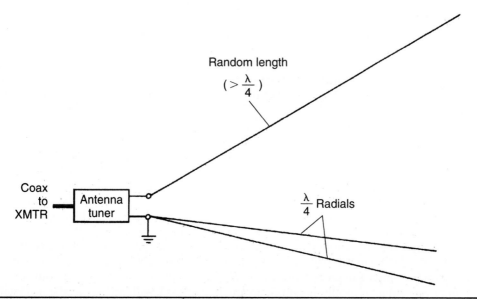

FIGURE 10.2 Radials improve the "ground" of random-length antenna.

chassis of the ATU (or the transmitter if no ATU is used) and run outside. Of course, this effectively converts your longwire into an off-center-fed wire of unpredictable performance, but it's far better than having no defined ground at all.

If you intend to use your longwire on more than one band, a single radial is of limited help. However, at least one company (MFJ) sells a tuner for the *artificial* ground. The MFJ model *MFJ-931* artificial RF ground is installed between the radial wire and the ATU or transmitter chassis ground connection, as shown in Fig. 10.3. With every band change, the tuning controls will need to be adjusted for maximum ground current as indicated by the built-in meter. Keep in mind, however, that use of an artificial ground—with or without a tuner—falls in the category of trying to make the best of a bad situation.

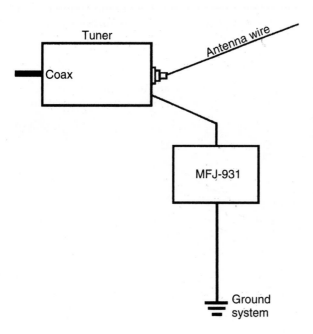

FIGURE 10.3 MFJ ground line tuner installation.

Longwire Pattern and Gain

As the number of wavelengths in an end-fed wire increases—either by using a fixed-length wire on a higher frequency or by lengthening the wire while keeping the operating frequency fixed—both the pattern and the direction of maximum gain change, as well. At $\lambda/2$ the pattern of an end-fed wire is the broadside "doughnut" of a $\lambda/2$ dipole, perhaps very slightly skewed as a result of being fed at one end instead of in the center. At a length of λ, the doughnut-shaped figure eight pattern has broken into a four-lobed cloverleaf with peaks at 45 degrees from the antenna's axis, and the maximum gain of the antenna (in the center of each lobe) has increased by almost 1 dB. As the frequency is further increased, the pattern breaks into more and more lobes, such as that depicted in Fig. 10.4 for a 2λ longwire, but the broadest of these maintain a gradual "march" toward the axis. At the same time, the peak amplitude of these main lobes gradually increases, reaching an additional 5 dB compared to the $\lambda/2$ wire at a length of 4λ, which corresponds to using an 80-m 135-ft $\lambda/2$ dipole on 10 m. Figure 10.5 plots the increase in gain of the main lobe relative to a $\lambda/2$ dipole as a function of longwire length.

Regrettably, most longwires are fixed in one location; if the longwire is your only HF antenna, having the direction of peak signal gradually precess from broadside to nearly 90 compass degrees away while changing frequency may not suit your operating needs. More typically, however, an amateur or SWL with acreage available installs multiple longwires for a preferred band. An ideal configuration might have the fed end of all the wires located at a common point where an ATU is located and can be switched easily from one wire to another.

A rarely mentioned problem with very long wires: Electrostatic fields build up a high-voltage dc charge on longwire antennas! Thunderstorms as far as 20 mi away are known to produce substantial charge build-up on long antennas, capable of causing serious damage to a receiver or its operator! A simple solution is to connect a transmitter-rated *RF choke* (RFC) between the antenna terminal and the ground connection of the ATU. In a pinch, a dozen 10-MΩ, 2W resistors in parallel can be substituted.

Nonresonant Longwire Antennas

The resonant longwire antenna is a *standing wave antenna* because it is unterminated at the far end. A signal propagating from the feedpoint toward the open end will be reflected back toward the transmitter from the open end; the resulting superposition of

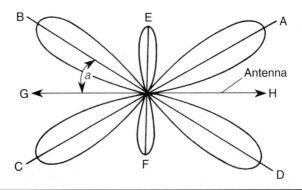

FIGURE 10.4 Radiation pattern of longwire antenna.

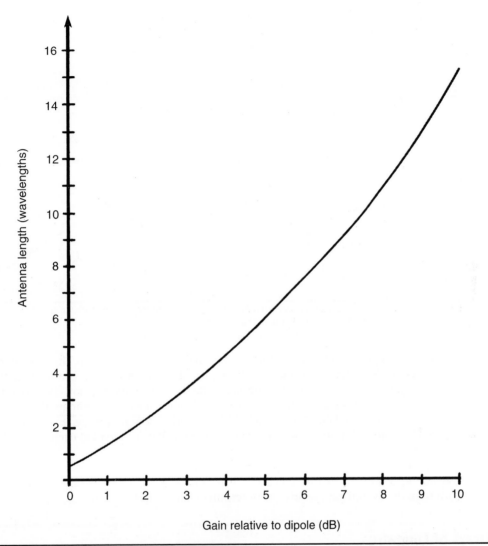

FIGURE 10.5 Antenna length versus gain over dipole.

the forward and reflected waves sets up stationary (i.e., "standing") currents and voltages along the wire, similar to the process described in Chap. 3.

In contrast, a *nonresonant longwire* is terminated at the far end in a resistance equal to its characteristic impedance. (See Fig. 10.6.) Thus, the outward-bound wave is absorbed by the resistor, rather than being reflected. Such an antenna is called a *traveling wave antenna*. Terminating resistor R_1 is selected to be equal to the characteristic impedance Z_0 of the antenna (i.e., $R = Z_0$). When the wire is 20 to 30 ft above average ground, Z_0 is about 500 to 600 Ω.

The radiation pattern for the terminated longwire is a unidirectional version of the multilobed pattern found on the unterminated longwires. The orientation of the lobes

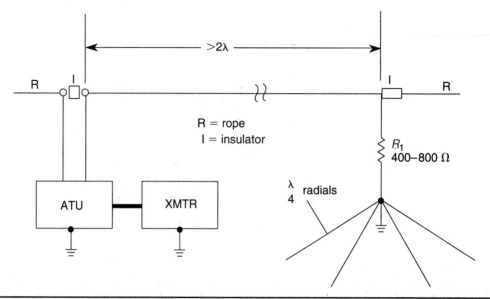

Figure 10.6 Longwire (greater than 2λ) antenna is terminated in a resistance (typically 400 to 800 Ω). Radials under the resistor improve the grounding of the antenna.

varies with frequency, even though the pattern remains unidirectional. The directivity of the antenna is partially specified by the angles of the main lobes. It is interesting to note that gain rises almost linearly with $n\lambda$, while the directivity function changes rapidly at shorter lengths (above three or four wavelengths the rate of change diminishes considerably). Thus, when an antenna is cut for a certain low frequency, it will work at higher frequencies, but the directivity characteristic will be different at each end of the spectrum of interest.

Of course, if a terminated longwire is used for transmitting, approximately half the transmitter power will be dissipated in the termination resistor, so one should subtract 3 dB from the overall gain relative to unterminated antennas.

Arrays of Longwires

Longwire antennas can be combined in several ways to increase gain and sharpen directivity. Two of the most popular of these are the *vee beam* and the *rhombic*. Both forms can be made in either resonant (unterminated) or nonresonant (terminated) versions. Since a 1λ or longer longwire is itself an array, these antennas can be considered to be *arrays of arrays*.

Vee Beams

The *vee beam* (Fig. 10.7) consists of two equal-length longwire elements (Wire 1 and Wire 2), fed 180 degrees out of phase and oriented to produce an acute angle between them. The 180-degree phase difference is provided automatically by connecting the two wires of the vee to opposing sides of a single parallel conductor feedline.

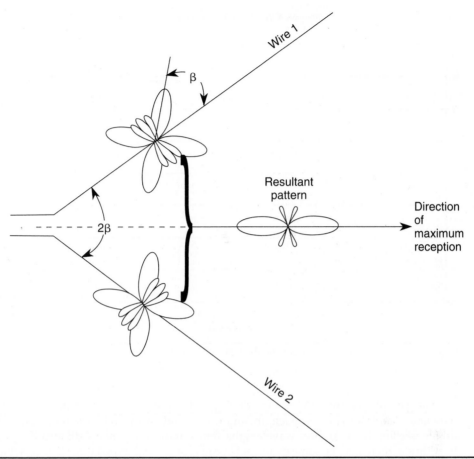

FIGURE 10.7 Radiation pattern of vee beam antenna consists of the algebraic sum of the two longwire patterns that make up the antenna.

  The unterminated vee beam of Fig. 10.7 has a bidirectional pattern that is created by summing together the patterns of the two individual wires. Proper alignment of the main lobes of the two wires requires an included angle, between the wires, of twice the radiation angle of each wire. If the radiation angle of the wire is β, then the appropriate included angle is 2β. To raise the pattern a few degrees, the 2β angle should be slightly less than these values. It is common practice to design a vee beam for a low frequency (e.g., 75-/80- or 40-m bands), and then to also use it on higher frequencies that are harmonics of the minimum design frequency. A typical vee beam works well over a very wide frequency range *only* if the included angle is adjusted to a reasonable compromise. It is common practice to use an included angle that is between 35 and 90 degrees, depending on how many harmonic bands are required.

  Vee beam patterns are usually based on an antenna height that is greater than a half-wavelength from the ground. At low frequencies, such heights may not be possible, and you must expect a certain distortion of the pattern because of ground reflection effects.

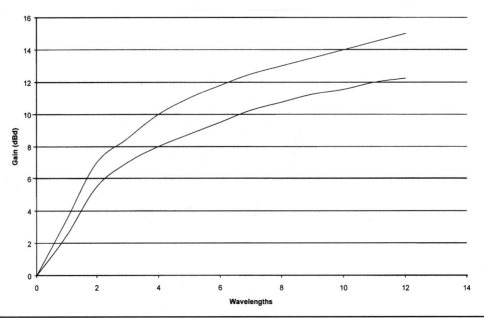

FIGURE 10.8   Gain versus length of vee beam and rhombic beam antennas.

The gain of a vee beam is about 3 dB higher than the gain of the single-wire long-wire antenna of the same wire length, and it is considerably higher than the gain of a dipole (see Fig. 10.8). At three wavelengths, for example, the gain is 7 dB over a dipole. In addition, there may be some extra gain because of mutual impedance effects—which can be about 1 dB at 5λ and 2 dB at 8λ.

Nonresonant Vee Beams

Like single-wire longwire antennas, the vee beam can be made nonresonant by terminating each wire in a resistance that is equal to the antenna's characteristic impedance (Fig. 10.9). Although the regular vee is a standing wave antenna, the terminated version is a traveling wave antenna and is thus unidirectional. Traveling wave antennas achieve unidirectionality because the terminating resistor absorbs the incident wave after it has propagated to the end of the wire. In a standing wave antenna, any energy reaching the far end is reflected back toward the source, so it can radiate oppositely from the incident wave.

Rhombics

The *rhombic antenna*, also called the *double vee*, consists of two vee beams positioned to face each other with their corresponding tips connected. The unidirectional, nonresonant (terminated) rhombic shown in Fig. 10.10 develops approximately the same gain and directivity as a vee beam of the same size. The nonresonant rhombic has a gain of about 3 dB over a vee beam of the same size (see Fig. 10.8 again).

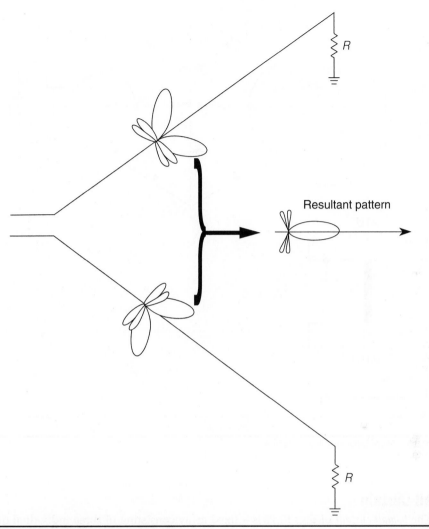

FIGURE 10.9 Nonresonant vee beam antenna.

The layout of the rhombic antenna is characterized by two angles. One-half of the included angle of the two legs of one wire is the *tilt angle* (φ), while the angle between the two wires is the *apex angle* (θ). But the two are not independent; from Fig. 10.10 it should be obvious that θ/2 = 90 − φ. A common rhombic design uses a tilt angle of 70 degrees, a length of 6λ for each leg (two legs per side), and a height above the ground of 1.1λ. θ for this antenna must necessarily be 40 degrees.

The termination resistance for the nonresonant rhombic is 600 to 800 Ω, and it must be noninductive across the entire range of operating frequencies. For transmitting rhombics, the resistor should be capable of dissipating at least one-third of the average power of the transmitter. For receive-only rhombics, the termination resistor can be a 2W carbon composition or metal-film type. Such an antenna works nicely over an octave (2:1) frequency range.

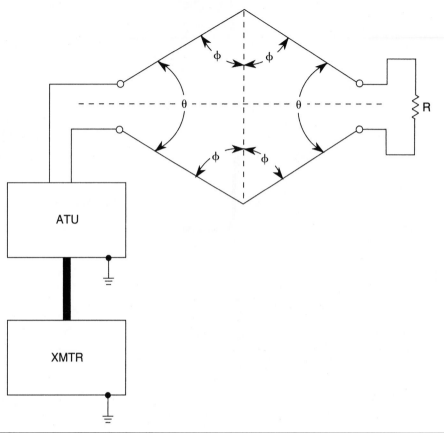

FIGURE 10.10 Terminated rhombic antenna.

Bobtail Curtain

The *bobtail curtain* of Fig. 10.11 is a fixed array consisting of three individual quarter-wavelength vertical elements in a line; adjacent elements are spaced a half-wavelength apart and attached electrically to a horizontal wire across their tops that serves both as a phasing wire and as the other half of an L-shaped dipole. The array is fed at the bottom end of the center element. This is a high impedance point, so a parallel-tuned tank circuit or an ATU capable of matching very high impedances is required.

The array must be kept high enough in the air that the bottoms of the λ/4 vertical elements are some distance above ground—ideally high enough in the air that humans and animals cannot touch their high-voltage tips, although, clearly, the feed wire to the center element may require special protection. A bobtail curtain for 80 m should thus be suspended about 75 or 80 ft up; with catenary effects causing the phasing line to sag, end supports of at least 100 ft are appropriate.

RF energy introduced at the bottom of the center element splits and goes equally to the two end elements when it reaches the phasing line at the top. Since each end element is λ/2 from the center element, the currents of all three elements are in phase, although the current in the center element is twice that of either of the others. (This is an

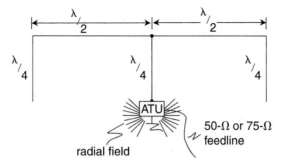

FIGURE 10.11 Bobtail curtain.

example of a 1:2:1 binomial array, as discussed in Chap. 5.) Since the currents feeding the two end elements are traveling in opposite directions along the top, and since the current to either end element goes through a reversal at the middle of the horizontal half-section, the horizontal radiation is partially canceled out.

Because all three elements are in phase, the principal lobe is *broadside* to the plane of the three radiating elements. For maximum signal north and south, for instance, the array should be suspended between supports on an east-west line.

The radiating elements are λ/4 monopoles and normally would require a low-resistance ground path for their return currents. However, if each vertical element is visualized as one side of a λ/2 dipole and half the horizontal distance along the phasing line to the adjacent element is seen as the other side of the dipole (or as an elevated groundplane), the vertical elements themselves have no particular need for a good ground system. However, unless there is an excellent RF ground at the base of the middle element for the matching network to connect to, there is no low-loss return path for RF currents flowing to/from the ground end of the matching network, and excessive transmitter power will be wasted in some vague but very lossy return path. Thus, the matching network at the base of the middle element may require a good system of radials. Since radials lying on the surface of the earth are not resonant, their exact length is not particularly critical. It might be helpful, depending on the quality of your ground, to put radial fields under the end elements, as well.

The bobtail curtain has an extremely low angle of radiation. At one time the author had an 80-m bobtail curtain aimed north and south at his home in the northeastern United States; it was a "killer" antenna for very long haul contacts over the north pole into the far reaches of Asiatic Russia and the western Pacific region—especially Indonesia—but it was totally useless for anything much closer.

Half-Square

If one end element of the bobtail curtain of Fig. 10.11 and its associated phasing line are removed, the resulting antenna is known as a *half-square*. Now we have two vertical radiators joined by a λ/2 horizontal phasing line. As with the original bobtail curtain, the antenna is fed at the bottom of one element—a high impedance point. (Keep in mind that both the fed end and the far end are high-voltage points.) We can consider that element and half the horizontal wire as constituting an L-shaped λ/2 dipole feeding a second λ/2 dipole. But since the current in consecutive λ/2 sections reverses, the instantaneous current in the second vertical element is always pointed in the same direction—up or down—as that of the first.

Thus the half-square is a two-element broadside array. If it is hung from east and west supports, maximum radiation is north and south. Off either end, contributions by the two vertical radiators are 180 degrees out of phase, thanks to the $\lambda/2$ spacing between them, so the array has a null to the east and west. Because the currents in the two halves of the horizontal wire are also out of phase, they mostly cancel and the array exhibits relatively little high-angle radiation.

Over average ground, the half-square has about 4 dB additional gain broadside to the axis of the antenna compared to a single $\lambda/4$ vertical, and the peak of the main lobe occurs at a lower elevation angle (20 degrees versus 23 degrees). As before, each $\lambda/4$ vertical monopole can utilize half the horizontal wire as a radial, but overall system efficiency and the specifics of the matching network may dictate the use of radials directly connected to the ground and chassis of the ATU that feeds the vertical wire. Adding radials under both vertical elements adds about 0.4 dB to the low-angle broadside gain over average earth.

20-m ZL-Special Beam

The antenna shown in Fig. 10.12 is a close relative of the Yagi beam. It consists of a pair of folded dipoles mounted approximately 0.12 wavelengths apart. Each element is 30.5 ft in length, and the spacing between the two is 7.1 ft. Each element can be made of two parallel wires separated with spacers or from 300-Ω television-type twin lead. Alternatively, if the antenna is to be rotatable, wires inside small PVC tubes supported at their ends by a truss assembly can be used, as can spaced aluminum tubing.

The two $\lambda/2$ folded dipole elements of the ZL-special are fed 135 degrees out of phase with respect to each other. The feedline is connected to one of the dipoles directly; the other is fed through a length of 300-Ω twin lead that has an electrical length of about 45 degrees ($\lambda/8$) and has been given a half-twist to add another 180 degrees of phase shift. Its physical length will depend on the velocity factor v_F of the particular line used.

The feedpoint impedance is on the order of 100 to 150 Ω, so the antenna is a good match for either 52-Ω or 75-Ω coaxial cable if a 2:1 impedance matching transformer is used.

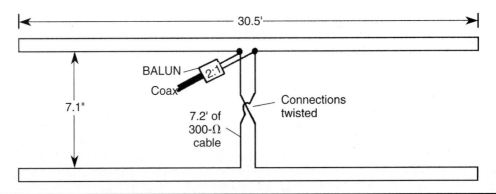

FIGURE 10.12 ZL-special beam antenna.

CHAPTER 11

Vertical Arrays

Despite its longevity, the vertical antenna is either praised or cursed by its users, depending upon their experiences with it. At HF in particular, "DXability" is often the criterion for judging the antenna's performance. As amateurs and others have come to appreciate the importance of a good RF ground system to the vertical's performance, in recent years the vertical has come into its own on the 1.8- and 3.5-MHz bands because it provides superior low-angle radiation compared to the low dipoles that are the only alternative for most users.

But a vertical antenna is *omnidirectional* in its *azimuthal* (compass rose) coverage; that is, it transmits and it "hears" equally well in all directions. Thus, a single vertical exhibits two weaknesses:

- It puts far more RF energy out in directions that are of no use (at that particular moment, at least) than it does in the desired direction.

- It does not have any way of eliminating noise coming from all 360 degrees around the compass when trying to receive a weak signal from a specific heading.

With respect to the first point, there are only three things the user with a single vertical can do to improve his/her signal at a specific faraway receiving location:

- Increase the transmitter power to the antenna.

- Improve the antenna efficiency by reducing resistive losses, especially by making sure an adequate system of radials is installed at the base of the antenna.

- Increase the electrical length of the vertical radiator (up to a maximum of $5\lambda/8$). This causes the low-angle field strength for a given transmitter power to increase because high-angle radiation is being reduced as the vertical is lengthened.

With respect to the second point, we note:

- Atmospheric band noise arrives from compass headings where the band is "open". That is, if it's shortly after sunset at the site of the vertical antenna, atmospheric noise on the MF and lower HF bands is likely coming only from points east of the antenna, where darkness and ionospheric propagation are prevalent. To the west, daylight absorption in the lower layers of the ionosphere substantially reduces the noise levels at the vertical. This is true of the noise ("QRN" in the long-established radio shorthand known as *Q-signals*) from distant thunderstorms.

- Nearby thunderstorms, on the other hand, can be heard equally well regardless of their location relative to the vertical, since the propagation mode is likely direct line of sight from clouds in the region to the antenna.

- Man-made noise ("QRM") is usually (but not always) nearby and unaffected by ionospheric propagation conditions. Today most man-made QRM originates in the high-speed microprocessors and digital circuits that infest all the appliances and electronics gear in our homes and offices or those of our neighbors. In general, we have little or no control over the direction of these sources from our vertical, so we can assume that their compass headings are likely to be randomly distributed.

Directivity and Phasing

We can improve our ability to hear and to be heard on any frequency—but especially on the MF and lower HF bands—where we are using a vertical antenna by operating two or more in an *array*. As we saw in Chap. 5, arrays can provide increased field strength in some directions at the expense of strength in other directions. This holds for both transmitting and receiving, so an array of vertical antennas can help overcome all but one of the limitations itemized in the preceding bullet list. (An array will not help us reduce atmospheric noise or QRM coming from the same direction as a weak signal we are trying to copy.)

Most AM broadcast stations have used arrays of vertical antennas for decades. Consequently, the best radio engineers of the past century have collectively designed and evaluated far more combinations of verticals than any one author could do on his or her own in a lifetime, and there is much to be learned and borrowed from the AM broadcast band favorites. Many standard patterns for two-, three-, four-, and five-vertical arrays dating from before World War II exist in the literature. Most of the more complicated ones are irrelevant to anything other than the AM broadcast band, but the simpler combinations have many virtues useful to amateurs and others.

Of course, amateurs and shortwave or broadcast band listeners usually have a requirement that AM broadcast stations do not: Most arrays must be designed so they can be switched or *steered* to more than one compass heading.

To do the subject of vertical arrays full justice would require a separate book at least as big as this one. So this chapter will serve primarily as an overview of some simple but effective arrays that can be formed with the proper placement and feed systems.

In general, an array of verticals involves two or more antennas in close proximity, each fed at a specified phase angle and amplitude relative to a common reference in order to produce the desired radiation pattern. Often the common reference is simply the RF applied to one of the antennas in the array that has been selected to serve as the reference point.

Two-Element Array

Figure 11.1 shows the possible patterns for a pair of vertical antennas spaced a half-wavelength (180 degrees) apart and fed *in phase* or (180 degrees) *out of phase*. Although other phasing relationships between the drive currents to the two antennas are possible, we'll stick with these because of the simplicity of the networks required to feed them.

When the two verticals (A and B) are fed in phase with equal currents, the radiation pattern of Fig. 11.1A (idealized here) results. It is a bidirectional figure eight that is maximum *broadside* to a line drawn between the two antennas. If there are no nearby obstructions and the ground around the two verticals is reasonably flat and featureless, and if the drive currents to the two antennas are well matched in both amplitude and phase, a deep null is formed along the axis of the array—i.e., along the line drawn between the elements A and B.

If the phase of the drive current to *one* of the elements is shifted by 180 degrees, the pattern rotates 90 degrees (a quarter of the way around the compass) and now exhibits directivity along the line of the centers (A-B), as depicted in Fig. 11.1B. This is often called an *end-fire* pattern or an *end-fire array.*

Note that the forward gains and the exact shapes of the broadside and end-fire patterns are seldom the same.

The simplest way to make sure the amplitudes of the feed currents to the two array elements are identical is to bring a common feedline from the transmitter to a point midway between the two elements and then feed each element from there. Figure 11.2A shows the feedline configuration for driving the array elements in phase; note, in particular, that lengths L_1 and L_2 must be the same, and ideally they should be cut from the same roll of coaxial cable or hardline so that their impedances and velocities of propagation are as closely matched as possible.

To switch the two-element array to its end-fire mode of Fig. 11.1B an extra 180-degree phase shift must be added to the drive signal delivered to one or the other (but not both) of the two elements. Although this phase shift can be provided with a discrete component LC network, a simple solution is to add an extra $\lambda/2$ section of the same cable in the line to one element. When doing so, it's important to include the *velocity factor* (v_F) of the cable in the calculation. v_F is a decimal fraction on the order of 0.66 to 0.90, depending upon the specific transmission line used.

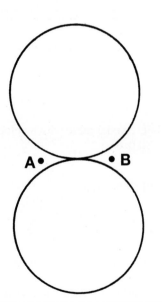

FIGURE 11.1A Pattern of two radiators fed in phase, spaced a half-wavelength apart.

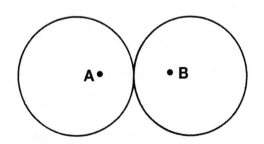

FIGURE 11.1B Pattern of two radiators fed out of phase, spaced a half-wavelength apart.

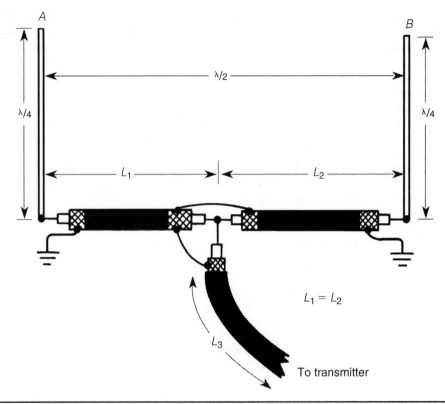

FIGURE 11.2A   Feeding a phased array antenna in phase.

Its effect is to cause a line that is electrically $\lambda/2$ to be much shorter than $\lambda/2$ physically. In particular:

$$L_{\text{PHYSICAL}} = v_F L_{\text{ELECTRICAL}} \tag{11.1}$$

Example 11.1   Assuming the use of standard RG-8U coaxial cable with a velocity factor of 0.67, a half-wavelength section of cable for 3.6 MHz is

$$L_{\frac{\lambda}{2}\text{PHY}} \text{ (feet)} = \frac{492\, v_F}{F\,(\text{MHz})} = 91.57\ \text{ft} = 91\ \text{ft}\ 7\ \text{in} \tag{11.2}$$

◆

If we are interested only in the end-fire pattern of Fig. 11.1B, we can feed the array as shown in Fig. 11.2B, where it is understood that the lengths shown for the segments of coaxial cable are *electrical* lengths, not physical lengths. If, instead, we wish to be able to switch back and forth easily between broadside and end-fed modes, we can use the circuit of Fig. 11.3. Here two convenient, but equal, lengths of coaxial cable (L_1 and L_2) are used to carry RF power to the antennas. One segment (L_1) is fed directly from the

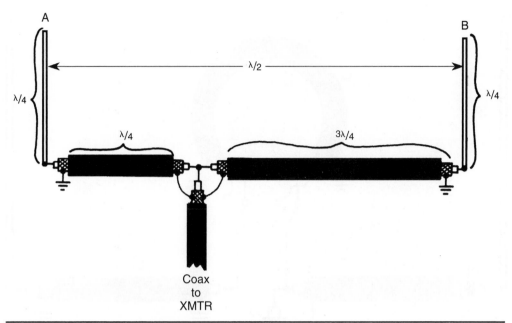

FIGURE 11.2B Minimum length feed for a phased array antenna.

transmitter's coaxial cable (L_3), while the other is fed from a phasing switch. The phasing switch is used to either bypass or insert a phase-shifting length of coaxial cable (L_4).

In principle this technique allows the selection of varying amounts of phase shift from 45 degrees to 270 degrees. However, at other than 0 degrees and 180 degrees, the currents no longer split equally between the two elements because the element feedpoint impedances are not the same. In general, for angles other than 0 degrees or 180 degrees, more complicated feed systems, including *current-forcing* techniques, are employed.

When unequal lengths of transmission line go to the elements of an array, unequal amplitudes are a likely by-product. The most noticeable effect for small differences, such as would occur when adding a $\lambda/2$ section of cable to implement the end-fire mode, will be to cause the nulls to fill in and lose some of their depth.

With the addition of a *phasing transformer,* a two-element vertical array can be remotely switched between broadside and end-fire modes while preserving balanced drive currents at the two elements. Figure 11.4 shows how a two-element array is fed through such a two-port phasing transformer. The phase-reversing switch S_1 can, of course, consist of coaxial relays controlled from the transmitter end of the transmission line. The transformer itself is made from a 1:1 toroidal balun kit such as those available from Amidon Associates and others.

Wind the three coils in trifilar style, according to the kit instructions. The dots in Fig. 11.4A show the "sense" of the coils, and they are important for correct phasing; call one end the "dot end" and the other end the "plain end"—and mark them differently—to keep track of them. If the dot end of the first coil is connected to J_3 (where the feedline from the transmitter or receiver attaches), connect the dot end of the second coil to the 0-degree output (J_1, which goes to array element A). The third coil is connected to J_2 through a DPDT low-loss RF relay or switch rated for the power levels involved. Al-

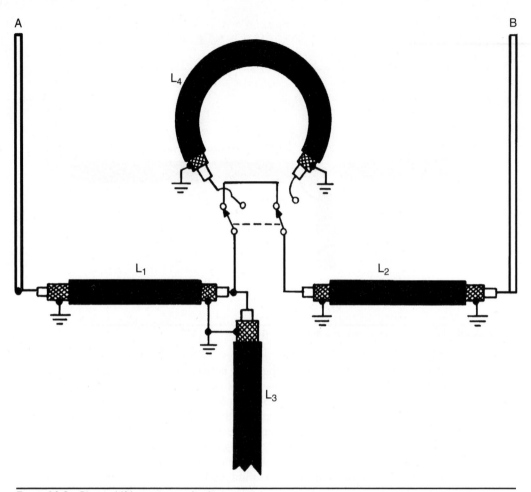

FIGURE 11.3 Phase-shifting antenna circuit.

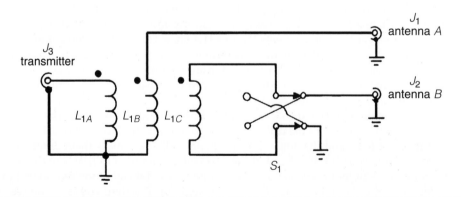

FIGURE 11.4A Phasing transformer circuit.

though the transformer adds a small amount of loss, it is applied equally to both elements, so the amplitudes of their drive currents remain equal regardless of which mode the array is operated in.

Be very sure, as shown in Fig. 11.4B, the coaxial cables to the individual elements are identical lengths from the point where they connect to J_1 and J_2 on the transformer housing to the base of either element.

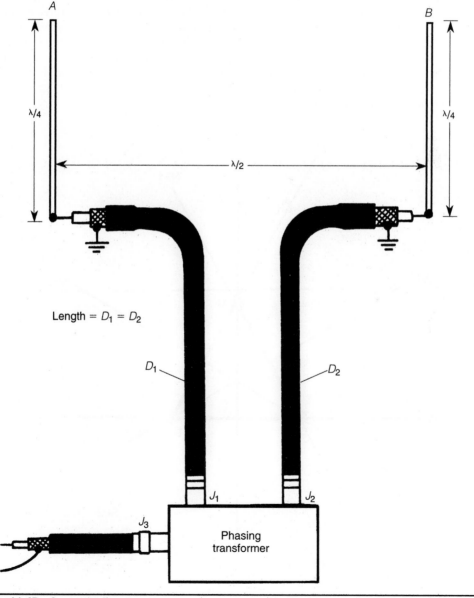

FIGURE 11.4B   Connection to antennas.

360-Degree Directional Array

The phased vertical antenna concept can be used to provide round-the-compass steering of the antenna pattern. Figure 11.5 shows how three λ/4 verticals (arranged in a triangle that is a half-wavelength on each side) can be used to provide unidirectional and bidirectional patterns through combinations of in-phase, out-of-phase, and grounded elements. For a specific pattern, any given element (A, B, or C) is grounded ("passive"), fed at 0-degree phase, or fed with 180-degree phase shift.

Table 11.1 lists for each compass direction the drive to each of the three elements, the maximum forward gain of that configuration and the elevation angle at which it occurs, and a rough *front-to-back* (F/B) number. The figures in Table 11.1 are specifically for a three-element 80-m array with identical 66-ft verticals spaced 136 ft (λ/2) apart. The design frequency is 3.6 MHz. Each element has fifteen 65-ft radials at its base. Any time an element is labeled "Passive" in the table, it is unfed and directly grounded to the radial field beneath it.

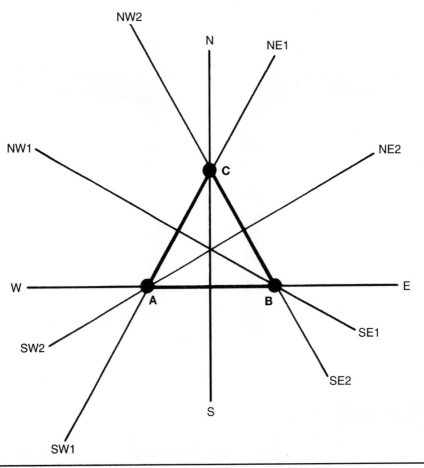

FIGURE 11.5 Three-element phased array.

A Phase	B Phase	C Phase	Heading (Degrees)	Gain (dBi)	Elev.	F/B (dB)
0°	0°	Passive	S (180°)	5.4	28°	3.8
Passive	0°	0°	NE2 (060°)	5.4	28°	3.8
0°	Passive	0°	NW1 (300°)	5.4	28°	3.8
0°	180°	Passive	E&W (090° and 270°)	3.1	23°	0
0°	Passive	180°	NE1 & SW1 (030° and 210°)	3.1	23°	0
Passive	180°	0°	SE2 & NW2 (150° and 330°)	3.1	23°	0
0°	Passive	Passive	SW2 (240°)	4.2	28°	10
Passive	0°	Passive	NW1 (300°)	4.2	28°	10
Passive	Passive	0°	N (000°)	4.2	28°	10
0°	0°	180°	E&W (090° and 270°)	4.8	23°	0
0°	180°	0°	SE2 & NW2 (150° and 330°)	4.8	23°	0
180°	0°	0°	NE1 & SW1 (030° and 210°)	4.8	23°	0
0°	0°	0°	Omnidirectional	1.8–2.0	44°	0

TABLE 11.1   Pattern Gains for Three-Element Equilateral Array of Fig. 11.5.

In practice, at any specific location the triangular *footprint* of the three verticals can be rotated up to 60 degrees in either direction relative to the compass headings shown to orient the best patterns where they can be of most help. (Notice that each type of pattern repeats every 120 degrees.) Some of the patterns listed in the table have little to recommend them, but, as a general rule, most amateur DXers on the MF and HF bands find that breaking the 360-degree compass rose into six 60-degree directions is more than adequate.

If space is at a premium, or if multiband operation of an array of *trap verticals* is desired, the same equilateral triangle can be used with shorter spacings between the verticals. For years, Hy-Gain published an application note that detailed how to feed three of their *18-AVQ* trap verticals spaced $\lambda/8$ apart at the lowest frequency (80 m) they covered. One of the difficulties with spacing other than $\lambda/2$ is that 180-degree phase shifts in drive current are no longer matched by the phase shift undergone by the radiated wave as it heads from one element to another; instead, spatial phase shifts of 45 degrees and 90 degrees are involved on 80 m and 40 m, respectively, and multiples of 360 degrees (i.e., in phase) on bands above 20 m. Nonetheless, directivity and gain are possible with such a system.

Four-Square Array

Another popular way to obtain directivity that can be switched to multiple compass headings is the *four-square* (or *4-square*) array. In its original implementation, four $\lambda/4$ vertical monopoles are located at the corners of a square that is $\lambda/4$ on a side. Drive currents of equal amplitudes are applied to all four elements, but with three different phases. To develop maximum gain along a diagonal of the square, if the rear vertical on the diagonal is defined to have zero phase shift, the front vertical (i.e., the one in the direction of maximum gain) is –180 degrees and the two side verticals are both –90 degrees. Obviously, all aspects of the array's performance can be identical every 90 degrees around the compass, simply by switching the phase of the drive current going to each specific vertical.

An amateur DXer in the northeastern United States would likely orient his or her 4-square so as to put the center of the four switchable pattern lobes at 60-, 150-, 240-, and 330-degree compass headings to favor Europe/North Africa, South America, Australia/New Zealand, and the Far East, respectively.

Other variations of the 4-square utilize somewhat different combinations of phase shifts and/or smaller spacings between elements. In most cases, 4-squares are constructed with ground-mounted monopoles and extensive radial fields, but today many employ elevated radial systems. If the user has the means to "force" drive currents of arbitrary phase and amplitude relationships to the four elements, there's virtually no limit to the number of customized azimuth patterns that can be obtained.

Receive-Only Vertical Arrays

Much work has been done in the past few years on phased arrays for receiving at MF and the lower HF bands. In stark contrast to the challenges of weak signal reception at VHF and above, the abundance of atmospheric noise below, say, 10 MHz gives the system designer additional leeway in selecting techniques for digging weak signals out of the background noise. At these low frequencies, receive-only arrays can differ from their transmitting antenna counterparts in two important ways:

- Because receive-only arrays are not used for transmitting, antenna *efficiency* is of only secondary importance; as a result, array elements much shorter than λ/4 and having very low feedpoint impedances as well as inexpensive tubing and support requirements can be employed, and preamplifiers (if needed) added at the array.

- Because no transmitted RF energy is directly applied to the elements, there is no need for high-voltage or high-current components that are normally required for transmit applications.

Over the past decade or two, leading 160-m and 80-m DXers fortunate enough to have the necessary acreage have designed and built *six-circle* and *eight-circle* arrays for these bands. Typical of these arrays, which provide outstanding receive directivity and *signal-to-noise ratio* (SNR) on these notoriously weak-signal frequencies, is the eight-circle for 160 m designed by Tom Rauch, W8JI. In this design, the output signals from four of eight short vertical elements equispaced around the circumference of a circle are combined with appropriate relays and phasing lines to provide more than 7 dB gain in the desired direction, relative to a single such element. Because the circle of elements is *eight-way symmetrical* around the compass rose, the main lobe can be switched in 45-degree increments.

At any given switch position (i.e., compass heading), only four of the eight elements are active: Two adjacent elements on one side of the circle form an end-fire two-element array, as do two adjacent elements directly opposite, on the far side of the circle. These two end-fire arrays are then combined in a two-element broadside array, thus enjoying high forward gain *and* deep nulls to the rear, thanks to the principle of *pattern multiplication*. Both the eight-circle and the four-element combined end-fire/broadside array building block are described in detail by John Devoldere in his excellent book *ON4UN's Low-Band DXing*, published by ARRL.

Although circle arrays for a given frequency range require less acreage than four bidirectional Beverage antennas, an absolute minimum for a 1.8-MHz eight-circle is about three acres, assuming ideally shaped property. Recently, John Kaufman, W1FV, has designed and published (through ARRL) a compact dual-band version of a *nine-circle* array requiring a very modest 140-ft-diameter footprint (corresponding to slightly more than a half-acre). The W1FV array consists of eight elements around the periphery of the circle, surrounding a single element at its center. It functions as a three-element in-line array that—just like the eight-circle—can be switched in 45-degree heading increments. The 160-m design acquits itself quite nicely on 80 m, as well, thus eliminating the need for the added costs (and acreage,) of a separate 80-m receiving array.

As we have said many times throughout this book, there is no "free lunch". High-performance arrays of short receiving elements are precision designs, requiring close matching of signal amplitudes and differential phase shifts from element to element. Poor feedline installations and sloppy grounding techniques, among other things, can lead to *common-mode* signal pickup and other ills that will easily nullify the directivity and SNR improvements possible with these antennas.

Grounding

It goes without saying that no ground-mounted or ground-plane vertical will perform very well without an adequate ground system beneath it. This is especially true for *arrays* of verticals, as well. Each element of an array of vertical monopoles should have its own extensive ground system (the W1FV array is an exception to this), whether composed of sheets of copper flashing, two dozen or more radials each generally as long as the vertical is tall, or saltwater. If radials or other conductors are employed, they should not be allowed to overlap unless they are insulated or unless all points of possible contact have been soldered or mechanically joined. Intermittent or oxidized joints can raise the background noise level when receiving and result in the radiation of spurious out-of-band signals when transmitting. Refer to the grounding information in Chaps. 9 and 30 for a more extensive discussion of grounding techniques.

CHAPTER **12**

The Yagi-Uda Beam Antenna

A s we have discussed in the past few chapters, antenna arrays deliver an increased signal to or from one or more favored directions at the expense of signal in other directions. We have seen examples of both *all-driven* (or *driven*) arrays and *parasitic* arrays. In the latter, one or more elements of the array are not directly connected to the feedline; instead, they cause RF energy to be *redirected* through careful choice of certain of their characteristics—in particular, their length relative to a half-wavelength at the operating frequency and the spacing between them and adjacent elements of the array.

Arguably the most popular parasitic array in common use today is the *Yagi-Uda beam antenna* or, more commonly, the *Yagi* (thus showing the advantage of having your name appear first on research papers and book covers). Amazingly, the Yagi has been with us for 85 years, having been first described publicly by its inventors in 1926! The Yagi is inherently a balanced antenna, making it a natural for horizontal polarization, but many tens of thousands of vertically polarized Yagis for the 11-m citizens band have dotted the landscape here in the United States since the 1960s. Nor is it difficult to recognize certain parasitic arrays of verticals used by broadcast stations and amateurs as ground-mounted equivalents of Yagi "half-elements".

Despite its age, the Yagi was not popularly employed on the high frequencies until after World War II, when the availability and affordability of aluminum tubing in appropriate lengths, diameters, and thicknesses made the backyard assembly and erection of these arrays feasible for the average radio amateur. Before then, only dedicated DXers seeking "that extra edge" made Yagis from whatever was available to them. In fact, the first full-size rotary three-element 20-m Yagi seen by one of the authors was made of cast-iron pipe; its owner reported it weighed 300 lb! It was mounted on a heavy-duty tripod on the roof of his house, directly over his radio room, and rotated by hand from inside the room. Other builders of early Yagis stapled ordinary copper wire to structural frameworks made from cheap, easily obtainable materials, such as wood two-by-fours.

Certainly there were many other horizontal HF arrays in use prior to 1940. But most were usually fixed in position and had only one or two favored directions. The W8JK two-element driven array was very popular—so much so that a few enterprising amateurs had designed and built rotatable versions. But by 1950 it was clear that the Yagi-Uda design with the need to directly feed only one element, coupled with the use of aluminum for the entire structure (originally described as the "plumber's delight" method by the late Bill Orr, then-W6SAI, in a February 1949 *QST* article), provided the basis for a lightweight and economical beam antenna on 14 MHz and above that could be raised to the top of a tower by a single person. There it could be rotated through the entire 360-degree azimuthal horizon with not much more than a heavy-duty TV rotor.

What makes the Yagi so popular an antenna? In brief: gain, directivity, simplicity of construction, single-element feedpoint—and all in a very small space. A three-element Yagi for 20 m can, in the footprint of a 50-ft-diameter circle, provide a more constant forward gain rotatable through all compass directions than can a half-dozen fixed rhombic wire antennas, each 400 ft long on its longer axis. Contrast the real estate requirements and the construction simplicity of a single tower supporting a single Yagi to those six rhombics needing as many as 24 supports!

By the mid-1950s, manufacturers such as Hy-Gain, Mosley, and Telrex were offering two- and three-element Yagis not just for individual high-frequency bands, but for three bands in one assembly employing a single coaxial feedline! These latter products were made possible by the use of *traps* in their elements, as we will discuss later in the chapter.

Today, common usage often has us referring to the Yagi as a *beam*. Example: "The antenna here is a three-element beam" or "The antenna here is a three-element triband beam". In general, whenever you hear a physical antenna referred to as a "beam", you can almost always interpret that to mean "Yagi". Whereas terms such as *beamwidth* can pertain to the characteristics of *any* type of antenna, the "beam" antenna is invariably synonymous with the "Yagi".

Use of a directional beam antenna provides several advantages over a more omni-directional antenna, such as a dipole or random length of wire:

- The beam antenna provides an apparent increase in radiated power because it focuses the available transmitter power into a single (or limited range of) direction(s). The doughnut shape that is a dipole's radiation pattern in free space has a gain over an isotropic radiator of approximately 2 dB in its best directions, broadside to the axis of the dipole. Add one additional element and the focusing becomes nearly unidirectional by increasing the *effective radiated power* (ERP) another 3 dB or so in the desired direction. Or add two elements to the dipole, and see an ERP increase of 6 dB over that same dipole in the desired direction. That's equivalent to swapping out a 25W transmitter for a 100W unit!

- Antennas are generally reciprocal devices, so they will work for receiving applications much as they do for transmitting. When aimed at a desired signal source, the beam antenna increases the amplitude of the received signal available at the input of the owner's receiver relative to the internal noise of the receiver itself. This is of great importance on the VHF and UHF bands, and not so important on the HF bands.

- The directivity of the beam antenna is usually marked by one or more sharp nulls in both the azimuthal and the elevation response patterns. This attribute forms the basis for some popular VHF and UHF direction-finding techniques.

- The receiving pattern of a Yagi substantially knocks down the strength of signals and band noise coming in from compass directions to the sides and rear of the desired signal path, thus effecting an improvement in the received *signal-to-noise ratio* (SNR). This effect is important and very useful at HF.

- Since the main forward lobe is relatively broad and the null off the sides quite sharp, small changes in beam heading to put interfering signals in the deepest part of the null may provide as much as 30 or 40 dB rejection of an unwanted signal—often with less than 1 dB reduction in the strength of the desired signal—a characteristic extremely useful on today's crowded HF bands.

All in all, if you are considering an upgrade to your station and your funds are insufficient to provide both an amplifier and a better antenna system, a good rule of thumb is to first spend your money on antennas—not on the amp—because the antennas will help you with *both* receiving and transmitting.

Theory of Operation

Figure 12.1 shows the free-space pattern (viewed from above) typical of the Yagi antenna at low *takeoff* and *arrival angles.* The antenna, located at point *P*, fires signals in all directions but the beam-forming pattern created by the physical configuration of its elements causes the relative strength of the signal to be as shown by the solid line envelope in the figure, with the direction of maximum signal strength indicated by the arrow at the top. The *beamwidth* of the antenna is defined as the angle α between the points on the main lobe that are 3 dB down from the peak at point *C*.

A perfect beam antenna will have only the main lobe or "beam", but that situation occurs only in dreams. All real antennas have both *side lobes* and *back lobes,* also shown in Fig. 12.1. These lobes represent not only wasted power transmitted in the wrong direction during transmission but also opportunities for interference while receiving. The goal of the antenna designer is to increase the amplitude of the main lobe while decreasing that of the side lobes and back lobes. Fortunately, it is often possible to effect a substantial reduction in the side lobes and back lobes with only a slight reduction in main lobe (forward) gain or other desirable parameters such as wide bandwidth or matching the transmission line impedance.

Figure 12.2 shows in schematic form the basic three-element *Yagi-Uda* antenna. With rare exception, all elements of a Yagi lie in the same plane, are parallel to each other, and have their centers all in a straight line. In a rotatable Yagi, these constraints are best met by attaching the centers of all elements to a common *boom.* Depending on the specific design approach taken, the boom may serve as an electrical connection, as well.

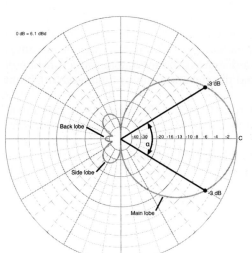

Figure 12.1   Pattern of a typical beam antenna.

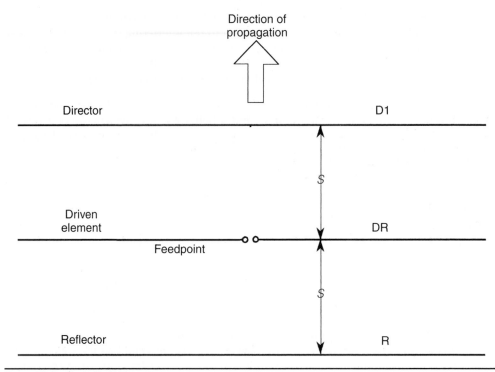

FIGURE 12.2   Basic Yagi-Uda antenna.

In addition to having a *driven element*—which is similar to a standard half-wavelength dipole fed at or near the center by any of a number of different means that we will discuss later—Yagi antennas employ either or both of two additional types of elements: *reflectors* and *directors*. These are called *parasitic elements* because they are not directly connected to the feedline but instead receive energy radiated from the driven element and then *reradiate* it. The combination of this reradiated energy from the parasitic elements with the original radiation from the driven element causes peaks and nulls in the radiation field at a distance from the antenna as a function of both the *azimuthal* (or *compass*) *heading* and the elevation angle of the receiving point from the antenna.

The mathematical equations to support and "prove" the operation of Yagi antennas are beyond the scope of this book, but here's a short *qualitative* description of what goes on:

The driven element of a Yagi is, for all intents and purposes, nothing more than a half-wave dipole. RF energy fed to it via the feedline connected at its center terminals creates RF currents and magnetic fields that radiate from the element, as we learned in Chap. 3. When these radiation fields reach a parasitic element (either a reflector or a director), they *induce* fields and currents in that element because it is, after all, a conductor. But because each parasitic element is deliberately made shorter (when used as a director) or longer (when used as a reflector) than an exact half-wavelength at the frequency of operation, the induced fields do not establish the same amplitude and phase relationships on the parasitic element as exist in the driven element.

These induced currents make each parasitic element into a radiator in its own right. That is, each director or each reflector now radiates an RF field just as the driven element does. However, the RF radiation from a reflector, for instance, has a different phase from that of the driven element, and it originates at a different point in space than the driven element occupies. The radiated fields created by the induced fields and currents in the parasitic elements travel in all directions into space. And at every point in space, they combine linearly and vectorially (i.e., in both amplitude and phase) with the original radiated fields from the driven element. Because the radiation from each element has its own unique phase and amplitude, the resultant is a radiation pattern that exhibits peaks and nulls depending on where in the space beyond the Yagi the receiving antenna is located.

Up to this point in the discussion, there is nothing particularly special about what we have described. For instance, if you hang a 20-m dipole in your attic along your ridgeline, you have probably created a multielement parasitic array comprised of your dipole and any house wiring circuits running around the attic. Of course, you don't have much control over the distance to the house wiring or the length of these branch circuits, but they *are* electrical conductors and they can have a major effect on the radiation pattern and input impedance of your dipole.

The "magic" worked by Yagi and Uda was that they derived equations (without benefit of electronic calculators or high-speed computers, it should be noted) capable of accurately predicting the radiation patterns that would result from various physical arrangements of elements, ultimately arriving at the configuration that now bears Yagi's name. From these equations and generally accepted engineering objectives for a directional antenna (high forward gain, high front-to-back ratio, wide bandwidth, reasonable input impedance, etc.), they then provided formulas for calculating practical and manually optimized dimensions for element diameter, length, and spacing.

As a matter of definition, if a *reflector* is used, it is placed "behind" the driven element—where "behind" is taken to mean on the side of the driven element *away* from, or *opposite*, the direction of desired maximum signal radiation—and it is typically a few percent longer than a half wavelength *in the conductor* at the operating frequency.

If a *director* is used, it is placed in "front" of the driven element (i.e., on the side of the driven element that is in the direction of desired maximum radiation). In a traditional Yagi design, the director nearest the driven element is typically about 3 percent shorter than a half wavelength *in the conductor*.

Spacings between adjacent elements of a Yagi typically are between 0.1 λ and 0.25 λ, where λ is the wavelength *in air*. The exact spacings depend on what the designer intends to optimize. For example, if weight, turning radius, and wind load are more important than bandwidth, a three-element Yagi can fit on a boom that is only 0.2 λ in overall length, but the designer and user will find that other attributes, such as standing wave ratio (SWR) bandwidth, front-to-back ratio, and that last ounce of forward gain, have been compromised.

The basic Yagi is a *planar* antenna, with all elements lying in the same plane and sharing the same orientation. Although there is no fixed rule regarding the number of either reflectors or directors, it is common practice to use a combination of a single reflector and one or more directors, in addition to the driven element. Thus, from the definitions and "rules" of the preceding two paragraphs, the most common three-element beam consists of a driven element flanked by a reflector at one end and a director at the other end of a boom supporting all three. Most four-element Yagis have two directors, but in recent years a different kind of four-element Yagi has made its appearance. This new design uses two driven elements to provide maximum gain and directivity over a wider bandwidth. In general, it can be considered a very wideband three-element Yagi.

Over the years, there have been commercial two-element Yagi antennas designed and sold that employed *either* a reflector or a director in combination with the driven element—especially for the 40- and 30-m bands. If a reflector is chosen, peak performance (i.e., forward gain and front-to-back ratio) will increase very rapidly from the *lower*-frequency side of the operating range of the beam, then fall off slowly as the frequency is increased through the operating range. If a director is chosen, the opposite is true. All major manufacturers of two-element HF Yagis at the time of this writing offer only the reflector + driven element version.

Three-Element Yagi Antennas

For a three-element HF Yagi with full-size elements, there are six important dimensions:

- Length of the driven element (DE)
- Length of the reflector (R)
- Length of the director (D)
- Diameter of the elements (d)
- Centerline spacing between the driven element and the reflector (R-DE)
- Centerline spacing between the driven element and the director (DE-D)

Of course, what appears in the preceding list as a single dimension (d) may be *many* dimensions, depending on whether the elements are constructed of *nested* or *telescoping* tubing of more than one outside diameter (OD). Also, at VHF and UHF (and even for some HF designs) the specific dimensions of the boom-to-element attachment method need to be incorporated in the design calculations. Finally, note that the sum of all the interelement spacings (measured to the centerline of each element) plus the width of one boom-to-element bracket determines the minimum required boom length for a specific design.

All six of these dimensions interact with each other. That is, if we have a Yagi design that meets our specifications for forward gain, front-to-back ratio, and SWR at 14.1 MHz, for instance, a change in the spacing between the reflector and the driven element may well worsen the performance of any or all of those parameters at 14.1 MHz. If we are lucky, we may be able to find another combination of element lengths that allows our Yagi to meet or exceed the performance at that frequency, but it's also quite possible that we won't.

As you might guess from the inclusion of element diameter in the preceding list, the dimensions for a multielement Yagi constructed of #12 copper wire are not likely to be identical to those for the same beam with elements of 1-in aluminum tubing. In fact, because of the close coupling between elements, multielement Yagis are far more sensitive to changes in all dimensions than simple dipoles, as we can see with the next two examples.

Example 12.1   A $\lambda/2$-dipole for 20 m constructed from #12 copper wire is trimmed for a purely resistive feedpoint impedance at 14.1 MHz. The resulting length of the dipole

is 33 ft 3½ in. When the wire is replaced with 1-in aluminum tubing, the dipole resonates at 33 ft 2½ in. The difference, 1 in, is probably inconsequential to the overall operation of the dipole.

Example 12.2   A three-element Yagi of a certain design and having elements of 1-in
aluminum tubing meets the following specifications at 14.1 MHz:

- Forward gain = 8.25 dBi
- Feedpoint impedance = $13.8 + j0$ Ω
- Front-to-back ratio = 31 dB
- Front-to-side ratio = 25 dB

With the use of a simple 4:1 balun, the input impedance becomes 55 Ω, and the 2:1
VSWR bandwidth using 50-Ω coaxial cable is 220 kHz.

Now suppose we replace all the aluminum elements with #12 copper wire of exactly
the same length. The performance of the resulting beam at 14.1 MHz becomes:

- Forward gain = 7.5 dBi
- Feedpoint impedance = $24.2 - j17.0$ Ω
- Front-to-back ratio = 16.3 dB
- Front-to-side ratio = 16.3 dB

In short, every important performance parameter of our three-element beam has been
compromised by blindly ignoring the effect of element diameter on the design
dimensions. And simply to obtain a reasonable SWR at the design frequency, we must
switch to a 2:1 balun to obtain a transformed input impedance of $48.4 - j34$ Ω. This
corresponds to an SWR of 1.9:1, but it is the best we will see unless we adjust the driven
element for zero reactance at 14.1 MHz.

If we launch a more thorough attempt to bring the wire beam up to the level of
performance of the original beam, we find we can get closer, but we're not able to
perfectly equal the original. For comparable front-to-back (F/B) and front-to-side (F/S)
ratios and a feedpoint impedance that is a good match for the 4:1 balun, we can obtain
about 8 dBi forward gain. But the resulting 2:1 SWR bandwidth is only 125 kHz—one
consequence of making the element diameter much smaller. In the process, we have
had to lengthen the director by 3 in and slightly shorten the reflector for this particular
design.

◆

Element diameter comes into play another way, as well. Once a Yagi design has
been developed for one frequency or band, that same design can be "translated" to
other frequencies by scaling *all* the important dimensions—including element diameter—in accordance with the change in wavelength. Unfortunately, in the process of scaling a good design to another center frequency, the resulting element diameter(s) may be

unacceptable: they may be mechanically too fragile or too heavy, but most likely they're just plain unobtainable. If *all six* of the listed dimensions are not scaled by the same factor, performance on the scaled band(s) may well be less impressive than expected. A practical approach is to use simple scaling to get into the ballpark with dimensions for the new band based on the original band, then use antenna modeling software (Chap. 25) or experimentation to fine-tune all the dimensions so as to optimize performance around standard aluminum tubing diameters or copper wire gauges.

Many different Yagi designs exist. In fact, the potential number is infinite, limited only by users' willingness to experiment or to use modeling software to examine various dimensional configurations. One reason for having many different designs is that usually one or more characteristics of the antenna must be compromised. Thus, one design may be developed for a three-element Yagi with wide SWR bandwidth, while another may favor a high F/B ratio over a narrow bandwidth centered on the design frequency.

Some designs are based on mechanical tradeoffs rather than electrical performance specifications. One such commonly encountered constraint is total boom length. (A manufacturer might choose a shorter boom length to keep material costs and antenna prices down. An amateur backyard experimenter might choose a shorter boom length because of limitations of his rotator or proximity of a nearby tree.)

For obvious reasons, boom lengths of 20 ft or slightly less are very popular. In particular, 3-in irrigation pipe makes a very strong boom. (But there are many commercial Yagis for those bands that butt two suitably thick-walled 10-ft aluminum tubes together to form a 20-ft boom—usually with interior and/or exterior reinforcing sections for some distance on either side of the junction.)

In the 85 years since the publication of the original Yagi-Uda paper, practicing scientists and engineers, along with amateur experimenters, have continued to improve the design. While much of Uda's later work was specifically aimed at the burgeoning VHF consumer television market, shortwave broadcasters and the U.S. military joined hams in bringing the Yagi to the HF bands. Here is a summary of some of the most important things to keep in mind when working with Yagi antennas:

- Exact interelement spacings are not critical on a Yagi; they are typically 0.1 to 0.3 wavelengths for popular proven designs. Once established for a design, however, changes to spacings interact with element lengths and do affect Yagi performance.

- Wider element spacings (for a given number of elements) provide greater SWR bandwidth and more constant forward gain and F/B ratio across the entire band. Of course, wider element spacings translate to longer booms!

- Maximum forward gain and maximum F/B ratio seldom occur at the same frequency; in general, longer boom lengths (for a given number of elements) make it easier to optimize multiple parameters over a wider common range of frequencies.

- As a glittering generality, Yagi gain is proportional to boom length. Stuffing additional elements onto a fixed boom length usually worsens performance.

- A reflector 3 percent longer than a half-wavelength *in the conductor* is a good starting point. (Said another way, start with a reflector length that is no longer than $\lambda/2$ in free space or air.)

- A first director 3 percent shorter than a half-wavelength in the conductor is a good starting point.

- Most HF Yagis are constructed from telescoped lengths of aluminum tubing to minimize mechanical stresses on the inner portions of the elements and the overall load on the boom. We (somewhat ambiguously) say the elements (and often the booms, too) are *tapered* in diameter; more precisely, they're *stepped-diameter* elements. Stepping (or tapering, if you must) alters the electrical lengths of the elements for a given physical length and must be taken into account in the design and modeling process.

- Most VHF and UHF Yagis use elements of a single diameter, so tapering effects are not an issue. However, the exact method of passing the center of the element through, over, or around the boom can have a significant effect on element electrical length and Yagi performance at those frequencies.

- All other things being equal, wire elements will be longer than elements made from larger-diameter tubing, but stepped-diameter elements are longer than single-diameter ones.

- The length of the driven element plays a relatively minor role in optimizing a Yagi's unique attributes (forward gain, F/B and F/S ratios, resistive part of the feedpoint impedance); in most designs, in fact, the driven element length is adjusted following all other adjustments, to make the reactive part of the feedpoint impedance go to zero.

- The input impedance at the center of a typical three-element Yagi is substantially less than that of a simple half-wave dipole, even in free space; 10 to 30 Ω is not uncommon, and the use of a 4:1 balun (for the lower end of the feedpoint impedance range), a stub, or a series-section transformer (for the upper end of the range) is often needed to provide a better match to the transmission line. (See Chap. 4.)

- Even though it is inherently a balanced antenna, a Yagi can be directly fed with an unbalanced feedline such as coaxial cable with little impact on forward gain; there may be some (minor) loss of symmetry in the radiation pattern, and it will probably be necessary to add feedline chokes at one or more locations along the cable to minimize common-mode pickup and radiation by the feedline.

- Other methods of feeding a Yagi include the gamma match (also unbalanced), the T-match (a balanced gamma match), and the hairpin match (a balanced stub-matching technique; a variant of the hairpin is used on many Hy-Gain beams under the trade name "Beta match"). Often, a lumped-component *balun* of the appropriate turns ratio and located right at the driven element feed terminals is the simplest approach.

- The Yagi has been characterized as an antenna that "wants to work"; there have been numerous anecdotal stories of how "amazingly" well long-boom, multidirector Yagis have continued to perform even though ice and/or wind have ripped one or more half-elements from the antenna.

- Many commercial Yagi models with shortened elements (to reduce wind load, rotational torque, and weight) are available. The most common methods for shortening elements include traps, loading coils, and linear loading wires.

- Perched at the top of a support, Yagi antennas are a prime target for lightning discharges and wind-driven precipitation static. In addition, Yagis on a tower may (inadvertently or deliberately) be part of a capacitive top-loading system for 160 or 80 m when the tower is used as a transmitting vertical on either of those bands. The Yagi designs that best address these concerns have both sides of all elements grounded to the boom—either directly or through the matching network.

Yagi Antennas with Four or More Elements

As additional elements are added to a Yagi, they typically are directors. For decades, until the advent of the dual-driven element variant previously described, a four-element Yagi almost automatically meant a reflector at one end of the boom, followed by the driven element and the *first* and *second* directors, in that order. Similarly, a five-element Yagi usually employed three directors. "Why not multiple reflectors?" you may ask. The nonmathematical answer is this: If the first reflector is properly designed and positioned on the boom relative to the driven element, virtually all of the energy radiated off the "back of the beam" (i.e., beyond the reflector) will be canceled by the phased combining of radiation from all the elements. Thus, there will be very little net RF passing beyond the reflector and, hence, very little useful work for a second or third reflector to do! Nonetheless, some very long, high-performance VHF and UHF Yagis use three reflectors mounted in a plane orthogonal to the plane of the other elements.

Yagi Construction

In this section we will briefly discuss some simple construction methods and inexpensive materials for Yagis that can be built relatively easily—i.e., *practical* ones. It is assumed that most readers who want a triband multielement Yagi will prefer to buy a commercial product, rather than build a homebrew model, because the *traps* or multibanding circuits employed likely involve the use of construction techniques not available to most hobbyists. The projects in this chapter are meant to be within the reach and capabilities of most readers and assume the availability of little metalworking equipment in the home workshop beyond drill, screwdriver, nutdriver, and hacksaw or tubing cutter.

As the design frequency of a rotatable Yagi drops below 30 MHz or so, mechanical issues come to the fore. In particular, reliable element self-support dictates the use of multiple tubing diameters for the elements, sometimes in conjunction with truss systems. It's not uncommon, for instance, to find as many as five or six different tubing diameters in each element of a 20-m beam. And, as element lengths increase, the strength of the associated boom-to-element clamps (brackets) becomes a critical design matter, as well.

For designs requiring more extensive workshops, the most practical—and the least expensive—approach for many of us is to start by buying one or more used Yagis, if for no other reason than to have a stock of suitably premachined parts and element sections of many different diameters. Don't overlook the parts value of a "junked" beam from a fellow hobbyist or your local CATV system's head-end! Lacking such sources, a few specialty components (certain types of boom-to-element brackets, for instance) are probably best obtained from the parts department of one of the antenna manufacturers or from the multiline antenna and tower retailers listed at the end of this book. But, just as it's far less expensive to buy a car fully assembled than it is to buy and assemble all the individual parts that go into one, don't plan on "rolling your own" economically unless you have lots of aluminum tubing and flat stock already on hand.

Rotatable Dipole

The simplest "Yagi" antenna is not usually considered to be a beam antenna at all, but the *rotatable dipole* does have directivity, and those amateurs fortunate enough to have a rotatable 80-m dipole (!) up 130 ft or more certainly feel like they have a full-fledged beam compared to the rest of us. So, for purposes of exploring basic Yagi construction techniques, the rotatable dipole serves as a one-element Yagi because it is, in fact, the driven element of the multielement Yagi beams to be discussed later.

As we saw in Chap. 5, the *half-wave dipole* is a bidirectional antenna with a figure eight pattern (when viewed from above). It is usually installed horizontally, although vertical half-wavelength dipoles on the 27-MHz citizens band and above are often found. Rotating a horizontal $\lambda/2$ dipole for the MF and lower HF bands around the compass is more of a mechanical design challenge than it is an electrical one, and it is certainly not for the faint of heart! Above 10 MHz or so, however, a rotatable dipole becomes practical—not only because the $\lambda/2$ element is a more manageable length but because the user doesn't need to lift it as high above ground to make it a reasonable performer. Specifically, the length of a halfwave dipole is approximately 16 ft on 10 m, 22 ft on 15 m, and 33 ft on 20 m; all of these antennas will perform well at a height of 30 ft or more if they are in the clear.

The starting point for determining the length of a $\lambda/2$ dipole constructed from copper wire is

$$L \text{ (feet)} = \frac{468}{F \text{ (MHz)}} \tag{12.1}$$

The length found from Eq. (12.1) is approximate because of end effects and other phenomena, so some "cut and try" may be required to arrive at a length that minimizes the reactive part of the feedpoint impedance. Also, because HF beams are seldom mounted high enough above ground or the roof of a support building to be treated as being in free space, the length that results in a resonant antenna (i.e., the feedpoint reactance is zero) will differ slightly from the free-space value. Although we have said this before, it is worth repeating here: Setting the length of a dipole to exactly the dimensions given by Eq. (12.1) is not one of the most important things to be concerned about.

Example 12.3   Find the length of a $\lambda/2$ dipole antenna for a frequency of 24.930 MHz in the 12-m amateur radio band.

Solution

$$L \text{ (feet)} = \frac{468}{F \text{ (MHz)}}$$

$$= \frac{468}{24.930} = 18.77 \text{ft}$$

The simplest way to feed this antenna is to break the dipole in the center and attach a 72-Ω coaxial cable there. The dipole then consists of two separate conductors, each of which is one-half the overall length (or, in the example given for the 12-m band, about 9.4 ft).

◆

Figure 12.3 shows a generic rotatable dipole design for use on 15, 12, and 10 m. On 12 and 10, the radiator half-elements are made from 10-ft lengths of ¾-in aluminum tubing cut to the lengths listed in Table 12.1, which are based on the dipole being mounted 1λ over average soil. For these bands, the tubing is mounted on "beehive" standoff insulators, which in turn are mounted on a 4-ft length of 2 × 2 lumber. The lumber should be protected with varnish or exterior polyurethane. In a real pinch, the elements can be mounted directly to the lumber (with hose clamps or long bolts and large washers that go all the way through the elements and lumber, for instance) without the insulators, but this is not the recommended practice—in part because any precipitation is likely to cause a leakage path and resulting impedance change across the two halves, thus changing the feedpoint impedance and measured SWR "seen" at the

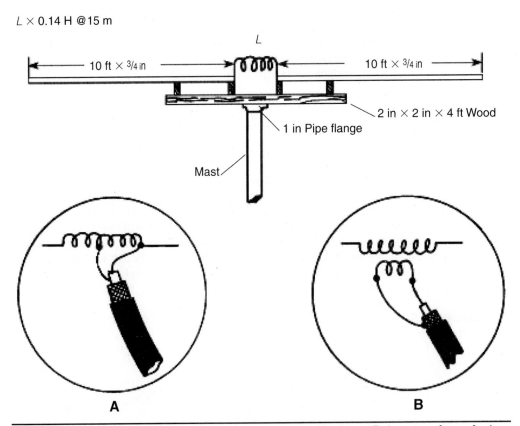

$L \times 0.14$ H @15 m

A   **B**

FIGURE 12.3   Rotatable dipole antenna. Inset A shows conventional feed; inset B shows transformer feed.

transmitter. The two ends of the element halves near the centerline of the dipole should be an inch or two apart, although the exact spacing is not critical in the HF bands. Connect one conductor of a coaxial cable or hardline to either half-element directly or through a 1:1 balun.

Band	Design Frequency	Total Length of One Side	¾-inch Tubing	⅝-inch Tubing
10 m	28.300 Mhz	8'3" (2.514 m)	8'3" (2.514 m)	
12 m	24.930	9'5⅜" (2.880 m)	9'5⅜" (2.88 m)	
15 m	21.150	11'1" (3.385 m)	10' (3.048 m)	1'1" (0.337 m)

TABLE 12.1   Rotary Dipole Half-Element Lengths

The mast is attached to the 2 × 2 lumber through any of several means. One simple method employs a 1-in pipe flange. These devices are available at hardware stores under the names *floor flange* and *right-angle flange*.

For the 15-m version, the 10-ft length of tubing is not long enough to be resonant anywhere within the band. The builder has a number of options available:

- Use it "as is", and tune out the reactive part of the feedpoint impedance with an ATU.

- Add a center loading coil to provide the "missing" inductance to make the dipole resonant.

- Add element "tips" to the 10-ft sections of tubing to make the element self-resonant on 15.

Although the gain of the assembly is not materially affected, the first option results in a VSWR on the 75-Ω feedline of roughly 4.0:1 across the entire band. (Even though the resistive part of the feedpoint impedance is an excellent match for 52-Ω cable, the high reactance makes the VSWR lower on 75-Ω cable.) This option should not be considered if any substantial transmit power (over, say, 100 W for normal cables) is planned.

For the second option, a 0.14-μH loading coil is mounted at the center, with its two ends connected to the two half-elements. The dimensions of the coil are four to five turns, 0.5-in diameter, 4-in length. For low power levels, the coil can be made of #10 (or #12) solid wire—and, for higher levels, ⅛-in soft copper tubing (often found in a coil inside a plastic bag in the plumbing section of the local hardware store).

If the center loading coil approach is chosen, there are two basic ways to feed the antenna, as shown in details A and B in Fig. 12.3:

- The traditional method is to connect the coaxial cable across half the inductor, as shown in detail A. Be sure the center of the inductor is connected to the shield, or grounded, side of the coaxial transmission line.

- *Link couple* the coil to the line through a one- to three-turn loop (as needed for impedance matching), as shown in detail B. One way to do this is with a toroidal inductor, which, coincidentally, lifts both sides of the dipole off the shield ground of the cable. (Whether isolation of the element from dc ground is good or bad is a discussion for another time.)

Even lower frequencies (than 21 MHz) can be accommodated by increasing the dimensions of the coil. The coil cannot be simply scaled because the relative length of the radiating portion of the antenna changes as the frequency changes. But it is possible to cut and try by adding turns to the coil, one turn at a time, and remeasuring the resonant frequency. Adding inductance to the coil will make the antenna usable on 17 m or 20 m in addition to 15 m.

For the home hobbyist, eliminating the need for a coil by making the λ/4 element halves the proper length may be a simpler construction project. We can do this by extending each tube with one of a smaller diameter. The added tube should extend back into the original tube 12 in or more to provide strength for the cantilevered portion of the outer tube and to leave some extra material for final adjustments to element length. Thus, for the example cited, a 24-in-long ⅝-in OD tube with 0.058-in wall thickness will "nest" inside the ¾-in element half to bring the total length of the half element to the required 11 ft 1 in. So-called adjacent sizes of aluminum tubing are designed to nest so that the smaller will be a tight slip-fit inside of the larger.

Depending on anticipated wind and ice loading, reinforcement of the ¾-in tubing in the area of the insulated standoffs and for some distance beyond may be appropriate. With standard 0.058-in wall aluminum tubing, the wind survival speed for the 15-m dipole is in the 70 to 80 mph range. Use any leftover ⅝-in stock to slide or drive a few feet of it into the feedpoint end of each ¾-in tube that will be supported.

When nesting aluminum tubing, *always* make sure the two mating surfaces are clean and smooth. For some of the common inside diameters, small wire brushes for cleaning the inside of copper plumbing fittings are useful. For smaller sizes, roll up a piece of fine-grit sandpaper and slide it in and out of the larger tube. Then be sure to coat the outer surface of the smaller tube with a thin film of an aluminum oxide suspension commonly found in electrical or plumbing supply stores and "big box" home supply stores under the brand names *Alumalox* and *Permalox*. Some manufacturers of Yagis with telescoping element sections include a small container of the material; one such trade name is *Penetrox*.

Nested tubes can be held together easily with one or two stainless steel sheet metal screws just behind each junction. Be sure to specify stainless steel, and be sure to get screws with a Phillips head. There is no worse way to spend a Saturday than trying to separate two pieces of tubing held together with a rusted screw whose slotted head has been worn off by futile attempts to twist it out!

Another way to hold nested tubes fixed with respect to each other is to slot the end of the larger tube and tighten a hose clamp over the slotted region after the smaller-diameter tube has been inserted the correct distance. This is the method used by some of the largest amateur antenna manufacturers over the years, but the author has never found this form of clamping to be sufficiently secure when an element becomes snagged on tree branches, guy wires, or halyards on its way up to its final mounting point. Even when this is the manufacturer's stock method of securing telescoping tubing sections, secure each joint with a stainless steel screw through both pieces of tubing. (You will first want to drill a pilot hole for each screw!) Alternatively, one major antenna manufacturer (Force 12) and many amateurs use *pop rivets* to hold element sections together. Current best practices on Internet discussion groups is that each joint should be secured with at least two pop rivets at right angles to each other.

The best procedure for completing a Yagi is to initially assemble the dipole or Yagi to recommended nominal dimensions with hose clamps and slotted tubes, attach the feedline (discussed later), raise the antenna up in the air, and check the VSWR at the transmitter end of the feedline before bringing the antenna back down to add the machine screws or pop rivets. Whether this is possible or practical in any given installation will depend heavily on the specifics of the site, the support, and the ground crew available. Sometimes a stepladder or garage roof can act as a temporary support. Another trick for minimizing the effect of nearby ground is to aim the Yagi straight up in the air with its reflector nearest the ground (although this is of no benefit for the rotary dipole). Remember, it's no crime to secure the initial dimensions with the screws, then remove the screws and redrill for a different element length, if necessary.

Plastic caps supplied for the element tips by many Yagi manufacturers are both a blessing and a curse. While keeping spiders and other insects out, they can trap moisture inside the element. Natural element sag is perfectly capable of draining moisture on its own, so a small piece of wire mesh, held in place with a small hose clamp, is a good way to prevent insects from setting up shop inside Yagi elements. Alternatively, drill small holes in the stock plastic caps and install them with the holes toward ground. Even the plastic caps should be held in place with good electrical tape or hose clamps because they are easily dislodged during the antenna raising process. Also, many of these caps tend to crack and lose their tight grip on the element when exposed to the sun over long periods.

If the rotatable dipole is constructed of full-length elements so that no center loading coil is employed, feeding it is simply a matter of connecting a 75-Ω coaxial cable (RG-6, RG-59, RG-11, or equivalent) to its center. The center conductor of the cable goes to one side of the element and the shield or braid goes to the other, using attachment techniques detailed in Chap. 28. As described before, the two element halves should be separated about 2 in by the supporting hardware. For someone with access to a wood lathe, a 2-ft length of hardwood dowel turned down to the inside diameter (ID) of the element tubing and treated with preservative (such as an exterior polyurethane paint) or other water repellant is an excellent center insulator. Strain relief for the cable must also be provided.

NOTE *As discussed in Chap. 3 and elsewhere, using a loading coil as a substitute for part of an antenna element reduces the element's effectiveness as a radiator. Remember, the difference between an efficient radiator and a discrete (lumped) circuit component is primarily in the physical configuration of the conducting material. When you coil up a portion of an antenna element into a small volume, you convert that portion of the radiator into a nonradiating lumped component. The deleterious effect on the antenna is greatest when this is done within the high-current section of the element. Therefore, whenever possible, loading of elements close to their centers should be avoided. Often, other forms of loading can attain the same reduction in length with far less decrease in radiated signal strength.*

Multielement Yagi Construction

Each element of a rotatable multielement Yagi antenna can be constructed in a manner similar to that used for the rotatable dipole described earlier. The only significant difference is that the parasitic elements (reflector and directors) do not have a feedline attached to them and the metal is continuous across the centerline of the element.

However, adding one or more parasitic elements to the rotatable dipole described previously introduces a number of issues that must be dealt with if the advantages of the parasitic array are to be realized:

As mentioned, most commercially available rotatable dipoles and multielement Yagis for the HF bands use nested tubing—sometimes relying on five or six different diameters that become progressively smaller the farther out on the element one gets from the boom. In antenna design and modeling lingo, these are known as *tapered* or *stepped-diameter* elements. A big advantage to tapering is that less strength is required to support long elements than if the element tubing were the same diameter all the way out to the tips. A disadvantage, however, is that tapering affects the accuracy of Eq. (12.1) and its "cousins" for directors and reflectors, so corrections must be made to the lengths of element halves to be sure of getting maximum gain or front-to-back ratio with a given beam design. One good way to solve that problem is to use one of the antenna modeling programs, such as *EZNEC*, that include stepped-diameter correction algorithms. We didn't worry about the effect when we added smaller-diameter tubing to the end of our rotatable 15-m dipole earlier because the dipole has such a broad resonance that the effect of stepped diameters is minimal. When one or more parasitic elements are involved, however, the bandwidth of the entire array is substantially narrower and its sensitivity to dimensional changes increases. As a result, the error created by ignoring changes in element diameter over the length of the element can be important.

A rotatable multielement beam requires a *boom* to support the elements in a common plane while holding them in fixed positions and orientations with respect to each other. The boom can be made of metal or wood. In the case of a metal boom, the driven element is frequently insulated from the boom, depending on the specific feed system chosen (see "Feeding the Yagi", later in this chapter), but the parasitic elements are often mounted directly to it. (The *XM240* from Cushcraft is an exception; both reflector and driven element are insulated from the metal boom.) In general, wood is a readily available material that is suitable for the smaller beams of 24 MHz and above. Below that frequency, the longer element lengths and increased spacings cause the weight of a satisfactorily strong wood boom to greatly exceed the weight of an equivalent aluminum boom. If wood is used for a small beam, however, it should be weatherproofed—either with multiple coats of an exterior polyurethane or by purchasing pressure-treated lumber.

Aluminum tubing suitable for use as a boom can be obtained from specialty metal dealers, usually found only in metropolitan areas. Another possible source, frequently overlooked, is a junked or damaged beam from a nearby amateur, perhaps located through a local radio club or craigslist ad. Local amateurs, SWLs, and CBers may also be a good source for boom-to-element brackets, too, although the outside diameter of the new elements at their centers may have to be beefed up to accommodate what's available. Alternatively, the price of brackets as replacement parts from antenna manufacturers may not be prohibitive, and at least one supplier markets a family of mounting brackets for those who like to "roll their own". If obtained from a distant supplier, it's generally a lot less expensive to ship a few brackets than it is long sections of aluminum tubing.

Aluminum boom strength depends on multiple factors:

- Diameter of the tube
- Thickness of the tube wall
- Grade of aluminum

Of course, aluminum is used for antenna booms and elements for the same reason it is used in airplanes: weight versus strength versus cost. Aluminum has other advantages that we often take for granted: It oxidizes in a fairly benign and easily reversible way, and it's easily cut by hand with a hacksaw or drilled with a hand drill, to name but two. Recognize, however, that aluminum is a broad umbrella for a variety of alloys prepared in many different ways. Aluminum tubing you find at the local hardware store may not have the strength you need, at least for larger Yagis. In his excellent book *Physical Design of Yagi Antennas,* David B. Leeson, W6NL (ex-W6QHS) warns readers that 6061-T6 and 6063-T832 are better choices for antenna builders than the more commonly stocked 6063-T5. If you must send away for aluminum tubing stock, see App. B for a list of suppliers.

Total boom length and the spacing of elements along the boom are important parameters in the design of any multielement Yagi. Boom lengths for three-element Yagis typically lie between 0.15 λ and 0.5 λ. As a "glittering generality", for a fixed number of elements, longer boom lengths are capable of wider SWR bandwidths and make it easier for the beam designer to achieve maximum gain and high F/B ratios across the same frequency range. But if narrower bandwidths and a slight reduction in maximum forward gain are acceptable to the owner, excellent performance can be obtained from beams with shorter booms.

Rotatable Three-Element 20-Meter Yagi

At frequencies below about 24 MHz, the elements of rotatable Yagis are best formed from multiple sections or segments of tubing having diameters that get smaller with distance from the boom or center of the element. Such a design approach can provide maximum wind and ice survivability with minimum weight elements. For 15 m, each half-element will likely consist of a (largest diameter) center section and perhaps three progressively smaller diameter sections. On 20 m, a center section and four to six progressively smaller diameters are common. Below 20 m, survivability issues usually lead to design approaches involving substantially more complicated methods (*trusses,* for example) of strengthening booms and elements alike, so 20 m represents just about the lowest frequency for which a simple design like the one presented here is mechanically adequate for the majority of wind environments.

Stepped-diameter correction of each element's length can be done manually, using a spreadsheet technique developed by Leeson and described in his book. Alternatively, antenna modeling software (such as W7EL's *EZNEC 5+*) that incorporates the Leeson correction can simplify the task. Once an element taper schedule such as that of Fig. 12.4A has been created based on *mechanical* (strength) considerations, the lengths and diameters of the various element sections are entered into the *EZNEC* wire table, and *EZNEC* automatically incorporates the Leeson corrections in its calculations to arrive at the correct *electrical* length for each element half. (Just be sure to first turn on the stepped-segment option!)

A convenient boom length for a rotatable three-element 20-m Yagi is 20 ft, since the boom can be a single length of 3-in irrigation tubing or a pair of 2-in OD × 10-ft aluminum tubes butted together with a short internal or external reinforcing tube and a boom-to-mast bracket of suitable dimensions, as depicted in Fig. 12.4B. Also shown is an alternative approach to constructing the boom from multiple sections of aluminum tubing if the longest that can be obtained conveniently are 6-ft sections. (See App. B for suppliers that routinely ship boom and element tubes by UPS.) Regardless of the exact

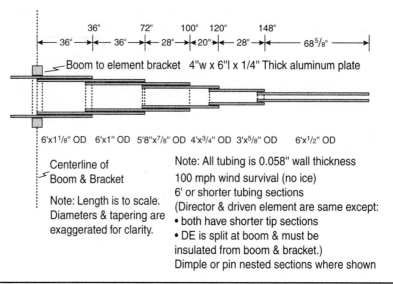

FIGURE 12.4A   Stepped-diameter 20-m Yagi reflector half-element.

approach, the boom tubing should be reinforced with properly nested reinforcing inserts at each point where U-bolts encircle the boom. For 2-in OD boom material with 0.120-in wall thickness, 0.175-in OD tubing makes excellent inserts. All inserts should be pinned to the outer tubes with two stainless steel bolts at right angles to each other.

Even after specifying a boom length there are countless ways to design a three-element Yagi. The designs of Table 12.2 are based on the following "wish list":

- Free-space forward gain of 8 dBi or greater across the design bandwidth
- Front-to-back ratios of 25 dB or more across the design bandwidth
- Driven element offset slightly from the boom-to-mast bracket location but still well within reach from the tower top or mast
- No need for boom or element truss supports
- Input impedance "well behaved" and compatible with simple matching schemes
- All element tubing less than 6 ft long for easy transportability and reduced scrap
- Lightweight low-profile elements with 100 mph wind survival (60 mph with ½-in ice)

Aluminum tubing of various diameters—all with 0.058-in wall thickness—is typically used in the construction of HF Yagis so that tubing in increments of ⅛-in OD can be nested (*telescoped*). Determining the length of each section is not a simple matter of "eyeballing" the element and selecting segment lengths that look about right. Instead, an understanding of mechanical design considerations and knowledge of materials strengths are required. Although not an easy read the first time through, Leeson's *Physical Design of Yagi Antennas* does a superb job of relating those matters to the design of

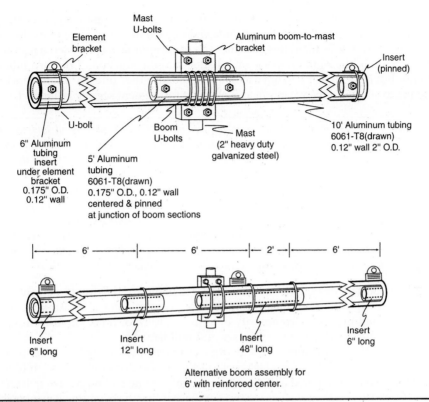

FIGURE 12.4B   20-ft Yagi boom with reinforced center section.

HF Yagi antennas and provides not just equations (for those who want to "roll their own") but also specific design examples (for those who simply want to build and install a known good design.)

Although it would have been possible to design this Yagi with fewer different diameters of tubing for elements—i.e., with more pronounced steps between nested tubes—the resulting design would have required special techniques for implementing a *double*

	CW	Mid	Phone
Reflector–DE spacing	100⅞	100⅞	100⅞
DE–director spacing	134⅜	134⅜	134⅜
Reflector tip length	216⅝	214⅜	213½
Driven element tip length	210	207⅝	206¾
Director tip length	199⅝	197¾	197

TABLE 12.2   Element Dimensions and Spacings (in inches) for Three-Element 20-Meter Yagi, Optimized for Three Different Frequency Ranges

diameter step at the junction of certain pairs of tubes. Some commercial antenna manufacturers accomplish this by *swaging* one end of each tube, but most of us do not have the tools or the skills to do this at home. The other approach is to insert a few inches of tubing having the right "in-between" diameter at the junction of the two tubing segments. But if we're going to have to obtain tubing for the "unused" diameters to avoid having to swage joints, we might as well use it for element segments, instead.

CAUTION *The half-element taper schedule of Fig. 12.4A has good wind and ice survival characteristics only because critical regions of the half-element are* double *thickness by virtue of inserting the next smaller diameter inside the larger tube for at least the distances shown in the figure. If you attempt to scrimp on tubing material and do not overlap tubes at least as much as shown in the drawing, your Yagi will fail at very low wind speeds! All telescoping joints and doubled thickness regions* must *be pinned so as to prevent relative slippage between any two tubes. Popular techniques for pinning include dimpling "hidden" reinforcing tubes with a punch, stainless steel sheet metal screws, and pop rivets. Unless you're extremely proficient at slitting the ends of tubing and properly applying hose clamps, using hose clamps on telescoping sections is not recommended by the author.*

An important design decision is whether to *shunt-feed* the driven element. If so, the element can be continuous metal tubing across its center point as it passes over or under the boom, but the added complexity of a *gamma match* or other shunt-feeding system may be a potential deterrent to the backyard builder.

Instead, this beam employs a split driven element, thus allowing the use of a simple feed technique, the 4:1 balun. This is not exactly a trivial decision because an 18-ft length of tubing, whether tapered or not, supported only at one end represents quite a *cantilevered beam*. The penalty for choosing this approach is that the boom-to-element bracket for the driven element must be long enough to allow proper spacing of four U-bolts (two properly spaced on each side of the driven element) along its length and the U-bolts and bracket must be strong enough to withstand the moment arm created by the force of the wind over the entire length of the element half. Also, insulating "tubes" must surround the element halves where they are grasped by the U-bolts, and the insulation must extend to just beyond the outer edges of the bracket. Purchase of appropriate boom-to-element insulating bracket assemblies from the antenna manufacturers or accessory suppliers listed in App. B may be the best option for the first-time builder of an HF Yagi.

For the director and reflector brackets, flat aluminum plates that are at least ¼-in thick, 3.5 in wide (the edge dimension that is parallel to the boom length) and 6 to 8 in long (the edge dimension parallel to the element length) should be fine for normal wind speeds. Alternatively, obtain *aluminum rectangular (bathtub) channel* of roughly the same dimensions for vastly increased strength.

Table 12.2 is used in conjunction with the element taper drawing of Fig. 12.4A to assemble each element of the 20-m Yagi. The table includes three separate sets of dimensions, each having a distinct focus:

- The "CW" column emphasizes performance across the 14.000 to 14.100 end of the 20-m band. In particular, when fed with 50-Ω transmission line, the SWR is less than 1.5:1 across that 100-kHz range and the F/B ratio exceeds 30 dB over the bottom 80 kHz or so.

- The "Mid" column is a design that maximizes SWR bandwidth while maintaining reasonably flat gain across the *entire* 20-m band. As a result, maximum gain at the design frequency is a fraction of a decibel less than in either of the other two designs, and F/B ratio can drop from a maximum of 30+ dB at 14.200 to 15 to 20 dB at the frequency extremes. The 50-Ω SWR is under 3.0:1 anywhere between 14.000 and 14.350.

- The "Phone" column is a design that keeps the SWR under 2.0:1 across the 14.150 to 14.350 range that corresponds to the full Amateur Extra license privileges in the United States. Front-to-back ratio is in excess of 30 dB at 14.250 and stays above 20 dB throughout the full 200-kHz segment.

The element tapering of Fig. 12.4A is used for this three-element Yagi. All half-elements of all three designs in Table 12.2 use identical taper schedules except for the length of the tip (½-in OD) sections, which is given in Table 12.2. Only the longest tip of all—the half-element for the "CW" Reflector—is shown in Fig. 12.4A. Thus, assembly of the six half-elements is identical, and uses exactly the same lengths of tubing, except for the final setting of the tips.

In all three designs, the input impedance is very close to 75 Ω at the design frequency (after passing through a 1:4 step-up balun or other transformer). Thus, the SWR bandwidths in systems employing 75-Ω feedlines are all noticeably wider than the stated 50-Ω bandwidths above. Free-space gain at the design frequency is approximately 8.5 dBi, typically varying ±0.25 dB throughout the useable bandwidth, but the designs have been tweaked for mounting 70 ft (approximately one wavelength) above flat ground consisting of "average" or "medium" earth.

Multimode DXers or contesters might choose the "CW" settings if the bulk of their voice-mode activity is apt to occur below 14.220. Alternatively, adding an adjustable ATU or fixed-component matching network at the equipment end of the transmission line should easily permit reduction of the SWR at the transmitter or amplifier to under 2.0:1 across the entire 20-m band. One setting of the ATU or matching network is sufficient—use a coaxial bypass relay or two-position switch to switch the matching network in and out of the circuit for the two halves of the band with either the "CW" or "Phone" design. With this approach, the main reason to choose the settings of one design over the other is to place the best F/B ratios in either the low or the high end of the 20-m band.

To recap, the six element halves of each design differ in only two ways:

- The lengths of the outermost sections, or *tips,* of the three elements vary according to the selected design frequency and which element they are part of (as given by Table 12.2).

- The director and reflector are continuous tubing as they cross the boom, while the driven element is split at the boom to allow connection of a 4:1 balun.

In general, where two tubes are telescoped, the length of internal "overlap" should be 3 in or greater. For a detailed discussion of techniques for securing telescoping sections, refer to the earlier section in this chapter on rotatable dipoles.

Wire Yagis

It is not necessary to use tubing or pipes for the antenna elements in order to obtain the benefits of the Yagi beam antenna. An example of a two-element wire beam is shown in Fig. 12.5. The wire beam is made as if it were two half-wavelength dipoles, installed

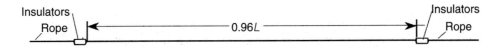

$$Z_o = 276 \log \frac{2S}{d}$$

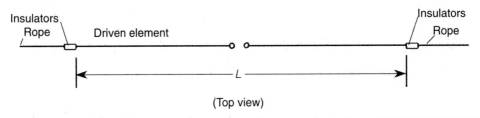

(Top view)

FIGURE 12.5   Wire beam especially useful for HF bands.

parallel to, and about 0.2 to 0.25 wavelengths apart from, each other. This particular set of dimensions is for a parasitic *director*, usually (but not always) made evident by the unfed wire being *shorter* than the driven element. Although the two-element wire beam is the most common, multielement wire beams are also possible. Some leading DXers and top-scoring contesters use wire Yagis of up to seven (!) elements fixed on compass

Band	80	40	20
Design frequency	3.600 MHz	7.100 MHz	14.125 MHz
Reflector–driven element spacing	38'3"	19'5"	9'9"
Driven element–director spacing	35'6"	18'0"	9'½"
Reflector length	139'4"	69'3"	34'7"
Driven element length	134'5"	67'10"	34'½"
Director length	129'9"	65'9½"	33'1"
Forward gain at design frequency	9.6 dBi	11.8 dBi	12.7 dBi
F/B ratio at design frequency	20 dB	25 dB	32 dB
F/S ratio at design frequency	20 dB	25 dB	24 dB
Z_{IN} at design frequency	85 Ω	22 Ω	17 Ω
Matching network	λ/4 Q-section	4:1 balun	4:1 balun
2:1 VSWR bandwidth	75 kHz (75 Ω)	100 kHz (75 Ω)	200 kHz (75 Ω)

TABLE 12.3   Three-Element Copper Wire Yagi Dimensions

headings of most interest or use to them. One of the most frequent uses of wire beams is on the lower bands (e.g., 40, 75/80, and 160 m), where rotatable beams are more expensive and often far more difficult (and expensive) to install at useful heights. Bear in mind, however, that the small element diameter of wire beams leads to a narrower bandwidth and a slight reduction in the maximum obtainable forward gain.

Dimensions for some three-element wire Yagis are provided in Table 12.3. In all cases, #12 uninsulated wire is used, and the Yagis are assumed to be about 70 ft above average ground. These beams give up about 0.5 dB of forward gain relative to ones made of aluminum tubing in order to get reasonable F/B and F/S ratios and a good match to either 50-Ω or 75-Ω feedlines through a 4:1 balun. Since the height chosen is a smaller percentage of a wavelength as the frequency decreases, there is greater ground loss at the lower frequencies and the elevation angle of peak forward gain increases. Using a 4:1 balun to bring the feedpoint impedance up to something compatible with commonly available coaxial lines, the 2:1 SWR bandwidth for a 75-Ω system is about 1.5 percent of the center frequency. The dimensions in this table are *not* correct for Yagi elements made of aluminum tubing and will be slightly off for different heights above ground.

Multiband Yagis

Largely because of the costs and complexities associated with erecting multiple beams high in the air, amateurs and professionals have developed a variety of approaches to covering more than one frequency from a single boom:

- Interlaced elements
- Trapped elements
- The log-periodic beam
- Adjustable (motorized) elements

In principle, beams with interlaced elements are easily understood; in practice, the electrical interactions and mechanical issues that are created by putting all these elements on a shared boom add to the design complexity and cost, and almost always lead to some compromises in performance. Years ago, most beams with interleaved elements were designed with manual calculations, supplemented by experimental results at test ranges. Today, a lot of the drudge work is handled by antenna modeling programs and the design of an effective multiband beam can be optimized far quicker.

One advantage of interlaced beams is that the user can opt to have a separate feedline for each band. In some applications (such as amateur multitransmitter stations), this is highly desirable. One disadvantage of interlaced beams is that they can be mechanical nightmares to construct, erect, and keep up in the air. Another disadvantage (common to *all* multiband beams, it should be noted) is that often the performance on one or more of the covered frequency ranges is compromised. Nonetheless, for some users they represent the best solution to a specific need.

Arguably the most popular of the multiband antennas is the *trap tribander* (sometimes a five-bander if provision has been made for the 12-m and 17-m WARC bands). Although most tribanders are sold into the amateur market, trapped tribanders designed for key shortwave broadcast bands or long-haul military HF comm links are also available.

Traps are nothing more than lumped-component parallel-resonant LC circuits inserted partway along both sides of each element, similar to the ones used in trap dipoles and shown schematically in Fig. 8.1A. In a 20-15-10 tribander there are two parallel-resonant circuits on each half element. The one closer to the boom is parallel resonant in or near the highest band of operation (10 m). On the next lower band (15 m), the increased reactance of the 10-m capacitor C1 essentially removes it from the circuit, while the 10-m inductor, L1, functions as a loading coil, allowing some shortening of the total element relative to a half wavelength.

The second trap in each half-element is parallel resonant near 15 m. In some commercial tribanders this circuit is housed in the same assembly as the 10-m circuit, but in others it is a foot or two farther from the boom, in a completely separate enclosure. Either way, it serves to decouple, or disconnect, the outermost part of the element from operation on 15 m.

On the lowest band (20 m), both C1 and C2 are essentially out of the circuit because of their high reactances, and the traps are electrically equivalent to two lumped-component inductors.

Because of these lumped-component inductors, each element half of a trapped beam is shorter than its counterpart in a conventional beam for the same frequencies. However, those same inductors may add weight to each half-element, as compared to the weight of the "missing" aluminum tubing in an equivalent untrapped (and longer) element.

One disadvantage of a multiband Yagi employing traps is that the spacing between adjacent elements is likely to be less than optimum on at least one of the bands covered by the beam. Nonetheless, a good antenna designer can often play with the effective element lengths on each band to partially counteract the lack of freedom in choosing unique interelement spacings for each band.

Traps always add RF energy loss to the antenna system. The exact amount of loss is hotly debated, but it has two components:

- Total signal strength at a distant point is reduced by the $I \times l$ contribution that would have come from each portion of the radiating element that has been replaced by a lumped-component, nonradiating, inductor.

- Resistive losses in the coils can be quantified indirectly if the *Q-factor* of the coils is known. High-Q coils will minimize such losses, but they are invariably heavier and costlier than the stock coils.

Bottom line: Trapped beams are extremely convenient. In return, they are reputed to give up, on average, somewhere between 0.5 and 2.0 dB compared to a trapless monobander of the same number of elements—part due to trap losses and part due to suboptimal element spacing on one or more bands. But depending on the user's operating interests, the additional losses may be insignificant—especially when weighed against the added real estate and dollar costs associated with multiple monobanders and separate supports, or the mechanical complexity and potential for interactions between beams mounted on the same support. Over the years, some outstanding contest scores and DXing accomplishments have been chalked up by operators using 20-15-10-meter tribanders. Especially popular for decades have been the Hy-Gain *TH-6* and *TH-7* family of trapped beams (now manufactured and sold by MFJ Products), the Mosley *TA-33* and *PRO*-series beams, and the Cushcraft *A-3S* and *A-4S*.

At this writing, the only *adjustable-element* Yagi on the market is the *SteppIR* family, manufactured by SteppIR Antennas, Inc. Small motors at the center of each element lengthen or shorten the elements of the company's two-, three-, or four-element Yagis in response to microprocessor commands from the radio room below. As a result, a single set of fixed-mounted elements can be optimized at all frequencies between 7 and 54 MHz, depending on the specific model purchased. Two- and three-element Yagis can almost instantly switch the direction of maximum forward gain by 180 degrees simply by commanding the parasitic elements to swap lengths, thus swapping their roles as reflector and director. Since the antenna dimensions do not have to be optimized for more than a single frequency at any one time, element spacing along the boom is virtually immaterial, although the SteppIR models that cover 6 m do include additional fixed-length parasitic elements for that band, interlaced with the motorized elements.

The SteppIR is clearly more expensive to manufacture and sell than a conventional monoband HF Yagi and, as you might expect, has a higher purchase price. But if you require a gain antenna capable of operation on any frequency between 7 and 54 MHz, it could well be the most economical approach when compared with the cost of multiband Yagis or a bevy of monobanders requiring multiple towers and/or stronger and taller masts. A bigger concern for some is the potential reliability and maintenance costs associated with having electronic circuitry located high in the air. Nonetheless, there are today many satisfied SteppIR users around the world.

Yagis with Loaded Elements

One of the most popular Yagi implementations on the market today is the monoband beam with shortened elements—especially on 30, 40, and 80 m, where beams with full-size elements pose mechanical challenges to the rotor and support system. Typically, elements are shortened in either of two ways:

- Lumped loading coils
- Linear loading wires

Keep in mind that while lumped loading coils may appear very similar to traps (especially when viewed from 50 to 100 ft below!), they differ from traps in that they are not meant to form a resonant circuit or act like a trap anywhere near the operating frequencies of the beam. A loading coil is simply replacing part of the radiating element with a lumped component inductor.

Like traps, loading coils have loss. The coils found on beams with shortened elements—notably the Cushcraft *40-2CD* and its successor, the *XM240*—are not high-Q coils. By comparison, the Force 12 *Delta* series of "shorty 40" loaded beams uses extremely high Q coils that are, not surprisingly, extremely bulky and heavy. Similar high-Q coils designed and sold by individuals are currently available; they are primarily of interest to a handful of amateurs who are building their own 40- and 75/80-m low-loss shortened Yagis.

But it is easy to overstate the practical effects of loss in the stock coils. The Cushcraft two-element 40-m beams have enjoyed an outstanding reputation for performance. One of the authors has owned four different brands of "shorty 40" two-element Yagis at one time or another during the past four decades—two with loading coils and two with linear loading wires—and has nothing but praise for the Cushcraft's electrical performance.

Two 40-m beams with linear loading currently on the market are the MFJ/Hy-Gain *Discoverer* series (in one-, two-, and three-element versions) and the M² *40M2LL* and *40M3LL* two- and three-element beams. The *Discoverer DIS-72* 2-element beam supersedes the venerable *402BA*. All these antennas insert an insulator midway out each shortened half-element; a loading wire runs from one side of the insulator back toward the boom, then back out to the other side of the insulator, thus effectively lengthening the element by folding a portion back on itself. Note, however, that linear loading wires are inherently low-Q components because their inductance is spread out over a large area and they have more loss resistance than high-Q coils for an equivalent shortening of the elements. They may also be subject to a slight (but temporary) shift in inductance with icing.

Feeding the Yagi

The feedpoint impedance of a dipole is on the order of 73 Ω in free space, although the actual impedance will vary above and below that figure for antennas within a wavelength of the earth's surface or other ground plane. The addition of parasitic elements properly spaced and adjusted for maximum forward gain and front-to-back ratio reduces the impedance even more. Often the feedpoint impedance of the antenna is too low—typically between 15 and 30 Ω—to be directly fed with coaxial cable, so some means of impedance matching is needed. For those designs, three feedpoint matching systems are in common use at HF:

- Balun
- Gamma match (and its cousin, the T-match)
- Hairpin match

Balun Feed

Some people feed the antenna through an impedance-matching *balun* transformer. However, not everyone appreciates that the balun for most three-element direct-feed Yagis needs to be a *step-down* configuration, *dropping* the 50- or 70-Ω impedance of the coaxial cable or hard line to one-fourth of either value, depending on the approximate input impedance of the specific antenna involved. See Chap. 4 for details.

Of course, the primary purpose of a balun—and the origin of its name—is to drive *bal*anced feedpoints, such as those found on most horizontal antennas, including dipoles and the driven elements of Yagis, with *un*balanced feedlines, especially coaxial cables. When used, the balun should be as close to the feedpoint as possible, such that the wires from the balun terminals to the feedpoint are almost nonexistent. Otherwise, the wire lengths need to be included in the calculations associated with the length of the driven element itself. Ideally, the "ground" terminal on the unbalanced side of the balun should be directly grounded to the boom of the Yagi at the balun, to minimize the impact of lightning-induced surges on the transmission line.

Gamma Match

Although our earlier discussion of the rotatable dipole—and later the driven element of multielement Yagi beams—was based on a λ/2 dipole split at the center to allow a direct *series* connection to a feedline, that is not the only possible way to feed the dipole.

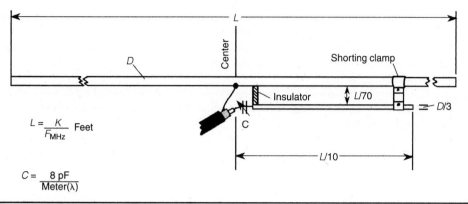

$$L = \frac{K}{F_{MHz}} \text{ Feet}$$

$$C = \frac{8 \text{ pF}}{\text{Meter}(\lambda)}$$

FIGURE 12.6 Gamma feed provides impedance matching.

The gamma match shown in Fig. 12.6 is one popular example of a *shunt*-feed system that attaches the feedline in *parallel* across a portion of the element that remains electrically continuous across its centerline. This approach is easy to describe: We are simply connecting the transmission line at a point (or *tap*) on the element where the impedance of the transmission line is comparable to the impedance of the standing wave of RF energy on the driven element (or dipole). We can best understand this by visualizing each half-element as a tappable resistor that is $\lambda/4$ in length. Thus, the role of the gamma match is to transform the impedance at some *tap point* along the element half to a resistive part that matches the Z_0 of the line and to simultaneously cancel out any leftover reactance.

To repeat for emphasis and clarity: The dipole or driven element of the gamma-matched Yagi is *not* broken in the center, as it is in the case of the simple series fed dipole. The outer conductor, or shield, of the coaxial cable is connected to the precise center point of the driven element, which is often directly grounded to the boom. The center conductor of the feedline is connected to one end of a capacitor that is in series with a component known as the *gamma rod*. If you are comfortable analyzing electronic circuits, one way to view the gamma rod is to think of it as link coupling to a zero-turn coil (the element). Early link coupling circuits used a moveable link to vary the "tightness" of coupling between the link and the primary resonating inductance in antenna tuners and plate tank circuits. Later antenna tuners used a fixed link but varied a capacitor in series with the link to ground—a configuration quite analogous to the gamma configuration of Fig. 12.5. Moving the shorting bar is equivalent to changing the number of turns on the link.

Here is one set of "cookbook" dimensions for gamma matching a Yagi to 50-Ω line:

- Distance from centerline of element to gamma rod shorting bracket: $L/10$
- Gamma rod diameter: $D/3$
- Spacing of gamma rod from driven element: $L/70$

where L = length of entire driven element, roughly equal to $\lambda/2$
        D = diameter of driven element

The first-time constructor of a gamma-matched driven element would be wise to make the gamma rod itself substantially longer than $L/10$, since these dimensions are only approximate, and both the position of the shorting bar and the setting of the capacitor will need to be varied by trial and error during the tune-up procedure described subsequently.

The capacitor in series with the center conductor of the coaxial cable has a value of approximately 8 pF per meter of wavelength at the lowest frequency in the band of interest or approximately

$$C \ (pF) = \frac{2400}{F \ (MHz)} \tag{12.2}$$

For any transmitter power levels more than a few watts, the capacitor must be a high-voltage transmitting variable type. In general, gamma match capacitors are either air or vacuum variables although those found on the older Cushcraft family of four-element monobanders for 20, 15, and 10 m consist of a fixed length of RG-8 coaxial cable with the outer jacket and shield braid removed so that just the center conductor and dielectric remain. This insulated wire is inserted into a portion of the hollow aluminum gamma rod, where the center conductor of the stripped coaxial cable forms one plate of the capacitor and the hollow gamma tube is the other.

Adjustment of the gamma match is accomplished in two steps:

- Adjust the length of the driven element (or dipole) for resonance, using a noise bridge, wattmeter, dip meter, impedance bridge, or other means.

- Alternately adjust the capacitor and the shorting bar/clamp for best impedance match between the antenna and the transmission line. Results of this tuning operation are best monitored at the antenna with a small portable analyzer (see Chap. 27) capable of measuring and displaying SWR or impedance curves over the frequency range of interest. The author uses a battery-powered AEA *HF Analyst* hauled to the top of his tower in a water bucket or hanging from his belt, but AEA and others have newer analyzers that are more compact and/or have more "bells and whistles" on them.

If a good match (as indicated by very low SWR) cannot be obtained with the initial spacing of the gamma rod from the element, write down the best SWR obtained at the design frequency and then increase the distance of the gamma rod from the element and go through the adjustment process again. If the SWR obtained this time is lower than the value written down, continue to increase the gamma rod spacing until no further improvement is possible. If, on the other hand, the SWR increased, reduce the spacing between the rod and the element and perform the adjustment procedure again. In some Yagi designs, it may also be necessary to shorten the driven element to get a match.

T Match

Not much needs to be said about the T match since it is nothing more than a balanced (or two-sided) gamma match. All the design considerations are identical. Some antenna experimenters prefer the T match because it presumably provides a more symmetrical input drive to the balanced driven element, but, in addition to requiring twice as many components, pattern skewing resulting from unbalanced drive to a dipole or driven element is hardly noticeable in practical usage. Actually, whatever advantage may accrue to minimizing pattern skew is probably negated by the loss in any lumped-component *balun* used to convert the unbalanced feed from coaxial cable to the balanced set of gamma circuits.

Hairpin Match

In some multielement Yagi designs, the feedpoint impedance at the terminals of the driven element lends itself to matching via a *stub*, which is a short piece of transmission line attached to the feedpoint terminals of the driven element and short-circuited at the far end. The driven element is split and insulated from ground on both sides, and the feedline attached somewhere along the stub between the feedpoint terminals and the short-circuit. Figure 12.7 is a sketch of a typical hairpin matching system. The exact point of attachment is one of the adjustments made when tuning the system for minimum SWR. Use of a stub is appropriate when the driven element impedance has a negative reactive (capacitive) component.

The hairpin has appeared under the trade name "Beta Match" on a number of MFJ/Hy-Gain beams over the years, including their line of *Long John* monobanders and the *TH-6DX* and *DXX* trap triband beams. (Strictly speaking, the Hy-Gain beta match differs slightly from a traditional hairpin because the former draws the boom into the stub calculations by having the hairpin wires straddle it.) On a commercially produced beam such as the Hy-Gain, the only adjustment necessary is to set the element lengths, the position of the feedline connection, and the position of the shorting bar to the dimensions shown in the instruction sheet for the center frequency you desire.

For a homebrew beam, tune-up is a two-step process similar to that of the gamma match:

- Adjust the length of the driven element (or dipole) for resonance, using a noise bridge, wattmeter, dip meter, impedance bridge, or other means. Then shorten each side of the element (to make it exhibit some capacitive reactance). A

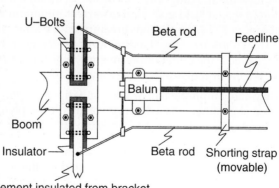

Figure 12.7   Beta match example.

reasonable starting point for HF hairpins is an initial reduction of approximately 1 in for every 3 m of operating frequency. Example: For a 15-m driven element, shorten each side by 4 or 5 in.

- Alternately adjust the distance of the feedline tap from the driven element and the shorting bar/clamp for best impedance match between the antenna and the transmission line. If you cannot obtain a good match, it may be necessary to go back to the first step and try another length for each side of the driven element.

The M² 40M2LLA two-element linear loaded monobander for 40 m uses a hairpin with the transmission line feeding a balun whose secondary leads connect directly to the driven element feedpoint terminals, so one degree of freedom (position of feedline tap along the shorted stub) is lost. Instead, the manufacturer's instructions effectively combine both of the preceding steps by having the user alternately adjust the position of the shorting bar and the lengths of the driven element tips in 1-in increments.

When designing your own hairpin match, the Smith chart (Chap. 26) can be a great help, allowing you to avoid heavy-duty number crunching by using its graphical techniques for stub matching. If you know the input impedance of the driven element, the stub-matching example will show you how to quickly obtain the hairpin dimensions and the point at which the feedline should be attached.

Moxon Beam

In Chap. 3 and again in Chap. 6 we took pains to emphasize that most of the signal strength developed at a faraway location from a dipole is a result of the high-current portions of the antenna. Thus, the standing wave of current in the outermost $\lambda/8$-section of each end of the dipole contributes little. That is not to say the outer sections of wire are useless and can be discarded—they aren't (useless), and they shouldn't be (discarded)! To the contrary, the outer ends of the dipole serve three very important purposes:

- They make the dipole structure resonant or nearly so when its total length is about a half-wavelength at the operating frequency, thus minimizing the reactive component of the feedpoint impedance.

- Because the standing wave of current on the antenna that causes efficient radiation to the far field goes from a current minimum to a current maximum in a distance of $\lambda/4$, the far ends are important because they "allow" the maximum current portions of the standing wave to emerge from the transmission line and appear near the center of the antenna, which is a structure physically designed to be a better radiator than the transmission line.

- They allow multiple half-wavelength dipoles to be connected together to form loops and other phased arrays of dipoles exploiting the principle that the standing wave of current reverses direction in each half-wave wire segment.

As noted in Chap. 6, one of the authors has had excellent results over many decades with 80-m (and, later, 160-m) dipoles configured for limited space by letting the outermost $\lambda/8$-section of each side drop down directly toward the ground. (See Fig. 6.17.) More recently, the same principle has been used by others to create two- and three-element Yagis for limited spaces by constructing the center half of each element out of rigid aluminum tubing and then dropping weighted lengths of #10 or #12 wire from the ends of each.

The two-element Moxon beam (Fig. 12.8), named for L. A. Moxon, G3XN, who first described it in 1993, is physically similar to an NVIS horizontal loop wherein the drooping ends of the two dipoles are pulled up to lie in the same plane as the centers of the elements but, unlike a loop antenna, the dipole ends are connected together mechanically but not electrically. Rather than being square, the outline of the Moxon beam as viewed from above is rectangular because the centers of the two bent dipoles are spaced less than λ/4 to optimize the antenna's performance as a parasitic array having one driven element and one reflector. Thus, each of the bent ends of the two elements must be somewhat less than λ/8 long and the center section of each element is perhaps 5λ/16 or more in length. An excellent summary of some practical Moxon wire beams for HF and VHF can be found on DK7ZB's Web site or at the Moxon Antenna Project site, www.moxonantennaproject.com.

Like all other two-element Yagis employing a reflector, forward gain, F/B, and SWR change more rapidly with frequency when approaching the design center of the beam from *lower* frequencies. Thus, dimensions and matching circuits should be optimized for the lower end of the band, since any of these performance characteristics will hold up respectably well higher up into the band. Of course, a Moxon beam constructed of tubing will have a wider bandwidth than one using wire.

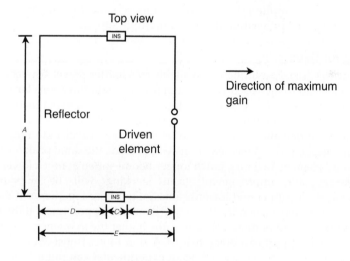

Top view

Direction of maximum gain

Reflector

Driven element

To cover the entire 20-meter band with a VSWR less than 1.6:1 into 50–Ω feedline:

    A: 296.0"
    B: 46.4"
    C: 6.0"
    D: 56.6"
    E: 109.0"

FIGURE 12.8 Moxon rectangle.

Some implementations of the Moxon utilize wires with spreaders (visualize the driven element of a cubical quad pointing straight up in the air), some use aluminum tubing for the element centers (i.e., the portions at right angles to the boom) and wires for the element ends, and others use aluminum tubing for the entire radiating structure. One recent implementation converted a Cushcraft *40-2CD* two-element shortened Yagi to a Moxon, thus trading loading coil losses in the original design for the possibly smaller losses of full-length elements with bent ends. A properly designed Moxon also boasts higher F/B ratio than the typical two-element Yagi.

The dimensions shown in Fig. 12.8 for a 20-m Moxon rectangle constructed with #12 wire yield a free-space forward gain of 6.14 dBi (or 4.0 dBd) in conjunction with F/B and F/S ratios of nearly 24 dB at the design frequency of 14.1 MHz. $Z_{IN} = 51 + j0.4 \, \Omega$ at the design frequency, and the SWR stays below 1.6:1 across the entire band. If all six segments of the 20-m design in the figure are scaled according to wavelength, the design should translate easily to other bands while retaining the 50-Ω resistive input impedance.

Stacking Yagi Antennas

There are two distinctly separate reasons to stack antennas on a single mast or tower:

- Desire for increased gain and pattern flexibility from interconnecting two or more antennas for the same frequency

- Need to put multiple bands on one support because of limitations (space, funds, zoning, etc.) preventing the use of multiple supports

Stacking for Gain

As we saw in Chap. 5, splitting the available transmitter power between two identical antennas can lead to increased signal strength in certain directions far from the antennas, at the expense of signal in other directions. Extensive experimentation and computer modeling have shown optimum HF or VHF stack spacings for a pair of identical beams mounted one above the other to be in the one wavelength range. Under ideal conditions, stacking two 3-element Yagis will provide the same peak forward gain as a single 6- or 7-element Yagi on a much longer boom—often at much lower support cost.

However . . . the improvement from stacking will be impaired—or even nonexistent—if the upper and lower beams are too close together or if the lower beam is too close to ground. Generally, stacking will be a disappointment if the lower beam is not $\lambda/2$ above ground or more. Of course, the higher the frequency, the more flexibility is possible on a support of a given height. A 70-ft tower is just about the shortest that can host a two-Yagi, 20-m stack, while an experimenter can put a 10-m stack at many different heights and spacings on the same tower.

To enjoy the benefits of stacked Yagis around the entire compass rose, the obvious (and expensive!) approach is a rotating tower—or a hybrid tower where only the portion from the lowest beam up rotates. But in many instances, coverage of a full 360 degrees is not a necessity, and the lower beam in the stack can attain as much as 300 degrees rotation by mounting its rotator on a short *outrigger* mast supported a foot or two from the nearest tower leg at both ends of the mast. Some modern rotator controllers, such as the Green Heron units, can be interconnected in a master/slave configuration so that the lower beam follows the commands for the upper through the more limited range. In an even simpler configuration, one or more lower beams can be fixed in position by being mounted directly to sides of the tower at the appropriate height(s). A DXer or contester in the northeastern United States might, for instance, attach beams aimed at 45 degrees (to maximize coverage into Europe) and 135 degrees (for South America and the southern half of Africa) directly to the tower at points roughly a wavelength below the top beam.

In practice, stack spacings on a guyed tower will be determined as much by guy wire attachment heights and element clearance distances as by any theoretical projections of stacking gain versus spacing. Luckily, exact stack spacing (*stacking distance*) is not critical.

Because the optimum stack spacing is expressed in terms of the number of wavelengths between the top and bottom beams in the stack, one might conclude that only monoband beams can be stacked with good results. That isn't true. Because the stacking gain curve exhibits such a broad peak as a function of spacing, multiband beams covering a 2:1 frequency span can be spaced as much as $3\lambda/4$ apart at the *lowest* frequency of operation, corresponding to a spacing of $3\lambda/2$ at the highest frequency. Thus, amateur HF tribanders or five-banders (when the 12- and 17-m WARC bands are included) covering a 2:1 frequency span can be stacked with very competitive results.

The normal assumption with stacking is that the feedline configuration to each beam in the stack is identical in length and impedance—a "T" connection, in other words, as shown in Fig. 12.9—thus applying roughly equal transmitter power in phase to the two Yagis. Enterprising HF DXers and contesters have found, however, that a useful adjunct to the two-beam stack is a coaxial cable relay control box with a four-position switch and associated transmission line phasing sections that allow the user to select *beams in phase* (BIP), *beams (180 degrees) out of phase* (BOP), *upper* beam only, or *lower* beam only. All of these choices are useful at one time or another not only because we are usually interested in communicating over a wide range of distances but also because the optimum elevation angle for HF ionospheric communications is always varying.

One advantage to using one or more fixed lower beams or having the upper and lower beams independently rotatable is that transmitter power (and receiving capability) can be split between two different directions at the same time. Thus, if looking to establish communications with two different parts of the world, the upper beam might be rotated in the direction of one continent while the lower beam is aimed at another.

Figures 12.10 and 12.11, provided by W. L. Myers, K1GQ, are plots of antenna

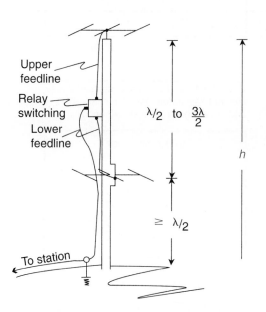

Upper & lower feedlines should be exactly the same length, measured from antenna feedpoint to switch box.

FIGURE 12.9   Feeding stacked Yagi antennas.

patterns obtained with *cocoaNEC*, antenna modeling software for the Mac written by Kok Chen, W7AY. In the two halves of Fig. 12.10, the dotted line in 12.10A is for a three-element HF yagi λ/2 above ground, and the solid line in 12.10A is for the same antenna spaced λ/2 higher, i.e., at one wavelength above ground. These plots correspond to feeding only the lower Yagi or only the upper Yagi of a two-Yagi stack, respectively. In Fig. 12.10B the dotted line corresponds to feeding power to the two antennas equally and in phase, while the solid line shows the pattern that results when the same power is fed to the two beams out of phase. Figure 12.11 is identical to Fig. 12.10 except that the antenna heights and spacing have been doubled in Fig. 12.11: The lower Yagi is now λ above ground and the upper Yagi is λ above the lower one, or 2λ above ground. Taken together, the heights of Figs. 12.10B and 12.11B show what to expect at the high- and low-frequency extremes (14 and 28 MHz) when stacking triband (or five-band) Yagis on a 70-ft tower. Perfect ground is assumed in the plots that follow.

Note that the maximum gain for the upper Yagi of Fig. 12.10A is 11.85 dBd, occurring at an elevation angle of about 14 degrees. In other words, the upper Yagi has a peak gain that is about 6 dB greater than that of a λ/2 dipole mounted at the same height above ground. The lower Yagi has a peak gain that is about 1 dB less, and it occurs at a higher elevation angle—about 24 degrees. (In other words, as we discussed in Chap. 5,

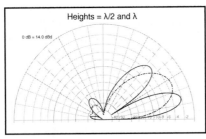

——— Upper @ λ
·—·— Lower @ λ/2

FIGURE 12.10A Patterns for individual three-element Yagi antennas at λ/2 and λ.

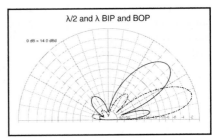

——— Beams out of phase (BOP)
·—·— Beams in phase (BIP)

FIGURE 12.10B Patterns for stacked three-element Yagi antennas fed in phase (BIP) and out of phase (BOP) at λ/2 and λ.

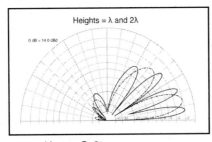

——— Upper @ 2λ
·—·— Lower @ λ

FIGURE 12.11A Patterns for individual three-element Yagi antennas at λ and 2λ.

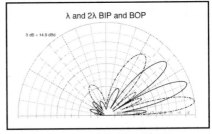

——— Beams out of phase (BOP)
·—·— Beams in phase (BIP)

FIGURE 12.11B Patterns for stacked three-element Yagi antennas fed in phase (BIP) and out of phase (BOP) at λ and 2λ.

the total gain at any given elevation and azimuth relative to some reference antenna, whether dipole or isotropic radiator, is the pattern of the antenna multiplied by the *array pattern*. In this case, the array pattern is formed in conjunction with the *image* Yagi located an equal distance *beneath* the ground; thus, the patterns of these two figures are those of a three-element Yagi multiplied by the ground reflection pattern for that specific antenna height. The peak boost provided by the ground reflection pattern over perfectly conducting ground would be 6 dB, occurring at an elevation angle that depends on the Yagi's height above ground; in general, the gain is somewhat less over real ground.)

Which antenna height is better? Certainly for very long distance communication, the upper Yagi has the benefit of slightly higher gain and noticeably lower *takeoff* angle for its main lobe. But at midday on the HF bands, the lower Yagi might well be the superior performer as DX signals shift to higher takeoff and arrival angles, especially if both stations involved in the communications are in North America, are in Europe, or even are attempting to communicate between eastern North America and western Europe. In particular, the lower Yagi is near the peak of its response curve at elevations around 30 degrees, where the upper Yagi has a deep null. The depth of the null will depend on the actual nature of the ground within a few wavelengths from the base of the tower (for flat terrain). Which antenna height is better? It depends. At one time or another, either one is better.

Now that we have pattern data for the two antennas, let's look at Fig. 12.10B to examine the effects of splitting our transmitter power between them by feeding them *in phase* (BIP) and *out of phase* (BOP). Here's what we observe:

- The peak of the broken curve (BIP) in Fig. 12.10B is 13.2 dBd, or 1.3 dB higher than the peak gain of the high Yagi of Fig. 12.10A. This *stacking gain* is less than the theoretical maximum of 3 dB because of the close spacings and because 3 dB can be attained only when the vertical elevation patterns of the two beams are identical (as in free space). Nonetheless, there are many communications scenarios when an extra 1.3 dB means the difference between success and failure.

- The elevation angle of the BIP main lobe (14 degrees) is the same as that of the upper beam alone.

- Feeding the two beams out of phase (BOP, solid line) creates a single main lobe similar to the main lobe of the lower Yagi alone but enhanced with 1.3 dB stacking gain and centered on a higher elevation angle (42 versus 24 degrees). As a typical day on the HF bands progresses, one might reasonably expect to progress from BIP through Lower to BOP on most transoceanic paths. Upper would be useful primarily for very long haul links at those installations where the lower beam was fixed on a different azimuthal heading and not able to contribute to forming the BIP pattern.

- Although the apparent benefit of stacking beams at $\lambda/2$ and λ above ground is "only" 1.3 dB compared to the peak pattern response of the upper beam alone, there are other benefits to stacking. BIP and BOP allow us to double the number of elevation angle ranges we can select with two beams. And often more important than stacking gain is the reduction in interference that comes from eliminating the secondary (upper) lobe of the higher Yagi by using BIP mode.

Figure 12.11 provides the same elevation pattern information for beams mounted at λ and 2λ above ground (corresponding to the 10-m band on the 70-ft tower of this example or, alternatively, stacked 20-m Yagis at 70 and 140 ft on a taller tower). Note that the solid curve (Upper) of Fig. 12.10A is now the dotted-line curve (Lower) of Fig. 12.11A. Because the two antennas are farther from ground and from each other (in terms of wavelengths) than in the previous example, the stacking gain has increased to about 2.0 dB. We also see more and thinner lobes the higher we go, but we should take the depth of the nulls with a grain of salt since the higher we go in frequency, the less perfect our actual ground is likely to be. And just like the previous pair of antennas, as we go through a typical HF day we should expect to progress from BIP to Lower to BOP, followed by the reverse sequence, for most transoceanic comm links.

Taken together, the two charts are useful in predicting the performance—both individually and together—of two 20-15-10 or 20-17-15-12-10 multiband beams mounted at 35 ft and 70 ft on a single tower. The curves of Fig. 12.10 then are representative of 20 m and those of Fig. 12.11 of 10 m. Before committing to those heights in a real installation, however, we would need to run plots for the three other bands to be sure there were no "surprises" awaiting us.

Also, keep in mind that these plots are based on three-element beams. Optimum stack spacings and heights above ground vary with the number of Yagi elements (and boom length), so be careful not to misapply these figures to other configurations.

From the antenna modeling results summarized in Figs. 12.10 and 12.11 we can conclude that having two antennas available at two different heights, selected singly or in combination, ensures a reasonable pattern gain at virtually any elevation angle useful for communicating over ionospheric "skip" distances. This, rather than any specific stacking gain, may be the real advantage of having higher and lower beams for the same band at HF.

Finally, don't underestimate the value of easy maintainability. The added benefit of providing full 360-degree rotation for the lower beam in a stack compared to, say, 300-degree coverage with one or more fixed or side-mounted beams may not be worth the extra cost and complexity. And the added gain of a four- or five-element Yagi compared to a simpler three-element beam may be completely nullified if something goes wrong with the driven element or matching network and you can't easily reach that far along the boom while buckled to the top of the tower.

Stacking for Limited Space

The other common use of the term *stacking* is when unrelated antennas for different frequency ranges or bands are mounted on the same mast or support structure. In this case, there is no intention to realize added gain; rather, the hope is that unrelated beams can be mounted near each other without adversely affecting either's performance. Unfortunately, there is no known way to assure the user in advance that such interaction won't occur except through the use of antenna modeling software in conjunction with extremely accurate antenna models and precise height and spacing measurements. Otherwise, all the user can do is fall back on anecdotes and experimental results reported by owners of specific combinations of stacked beams. While some manufacturers of commercial beams may specify a minimum stacking distance, the spacings they call for (typically on the order of $\lambda/2$ at the *lowest* frequency!) are ridiculously large for practical HF installations.

Instead, here are some commonsense guidelines that may be helpful in planning your installation:

- As a broad generality, lower-frequency beams (larger dimensions) will have a greater effect on higher-frequency beams (smaller dimensions) than the reverse.

- Try to avoid stacking antennas with full-size or trapped elements for harmonically related frequencies (such as 10 and 20 m). Especially avoid third harmonic combinations such as 40 m and 15 m. Instead, combine 15- and 20-m beams, or 10 and 15, or 12 and 17, etc. This prohibition is often not necessary for a beam with loading coils (such as the Cushcraft 40-2CD or XM240, or the Force 12 Delta family) or linear loading (such as the M² 40M2LL) because a lumped inductance that makes the elements resonant or nearly so on the design band will likely make the same element nonresonant on the harmonics.

- If it is absolutely necessary to stack two beams that interact, consider rotating one of the beams 90 degrees with respect to the other. However, be aware that the boom of one beam may well be resonant in the frequency range of the other; addition of a *current breaker* or other form of boom detuning may help.

- Choose the overall height of one or both beams, if possible, to create a ground reflection null in the elevation pattern(s) at 90-degree elevation (straight up). If your antennas are for long-distance work, you should already be striving for this.

- Accept the fact that some models from some manufacturers are inherently more susceptible to interaction than others.

- Put the lower-frequency beam *above* the higher-frequency one whenever the tower and mast strength allow it.

- Consider using a multiband beam (having traps or interlaced elements on a shared boom) in place of separate beams.

Examples of some stacking combinations that one of the authors has had good luck with over the years include:

- Hy-Gain *TH-6DXX* (20-15-10) and *402BA* (40) separated by 5 ft at 62 ft and 67 ft
- Cushcraft *40-2CD* (40) and *A-4S* (20-15-10) separated by 5 ft at 92 ft and 97 ft
- Cushcraft *20-4CD* (20) and *15-4CD* (15) separated by 5 ft at 125 ft and 130 ft
- Cushcraft *20-4CD* and *40-2CD* separated by 5 ft at 92 ft and 97 ft
- Cushcraft *20-4CD* and M² *40M2LL* (40) separated by 5 ft at 92 ft and 97 ft

In one case, the two beams (for 20 and 15) were not harmonically related; in all other cases, one of the two antennas employed loaded elements that kept it from being resonant on other bands.

Rolling Your Own

In years past, this book might have contained a greater number of charts displaying dimensions for a variety of different Yagi design philosophies and optimization preferences. However, there are many excellent resources already in print, as summarized here. Perhaps more to the point, however, the easy availability of inexpensive or free antenna modeling software makes it possible for any reader interested in designing and building a custom Yagi from scratch to "play with the numbers" and develop his/her own intuitions as to how key parameters (forward gain, F/B ratio, bandwidth, etc.) vary as dimensions are changed.

Further Reading

The designs and construction techniques of this chapter do not take into account the special demands of severe environments. The unusually high winds of mountaintops and coastal areas, the icing that must always be expected between midlatitudes and the poles, and saltwater mist for coastal and island installations all impose extra burdens on antennas and their supports. Further, the interplay between mechanical and electrical considerations becomes far more critical the longer the boom and elements are.

The definitive resource for understanding why and how to strengthen a Yagi is *Physical Design of Yagi Antennas,* by David B. Leeson, W6NL (ex-W6QHS), available from ARRL. In addition to explaining why and to what extent booms, elements, brackets, and masts should be reinforced and balanced, it is chock-full of valuable practical information—extending beyond purely mechanical considerations—for anyone interested in modifying commercial antennas or designing and building his/her own.

Another seminal work is *Yagi Antenna Design,* by James L. Lawson (deceased), ex-W2PV, ARRL, 1986. Here the principal focus is on the electrical design and performance of HF Yagis, and many practical families of designs with complete dimensions are provided for beams having between two and eight elements. Unfortunately, this book predates amateur access (in the United States, at least) to the so-called WARC bands at 30, 17, and 12 m.

Low-Band DXing, by John Devoldere, ON4UN, ARRL, fifth edition, 2011, is not primarily about Yagi antennas but does treat their design and performance on 80 and 40 m at some length. There is also an extended section on hairpin matching, including some explanatory calculations coupled with experimental techniques for obtaining the best match.

CHAPTER 13

Cubical Quads and Delta Loops

Two popular multielement types of antennas employ elements formed from wire loops having a total length of approximately one wavelength. The *cubical quad* employs *square* loops and the *delta loop* is built with *triangular* loops. Both antenna types are found in single-element and multielement varieties, and both are easily analyzed in terms of equivalent half-wave dipoles that make them up. The single-element version of each is a form of 1λ loop that was evaluated in Chap. 7.

Cubical Quad

The *cubical quad* was desgned in the mid-1940s at radio station HCJB in Quito, Ecuador. HCJB is a Protestant missionary shortwave radio station that delivers a booming signal worldwide from its high-altitude location. According to the story, HCJB originally used Yagi antennas to provide consistent signal strength in selected directions. In the thin air of Quito, the high voltages at the ends of the elements caused *coronal* arcing that resulted in instantaneous voltage standing wave ratio (VSWR) changes capable of shutting down or damaging the high-power transmitters and, over time, destroying the element tips themselves. Station engineer Clarence Moore designed the cubical quad antenna to solve this problem. Figure 13.1 shows the dimensions and construction techniques for a typical quad loop.

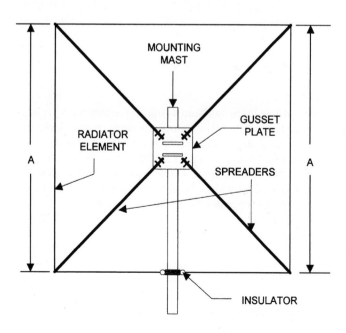

FIGURE 13.1 Quad loop antenna.

Because it is a full-wavelength antenna, with each side a quarter-wavelength long and fed at a current loop in the center of one side, the voltage maxima occur $\lambda/4$ from the feedpoint in either direction. At those maxima the wire loop is continuous (in contrast to the open-circuited element tips of a dipole or Yagi), thus substantially reducing the tendency to arc.

In addition to the reduction in coronal noise identified by Clarence Moore, the quad loop antenna is preferred by many people over a dipole for two other reasons:

- The quad loop requires a smaller footprint for a given frequency or band because it is only a quarter-wavelength long across the top or bottom (dimension A in Fig. 13.1).
- The loop form is reported by many to make it less susceptible to local electromagnetic interference (EMI) and precipitation static.

The quad loop of Fig. 13.1 can be used by itself with the loop orientation in any plane whatsoever. It can also be used as the basic building block for a multielement cubical quad. When fed at the center of a horizontal side, the single-element loop will exhibit a horizontally polarized figure eight azimuthal radiation pattern similar to a dipole's, but the elevation pattern will be squashed in amplitude at high angles. The reason for this can be seen by recognizing that the quad loop is actually a simple example of an *array*, which we discussed in Chaps. 5 and 10. In particular, the loop consists of two "bent" dipoles whose centers are spaced $\lambda/4$ apart. When fed as described (i.e., horizontally polarized) and compared to a $\lambda/2$ dipole having the same orientation as the fed leg of the loop, the quad loop exhibits about 1 dB greater broadside gain in free space. Factors keeping the quad from realizing full stacking gain of a two-element array are the closeness of the two bent dipoles and the mutual coupling effects between the two.

Just as a Yagi is formed from a $\lambda/2$ driven element in conjunction with reflectors and directors, a multielement cubical quad consists of a driven quad loop and one or more parasitic loops whose actual or electrical lengths have been adjusted slightly from a nominal $\lambda/2$ so that each acts as a reflector or a director. In practice, two-, three-, and four-element cubical quads are commonly found.

Quad versus Yagi

In the years since Clarence Moore's article there has been a running controversy regarding the performance of quads as compared to other beam antennas, particularly the Yagi. Some users claim that the two-element cubical quad has a forward gain of about 1.5 to 2 dB more than a two-element Yagi (with a comparable boom length between the two elements). In addition, some claim that the quad has a lower angle of radiation. Many feel that the quad seems to work better at low heights above the earth's surface but that the difference disappears at heights greater than a half-wavelength.

One difficulty with any of the comparisons is defining the "height" of a quad. When comparing quads to Yagis, for instance, do you put the height of the top wire on the quad loop at the same height as the Yagi or do you set the equivalent boom heights (i.e., the center of each array in the vertical plane) to be the same?

Another difficulty is deciding what performance measures are most important. Forward gain? Front-to-back (F/B) or front-to-side (F/S) ratio? VSWR bandwidth? Gain bandwidth? Total weight? Wind load? Boom length? Ease of installation? These are important choices to make early on because, as a general rule, in neither Yagis nor quads do maximum forward gain, F/B, F/S, and bandwidth occur at the same dimensions or frequency.

With the availability of good modeling software, we can get at some of the comparisons, but the ultimate decision as to which type of antenna is "better" will depend on how these various parameters are prioritized. To expose some of the tradeoffs, the author recently modeled (for 15 m) a quad loop against a dipole and then a two-element quad against a two-element Yagi using *EZNEC 5+*. All were fed and oriented for horizontal polarization. Subject to the assumptions in the author's model for each antenna, the results for a single element indicate:

- In free space a single-element square loop exhibits about 1.5 dB better broadside gain than a single bent dipole whose center has the same orientation as the fed leg of the loop. But since a bent dipole has lost about 0.5 dB of broadside gain compared to a standard $\lambda/2$ dipole that is horizontal across its entire length, the loop is only about 1 dB better than a conventional $\lambda/2$ dipole in free space.

- Compared to a conventional horizontally polarized $\lambda/2$ dipole at the same height above "average" ground as the top (horizontal) leg of a loop fed at the center of either the top or bottom leg, the loop exhibits lower broadside lobe gain until a height above ground of about 0.67λ. At that point the loop becomes better—and remains that way through at least a height of 2λ. The differences throughout that range are on the order of 0.5 to 1 dB. See Fig. 13.2.

- Throughout the entire range of heights from $\lambda/4$ to 2λ over average ground, the dipole has a slightly lower elevation angle (angle of peak broadside radiation). This is understandable, since approximately half of the RF energy feeding the loop is radiated by the lower bent dipole, thus contributing some higher-angle radiation to the loop's overall pattern. In fact, careful examination of Fig. 13.2 reveals that the elevation angle of the quad loop's peak radiation is the same as that of the dipole when the midpoint of the quad (halfway between the top and bottom horizontal wire segments) is at the same height as the dipole—i.e., when the top wire in the quad is $\lambda/8$ higher than the dipole. Another way of saying this is that the quad is a two-element array and its *net* ground reflection factor is based on the *average* height of the radiators in the array—i.e., the vertical center of the array.

At first blush, a two-element quad does enjoy about 1.2 dB forward gain advantage over a two-element Yagi at the same design frequency (in this case, 21.2 MHz) when both are modeled with #12 wire for their elements. It's possible to tweak the Yagi dimensions to reduce that advantage to somewhere between 0.5 and 1.0 dB at the expense of lower F/B and F/S ratios for the Yagi. However, in addition to element spacing and lengths, element diameter is also a factor. In particular, a "practical" rotatable Yagi will have elements of aluminum tubing instead of #12 wire supported by insulated spreader arms. This improves the bandwidth of the Yagi for a given boom length, all other factors being the same. Thus, after a few hours at the computer with *EZNEC 5+*:

- A two-element Yagi modeled with either #12 copper wire or 0.75-in aluminum tubing and 0.16λ spacing (roughly 7½ ft) between reflector and driven element exhibited 6.5 dBi (*decibels relative to isotropic*) forward gain and more than 10 dB

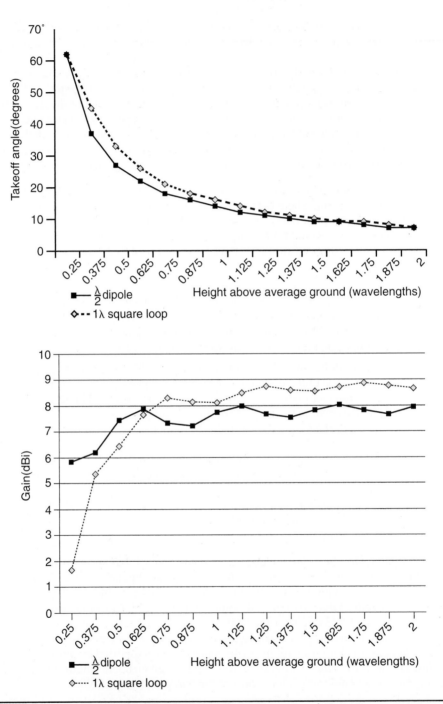

Figure 13.2 Gain (bottom) and takeoff angle (top) of 1λ quad loop versus λ/2 dipole over average earth.

F/B ratio at 21.2 MHz. Perhaps more important, however, the forward gain varied by less than 0.6 dB across the entire band and the VSWR never rose above 1.5:1 anywhere in the band when fed directly with 50-Ω cable (i.e., with no matching network required).

- A two-element quad modeled with 0.21λ (about 10 ft) spacing exhibited 7.6 dBi forward gain with F/B and F/S rejection comparable to the Yagi just described. Its input impedance at the design frequency was 120 Ω, which can be transformed down to 48 Ω with a $\lambda/4$ section of 75-Ω cable located at the feedpoint. However, there was substantial variation in the input impedance across the band, resulting in a much narrower VSWR bandwidth, and the F/B and F/S ratios showed a 2:1 variation, as well.

- At element spacings ranging between 0.11λ (about 5 ft) and 0.21λ it is possible to maximize *some* performance specs near the design frequency at the expense of others, but the two-element wire quad always shows a greater variation in input impedance (hence, VSWR bandwidth) and F/B or F/S ratios with frequency than does the two-element Yagi constructed of either copper wire or aluminum tubing.

- On the whole, it is easier to maintain F/B and F/S ratios in excess of 10 dB at the design frequency with the two-element quad, but these ratios show much greater variation across the band than do those of the Yagi.

- It is somewhat easier to adjust the Yagi driven element (in both the model and real life) by lengthening or shortening the element. Similar adjustments to the quad driven element affect all four legs unless the supporting frame is distorted.

Like the basic 1λ loop, one advantage of the cubical quad is *footprint*. The "wing-span" of each quad element is roughly half that of each corresponding element of a full-size Yagi, so the *turning radius* of a two- or three-element quad will almost always be somewhat smaller than that of a Yagi of the same number of elements. The advantage is somewhat less when the quad is compared with a Yagi employing traps, loading coils, or linear loading wires to shorten its elements.

Proximity to earth is no boon to the quad. With the top leg of each element at 1λ (e.g., 46 ft on 15 m), the two-element quad exhibits almost exactly 1-dB advantage over the Yagi at the same height over average ground, albeit at a slightly higher wave angle (15 degrees versus 14 degrees). As both antennas are brought closer to ground, the quad begins to lose its gain advantage, and by a height of $\lambda/2$ the Yagi actually enjoys a 0.5-dB advantage in forward gain, which occurs at 26 degrees versus the quad's peak at 29 degrees. Again, this is not surprising, since part of the power fed to the quad goes into the lower leg, where the greater proximity to earth causes it to favor relatively high wave angles.

Feeding the Quad

The driven element of a cubical quad can be fed in the center of a horizontal side (Figs. 13.1A and 13.3A), in the center of a vertical side (Fig. 13.3B), or at a corner (Fig. 13.3C). In all cases, the currents on opposing sides of the quad geometry are in phase; thus, maximum gain is along the axis of the loop—that is, on a line perpendicular to the plane of the loop. Thus, for all the quads of Fig. 13.3, maximum gain is into and out of the page.

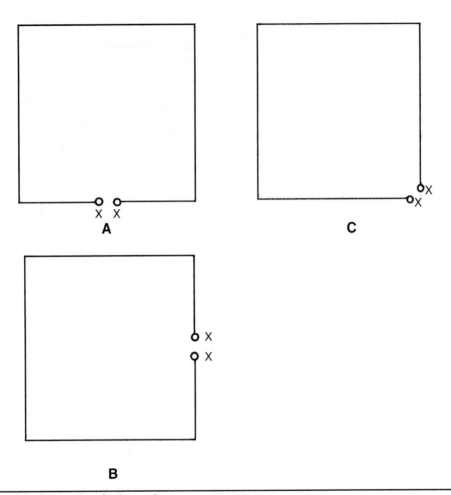

Figure 13.3   Feed options for the quad.

A loop located within a few wavelengths of ground or another reflecting surface is actually an array of *four* elements: the two original bent dipoles plus their two "image" antennas. As discussed in Chap. 5, when the bottom of the loop can be lifted to a height of at least a half-wavelength above the earth or other reflecting surface, it begins to make sense to utilize horizontal polarization by feeding it at the midpoint of the top or bottom side. In this manner, maximum utilization of a ground reflection factor that may be worth up to 6 dB at a distant receiving point is attained. For most amateurs, this suggests horizontal polarization at 14 MHz and above. At support heights below λ/2, however, the ground reflection factor for horizontally polarized antennas tends to put most of the radiation at very high elevation angles, so if long-distance communication is the user's objective there is merit to switching to vertical polarization by feeding the loop on one side or the other. Of course, if you wish to hedge your bets, or your quad loop is at that "difficult" height, you can feed it at a corner to obtain equal parts horizontal and vertical polarization.

There is virtually no difference in the pattern or gain of a quad loop caused by feeding the bottom leg instead of the top. The choice should be determined by the specifics of your installation, purely on the basis of simplicity, convenience, and cost. Feeding at the bottom usually results in a shorter feedline and its associated losses. *Boundary conditions* for the two bent dipoles that form a loop element limit its operation to the band for which it is designed. (That is, favorable broadside array gain of two bent dipoles having their ends connected requires the currents in opposing legs of the loop to be in phase.) Therefore, multiband cubical quad antennas are typically created the "brute force" way, by suspending one wire loop for each higher-frequency band inside the loop for the lowest frequency, thus giving rise to a "spiderweb" appearance. All bands share a common boom but the parasitic elements of one band may not be coplanar with those of any other band. Typically the driven element for each band is independently fed; separate feedlines can be run all the way back to the transmitter room or a remote mast-mounted coaxial switch capable of handling the expected transmit power can be employed at considerable savings in cable costs.

Construction Ideas

In its simplest form, a single-element quad loop antenna (single-band or multiband) is mounted to spreaders connected to a square gusset plate. At one time, spreaders were made by treating bamboo stalks with a preservative, but some quads have been built with aluminum tubing adorned with standoff insulators to support the wire loop at the ends of the four spreaders. Today most quads are built with fiberglass spreaders specifically manufactured for this market. A number of kits are advertised in amateur radio magazines and Web sites.

Details for a one-element gusset plate are shown in Fig. 13.4. The gusset plate can be made of aluminum like a Yagi boom-mast clamp, or it can be a strong insulating material such as fiberglass or ¾-in marine-grade plywood. The plate is mounted to the support mast using two or three large U-bolts (stainless steel or galvanized, to prevent corrosion). Spreaders are mounted to the gusset plate using somewhat smaller U-bolts.

Element spacing in a multiband quad poses a design problem analogous to that found in multiband trap Yagis. In principle, element spacings for a 10-m quad should be half the equivalent spacings for a 20-m quad if the two bands are optimized according to the same rules. Clever mechanical design, exemplified by the boom-spreader device—known as a *spider*—of Fig. 13.5, has solved this problem for the two- and three-element multiband quads. A four-element quad is a bit more complex, however. Nonetheless, some four-element quad designs employ the gadget of Fig. 13.5 in conjunction with three separate single-loop boom-spreader brackets spaced along the far end of the boom to support and properly position a fourth element (usually a second director) on each band. Alternatively, a "half-spider" that secures the spreaders for all second directors at some angle other than 45 degrees might work, depending on the spacings called for within a given design.

Figure 13.6 shows a schematic representation of a horizontally polarized two-element quad with elements spaced 0.11 to 0.22 wavelengths apart. The driven element is connected to the coaxial-cable feedline directly or through a $\lambda/4$ matching section of 75-Ω line. The other element is a reflector, so the total wire length of the loop is a few percent longer than that of the driven element. A tuning stub can be used to adjust the reflector loop for maximum gain or F/B ratio.

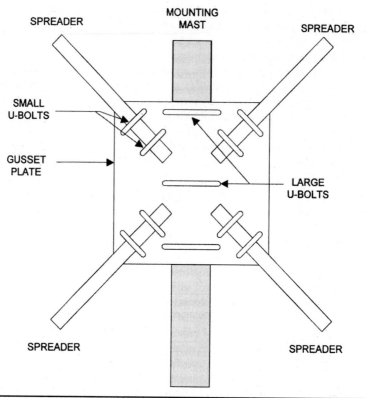

SPREADER

MOUNTING
MAST

SPREADER

SMALL
U-BOLTS

GUSSET
PLATE

LARGE
U-BOLTS

SPREADER

SPREADER

FIGURE 13.4   Quad loop gusset plate.

Because each element is a square loop, the actual length varies from the naturally resonant length by about 4 percent, but the forward gain and F/B ratios are more sensitive to the reflector dimensions than they are to the driven element length, which has more impact on the reactive part of the antenna input impedance. The overall lengths of the wire elements are given by the following formulas:

$$\text{Driven element:} \qquad L(\text{feet}) = \frac{1005}{F\,(\text{MHz})} \qquad\qquad (13.1)$$

$$\text{Reflector:} \qquad L(\text{feet}) = \frac{1030}{F\,(\text{MHz})} \qquad\qquad (13.2)$$

$$\text{Director:} \qquad L(\text{feet}) = \frac{975}{F\,(\text{MHz})} \qquad\qquad (13.3)$$

FIGURE 13.5   Multiband quad "spider".

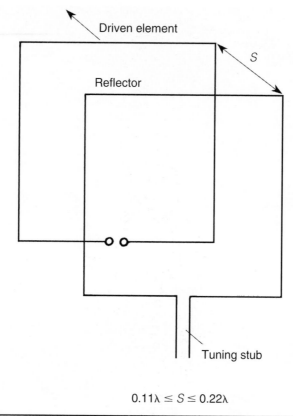

$$0.11\lambda \leq S \leq 0.22\lambda$$

FIGURE 13.6   Quad beam antenna.

One method for constructing a single-band, two-element quad is shown in Fig. 13.7. This particular scheme uses a 12- × 12-in wooden plate at the center, bamboo (or fiberglass) spreaders, and a wooden (or metal) boom. The construction must be heavy-duty in order to survive wind loads. A better solution might be to buy a quad kit consisting of spreaders and spider for either a two- or a three-element quad.

With most kits, more than one band can be installed on a single set of spreaders. The size of the spreaders is set by the lowest band of operation, so higher-frequency bands can be accommodated with shorter loops on the same set of spreaders. And while the use of a spider allows element spacing proportional to wavelength on multiple bands, there is no shortage of two-, three-, four-, and five-element quads that forsake proportional spacing and simply nest concentric loops for three, four, and five bands on spreaders that are perfectly vertical. Thus, there is no reason the construction technique of Fig. 13.7 can't be used to support loops for multiple bands as long as it is understood that VSWR bandwidth and the variation of other key performance parameters such as forward gain, F/B and F/S ratios across the operating band will be different on every band.

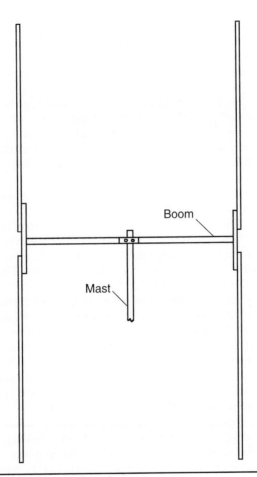

FIGURE 13.7 Single-band two-element quad on a boom.

Delta Loop

At frequencies below 14 MHz, spreaders and supports for fixed or rotatable multiele-ment quads become impractical for most amateurs. For some, the fixed *delta loop* is a practical form of multielement parasitic array for the 30-, 40-, and 80-m bands.

The delta loop is a triangular or three-sided 1λ loop. Although the ideal configura-tion for a delta loop is with the horizontal side of the triangle at the top, this is usually practical only at VHF. Instead, turning the loop upside down allows it to be suspended from a single high support in return for a slight reduction in performance. Rope or wire *catenaries* between tall trees are a popular way of supporting the delta loop, but a two-element 40-m delta loop can be suspended from the ends of a 20-ft aluminum boom attached near the top of a tower that is 70 ft or more tall. The approach has also been used on 80 m, but the boom length and minimum tower height requirements for that band typically lead to greater mechanical complexity, such as the use of trusses to strengthen the boom. Regardless of how it is supported, the two lower corners of the delta loop itself are usually tied off with nonconducting cordage to nearby trees or dedicated anchors.

Like the quad, the delta loop can be fed anywhere along the loop, with the ratio of horizontally to vertically polarized radiation changing accordingly. Feeding at the top or at the midpoint of the bottom wire results in a horizontally polarized loop. Moving the feedpoint halfway (or $\lambda/4$) along the total wire distance between those two points results in a vertically polarized loop. The feedpoint for vertical polarization is therefore $\lambda/4$ down from the top on either side; if the loop is an equilateral triangle, that's equivalent to $\lambda/12$ above either bottom corner, or ¼ the length of a side above one of the two bottom corners, as shown in Fig. 13.8.

Assuming the delta loop is an equilateral triangle, its total height is approximately $\lambda/(3.5)$, or 41 ft at 7 Mhz and 81 ft at 3.5 Mhz. For the loop to retain any semblance of its inherent free-space broadside gain, the bottom should be no closer to the earth than, say, $\lambda/8$, so in practice the top support for a delta loop should be at least 70 ft high for 40 m and at least120 ft high for 80 m.

Like Yagis and cubical quads, most users of a two-element delta loop prefer the parasitic element to be a reflector rather than a director. As with the quad, the reflector loop should be about 3 or 4 percent longer than the resonant length for the desired center frequency. At HF the earth will be close to the bottom of the loop, so experimentation to find the exact length that works best likely will be required. Some users use relays in protective enclosures at each loop to switch extra wire in and out of either or both of the two elements as a way of reversing the pattern or of having optimum lengths at two widely separated frequencies in the 3.5-MHz band.

Because one wavelength of wire is split among only three sides, instead of four, typically the bottom leg of an HF delta loop will be closer to the ground than the bottom leg of a quad for the same support height. However, in many cases, the delta loop is favored because the user has a single support that is higher than any other support—

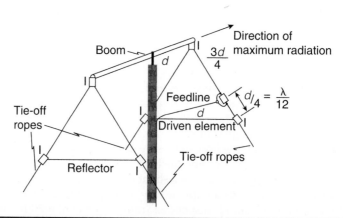

FIGURE 13.8 Vertically polarized two-element delta loop.

especially in the case of multielement delta loops, where the elements are often suspended from a long boom fixed-mounted near the top of the tallest tower in an antenna installation.

Make no mistake—because of its triangular shape and acute interior angles, the delta loop is a compromise compared to the square quad loop. Modeling the free-space difference between a square loop and an equilateral triangle indicates about 0.25 dB difference in the gain of the main lobe (in favor of the square loop) when the length of each loop is adjusted for a purely resistive feedpoint impedance at the design frequency. Nonetheless, the delta loop can do a passable job at providing low-angle gain on the lower HF bands.

Diamond Loop

The *diamond loop* consists of a square loop rotated around its broadside axis by 45 degrees. When fed at a top or bottom corner, as shown in Fig. 13.9, it delivers 1.5 dB more forward gain in free space than a delta loop designed with the same rules. The extra gain is the direct result of the greater spacing and better physical layout of the two inverted vees that form the array. However, that's not really a fair comparison, since the height of the diamond loop is *twice* that of the delta loop for the same frequency. Thus, the diamond loop is not usually a candidate for the lower HF bands, but single-element and multielement versions might warrant consideration as fixed arrays at 20 m and above if the user has a single tall support or can stretch a catenary line between two such supports.

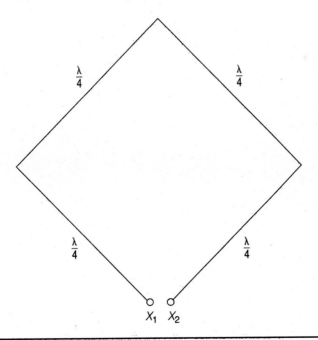

FIGURE 13.9   Horizontally polarized diamond loop.

Concluding Thoughts

At HF it is *very* difficult to erect *any* horizontally polarized antenna high enough to be able to forget about the effect of the earth below. In principle, the theoretical 6-dB gain that results from the ground reflection is worthwhile, but in practice the reflection gain from real earth is either far less than 6 dB or the antenna is so low (in terms of wavelengths) that the increased signal goes mostly into the ionosphere at very high angles, where it is often of no use, depending on your operating or listening interests. Similarly, the theoretical 3-dB gain from the second radiating element (the other half of the loop) is seldom achieved in the real world. Part of the reason for this is that in the case of 1λ loops and quad elements, the spacing of the opposing sides is only $\lambda/4$—leading to quadrature phasing and partial, rather than full, cancellation of the signal in unused directions. Equally important, the radiated field from the lower leg of a square loop is subject to the ground reflection pattern of a much lower height, so in general a loop's maximum gain is at a somewhat higher elevation angle than it is for the top leg alone.

In the author's opinion, unless one has some other reason for choosing the quad—such as the high-power corona discharge problem at HCJB—the mechanical complexity of the quad may outweigh its benefits. Often, more can be realized by analyzing the ground reflection factor for different antenna heights and then applying the same (or less) effort to putting a simple two- or three-element Yagi up at a height that yields the desired elevation pattern.

PART **V**

High-Frequency Antennas for Specialized Uses

CHAPTER 14

Receiving Antennas for High Frequency

Antennas are *reciprocal* devices. That is, just as transmitter power applied to the feedpoint of an antenna results in a field of a certain strength at a remote point, fields originating remotely can impinge upon that same antenna and create a received signal voltage and power at that same feedpoint.

Under most—but not all—circumstances, the best antenna to use for transmitting is also the best antenna to use for receiving. Sometimes, however, the use of a separate receiving antenna (or antennas) can lead to better reception and an overall improvement in communications reliability. This is especially true for the very low frequency (VLF), low frequency (LF), medium frequency (MF), and lower high-frequency (HF) ranges of the radio spectrum, for any of the following reasons:

- Dimensions and construction costs of effective and efficient transmitting structures (including the radials beneath them) are relatively large; the vast majority of transmitting antennas in use below 10 MHz consist of a single radiating element fixed in one position and exhibiting a nearly omnidirectional pattern around one axis. As a result, these transmit antennas provide very limited rejection of interfering signals from directions other than that of the desired signal.

- Most communications receivers have more than enough sensitivity below 10 MHz. That is, atmospheric noise is almost always many decibels above the noise floor of today's receivers. Consequently, receive antenna efficiency is relatively unimportant, and high-performance multielement directive receiving array designs capable of being "steered" to multiple headings around the compass can be built on much smaller land parcels and at substantially reduced cost compared to transmitting arrays providing comparable received signal-to-noise ratios.

- The ability to copy weak signals is often limited by atmospherics (lightning-generated noise, or QRN) or local noise sources that are stronger at the lower frequencies and often have different arrival angles or wave polarization than the desired signal(s). Consequently, it is often easier and more important to null out the noise with an easily steerable receiving antenna than it is to boost the incoming signal enough to override the noise.

- Two identical phase-locked receivers (such as are found in some of today's transceivers)—each connected to its own receiving antenna or one connected to a receiving antenna and the other to the transmit antenna—provide enhanced ability to copy weak signals in the presence of signal fading (QSB), ionospheric skew paths, and/or polarization shift.

Spurred by accelerating worldwide interest in the 160-m band as *long range navigation* (LORAN) systems were decommissioned and frequencies became available to amateurs, the decade following publication of the fourth edition of this handbook has been a fertile period for development of receiving antennas for that band. Thanks in large measure to the use of the Internet to rapidly disseminate new design concepts and experimental results, "folklore" about antenna performance has largely been replaced with a solid body of good science. Today the low-band receiving enthusiast—whether radio amateur, broadcast band listener, or shortwave listener—has available an arsenal of antenna types to enhance his/her receiving capabilities. They include:

- Beverage and Snake
- Multiturn loop
- EWE
- K9AY loop
- Pennant
- Flag
- Short verticals
- Longwires

Many of these antenna types are employed not only singly but often in phased configurations with duplicates of themselves. Some of these antennas cannot be analyzed by viewing them as derivations of the half-wave dipole. All except the longwire are generally unsuitable for transmitting purposes for multiple reasons:

- Low radiation resistance (difficult to match)
- Very low efficiency
- Easily destroyed by even modest power levels

Because of their inefficiency, these antennas generally present a signal level to the receiver input terminals that is anywhere from 10 to 40 dB *below* signal levels typically delivered by a transmit antenna during receiving periods. But because the atmospheric noise level at MF and lower HF frequencies is so high, receiver sensitivity, per se, is seldom an issue and preamplifiers are only occasionally required.

Beverage or "Wave" Antenna

The *Beverage* or *wave antenna* is considered by many people to be the best receiving antenna available for very low frequency (VLF), AM broadcast band (BCB), medium-wave (MW or MF), or tropical band (low HF region) DXing.

In 1921, Paul Godley, who held the U.S. call sign 1ZE, journeyed to Scotland under sponsorship of the American Radio Relay League (ARRL) to erect a receiving station at Androssan. His mission was to listen for amateur radio signals from North America. As a result of politicking in the post–World War I era, hams had been consigned to the supposedly useless shortwaves ($\lambda < 200$ m), and it was not clear that reliable international communication was possible. Godley went to Scotland to see if that could happen; he reportedly used a wave antenna (today, called the *Beverage*) for the task. (And, yes, he was successful—23 North American amateur stations were heard! As the cover of *QST*, the monthly journal of the ARRL, for March 1922 trumpeted, "We got across!!!")

The Beverage was used by RCA at its Riverhead, Long Island (New York), station in 1922, and a technical description by Dr. H. H. Beverage (for whom it is named) appeared in *QST* for November 1922, in an article entitled "The Wave Antenna for 200-Meter Reception". It then virtually disappeared from popular view for decades as amateurs, military, and commercial services moved higher and higher in frequency, until removal of LORAN restrictions on the 160-m band by the United States and other countries in the 1960s and 1970s caused an upswing of interest in DXing on that band. In 1984, an edited and updated version of the 1922 *QST* article reintroduced active amateurs to the merits of the antenna.

The Beverage is a terminated longwire antenna, usually greater than one wavelength (Fig. 14.1), although some authorities and many users (including the author) maintain that lengths as short as $\lambda/2$ can be worthwhile. The Beverage provides good directivity but is not very efficient—in part because roughly half of the received energy is dissipated in a termination resistor at one end. As a result, it is seldom used for transmitting. (This is an example of how different attributes of various antennas make the law of reciprocity a sometimes unreliable guide to antenna selection.) Unlike the regular longwire discussed in another chapter as a multiband transmitting antenna, the Beverage is intended to be mounted close to the earth's surface (typically $< 0.1\lambda$); heights of 8 to 10 ft are most commonly employed so that two- and four-footed mammals are less apt to run into the wire, but lower heights may have the advantage of minimizing stray pickup from the vertical "drop" wire at either end. At greater heights, the Beverage loses its relationship to the (lossy) ground beneath it and its pattern becomes more and more that of a traditional end-fed terminated wire.

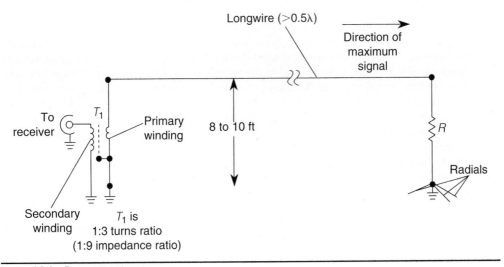

FIGURE 14.1 Beverage antenna.

When deployed over normal (i.e., lossy) ground, the Beverage antenna responds to vertically polarized waves arriving at low angles of incidence. These conditions are representative of the AM broadcast band, where nearly all transmitting antennas are vertically polarized, and of incoming long-haul DX signals on the amateur bands. In addition, the ground- and sky-wave propagation modes found in these bands (VLF, BCB, MF, and low HF) exhibit relatively little variation with time.

The Beverage works best in the low-frequency bands (VLF through 40 m or so). Beverages are not often used at higher frequencies because high-gain rotatable directional arrays—notably, Yagis and cubical quads—have a much smaller footprint than a set of six or so Beverages would require, are highly efficient, have excellent gain in their main lobe, are relatively easy to erect to a useful height at a modest cost, and can also be used for transmitting. In addition, the likelihood of "corrupting" the Beverage's unidirectional pattern with horizontally polarized signals increases because the usual height of the Beverage above ground begins to be a sizeable fraction of a wavelength at the higher frequencies.

The Beverage performs best when erected over moderately or poorly conductive soil, even though the terminating resistor needs a good ground. One wag suggests that sand beaches adjacent to salty marshes make the best Beverage sites (a humorous oversimplification). Figure 14.2 shows why poorly conductive soil is needed. The incoming ground wave (broadcast band) or low-angle (long-haul DX) E-field vectors arrive essentially perpendicular to the earth's surface. Over perfectly conducting soil, the vertical waves would remain vertical. But over imperfectly conducting soil, the field lines tend to bend near the point of contact with the ground. (To say it another way, a perfectly conducting ground cannot support a voltage or E-field between any two points on the ground plane. Hence, the *only* E-field that can exist immediately above a perfect ground is vertically polarized.)

As shown in the inset of Fig. 14.2, the incoming wave delivers a horizontal component of the E-field vector along the length of the wire, thus generating an RF current in the conductor. Signals arriving from the unterminated end of the wire build up a voltage on the wire as they travel, or "accumulate", down the wire, but that voltage is absorbed, or dissipated, in the termination resistor. Signals arriving from the termination end, however, travel to the feedline end, where they are collected by the step-down transformer or the feedline itself. If the Beverage is properly terminated by the transformer in conjunction with the feedline impedance, those signals simply keep moving through the feedline to the receiver, never to reappear. If there is a mismatch at the junction of the Beverage wire and the feedline or at the receiver itself, a portion of the signal headed for the receiver will be reflected back onto the Beverage wire, but some signal still continues to the receiver. The reflected portion continues to travel along the Beverage until it arrives at the termination resistor, where it is absorbed.

Are there optimum lengths for a Beverage? Some Beverage experts suggest specific lengths that provide "natural" cancellation of signals (called a *cone of silence*) off the back of the antenna pattern for a given band of frequencies. Some sources state that the length can be anything greater than or equal to 0.5λ, yet others say greater than or equal to 1λ is the minimum size. One camp says that the length should be as long as possible, while others say it should be close to a factor called the *maximum effective length* (MEL), which is

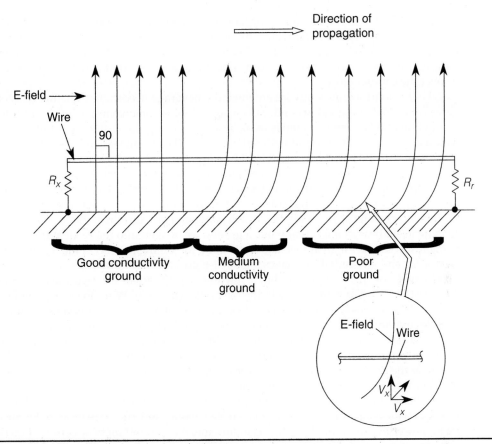

FIGURE 14.2   Wave tilt versus ground conductance.

$$\text{MEL} = \frac{\lambda}{4\left(\dfrac{100}{K} - 1\right)} \tag{14.1}$$

where MEL = maximum effective length, in meters
     λ = wavelength, in meters
     K = velocity factor, expressed as a percent

Misek, who may well be the leading proponent of the Beverage antenna, uses numbers like 1.6λ to 1.7λ over the 1.8- to 7.3-MHz region, and 0.53λ to 0.56λ on frequencies lower than 1.8 MHz. Dr. Beverage was once quoted as saying that the optimum length is 1λ. Perhaps the most important point to take away from this discussion is that the exact length is not critical. The author has employed Beverages of various lengths for over a quarter-century and has found little to recommend lengths in excess of 500 to 550 ft (1λ on 160 m) for use of frequencies between 1 and 7 MHz. Some of his current Beverages are in the 250- to 300-ft range; they have made it possible to pull weak stations out of the noise on 160 and 80 m on so many occasions it is hard to believe there is much improvement left to be gained by lengthening them. Of course, the longer a Beverage is, the narrower the main lobe—so in principle the greater the number of them you need to cover the compass rose.

Like the nonresonant longwire antenna, the Beverage needs a termination resistor that is connected to a good ground. This requirement is somewhat in conflict with the assumption that Beverage antennas work best over lossy ground, but the solution is to make a good *artificial* RF ground with radial wires at the terminated end. As in the longwire case, insulated or bare wires, at least 0.2λ long on the lowest frequency of interest, make the best radials. However, a substantial improvement in the ground is possible using wires only 15 to 20 ft in length (which is much less than λ/4 at the frequencies of interest), atop or just below the surface of the soil. Some experienced Beverage users recommend just one or two radials for each band, laid out at right angles to the Beverage wire itself. Bear in mind that while a ground rod driven into the earth at the terminating resistor makes a great mechanical tie point for joining the end of the Beverage to the resistor and any radials, it makes a lousy RF ground. The author has had good results (maybe not "perfect" results, but . . .) with two 0.1λ radials for each amateur band of interest, oriented at right angles to the Beverage wire itself. Wires in close proximity to ground—even lossy ground—should be thought of as providing nonresonant capacitive coupling rather than resonance, so their exact length is not important.

In addition to the radials and ground rod, Misek also recommends using a wire connection between the ground connection at the termination resistor and the ground connection at the receiver transformer. According to Misek, this wire helps to stabilize the impedance variations at higher frequencies.

The basic single-wire Beverage antenna of Fig. 14.1 consists of a single conductor (#18 to #8 wire gauges are most common) erected about 8 to 10 ft above ground. Because the RF power in a receiving Beverage is minuscule, the size and metallic composition of the wire are primarily set by *mechanical* (strength) and *metallurgical* (corrosion lifetime) considerations, rather than electrical. In contrast to most antennas, the greatest threat to Beverage longevity comes from dead trees and limbs falling on them or animals running into them. #14 THHN copper wire from building supply outlets is probably the most consistently economical source, although the antenna may have to be retensioned periodically. Some have used copper-clad steel or electric fence wire, but the author has found that the use of alternative metals or bimetallic wires often brings with it a new set of problems greater than the slight stretchiness of softdrawn copper wire.

Some Beverages are unterminated (and, hence, bidirectional), but most are terminated at one end in a resistance R equal to the antenna's characteristic impedance, Z_0. For the heights discussed, Z_0 is likely to be between 300 and 800 Ω. The receiver end should also be terminated in its characteristic impedance, but ideally requires an impedance-matching transformer to reduce the antenna impedance to the 50-Ω or 75-Ω standard impedance used by most modern HF transceivers and receivers.

Installation of the Beverage antenna is not overly critical if certain rules are followed:

- For safety of humans and animals, install the antenna at a height of 8 to 10 ft above ground. Check its entire length frequently; because they are so low, Beverages have an annoying habit of collecting fallen branches or entire trees!

- Use a compass to run the Beverage wire in a reasonably straight line for the distance and beam heading you have selected. (The iPhone has a wonderful compass app that is free!) Typical Beverage lengths range from 250 to 1200 ft. If space permits, choose your Beverage location to maximize the distance between it and other low-band antennas, towers, power lines, buildings, and nearby sources of electrical noise (including your neighbor's electrical heating pad and all the wall adapters in your own home as well as his).

- Support the wire as necessary (probably every 25 to 100 ft, depending on wire type). Beverages in the open can be mounted on treated 4 × 4 posts, but a potentially less expensive approach is to drive a few feet of an inexpensive steel rod (*rebar*, for instance) partway into the ground and drop an 8-ft length of PVC pipe over the rod. (Drill or slot the top of the PVC to hold the wire in place.)

- In wooded areas *insulated* wire can be draped over tree limbs of the appropriate height, or electric fence insulators obtained at farm supply stores can be nailed into the tree trunk. (Never wrap a wire or rope around a tree trunk or branch!) If the preferred path for your Beverage takes it *between* trees, run nylon twine between two trees straddling the desired path and hang the Beverage from the middle of the twine.

- Keep the wire a relatively constant height above ground. In other words, follow the slope of the terrain. The farther your Beverage wire is above ground, the less it is able to discriminate between desired low-angle signals and undesired higher-angle signals.

- Drop the vertical end sections straight down unless it's more convenient to bring them down gradually through some horizontal distance.

- Drive a 6- or 8-ft ground rod at the termination end and attach one or two radials, 50 to 200 ft long, at right angles to the Beverage. The termination end is the end from which signals will be received. In other words, if the termination resistor is at the western end of a Beverage wire running east and west, the Beverage will favor signals coming from the west. Excellent 8-ft galvanized steel rods and mating brass clamps are available from many electrical supply stores and utility supply houses such as Graybar and others.

- Connect a termination resistor (see text below) from the bottom end of the drop wire at the termination end of the Beverage to a clamp on the ground rod there.

- Drive another ground rod and lay out a separate set of radials at the opposite end of the wire. This is the feedpoint of the Beverage.

If you have a junk box full of old resistors, a 470-Ω 2W carbon composition resistor makes a great Beverage termination. If you have only larger resistance values, connect two or more in parallel; the greater the installed wattage, the less susceptibility to damage from lightning-induced surges. The exact value of resistance is not critical, but the *type* of resistor is. Many of today's resistors, especially the wire-wound ones, are not suitable for this application because they are intended for use in dc or audio circuits and exhibit too much inductive reactance at RF. A suitable alternative to carbon composition is a thick-film resistor, such as Ohmite's TCH line.

Once your Beverage is installed, a good preliminary test is to measure the dc resistance between the unterminated drop wire and the ground rod at that end of the Beverage. With average ground characteristics, R_{TOTAL} (the sum of the termination resistance plus the ground path between the meter and the grounded end of the termination) will probably range from 1.5 R_{TERM} to $4R_{TERM}$, depending on the length of the Beverage and the characteristics of the ground beneath it. The primary value of this test is simply to prove continuity through the Beverage wire and termination resistor into the ground. This test should be performed when the installation is new, and the value of R_{TOTAL} immediately recorded for future reference.

If you have the necessary test equipment, consider "sweeping" the impedance of the Beverage at its feedpoint over the frequency range of interest while you adjust the value of the terminating resistor for flattest response over the useful range. (Temporarily use a noninductive potentiometer in place of the permanent terminating resistor.) Make sure you terminate or match the feedpoint end of the Beverage according to the instructions with the test equipment. This is an excellent use of a *vector network analyzer* (VNA) of the type discussed in Chap. 27.

An inexpensive Beverage feedline is 75-Ω RG-6 coaxial cable purchased by the 500- or 1000-ft spool from a building supplies store.

When using either 50-Ω or 75-Ω coaxial cable for the feedline, a 9:1 *step-down transformer* at the Beverage feedpoint is a smart addition. (Do *not* put the transformer at the receiver end of the feedline!) Figure 14.3 shows how to build an appropriate transformer using a ferrite core and a *trifilar* winding. Type 31 or Type 43 materials should work well, but many mixes will work nearly as well. Ideally, the ground for the shield of the coaxial feedline at the transformer needs to be kept isolated from the grounded end (winding terminal a_2 in Fig. 14.3) of the transformer winding that connects to the Beverage, which would require a second, completely isolated coaxial cable output winding identical to winding a_1-a_2 on the same core and a second "ground" terminal, but the author has three "very satisfactory" short Beverages (each 250 to 300 ft) with no ground isolation at all that provide excellent signal-to-noise enhancement on 160 and 80 m most nights.

After connecting the feedline to the step-down transformer at the unterminated end of the Beverage, perform the same dc continuity test with an ohmmeter located at the receiver end of the feedline, thus checking the integrity of that portion of the signal path as well. Of course, you should see a very low value of R—no more than a few ohms, at most. The exact value will depend on the length of your feedline and the resistance of its conductors.

At this point, if a VNA or other suitable piece of test apparatus is available, a sweep of the entire antenna system from the receiver end of the feedline to the termination resistance at the far end of the Beverage wire can be an extremely valuable record should troubleshooting be necessary in the future.

Much has been learned and disseminated about the Beverage in the past decade. A book nearly as large as this one could be written about the antenna and all its subtleties and "picky details" that should be attended to. Nonetheless, "imperfectly" constructed Beverages can make a whole new layer

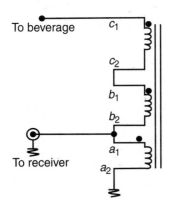

FIGURE 14.3 Trifilar-wound 9:1 step-down Beverage transformer.

of weak DX signals on the AM broadcast band or lower-frequency ham bands jump out of the background noise in your headphones. Yes, the author's 250-ft Beverages are "too short" to be of much use on 160 m. Yes, the author's Beverages should have matching transformers and isolated grounds. Etcetera, etcetera. But once the three most important compass headings were determined, each of the author's Beverages took no more than two or three hours to install and connect. And they work on 160, 80, 40, and the AM broadcast band! Could they be fine-tuned to work better? Yep! Do they work well enough? Yep! Did they cost a lot? Nope! In short, the Beverage is the epitome of a *practical* antenna.

The Bidirectional Beverage

With some clever transformer "magic", two Beverages, consisting of two parallel wires and capable of receiving from two separate compass headings 180 degrees apart, can share a common set of supports and a common feedline attached to the end of the wire pair that is nearer the receiver.

Figure 14.4 shows the circuit configuration that allows switching of the wires and the feedlines for two opposing directions. The parallel wires can be implemented in a variety of ways:

- Separate wires about a foot apart, supported by cross-arms atop wood or PVC supports
- Open-wire transmission line
- Two-conductor jacketed house wiring
- Surplus multiconductor cable

Transformers T_1 and T_2 are designed to transform the Beverage impedance down to the characteristic impedance, Z_0, of the coaxial feedline to the receiver. In operation, one of the two coaxial cables feeds the receiver while the other cable *must* be terminated in Z_0. This can be done at the receiver end, but, alternatively, to avoid running two transmission lines to the receiver, the resistive load switching and swapping of transformers feeding the receiver can be performed in a protective enclosure near the end of the Beverage that is attached to transformer T_1. In the latter configuration, the dc control signal

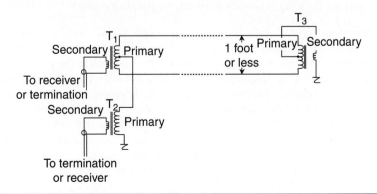

FIGURE 14.4 Two-wire reversible Beverage.

to switch a multipole double-throw relay in the enclosure can be multiplexed on the coaxial cable.

Assume first that the feedline attached to transformer T_1 is terminated in Z_0. Signals coming from the right side of the page build in voltage with respect to ground along both Beverage wires equally. At the left end of the wires, this common mode voltage appears across the entire primary winding of T_1, including its centertap, and drives the top end of the primary winding of T_2, whose secondary feeds the active coaxial cable to the receiver. Signals coming from the left side of the page build up equally on both Beverage wires from left to right, appearing on the entire primary of T_3, including its centertap, which drives one end of a secondary winding. The signal across the secondary of transformer T_3 now drives the two Beverage wires in *push-pull* or *differential* mode, and that signal propagates from right to left along the wires until it reaches the primary of T_1, where it is coupled to the secondary and dissipated in that winding's Z_0 termination. To receive signals from the opposite direction, the receiver is connected to the secondary of T_1, and the secondary of T_2 is terminated in Z_0.

The exact number of turns required on each transformer is dependent upon the specifics of each individual Beverage installation, including wire diameter and length, spacing between the two wires, height above ground, and ground conductivity. If the termination and cable switching is done near T_1 and T_2, care must be taken to isolate the ground end of the termination from the coaxial cable ground (through a separate pole on the relay), or the ground connecting the two must be virtually perfect.

Phased Beverages

Additional directivity and signal amplitude can be obtained by phasing two or more Beverages. At least two different methods of phasing are currently in use:

- Two identical Beverages, parallel to each other, spaced $\lambda/4$ or more apart, with no offset relative to the desired receiving direction. (That is, they can be thought of as two opposing long sides of a rectangle.) The outputs of the two wires are combined in phase before reaching the receiver.

- Two identical Beverages, parallel to each other and closely spaced (i.e., within a few feet), but offset somewhat in the desired receiving direction. (Similarly, picture the long sides of a very skinny parallelogram.)

A Beverage erected with two wires—parallel to each other, at the same height, spaced about 12 in apart (Fig. 14.5), with a length that is a multiple of a half-wavelength—is capable of *null steering*. That is, the rear null in the pattern can be steered over a range of 40 to 60 degrees. This feature allows strong, off-axis signals to be reduced in amplitude so that weaker signals in the main lobe of the pattern can be received. There are at least two varieties of the *steerable wave Beverage* (SWB).

If null steering behavior is desired, then a phase control circuit (PCC) will be required—consisting of a potentiometer, an inductance, and a variable capacitor all connected in series. Varying both the "pot" and the capacitor will steer the null. The direction of reception and the direction of the null can be selected by using a switch to swap the receiver and the PCC between port A and port B.

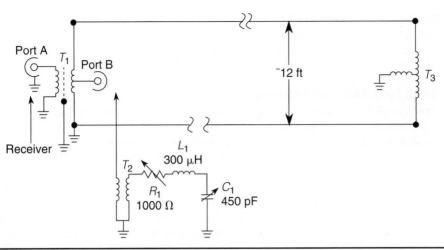

FIGURE 14.5 Beverage antenna with a steerable null.

Snake or BOG Antenna

The height above ground of a Beverage is the result of a compromise that invariably includes a need to locate the antenna high enough to avoid damage to or from humans and animals. As a result, the rejection of signals from undesired directions and signals with horizontal polarization is lessened.

The *Beverage on ground* (BOG) or *snake* antenna attempts to minimize unwanted signal ingress by eliminating the vertical *downlead* at each end of the Beverage. A side benefit is the elimination of all supports! The primary disadvantage is a noticeable reduction in received signal level stemming from the very close coupling of the antenna to the ground beneath it. Typically, a preamplifier must be added to a BOG or snake in order to develop signal levels sufficient to override the receiver noise floor. Ideally the preamp should be located at the junction of the Beverage and its feedline, rather than at the receiver end of the feedline.

Receiving Loops

Even in their smallest implementations, the preceding antennas require a fair amount of real estate for their proper operation. In addition to the space taken up by the antenna itself, there is a "hidden" requirement of another $\lambda/2$ or more distance between the receiving antenna and any transmitting antennas resonant on the same band(s). Further, it is wise to locate and orient Beverages so as to keep buildings containing computers, microprocessors, and other sources of RF noise *behind* the antenna—i.e., at the receiver, or unterminated, end. In general, a broadcast-band listener (BCL) or shortwave listener (SWL) contemplating the installation of Beverages for multiple compass headings will need a 10-acre or larger parcel.

Two entirely different classes of antenna have been of great use to listeners with limited space for antennas. The first of these is the traditional *small, multiturn loop* that has been around for years. The second is a whole new class of small, single-turn loops that emulate the directivity of large transmitting loops while exploiting size reduction design techniques uniquely available to receiving antennas that do not need to exhibit RF efficiency.

Small Loop Receiving Antennas

Radio direction finders and people who listen to the AM broadcasting bands, VLF, medium-wave, or the so-called low-frequency tropical bands are all candidates for a small loop antenna. These antennas are fundamentally different from the large (transmitting) loops of Chap. 7 and the multielement versions of Chap. 13. Large loop antennas typically have a total wire length of 0.5λ or greater. Small loop antennas, on the other hand, have an overall total conductor length that is less than about 0.1λ.

Earlier in this book we talked about the importance of proper physical configuration to efficient performance as a radiating antenna by a conductor. In particular, we pointed out that a half-wavelength of copper wire that is wound into a tight coil is a poor antenna and is best treated as a lumped-circuit inductor. Now we go a step further and note that a small loop antenna responds to the magnetic field component of the electromagnetic wave instead of the electrical field component. (Remember from high school science class that passing a bar magnet back and forth rapidly through the interior of a multiturn coil produces temporary voltages across the terminals of the coil. Consider the time-varying magnetic field in a transmitted wave as analogous to the movement of the bar magnet.)

Thus we see a principal difference between a large loop (total conductor length greater than 0.1λ) and a small loop when examining the RF currents induced in either loop when a signal intercepts it. In a large loop, the current is a standing wave that varies (sinusoidally) from one point in the conductor to another, much like a dipole, and we analyze its operation in much the same way we have examined dipoles; in the small loop antenna, the current is the same throughout the entire loop.

Since the electric and magnetic fields of a radiated wave in free space are at right angles to each other and to the direction the wave propagates through space, and since we are primarily listening for vertically polarized signals with loops, the magnetic field is oriented horizontally, or nearly so.

The differences between small loops and large loops show up in some interesting ways, but perhaps the most striking difference is found in the directions of maximum response—the main lobes—and the directions of the nulls. Both types of loops produce figure eight patterns but at right angles with respect to each other. The large loop antenna produces main lobes that are *orthogonal*—i.e., at right angles or "broadside"— to the plane of the loop. Nulls are off the sides of the loop. The small loop, however, is exactly the opposite: the main lobes are off the sides of the loop (in the plane of the loop), and the nulls are broadside to the loop plane (Fig. 14.6A). Do not confuse small loop behavior with the behavior of the loopstick antenna. Loopstick antennas are made of coils of wire wound on a ferrite or powdered-iron rod. The direction of maximum response for the loopstick antenna is broadside to the rod, with deep nulls off the ends (Fig. 14.6B). Both loopsticks and small wire loops are used for radio direction-finding and for shortwave, low-frequency medium-wave, AM broadcast band, and VLF listening.

The nulls of a loop antenna are very sharp and very deep. If you point a loop antenna so that its null is aimed at a strong station, the signal strength of the station ap-

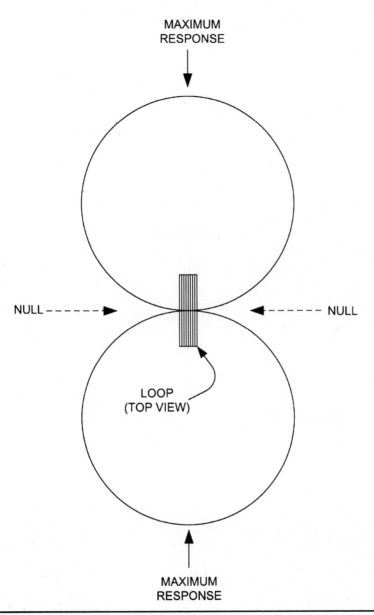

**MAXIMUM
RESPONSE**

NULL - - - - - ▶

◀ - - - - - NULL

**LOOP
(TOP VIEW)**

**MAXIMUM
RESPONSE**

FIGURE 14.6A   Small loop antenna.

pears to drop dramatically at the center of the notch. Turn the antenna only a few degrees one way or the other, however, and the signal strength increases sharply. The depth of the null can reach 10 to 15 dB on sloppily constructed loops and 30 to 40 dB on well-built units (30 dB is a very common value), and there have been claims of 60-dB nulls for some commercially available loop antennas. The construction and uniformity of the loop are primary factors in the sharpness and depth of the null.

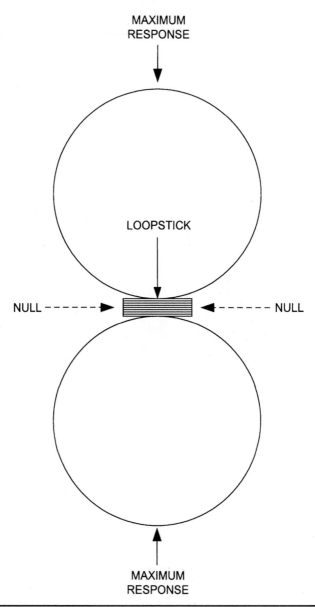

FIGURE 14.6B Loopstick antenna.

Radio direction-finding (RDF) has long been a major application for the small loop antenna, especially in the lower-frequency bands. Chapter 24 is devoted to the use of loops and other antenna types specifically for RDF purposes.

Today, the use of these small loops has been extended into the general receiving arena, especially on the low frequencies (VLF through lower HF range). In this application they are often physically quite large (compared, for instance, to the receiver they are apt to be used with) and require many turns of wire to develop the necessary signal amplitudes and notch depth. Receiving loops have also been used as high as VHF and are commonly used in the 10-m ham band for hidden transmitter hunts.

Let's examine the basic theory of small loop antennas and then take a look at some practical construction methods.

Grover's Equation

Grover's equation (Grover, 1946) seems closer to the actual inductance measured in empirical tests than certain other equations that are in use. This equation is

$$L = \left(K_1 N^2 a\right) \times \left\{ \ln\left[\frac{K_2 aN}{(N+1)b}\right] + K_3 + \left[\frac{K_4(N+1)b}{aN}\right] \right\} \tag{14.2}$$

where L = inductance, in microhenrys (µH)
  a = length of a loop side, in centimeters (cm)
  b = loop width, in centimeters (cm)
  n = number of turns in loop

K_1 through K_4 are shape constants and are given in Table 14.1. *ln* is the natural logarithm of this portion of the equation; it is typically obtained from a table or scientific calculator. (See App. A for information on natural logarithms.)

Shape	K_1	K_2	K_3	K_4
Triangle	0.006	1.155	0.655	0.135
Square	0.008	1.414	0.379	0.33
Hexagon	0.012	2.00	0.655	0.135
Octagon	0.016	2.613	0.7514	0.0715

TABLE 14.1 Shape Constants

Air Core Frame ("Box") Loops

A wire loop antenna is made by winding a large coil of wire, consisting of one or more turns, on some sort of frame. The shape of the loop can be circular, square, triangular, hexagonal, or octagonal, but for practical construction reasons the square loop is most popular. With one exception, the loops considered in this section will be square, so you can easily duplicate them.

The basic form of the simplest loop, as shown in Fig. 14.7, is square, with sides of length *A*. The width of the loop (*B*) is the distance from the first turn to the last turn in the loop or, alternatively, the diameter of the wire if only one turn is used. The turns of the loop in Fig. 14.7 are *depth-wound*—meaning that each turn of the loop is spaced in a slightly different parallel plane—and spaced evenly across distance *B*. Alternatively, the loop can be wound such that the turns are in the same plane (known as *planar winding*). In either case, the sides of the loop (*A*) should be not less than five times the width (*B*). There seems to be little difference in performance between depth- and planar-wound loops: the far-field patterns of the different shape loops are nearly the same if the re-

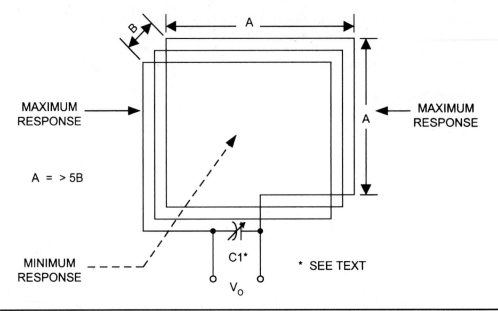

FIGURE 14.7   Simple loop antenna.

spective cross-sectional areas (πr^2 for circular loops and A^2 for square loops) are less than $\lambda^2/100$.

The reason a small loop has a null when its broadest aspect is facing the signal is simple, even though it seems counterintuitive at first blush. In Fig. 14.8 we have two identical small loop antennas at right angles to each other. Antenna A is in line with the advancing radio wave, whereas antenna B is broadside to the wave. Each line in the wave represents a line where the signal strength is the same, i.e., an *isopotential* line. When the loop is in line with the signal (antenna A), a difference of potential exists from one end of the loop to the other and current can be induced in the wires. When the loop is turned broadside, however, all points on the loop are on the same potential line, so there is no difference of potential between segments of the conductor. Thus, little signal is picked up (and the antenna therefore sees a null).

The actual voltage across the output terminals of an untuned loop is a function of the angle of arrival of the signal α (Fig. 14.9), as well as the strength of the signal and the design of the loop. The voltage V_o is given by

$$V_o = \frac{2\pi ANE_f \cos{(\alpha)}}{\lambda} \qquad (14.3)$$

where V_o = output voltage of loop
   A = area of loop, in square meters (m^2)
   N = number of turns of wire in loop
   E_f = strength of signal, in volts per meter (V/m)
   α = angle of arrival of signal
   λ = wavelength of arriving signal

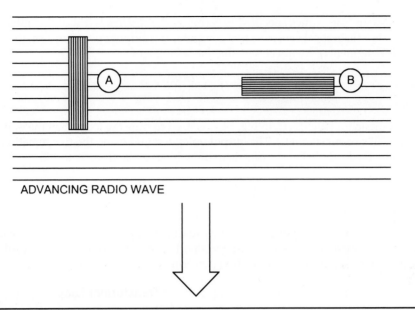

ADVANCING RADIO WAVE

FIGURE 14.8 Two small loop antennas at right angles to each other.

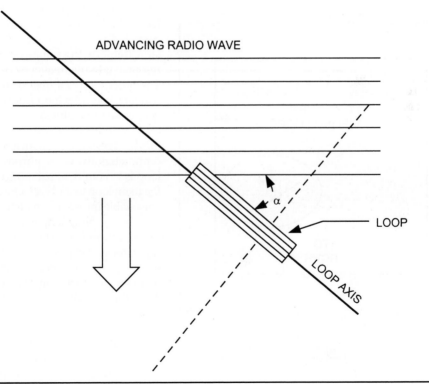

ADVANCING RADIO WAVE

LOOP

α

LOOP AXIS

FIGURE 14.9 Untuned loop antenna at an angle to received wave.

Loops are sometimes specified in terms of the *effective height* of the antenna. This number is a theoretical construct that compares the output voltage of a small loop with a vertical segment of identical wire that has a height of

$$H_{eff} = \frac{2\pi NA}{\lambda} \tag{14.4}$$

If a capacitor (such as C_1 in Fig. 14.7) is used to tune the loop, then the output voltage V_o will rise substantially. The output voltage found using the first equation is multiplied by the loaded Q of the tuned circuit, which can be from 50 to 100:

$$V_O = \frac{2\pi ANE_f Q \cos{(\alpha)}}{\lambda} \tag{14.5}$$

Even though the output signal voltage of tuned loops is higher than that of untuned loops, it is nonetheless low compared with other forms of antenna. As a result, a loop preamplifier usually is needed for best performance.

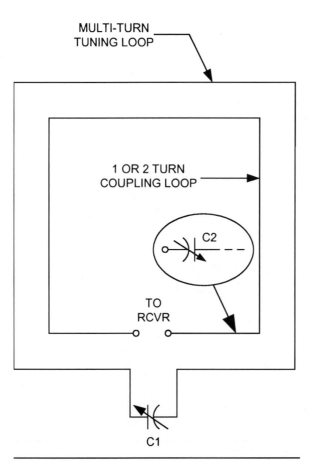

MULTI-TURN TUNING LOOP

1 OR 2 TURN COUPLING LOOP

C2

TO RCVR

C1

Figure 14.10   Transformer loop antenna.

Transformer Loops

It is common practice to make a small loop antenna with two loops rather than just one. Figure 14.10 shows such a transformer loop antenna. The main loop is built exactly as previously discussed: several turns of wire on a large frame, with a tuning capacitor to resonate it to the frequency of choice. The other loop is a one- or two-turn *coupling loop*. This loop is installed in very close proximity to the main loop, usually (but not necessarily) on the inside edge, not more than a couple of centimeters away. The purpose of this loop is to couple signal induced from the main loop to the receiver at a more reasonable impedance match.

The coupling loop is usually untuned, but in some designs a tuning capacitor (C_2) is placed in series with the coupling loop. Because there are many fewer turns on the coupling loop than on the main loop, its inductance is considerably smaller. As a result, the capacitance to resonate is usually much larger. In several loop antennas constructed for purposes of researching this chapter, the author found that a 15-turn main loop resonated in the AM BCB with a standard 365-pF capacitor, but the two-turn coupling loop required three sections of a ganged 3×365-pF capacitor connected in parallel to resonate at the same frequencies.

In several experiments, the loop turns were made from computer ribbon cable. This type of cable consists of anywhere from 8 to 64 parallel insulated conductors arranged in a flat ribbon shape. Properly interconnected, the conductors of the ribbon cable form a continuous loop. It is no problem to take the outermost one or two conductors on one side of the wire array and use them for a coupling loop.

Tuning Schemes for Loop Antennas

Loop performance is greatly enhanced by tuning the inductance of the loop to the desired frequency. The bandwidth of the loop is reduced, which reduces front-end overload from strong off-frequency signals. Tuning also increases the signal level available to the receiver by a factor of 20 to 100 times. Although tuning can be a bother if the loop is installed remotely from the receiver, the benefits are well worth it in most cases.

Various techniques for tuning are detailed in Fig. 14.11. The parallel tuning scheme, which is by far the most popular, is shown in Fig. 14.11A. In this type of circuit, the capacitor (C_1) is connected in parallel with the inductor, which in this case is the loop. Parallel-resonant circuits present a very high impedance to signals on their resonant frequency and a very low impedance to other frequencies. As a result, the voltage level of resonant signals is very much larger than the voltage level of off-frequency signals.

The series-resonant scheme is shown in Fig. 14.11B. In this circuit, the loop is connected in series with the capacitor. Series-resonant circuits offer a high impedance to all frequencies except the resonant frequency (exactly the opposite of the case of parallel-resonant circuits). As a result, current from the signal will pass through the series-resonant circuit at the resonant frequency, but off-frequency signals are blocked by the high impedance.

There is a wide margin for error in the inductance of loop antennas, and even the complex equations that appear to provide precise values of capacitance and inductance for proper tuning are only estimates. The exact geometry of the loop "as built" determines the actual inductance in each particular unit. As a result, it is often the case that the tuning provided by the capacitor is not as exact as desired, so some form of compensation is needed. In some cases, the capacitance required for resonance is not easily available in a standard variable capacitor, and some means must be provided for changing the capacitance. Figure 14.11C shows how this is done. The main tuning capacitor can be connected in either series or parallel with other capacitors to fine-tune the overall value. If the capacitors are connected in parallel, the total capacitance is increased (all capacitances are added together). If the extra capacitor is connected in series, however, then the total capacitance is reduced. The extra capacitors can be switched in and out of a circuit to change frequency bands.

Tuning of a remote loop can be a bother if it is done by hand, so some means must be found to do it from the receiver location (unless you enjoy climbing into the attic or onto the roof). Traditional means of tuning called for using a low-rpm dc motor, or stepper motor, to turn the tuning capacitor. Once upon a time, the little 1- to 12-rpm motor used to drive rotating displays in retail store show windows was popular. But this approach is not really needed since the advent of *varactors*.

A varactor is a reverse-biased diode whose junction capacitance is a strong function of the applied dc bias voltage. A high voltage (such as 30 V) drops the capacitance, whereas a low voltage increases it. Varactors are available with maximum capacitances of 22, 33, 60, 100, and 400 pF. The latter are of great interest to us for this application because they have the same range as the tuning capacitors normally used with loops.

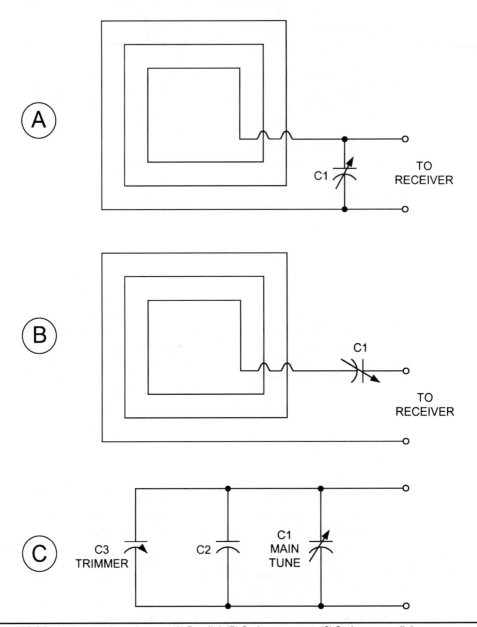

FIGURE 14.11   Various tuning schemes. (A) Parallel. (B) Series resonant. (C) Series or parallel.

Figure 14.12 is the schematic of a remote tuning system for loop antennas. The tuning capacitor is a combination of a varactor diode plus two optional capacitors: a fixed capacitor (C_1) and a trimmer (C_2). The dc tuning voltage (V_T) is provided from a fixed dc power supply ($V+$) located at the receiver end. A potentiometer (R_1) is used to tune the loop by setting the voltage going to the varactor. A dc blocking capacitor (C_3) keeps the dc tuning voltage from being shorted out by the receiver input circuitry.

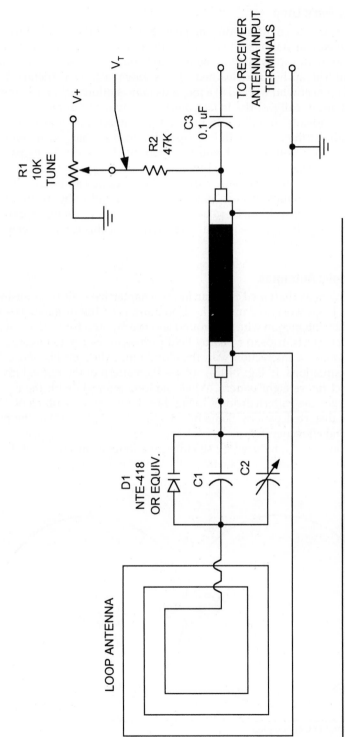

Figure 14.12 Remote tuning scheme using varactor diode.

The Sports Fan's Loop

Okay, sports fans, what do you do when the best game of the week is broadcast only on a low-powered AM station and you live at the outer edge of their service area, where the signal strength leaves much to be desired? You use the sports fan's loop antenna, that's what! The author first learned of this antenna from a friend—a professional broadcast engineer who worked at a religious radio station that had a pipsqueak signal but lots of fans. It really works; in fact, one might say it's a "miracle".

The basic idea is to build a 16-turn, 60-cm² tuned loop and then place the AM portable radio at the center with its loopstick aimed so that its null end is broadside of the loop. When you do so, the nulls of both the loop and the loopstick are in the same direction. The signal will be picked up by the loop and then coupled to the radio's loopstick antenna. Sixteen-conductor ribbon cable can be used for making the loop. For an extra touch of class, place the antenna and radio assembly on a dining room table lazy Susan to make rotation easier. A 365-pF tuning capacitor is used to resonate the loop. If you listen to only one frequency, this capacitor can be a trimmer type.

Shielded Loop Antennas

The loop antennas discussed thus far in this chapter have all been unshielded types. Unshielded loops work well under most circumstances, but in some cases their pattern is distorted by interaction with the ground and nearby structures (trees, buildings, etc.). In the author's tests, trips to a nearby field proved necessary to measure the depth of the null because of interaction with the aluminum siding on his house. Figure 14.13 shows two situations. In Fig. 14.13A we see the pattern of the normal free-space loop, i.e., a perfect figure eight pattern. When the loop interacts with the nearby environment, however, the pattern distorts. In Fig. 14.13B we see some filling of the notch for a moderately distorted pattern. Some interactions are so severe that the pattern is distorted beyond all recognition.

Interaction can be reduced by shielding the loop, as in Fig. 14.14. Loop antennas operate on the magnetic component of the electromagnetic wave, so the loop can be

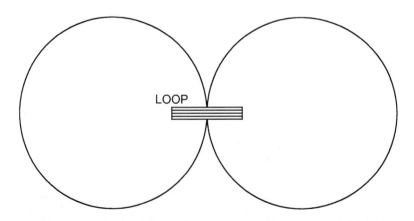

FIGURE 14.13A   Normal free-space loop.

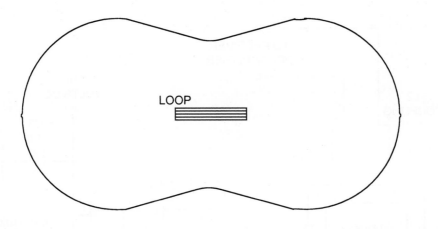

FIGURE 14.13B   Filling of the notch.

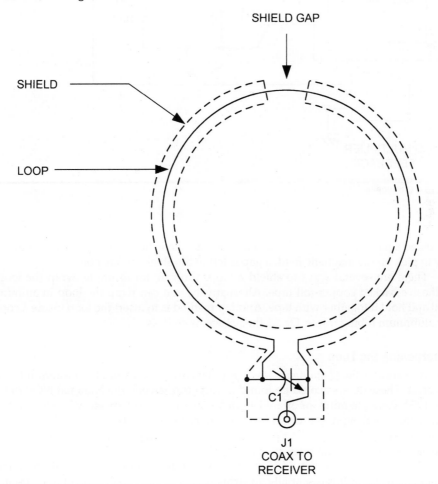

FIGURE 14.14   Shielding the loop.

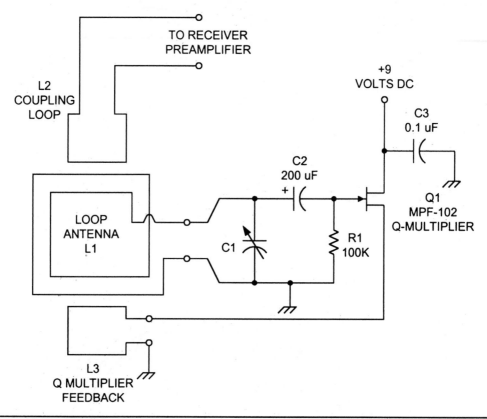

FIGURE 14.15 Q-multiplier.

shielded against E-field signals and electrostatic interactions. In order to retain the ability to pick up the magnetic field, a gap is left in the shield at one point.

There are several ways to shield a loop. You can, for example, wrap the loop in adhesive-backed copper-foil tape. Alternatively, you can wrap the loop in aluminum foil and hold it together with tape. Another method is to insert the loop inside a copper or aluminum tubing frame. Or . . . the list seems endless.

Sharpening the Loop

Many years ago, the *Q-multiplier* was a popular add-on accessory for a communications receiver. These devices could be found in certain receivers (the National NC-125 from the 1950s comes to mind) or offered as an outboard receiver accessory by Heathkit, and many construction projects could be found in magazines and amateur radio books. The Q-multiplier improves receiver performance by increasing the sensitivity and reducing the bandwidth of certain receiver stages.

The Q-multiplier of Fig. 14.15 is an active electronic circuit placed at the antenna input of a receiver. It is essentially an Armstrong oscillator that does not quite oscillate. These circuits have a tuned circuit (L_1/C_1) at the input of an amplifier stage and a feedback coupling loop (L_3). The degree of feedback is controlled by the coupling between L_1 and L_3. The coupling is varied by changing both how close the two coils are and their relative orientation with respect to each other. Certain other circuits use a series potentiometer in the L_3 side that controls the amount of feedback.

The Q-multiplier is adjusted to the point that the circuit is just on the verge of oscillating. As the feedback is backed away from the threshold of oscillation—but not too far—the bandwidth narrows and the sensitivity increases. It takes some skill to operate a Q-multiplier, but it is easy to use once you get the hang of it and is a terrific accessory for any loop antenna.

Loop Amplifier

Figure 14.16 shows the circuit for a practical loop amplifier that can be used with either shielded or unshielded loop antennas. It is based on *junction field-effect transistors* (JFETs) connected in cascade. The standard common-drain configuration is used for each transistor, so the signals are taken from the source terminals. The drain terminals are connected together and powered from the +12V dc power supply. A 2.2-µF bypass capacitor is used to put the drain terminals of Q_1 and Q_2 at ground potential for ac signals while keeping the dc voltage from being shorted out.

The two output signals are applied to the primary of a transformer, the centertap of which is grounded. To keep the dc on the source terminals from being shorted through the transformer winding, a pair of blocking capacitors (C_4, C_5) is used.

Input signals are applied to the gate terminals of Q_1 and Q_2 through dc blocking capacitors C_2 and C_3. A pair of diodes (D_1, D_2) keeps high-amplitude noise transients from affecting the operation of the amplifier. They are connected back to back in order to snub out both polarities of signal.

Tuning capacitor C_1 is used in lieu of a capacitor in the loop; it resonates the loop at a specific frequency. Its value can be found from the equation given earlier.

The transistors used for the push-pull amplifier (Q_1, Q_2) can be nearly any general-purpose JFET device (MPF-102, MPF-104, etc.). A practical approach for many people is to use transistors from service replacement lines, such as the NTE-312 and NTE-316 devices.

Special Problem for VLF/LF Loops

A capacitance is formed whenever two conductors are side by side. A coil exhibits capacitance as well as inductance because the turns are side by side. Unfortunately, with large multiturn loops, this capacitance can be quite large. The *distributed capacitance* of the loop self-resonates with the inductance. Loop antennas do not work well at frequencies above their self-resonant frequency, so it is sometimes important to raise the self-resonance to where it does not affect operation at the desired frequencies.

Figure 14.17 shows one way to raise the self-resonant point. The turns on the loop are broken into two or more groups separated by spaces. This method reduces the effective capacitance by placing the capacitances of each group of wires in series with the others. The effective capacitance of a series string of capacitors is always less than the value of the smallest capacitor.

$$\frac{1}{C_{TOTAL}} = \frac{1}{C_1} + \frac{1}{C_2} + \dots + \frac{1}{C_N}$$

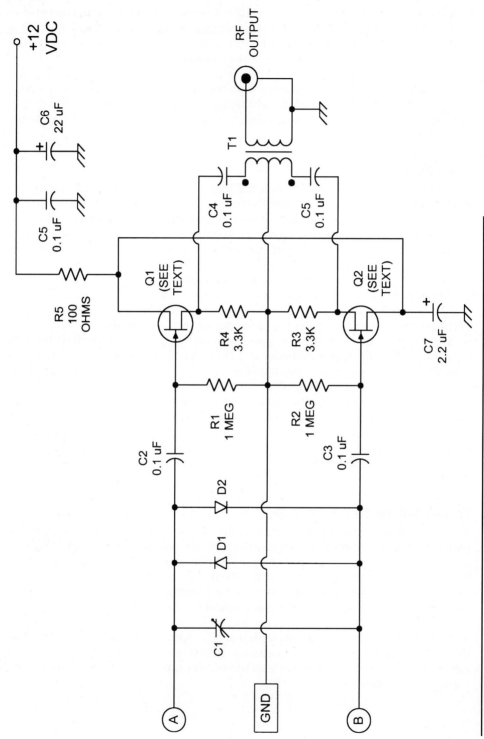

Figure 14.16 Practical loop amplifier.

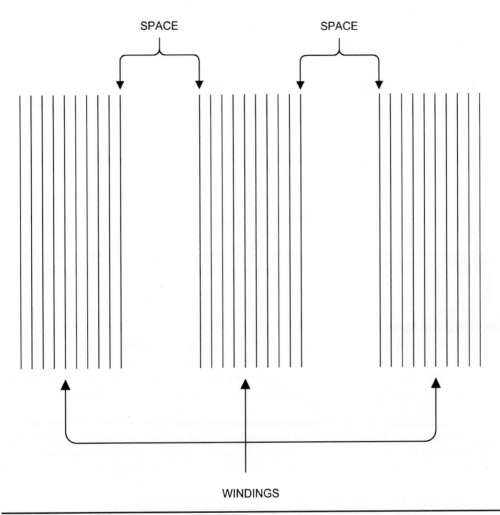

SPACE   SPACE

WINDINGS

FIGURE 14.17   Raising the self-resonant frequency.

Coaxial-Cable Loop Antennas

One of the more effective ways to make a shielded loop is to use coaxial cable. Figure 14.18 shows the circuit of such a loop. Although only a single-turn loop is shown, there can be any number of turns. One reader made a 100-kHz LORAN (a navigation system) loop using eight turns of RG-59/U coaxial cable wound with an 8-ft diameter.

Note the special way that the coaxial cable is connected. This method is called the *Faraday connection* after the fact that the shield of the coax forms a *Faraday shield*. At the output end, the center conductor of the coaxial cable is connected to the center conductor of the coaxial connector, and the cable shield is connected to the connector ground/ shield terminal. At the other end of the loop, the shield is left floating, but the center conductor is connected to the shield *at the connector end,* not at any other point.

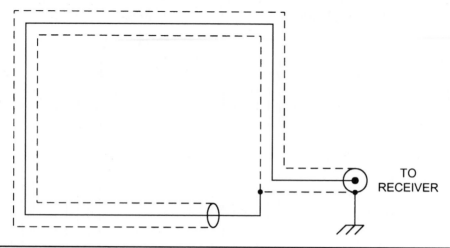

FIGURE 14.18   Coaxial-cable shielded loop.

The EWE Antenna

In 1995 Floyd Koonce, WA2VWL, published details of his EWE antenna, which provides MF and low HF receiving directivity in a much smaller footprint than a Beverage antenna. The acronym EWE is not only descriptive (Earth-Wire-Earth) but a play on words, as well, since the antenna has the shape of an inverted "U" (in response to which the author can only say "Bah!").

The EWE of Fig. 14.19 functions much as if it were a two-element phased array of short verticals. However, like the Beverage, the EWE is a *traveling wave* antenna. Directivity is a result of phase and amplitude differences between the two vertical segments; these differences, in turn, originate in the length of the horizontal wire connecting the vertical runs, the value of the termination resistor, and the fact that a current flowing upward in one vertical segment flows downward in the other.

Unlike the Beverage, the EWE receives from the end of the antenna *opposite* the termination resistor. Because the antenna is not a resonant device, it is relatively broadband in operation, although the optimum value of termination resistance varies with frequency (and with the exact characteristics of the ground beneath the antenna). The value of the termination is best set by tuning for a null off the back of the antenna by listening to appropriate stations in the AM broadcast band or on 160 m. Typical values for the termination resistor are in the 700- to 900-Ω range. The input impedance is somewhat lower and is a reasonably good match to 50-Ω coax and receiver inputs via a 9:1 transformer.

Although the EWE requires a much smaller space than a Beverage antenna, to achieve its full po-

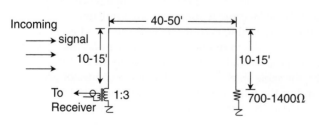

FIGURE 14.19   EWE antenna.

tential it needs to be kept away from other conductors. This constraint can substantially increase the actual space required by the EWE.

Pennants, Flags, and the K9AY Loop

The appearance of the EWE inspired other experimentally inclined radio amateurs to develop additional low-band receiving antennas for small spaces. Subsequent to publication of the EWE, at least three new single-turn terminated loop designs came into being. As is the case with the EWE, for those lacking the necessary space for a set of Beverage antennas, these loops provide an alternative, although those fortunate enough to have the space for both generally report hearing a little bit better with their Beverages.

Directivity of this family of loops is based on constructive versus destructive combining of the E- and H-fields that constitute an arriving radio wave. An arriving E-field in the plane of the loop induces a voltage in the loop just as it would in any other short (relative to wavelength) wire. The associated H-field at right angles to the loop plane induces a current in the loop that becomes a voltage across the terminating resistor. When these fields arrive from one direction, the two voltages sum; when arriving from the opposite direction, they subtract.

Figure 14.20 depicts the K9AY terminated loop. Exact shape of the loop is not critical, but the bottom leg must come near earth for the termination and transformer ground connections. The total length of wire in the loop is about 85 ft, and a single support only 25 ft tall is required. The feedpoint impedance is similar to that of a typical Beverage, so a 9:1 transformer is required. Like the EWE and unlike the Beverage, however, the direction of maximum signal pickup is *opposite* the terminated end.

The pattern of the K9AY loop is unidirectional. By using relays to switch the termination and feedpoint, one loop can be switched between two opposing directions. Two loops, at right angles to each other and similarly switched, can provide four directions from a single fixed mast.

Like the other members of the pennant and flag family, the K9AY loop is quite sensitive to detuning from nearby conductors. As a result, some of its impressive space savings are not always fully realizable if the user also has transmitting antennas and towers nearby. Proper operation of the K9AY is fairly sensitive to the exact value of the termination resistor and depends on both the operating frequency and the actual characteristics of the ground beneath the antenna. Many implementations use a remotely controlled termination resistor to optimize the depth of the null to the rear as the received frequency is changed and to compensate for seasonal variations in soil characteristics.

In many operating scenarios, the value of this antenna is likely to be more for its null off the rear than for any forward directivity it exhibits. But that's quite appropriate for much of the weak-signal work that occurs on the 160- and 80-m amateur bands, or when broadcast band DXing.

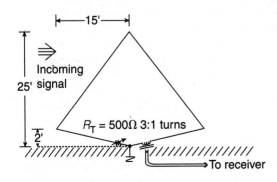

FIGURE 14.20 K9AY loop.

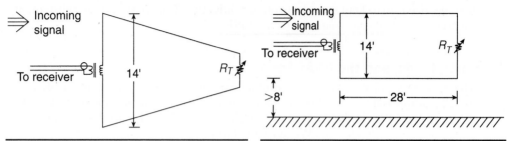

FIGURE 14.21A   Pennant antenna.

FIGURE 14.21B   Flag antenna.

The pennant (Fig. 14.21A) was developed by Jose Mata, EA3VY, and Earl Cunningham, then-K6SE (deceased), to overcome the EWE's sensitivity to the characteristics of the ground beneath it. The resulting pennant design exhibits stability of pattern and performance over a wide variety of grounds and can be raised (on a mast) to a wide range of heights. The favored design also tweaks dimensions and terminating resistor value to arrive at a purely resistive feedpoint impedance for optimum matching with a 9:1 transformer.

Just as the pennant is a triangular receiving antenna, the flag (Fig. 14.21B) is a rectangular one. Because the capture area of a flag is somewhat larger than that of a pennant, the flag antenna provides somewhat greater output voltage to the feedline for a given signal than does the pennant.

Unlike the Beverage, all these antennas receive *away* from the terminating resistor. All must be mounted on, or supported by, nonmetallic members (such as PVC pipe) and must not be located near any metallic structures—especially those having dimensions in the vertical plane that are a sizeable fraction of a wavelength at LF through lower HF frequencies—unless steps are taken to detune those structures at the frequencies of interest while receiving.

Many amateurs and BCLs have employed multiple phased pennants and flags to obtain even greater improvements in received signal-to-noise ratio. An Internet search on these antenna types will return many detailed and useful references.

Longwire Antennas

No discussion of HF receiving antennas would be complete without mention of the *longwire* antenna. This is often the first external antenna ever used by a broadcast band DXer or SWL. Its primary attribute is simplicity. Its primary disadvantage is that it has a pattern that is either unpredictable or shifts from broadside to axial with increasing frequency. However, the beauty of this antenna is that if it is at least 100 ft long, it can provide hours of enjoyable listening to anything from VLF airport beacons through the AM broadcast band and on up into the amateur and international shortwave broadcast bands. The author has even used a longwire with receivers (such as small portables) that had no provision for connecting to an external antenna. Simply bring the end of the longwire near the receiver—perhaps wrapping it around the receiver enclosure a few turns.

Unfortunately, that same "one size fits all" versatility means the longwire rejects nothing, and strong off-frequency signals can overload just about any receiver unless some form of selectivity is applied at its front end. Even a high-end receiver will benefit from having a low-power antenna tuner (ATU) inserted in the path between it and a longwire antenna.

The longwire antenna is covered at length in Chap. 8 ("Multiband and Tunable Wire Antennas") as a transmitting antenna, but some of the installation considerations there are of little importance for a receive-only installation. In particular, when used for transmitting a longwire must be operated against a good RF ground system, but if used only for receiving, the grounds associated with the receiver's power cord are probably sufficient at all but the highest frequencies.

Because the longwire enters the structure where the receiver is located, it tends to faithfully pick up signals from any of the incidental radiators nearby: wall adapters, Ethernet circuits, computer power supplies, etc. In today's electromagnetic environment, this may turn out to be its greatest shortcoming.

CHAPTER 15

Hidden and Limited-Space Antennas

One of the most significant impediments to amateur radio operators and short-wave listeners is the space available for their antennas. In many thousands of other cases, especially in the United States, the limitation is less one of space than of regulators. More and more subdivisions are built with covenants attached to the deed that prohibit the buyer from installing outdoor antennas. Once limited to town house developments, these onerous covenants (known as *CC&Rs*) are now routinely placed on single-family dwellings as well. The problem has reached such proportions that during 2010 a bill was introduced in the U.S. Congress that—if it had become law—would have required homeowner associations, property managers, and building owners to extend reasonable accommodation to residents and tenants.

In the meantime, however, this chapter is for all those who continue to be thwarted in their attempts to put up an antenna of any kind—especially antennas for the HF bands. Here we will examine some of the alternatives available to those readers who have either a limited-space situation (such as a small city lot) or are unable to move out of a subdivision where there are absolute prohibitions against outdoor antennas. The suggestions contained in this chapter are not universal, and indeed the authors recommend that you *adapt*, as well as *adopt*, these recommendations to your own situation, and come up with some of your own. In short, creativity within the constraints of the laws of physics governing radio antennas is encouraged.

Hidden Antennas

A hidden antenna is one that is either completely shielded from view or disguised as something else. We also include in this category antennas that are in *semiopen view,* but which are not obvious (except to the trained and diligent eye). For example: Some people have had success staying on the air with "hidden" longwires made of very fine wire (#26 enameled wire is popular). Installed in the open, high off of the ground (as in an apartment installation), these antennas often escape detection by anyone other than their installer.

The dipole is a popular antenna with both shortwave listeners and amateur radio operators. As you learned in Chap. 5, the dipole is usually horizontal, usually made of wire or tubing, usually a half-wavelength long, usually erected outdoors, and usually fed in the center (ideally) with open-wire line or 50- or 75-Ω coaxial cable. But there's nothing magic about any of those "usually" descriptions. So if your HF antenna needs to be hidden, here are some ideas to help make an untenable situation at least tolerable:

- Hang your dipole from the rafters or *carrying beam* in the attic of your residence. That's the highest place, and it's out of the way of family members and landlords.
- Feed your wire at one end instead of in the middle, if need be. Turn it into an end-fed Zepp or just a simple longwire. Be aware that you will need a good ground system attached to your transmitter for maximum effectiveness of the longwire.
- Bend the ends if you run out of attic space. Let them dangle, or let them come down the rafters at an angle. See Fig. 15.1.
- Use traps to get multiple bands out of your wire—especially if you're using coaxial cable as your feedline.
- If you've never used one before, add a full-range antenna tuning unit (ATU) so you can load up your attic wire on virtually any frequency you want.

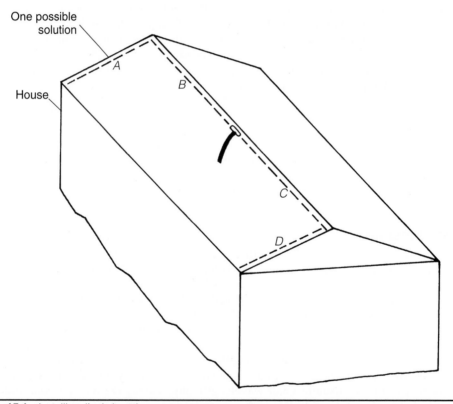

FIGURE 15.1   Installing dipole in attic.

As we have mentioned many times, the length of a λ/2 dipole is given approximately by

$$L_{ft} = \frac{468}{F_{MHz}} \tag{15.1}$$

If you do some quick calculations you will find that antennas for the 10-, 13-, 15-, and (possibly) even 18- and 20-m bands, will fit entirely inside the typical town house attic. This statement is also true of the 11-m citizens band antenna. But what about the lower-frequency bands?

Figure 15.1 hints at a possible solution for the lower frequencies (specifically, 80, 40, and 30 m) in a town house or single-family residence where the radio enthusiast has access to the highest interior space. In a sticky situation we can install a dipole with its arms bent in conformity with the space available. In this example, only one of many possible methods for accomplishing this job is shown. Here each quarter-wavelength section is composed of two legs, *AB* and *CD*, respectively. Ideally, segments *B* and *C* are the longest dimensions. Another method is to reverse the direction of one end leg—say, for example, *D*—and run it to the other corner of the building over the peak of the roof. In reality, you might not be able to use the exact layout shown—so ad lib a little bit.

How about performance? Will the constrained dipole of Fig. 15.1 work as well as a regular dipole installed a wavelength or two off the ground and away from objects? Of course not—especially for DXing, where low radiation angles make a big difference. But that is not the problem being solved; getting on the air *at all* is the problem at hand. You will find that the pattern of the constrained dipole is distorted compared with that of the regular dipole. In addition, the feedpoint impedance is not going to be 50 or 73 Ω (except by some fluke), so you will be required to use an antenna tuner of some kind.

Although wood (as in rafters and studs) is an insulator, the wire used in the constrained dipole (or other forms of attic antenna) should be mounted on TV-type screw-in standoff insulators if any significant output power is employed. Almost any store that sells TV antennas or accessories will have them. These standoff insulators are also available in many local hardware stores and department stores that sell TV antennas, as well as at electronics parts suppliers.

CAUTION Do not simply tape or staple the wire to the wooden underside of your roof. The reason is simple: In all transmitting antennas, voltages can get high enough—not only at the ends but at many multiple locations along the wire—to produce corona effects. The resulting arcing can easily start a fire!

Another alternative for the attic antenna is the nonresonant loop shown in Fig. 15.2. Although presented as a top view, the loop can be installed in any configuration that is compatible with the available space. In fact, the best performance will be bidirectional when the loop is installed vertically. Again, use standoff insulators and insulated wire for the installation. Ideally, the giant loop is fed with parallel line and tuned with a balanced ATU, but it is no crime to feed it with coaxial cable. As was true with the constrained dipole, the performance is not to be equated with the performance of higher antennas outside but is far superior to the prospect of not being on the air at all!

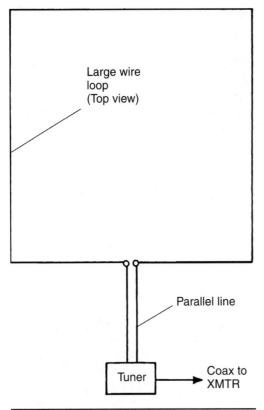

Keep in mind that indoor antennas are usually inferior receiving antennas because of their proximity to a multiplicity of noise sources: noisy power wiring in the walls of the structure, microprocessor waveforms, Ethernet drivers, wall adapters for myriad consumer electronics devices, and the like. If you are plagued by noise that comes from a single direction, you might consider the use of a noise-canceling accessory, such as those made by ANC or MFJ, for your receiver or transceiver. These devices pick up the noise on a *sense antenna* that is often nothing more than a short piece of wire and then change the phase of the detected noise signal so it cancels the noise on your main antenna. The noise canceler approach works best when the offending noise comes from a single source and/or a specific direction.

For those who can entertain the possibility of hidden antennas outdoors, another ploy is the old "flagpole trick" shown in Figs. 15.3A and 15.3B. Some housing developments allow homeowners to promote their patriotism by installing flagpoles—and flagpoles can be antennas in disguise. In the most obvious case, you can install a brass or aluminum flagpole and feed it directly from an ATU. For single-band operation, especially on the higher frequencies, you can delta-feed the "flagpole" unobtrusively and call your flagpole a vertical antenna. But that is not always the best solution.

FIGURE 15.2 Large loop antenna.

Figures 15.3A and 15.3B show two different methods for creating a flagpole antenna; both depend on using white PVC plumbing pipe as the pole. The heavier grades of PVC pipe are self-supporting to heights of 16 to 20 ft, although lighter grades are not self-supporting at all (hence, are not useable).

Figure 15.3A shows the use of a PVC flagpole in which a #12 (or #14) wire is hidden inside. This wire is the antenna radiator. For some frequencies the wire will be resonant, and for others it will surely be nonresonant. Therefore, an ATU is used either at the base of the antenna or inside at the transmitter. If the wire is too long for resonance (as might happen in the higher bands), place a capacitor in series with the wire. Multiple settings may be required, so use a multisection transmitting variable that has a total capacitance selectable to more than 1000 pF. Alternatively, use a vacuum variable capacitor of the same range.

In cases where the antenna is too short for resonance, as will occur in the lower bands, insert an inductance in series with the line to "lengthen" it. Alternatively, use an L-section tuner at the feedpoint.

A good compromise situation is the use of a 16-ft length of "flagpole" pipe with a 16-ft wire embedded inside. The 16-ft wire is resonant at 20 m, so it will perform like a quarter-

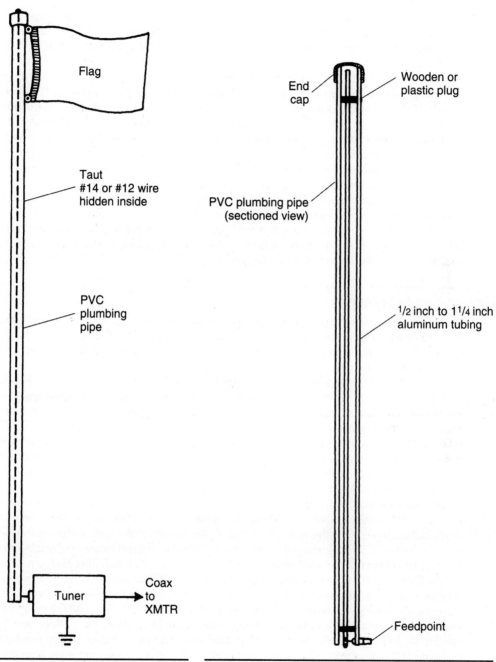

Flag

Taut
#14 or #12 wire
hidden inside

PVC
plumbing
pipe

Tuner

Coax
to
XMTR

End
cap

Wooden or
plastic plug

PVC plumbing pipe
(sectioned view)

$1/2$ inch to $1^1/4$ inch
aluminum tubing

Feedpoint

FIGURE 15.3A   Flagpole antenna made
from thick-walled PVC pipe.

FIGURE 15.3B   PVC flagpole hiding an aluminum tubing
vertical.

wave grounded vertical monopole antenna on that band. The tuner will then accommodate frequencies above and below 20 m. Remember: A monopole will not work well at all without an appropriate ground or system of radials beneath it. Radials should be relatively easy to hide, although you may have to wait until midnight to lay them in the thatch of the lawn surrounding your flagpole! (If caught in the act, you can always claim that the radials are part of the lightning discharge system associated with your flagpole.)

Another alternative is the version shown in Fig. 15.3B. In this case the wire radiator is replaced with a section of aluminum tubing. A wooden or plastic insert is fashioned with a drill and file to support the aluminum tubing inside the PVC tubing. One way to make the support is to use a core bit in an electric drill to cut out a disk that fits snugly inside the PVC tubing. Rat out the center hole left by the core bit pilot to the outside diameter of the aluminum tubing. The support can be held in place with screws from the outside or simply glued in place. One advantage to using tubing instead of a wire is a modest increase in the bandwidth of the vertical.

Coming up with the right concept for operating with a hidden antenna can pose a serious challenge. But with some of these guidelines and a little creativity, you should be able to get on the air and enjoy your hobby.

Limited-Space Antennas

Many people live in situations where it is permissible to install an outdoor antenna, but it is not practical to install a full-size antenna. Here we examine some of the options open to those with limited space for amateur radio, citizens band (CB), or shortwave listener (SWL) antennas.

Once again we return to the simple dipole as the basis for our discussion. Figure 15.4 sketches several alternatives for installing an outdoor antenna in a limited space. In Fig. 15.4A, the slanted dipole antenna uses the standard dipole configuration, but one end is connected to the high point of the building, while the other is anchored near the ground. The coaxial cable is connected to the midpoint of the antenna in the usual manner for regular dipoles. If the end of the dipole is within reach of people on the ground, someone may get a nasty RF burn if the antenna is touched while you are operating. Take precautions to keep people and pets away from any part of your antenna.

Another method is the vee dipole shown in Fig. 15.4B. In a regular dipole installation, the ends of the antenna are along the same axis. (In other words, they form an angle of 180 degrees.) In the example of Fig. 15.4B, however, the angle between the elements is less than 180 degrees but greater than 90 degrees. In some cases we might need to bend the elements partway along their length, rather than installing them in a vee shape. Figure 15.4C shows an angled or *bent dipole* with four segments. In all three examples, Figs. 15.4A through 15.4C, you can expect to find that the length needed for resonance will vary somewhat from the standard $468/F(\text{MHz})$ value, and the feedpoint impedance is very likely to be something other than 73 Ω. Also, the pattern will be somewhat distorted by any house wiring, aluminum siding, or other nearby conducting materials with at least one dimension roughly comparable to $\lambda/4$ or more. Nonetheless, although these antennas may not work quite as well or as predictably as a high dipole clear of nearby objects, the performance can be surprisingly good.

Another limited-space wire antenna is the (so-called) half-sloper shown in Fig. 15.4D. Although single-band versions are often seen, the example shown is a multiband version. Resonant traps result in relatively low SWR across multiple frequency ranges,

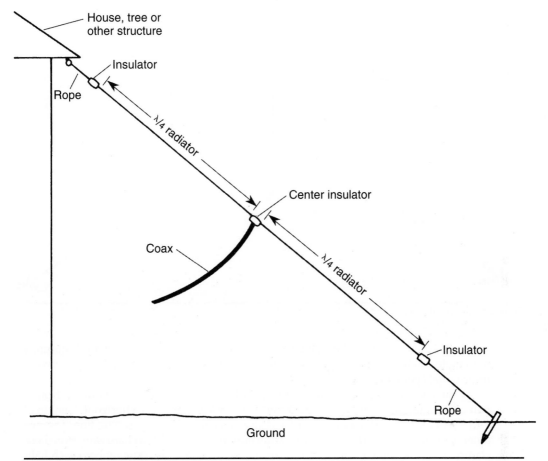

FIGURE 15.4A   Sloping dipole.

thus allowing the use of coaxial cable for the feedline and potentially eliminating the need for an ATU. In contrast to center-fed dipoles, this antenna is a sloping *monopole* that is strongly dependent upon a good RF ground at the feedpoint for reasonable transmit efficiency.

Figure 15.5 shows another antenna that's useful for limited-space situations. Although it is easily constructed from low-cost materials, the antenna is also sold by several companies under various rubrics including "cliff-dweller", "apartment house", "town house", or "traveler's" antenna. It consists of a 4- to 16-ft section of aluminum or copper tubing. Some of the commercial antennas use telescoping tubing that can be carried easily in luggage. As with the longwire, this windowsill antenna is tuned to resonance with an L-section coupler. Again, a good ground (or radial system) will greatly improve the performance of this antenna, which is capable of working DX for you—even on the lower frequencies.

Figure 15.6 shows the use of a mobile antenna for a fixed or portable location. Simply mount a mobile whip, such as the *Hustler,* on the windowsill or other convenient mount. Since this, too, is a vertical monopole, an adequate ground system is essential to

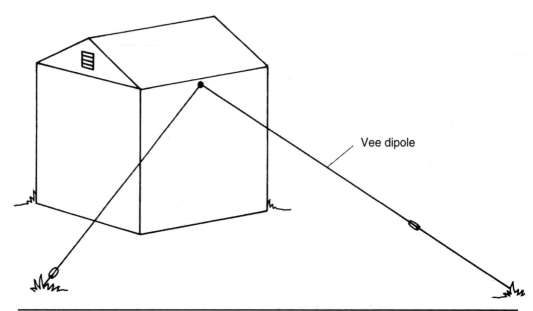

Figure 15.4B   Vee dipole.

this antenna, just as it is for the longwire and windowsill antennas. At a minimum, a monopole should have a counterpoise or elevated ground consisting of at least two or three radials per band.

A directional *rotatable* dipole is shown in Fig. 15.7. This antenna is made from a pair of commercial mobile antennas connected "back-to-back" on a horizontal length of 1-in × 2-in lumber and fed in the center with a coaxial cable or open-wire line. The two antennas can be clamped to the wood center support with standard hardware store hose clamps or strong cable ties. The tips of the mobile antenna determine where each half-

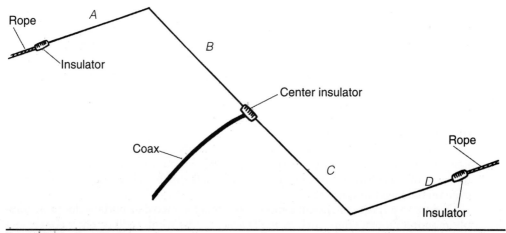

Figure 15.4C   Bent dipole.

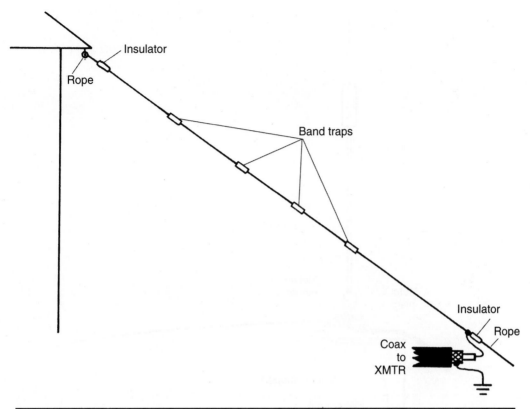

FIGURE 15.4D   Sloping trap vertical.

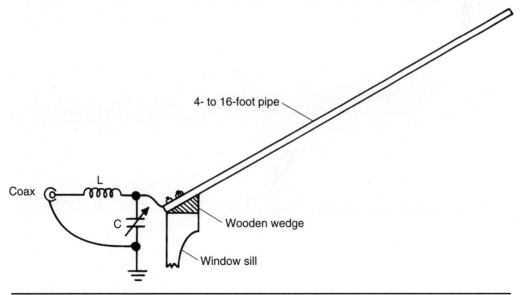

FIGURE 15.5   Window sill antenna.

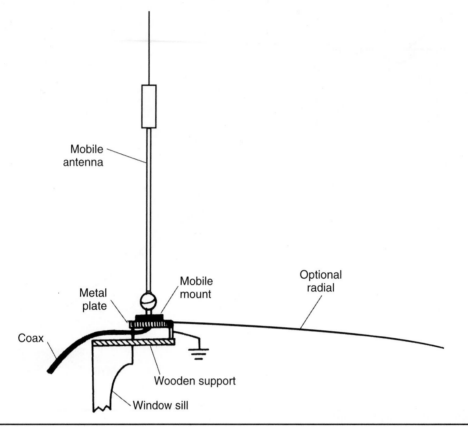

Figure 15.6   Use of mobile antenna as a windowsill antenna.

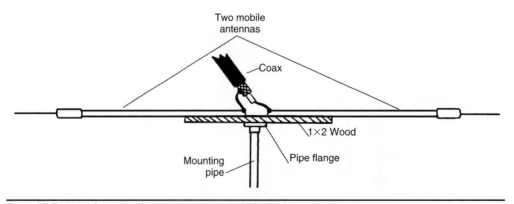

Figure 15.7   Use of two mobile antennas as a rotatable dipole.

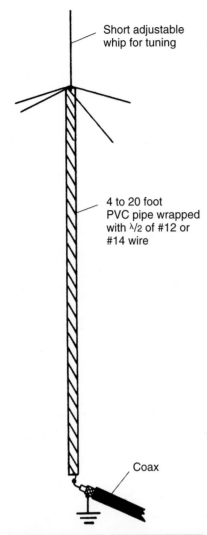

Short adjustable whip for tuning

4 to 20 foot PVC pipe wrapped with λ/2 of #12 or #14 wire

Coax

FIGURE 15.8A Helically wound vertical antenna.

element is resonant, and you must be sure to adjust both sides of the antenna to the same length. On HF these are extremely short dipoles for the frequencies involved, so their tuning will be very sharp, their useable bandwidth very narrow, and their efficiency rather low. One advantage to using them in a horizontally polarized configuration, however, is that no RF ground system is required. Another advantage, particularly if you live in a high-rise condo or apartment building with your own balcony and can lay the completed assembly out on a nonmetallic table or other support, is they then pick up as much as 6 dB ground reflection gain at a decent height. Try it—the results may surprise you!

Two examples of *helically* wound antennas are shown in Fig. 15.8. In this type of antenna, an insulating mast is wound with a half-wavelength of antenna wire. The mobile HF whips from Mosley Antennas utilize this technique. The overall length of the antenna is considerably less than a half-wavelength except at the highest frequencies. In order to dissipate the high voltages that tend to build up at the ends of these antennas when operated at higher power levels, a capacitance hat is often used. These "hats" can be disks (pie tins work well) or conducting rods about 16 to 24 in long. The version shown in Fig. 15.8A is a vertical monopole antenna, and like other such verticals it must be installed over either a good ground or a counterpoise ground. The version shown in Fig. 15.8B is a horizontally polarized dipole. Keep in mind that helical whips are somewhat less efficient than other mobile loading coil schemes because the helical coil uses much more wire than a lumped-component coil to achieve the same loading. The extra wire equates to added resistive power loss (heating), which may give the (erroneous) impression of wider SWR bandwidth.

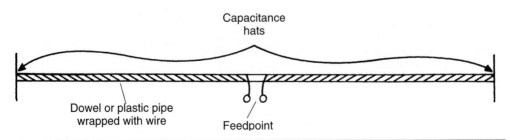

Capacitance hats

Dowel or plastic pipe wrapped with wire

Feedpoint

FIGURE 15.8B Helically wound dipole antenna.

CHAPTER 16

Mobile and Marine Antennas

Mobile operation of radio communications equipment dates back nearly as far as "base station" operation. From the earliest days of wireless, radio buffs have attempted to place radio communications equipment in vehicles. Unfortunately, two-way radio was not terribly practical until the 1930s, when the first applications were amateur radio and police radio (which used frequencies in the 1.7- to 2.0-MHz region). Over the years, land mobile operation has moved progressively higher in frequency for practical reasons. The higher the frequency, the shorter the wavelength and therefore the closer a mobile whip can get to being a full-size antenna. On the 11-m citizens band, for example, a quarter-wavelength whip antenna is 102 in long, and for the 10-m amateur band only 96 in long. At VHF and above, full $\lambda/4$ mobile antennas are practical. Partially as a result of this, much mobile activity takes place in the VHF and UHF region.

Interestingly, even as land mobile moved higher in frequency with the passing decades, amateur radio mobile activity has moved both higher and lower, with ever-increasing interest in overcoming the challenges that very long wavelengths present. Today, enterprising amateurs in the United States are working transoceanic DX on 160 m during their daily commutes!

Mobile operation in the amateur 144-, 220-, and 440-MHz bands is extremely popular because of several factors, not the least of which is the small size (and cost) of efficient vehicle-mounted verticals (or *whips*) for these bands.

Mobile Antennas for VHF and UHF

At 2 m and above, the roof of even a small coupe has enough flat or nearly flat surface area to act very much like an ideal ground plane—provided, however, that the antenna is located near the middle of the roof, and that the shield, or grounded, conductor of the coaxial feedline makes a clean, low-resistance connection to the roof at the base of the antenna. If that is done, a simple 19-in piece of tinned copper wire or plated spring steel will work as well as any store-bought whip on 2 m. Over the years, the author has had 2-m whips installed by a commercial two-way radio shop on more than one occasion, but has also done it himself a few times. If you are comfortable with snaking RG-58 or RG-8X cable between the roof and the liner of your vehicle and know how to install an old-fashioned chassis-mount SO-239 (drill a pilot hole for a chassis punch of the appropriate size, such as those made by Greenlee), you can have a 2-m mobile antenna that is hard to beat! (Don't forget to seal against rain around the connector. Use a rubberized gasket or a caulking compound.)

But suppose your vehicle is leased? Or has a ragtop—in other words, it's a convertible? Or there's only a few inches of clearance between the top of your sport utility vehicle (SUV) and your garage door? Then you will need to consider alternative locations for your antenna.

For VHF or UHF whips, the next best location is usually the center of the trunk lid—assuming, of course, that your vehicle *has* a trunk! Even though the roof of a car will be well above the base of the VHF or UHF whip, this isn't necessarily as bad for your signal or your antenna pattern as you might think. For one thing, in the typical sedan or convertible, anything mounted on the trunk lid has a clear horizontal shot (through the car's windows) in many front-facing directions. Secondly, *collinear* (see Chaps. 9 and 20) whips for VHF and UHF are relatively inexpensive and can bring at least one element of the collinear up above the roofline for a full 360-degree "view". A secondary advantage of locating the mobile antenna at the rear of the vehicle is that in many installations noise pickup from the various ignition components and microprocessors in the front will be less.

If a roof-mounted or trunk lid–mounted whip is not an option (because of garage doors or other constraints), sometimes a VHF or UHF whip can be mounted on top of a front fender (directly above a wheel well, in other words) or along the side or rear of the hood, in the gap between the hood and other vehicle body sheet metal sections. In other cases—especially SUVs—whips can be mounted to the oversize outside rearview mirror brackets found on larger vehicles, just as over-the-road drivers of those "big rigs" have their CB antennas mounted on their mirror brackets. Bear in mind, however, that with these latter locations, the overall symmetry of the ground plane is compromised.

Finding an acceptable ground plane on a fiberglass vehicle such as the Corvette is a more difficult task. However, one advantage to having a Corvette (actually, the author can think of *many*!) is that low whips, such as bumper-mounted ones, will encounter less loss of signal from massive pieces of metal near the whip. Further, it should be possible to get at least a *skeletal* ground plane from the frame and drivetrain members, provided care is taken to make sure all these parts are electrically bonded together and to the transmission line ground at the base of the whip.

If all else fails, bumper mounts, side fender mounts, and trailer hitch mounts can be used, even though the typical VHF or UHF whip (even the collinear) will be well below the roofline of the vehicle. However, for HF antennas, these last-named mounting techniques are actually the most common!

Mobile Antennas for HF and MF

Radiating a strong signal from a mobile installation is very much a frequency-dependent problem. While many vehicles provide a satisfactory ground plane and sufficient vertical clearance for a $\lambda/4$ or taller whip on 2 m and above, having a strong HF mobile signal is one of the most enduring challenges of CB and amateur radio (and some other services, as well). Even in the presence of a perfect ground plane, most mobile vertical monopoles below 30 MHz are such a small fraction of $\lambda/4$ that radiation efficiency and radiation resistance are extremely low by comparison to well-designed fixed station antennas.

The magnitude of the standing wave of current on a $\lambda/4$ monopole is generally approximated as a cosine curve having maximum amplitude at the base feedpoint and dropping to zero at the tip, where the boundary condition there requires that $I_{TOTAL} = 0$. If the vertical is significantly less than $\lambda/4$, a plot of current versus position along the line still starts at zero at the tip but is usually taken to be a straight line, reaching an I_{MAX} at the feedpoint. Although the signal strength at a distant receiving antenna is proportional to the total area under the current-versus-length curve of the transmitting antenna, in theory one could simply increase the drive current to make the $I \times l$ product the same as for a $\lambda/4$ monopole; in other words, one could greatly increase the slope of the curve. (If that weren't true, the peak gain of the infinitesimally short dipole of Chap. 3 would be much further below that of a $\lambda/2$ dipole than 0.6 dB!) Theory and practice are two different things, however, and most mobile HF antennas are at an immediate disadvantage for the following reasons:

- Since the total length l is substantially less than $\lambda/4$, the impedance at the feedpoint exhibits a very high capacitive reactance, which cannot be matched to obtain maximum power transfer to the short monopole except with very lossy inductors. (The extremely high circulating currents passing back and forth between the short monopole and the lumped inductor result in substantial I^2R losses.)

- The resistive part of the feedpoint impedance also decreases with decreasing length (relative to $\lambda/4$) and it becomes increasingly difficult—even with matching networks—to put the preponderance of the transmitter's RF output into the radiation resistance, R_{RAD}, of the antenna instead of losing it to ground losses and other dissipative losses in cables, connectors, etc.

In general, the key to HF or MF mobile installation success requires attention to certain details:

- Maximize the equivalent (or *effective*) length of the vertical radiator. Some things to do:
    - ○ Make sure the antenna has a top hat and a center loading coil, as described in the next secion. Both devices increase the effective length of the monopole, raising its feedpoint resistance and reducing the (capacitive) feedpoint reactance.
    - ○ Use simple trigonometric formulas (such as in App. A) to estimate the apparent $I \times l$ product for various possible mounting locations. (For instance, a low bumper mount may actually given you a greater $I \times l$ product than a roof mount, even though part of the high-current portion of the antenna is shielded by the car body—simply because the total l of the whip is able to be that much longer.) Each situation will be different because of the shape of the metal parts of your vehicle. When you do this analysis, treat the radius of the top hat as adding length to the top end of the whip; in other words, assume the current is zero at the outer ring of the top hat and work inward to the whip and then down the whip to the feedpoint. Assume the loading coil does not contribute to the radiated field (i.e., do not include any contribution to the $I \times l$ total corresponding to the vertical distance occupied by the coil) but *do* assume that it contributes to the effective length of the antenna.

- As the operating frequency is lowered, the amount of metal needed to form an adequate ground plane increases. Even at 6 m, the roof of most vehicles is not large enough to approach the ideal; as a general rule, as the operating frequency is lowered, ground losses in the mobile antenna system will increase. As with fixed or base station vertical monopoles, few antenna design and construction details are as important as those of the ground system directly beneath the vertical. In most cases (Chevy Corvettes and most recreational boats being prominent exceptions), the vehicle body is an oddly shaped mass of metal, and the most efficient mobile installations attempt to optimize antenna performance despite close proximity to that conductor. Some suggestions:

  ○ In general, the objective is to reduce RF losses in the vehicle's metallic mass to the smallest extent possible. Because the radiation resistance of the mobile monopole is so low, the quality of the ground beneath is even *more* important than it is for a fixed vertical.

  ○ Insofar as possible, locate the antenna so as to keep as much of the metal as possible *below* the primary radiating portion(s) of the antenna.

  ○ A superior mobile installation will have as many major metal components of the vehicle as possible electrically bonded together, as well as to the ground (shield) side of the transmission line at the base of the mobile antenna. All this bonding is best done when the vehicle is new, before underside steel has a very bad tendency to develop rusty surfaces. For many of us, the tools and materials necessary to do the job right are beyond the capabilities of our workbenches, and we are better off (if we can afford it) having it done by a good two-way radio installation shop.

- Similarly, losses in connections, connectors, transmission lines, antenna tuning units, and even the monopole itself can eat up most of the RF energy from the transmitter. Pay close attention to every link in the system.

- Don't forget the receiving side of the equation. Locate the mobile antenna as far as possible from vehicular microprocessors and ignition system components. Route any power and RF cables connecting your equipment to the antenna, the vehicle battery, and any other devices as far from other vehicular cabling as possible. Place properly chosen and constructed *common-mode chokes* on all cables entering and exiting the receiver or transceiver. Minimization of RFI from other sources throughout the vehicle is beyond the scope of this book, but quite a body of experimental successes has been accumulated; much of it is accessible on the Web by properly constructing your search request (or with the help of the eight-year-old next door).

Effective high-frequency (HF) mobile operation requires substantially different antennas than those commonly installed for VHF or UHF. Quarter-wavelength antennas are feasible on only the 10-, 11-, and 12-m HF bands. By the time the frequency drops to the 21-MHz (15-m) band, the antenna size must be approximately 11 ft long, much too long for practical mobile operation. Height limits imposed by utility lines, overpasses, parking garages, and tollbooths, to name a few, limit antenna length to about 9 ft; for many amateurs the 8-ft whip is most popular because it is resonant on 10 m. At all lower frequencies, the 8-ft whip is too short and it becomes capacitive, thus requiring an equal inductive reactance to cancel the capacitive reactance of the antenna at the chosen operating frequency.

The lower the operating frequency, the worse the problem becomes. In fact, HF and MF mobile operation suffers from a "double whammy" as the frequency drops: Not only is the radiating monopole getting far too short for the frequency, but the ground plane beneath it is becoming less and less effective, as well!

When the effects of imperfect ground planes and short vertical radiating elements are combined, it's not unusual to find that a typical HF mobile installation may have overall efficiencies of between 1 percent and 10 percent those of a conventional quarter-wavelength radiator married to an extensive ground plane!

Types of Short Mobile Antennas

Given the impossibility of having a $\lambda/4$ whip on the HF and MF bands while mobile, the most common compromise is to add a combination of discrete and/or distributed components to the mobile whip in hopes of increasing the feedpoint resistance and eliminating or at least minimizing the feedpoint reactance. But what kinds of components? And where do we place them in the existing antenna?

Figure 16.1 shows three basic configurations of coil-loaded HF antennas for frequencies lower than the natural resonant frequency of the antenna. In each case, the antenna is series-fed with coaxial cable from the base; point A is connected to the coaxial cable center conductor, and point B is connected to the shield and the car body, which serves as a (usually quite imperfect) ground plane. The system shown in Fig. 16.1A is base-loaded. Although convenient, simple to build, and the easiest to support mechanically, base-loaded verticals are among the least efficient radiators around and should be avoided if at all possible. Their inefficiency is a direct result of the high-current portion of the antenna element being replaced by a lumped-component loading coil, an approach we have eschewed in earlier chapters.

Figure 16.1B depicts a center-loaded whip, with improved current distribution. This configuration is by far the most common among commercially available HF mobile antennas, although the coil may be high enough and/or large enough to require special support.

Finally, we see the top-loaded coil system in Fig.16.1C. When accompanied by a capacitive *top hat*, this is by far the most efficient configuration, as has been proved time and time again at numerous mobile "shoot-outs" at hamfests around the world. It also is the most top-heavy of the three configurations, requiring the largest coil, the strongest mast, and sometimes even guy wires or guy rods high on the mast from anchors near the top of the vehicle.

Regardless of the specific configuration, in all three cases the purpose of the coil is to cancel most, if not all, of the capacitive reactance of the electrically short antenna at the operating frequency.

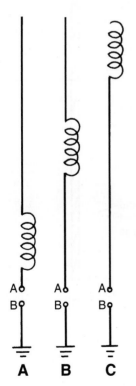

| A | B | C |

FIGURE 16.1 Loaded mobile antennas.

A modified version of the coil-loaded HF mobile antenna is shown in Fig. 16.2A. In this configuration the loading inductance is divided between two individual coils, L_1 and L_2. Coil L_1 is adjustable with respect to the antenna, while L_2 is fixed; coil L_2 is tapped, however, in order to match the impedance of the antenna to the characteristic impedance of the coaxial-cable transmission line. When tuning this antenna, two instruments are useful: a *field strength meter* and a *VSWR meter.* Placed at least a few wavelengths from the mobile installation, the field strength meter gives a *relative* indication of the signal radiated from the antenna, while the VSWR meter helps determine the state of the impedance match.

Variations on this theme are found in Figs. 16.2B and 16.2C. In the first variation, a helical whip, similar to the family of mobile whips available from Kenwood, is shown in Fig. 16.2B. In this configuration, the inductor is distributed along the length of a fiberglass antenna rod. Rather than being a cylindrical metal tube or rod, the conductor is a wire helically wound on a fiberglass shaft and encased in *shrink-wrap sleeving* to minimize abrasion damage. An adjustable-length tip section tunes the antenna to resonance if the mobile transmitter or transceiver does not have a built-in ATU. The length of this tip should be set with the help of a VSWR meter. Fixed station versions of this type of an-

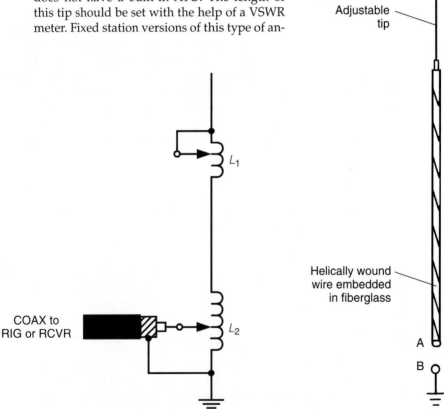

FIGURE 16.2A   Practical mobile antenna for HF bands below 10 m; L_1 resonates, L_2 matches impedance.

FIGURE 16.2B   Helically wound mobile antenna.

tenna are very popular on the 11-m citizens band, where they may be as short as 30 in or as long as 48 in.

The other variant is shown in Fig. 16.2C. In this case, the lower end of the radiator consists of a metal tube topped with a loading coil. An adjustable shaft at the top end is used to tune the antenna to resonance. Commercially manufactured forms of this approach are easily found; the New-Tronics *Hustler* product line, for instance, includes a family of lower rods, couplers, mounts, and adapters, and features resonator coils for both low and high power. The fixed lower rod is used on all bands, while the coil and adjustable *stinger* form a separate "resonator" for each HF band. Multiband operation of the *Hustler* can be accomplished by using a resonator mounting bracket such as the three-way unit shown in Fig. 16.2D. (But don't use it while moving; the stock *Hustler* bracket is aluminum, which can quickly fatigue and crack when subjected to buffeting while under way.) With no resonators at all atop it, the MO-1 or MO-2 fold-over shaft can be used by itself on 6 m, especially if the mobile HF rig has an internal antenna tuner (example: the Kenwood TS-480S/AT).

A common problem with all short coil-loaded mobile antennas is that they tend to be very high Q antennas. In other words, they are very sharply tuned and have a very narrow useable bandwidth; as little as 25-kHz change of operating frequency will detune the antenna significantly on 80 m. Although an antenna tuner at the transmitter will reduce the VSWR to a point that allows the transmitter to operate, that type of tuner does not fully address the problem. The actual problem is that the antenna is very short relative to a quarter-wavelength, and the feedpoint impedance of the antenna varies drastically with small changes in operating frequency. The only cure for this problem is to readjust the resonator's adjustable shaft as the band segment is changed. Unfortunately, this solution requires the opera-

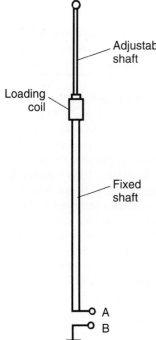

FIGURE 16.2C Top-loaded mobile antenna common in amateur radio.

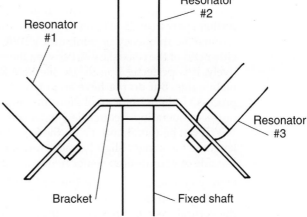

FIGURE 16.2D Multiband top-loaded antenna.

tor to get out of the vehicle in order to do a good job of retuning. In recent years more and more mobile operators have switched to using a motor-driven variable inductor for the loading coil. Today several manufacturers (High Sierra and Tarheel are two well-known and highly respected suppliers) offer both base-loaded and center-loaded coils and entire antenna assemblies that are either motor-driven or relay-selectable to permit frequency changing from inside the vehicle. Some models even feature a remote-control tilt to allow driving into the garage!

Arguably the best enhancement that one can make to any short HF whip is a ca-pacitive *top hat*. Usually consisting of a half-dozen or more radial *spokes* emanating from the top of the stinger to a circumferential support wire or tube, a top hat increases the electrical length of the entire assembly, thus bringing more of the natural high-current portion of the monopole out of the feedline and into the vertical element. Benefits include overall antenna efficiency improvements and increased VSWR bandwidth. Top hats for 21 MHz and below should be as large as you can make them; it's unlikely you'll exceed the dimensions at which they become an electrical detriment! Mobile HF antennas from some manufacturers have top hats with a 3- or 4-ft diameter; in many designs, guying of the upper half of the antenna is required.

Tuning HF Mobile Antennas

Tuning a typical mobile HF antenna is not particularly difficult, since there's very little you can adjust. Here's a sequence that works for center-loaded verticals with fixed-inductance loading coils (*resonators*):

- Initially set the whip or stinger to the manufacturer's recommended length. If you have no guidelines—or your mobile antenna is totally homebrew— set the total length to the maximum you can within the practical constraints (overpasses, utility lines, etc.) mentioned earlier.

- If your loading coils are not adjustable, attempt to find a length for the stinger (the portion of the whip *above* the loading coil) that produces a low VSWR back at the transceiver in the lower portion of your desired range of operating frequencies on a given band. While transmitting briefly on spot frequencies (no further apart than 25 kHz on 80 and 40) throughout the appropriate band, determine the frequency of minimum VSWR, if possible. (Use a VSWR meter at either end of the coaxial cable between the mobile rig and the antenna. Alternatively, it is perfectly fine to use the VSWR meter built into many of today's mobile rigs. But do not have an antenna tuner in the line at this stage of the process.) If there is not an obvious minimum VSWR somewhere in the desired range of operating frequencies, try a different stinger length. If a minimum VSWR can be found and it's on the *high* side of your chosen center frequency, lengthen the stinger slightly and remeasure. If the minimum VSWR is on the *low* side of your desired center frequency, shorten the stinger and remeasure.

- Once you have obtained the lowest VSWR you can by adjusting the length of the stinger, reintroduce any matching networks, starting with any at the base of the vertical. As a last resort, use the antenna tuner in your mobile rig to minimize the VSWR. Presumably this will coincide with maximum RF out of the vertical if you have been monitoring the field strength meter mentioned earlier. In short, about all you can do is attempt to match the antenna to the feedline and the transmitter, to maximize RF transfer and minimize any tendency by the transmitter to reduce power in the face of VSWRs higher than 1.0:1.

- At some point in this procedure you may find you can't make the stinger any shorter because it has bottomed out in the clamp at the top of the loading coil. If so, you may need to cut some of it off. Be careful! It's hard to *add* length if you cut it prematurely. In fact, if the minimum VSWR is not unreasonable (say, 1.5:1) with the original stinger stuffed all the way into the clamp, be satisfied. Do any further VSWR reduction with a matching network in the line or back at the transceiver. In short: Learn when to leave well enough alone!

In other mobile whip installations, there may be an adjustable (or tap-selectable) loading coil in addition to the adjustable stinger. Figure 16.2A takes this idea one step further with an adjustable or tapped center-loading coil in series with a tapped impedance-matching coil at the base. Alternatively, the upper coil is a fixed inductance, and the only adjustment to the upper part of the vertical is to the length of the adjustable resonator shaft (stinger). Ideally, a *dip meter* (see Chap. 27) is available to set the upper inductance of the stinger so as to roughly resonate the vertical at the desired operating frequency. Then—and only then—should the matching network at the base of the antenna (or back at the transceiver) be adjusted. With this configuration, it may also be possible to initially adjust the upper coil or stinger for maximum reading on a field strength meter, then follow up with adjustment of the base matching network for minimum VSWR on the transmission line back to the transceiver. However, tuning the upper coil and/or stinger for maximum field strength can lead to ambiguous or even erroneous results if the output power of the transceiver changes substantially in response to the changing VSWR it sees as the stinger wire or upper coil is adjusted.

Marine Radio Antennas

From its inception, radio communication considerably lessened the dangers inherent in sea travel—so much so, in fact, that the maritime industry took to the new "wireless telegraphy" earlier than any other segment of society. In the early days of radio, a number of exciting rescues occurred because of wireless. Even the infamous *S.S. Titanic* sinking might have cost less in terms of human life if the wireless operator aboard a nearby ship had been on duty. Today, shore stations and ships maintain 24-hour surveillance, and vessels can be equipped with autoalarm devices to wake up those who are not on duty. Even small pleasure boats are now equipped with radio communications, and these are monitored around the clock by the U.S. Coast Guard.

When operating type-accepted radio equipment on marine frequencies, the small shipboard operator can have a selection of either HF single-sideband communications or VHF-FM communications. The general rule is that a station *must* have the VHF-FM and can be licensed for the HF SSB mode only if VHF-FM is also aboard. The VHF-FM radio is used in coastal waters, inland waters, and harbors. These radios are equipped with a high/low power switch that permits the use of low power (1 W typically) in the harbor, but higher power when under way. The HF SSB radios are more powerful (100 W typically), and are used for offshore long-distance communications beyond the line-of-radio-sight capability of the VHF-FM band. Of course, a radio amateur can usually utilize the full panoply of modes, subject to the regulations of his home country and any other country whose coastal waters he enters. For radio regulatory purposes, the world has been divided into three regions, and the rules—including, for instance, allowable frequencies of operation—differ from region to region. Consequently, someone planning on using amateur radio equipment outside the coastal waters of his or her own country should be up to date on the pertinent regulations before embarking.

The problems of antennas on boats are the same as for shore installations but are aggravated by certain factors. Space, for one thing, is less on a boat, so most HF antennas are, like mobile antennas, quite short relative to the wavelength in use. Also, RF grounds are harder to come by on a fiberglass or wooden boat, so additional grounds must usually be provided.

A typical powerboat example is shown in Fig. 16.3. The radio is connected to a whip antenna through a transmission line (and a tuner on HF), while also being grounded to an externally provided ground plate. Over the years, shipboard radio grounds have taken many different forms. For example, the ground might be copper or aluminum foil cemented to the boat hull. Alternatively, it might be a bronze plate or hollow bronze tube along the centerline of the boat or along the sides just below the waterline. The ground is connected to the engine. Careful attention must be paid to the electrical system of the boat when creating external grounds to prevent electrolytic corrosion from inadvertent current flows.

On sailboats the whip antenna (especially VHF-FM) might be mast-mounted as shown in Fig. 16.4, or it might be an aluminum mast itself. Alternatively, a *stay* might serve double duty as an HF antenna. In any event, the same grounding schemes apply, including the additional option of a metal keel or metal foil over a nonmetallic keel.

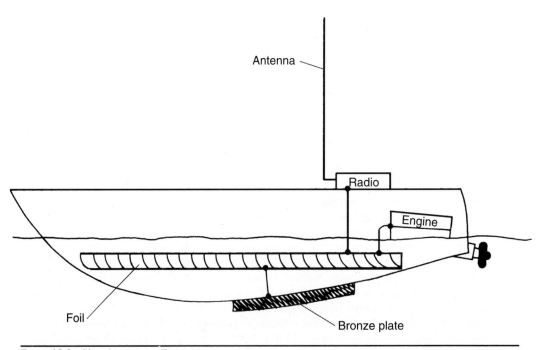

FIGURE 16.3 Motorboat grounding system.

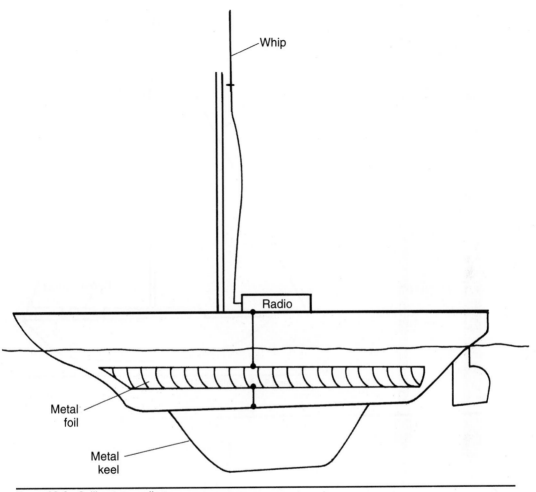

FIGURE 16.4   Sailboat grounding.

The whips used for boat radios tend to be longer than land mobile antennas for a given frequency because they are not subject to the same problems of overpass and tollbooth height. The VHF-FM whip (Fig. 16.5) can be several quarter-wavelengths and take advantage of *collinear* gain characteristics thereby obtained. Whips for the HF bands tend to be 10 to 30 ft in length and often look like trolling rods on powerboats. You will also see citizens band radios and whips on board.

Longwire antennas also find use in marine service. Figure 16.6 shows two possible types of installation. The antenna in Fig. 16.6A shows a wire stretched between the stern and bow by way of the mast. The antenna is end-fed from an antenna tuner or "line flattener". The longwire shown in Fig. 16.6B is similar in concept but runs from the bottom to the top of the mast. Again, a tuner is needed to match the antenna to the radio transmission line. Notice that the antennas used in this manner are actually not long-wires in the truly rigorous sense of the term but, rather, random-length antennas.

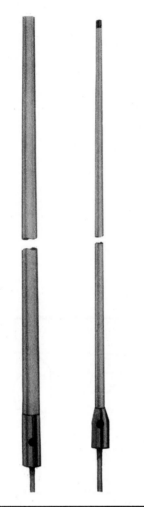

FIGURE 16.5 VHF-FM boat antenna. (*Courtesy of Antenna Specialist Company*)

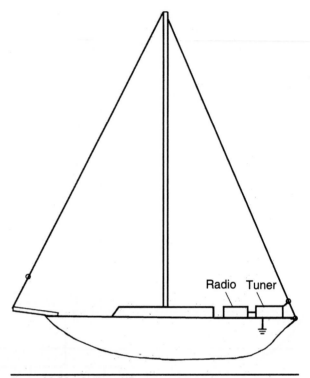

FIGURE 16.6A Sailboat HF antenna.

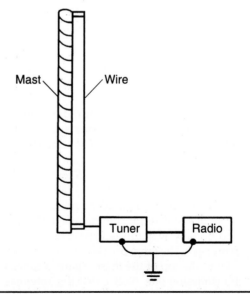

FIGURE 16.6B Mast antenna.

The tuner can be any of several designs, shown in Fig. 16.7. The reversed L-section coupler shown in Fig. 16.7A is used when the antenna radiator element length is less than a quarter-wavelength. Similarly, when the antenna is greater than a quarter-wavelength, the circuit of Fig. 16.7B is the tuner of choice. This circuit is a modified L-section coupler that uses two variable capacitors and an inductor.

Finally, we see in Fig. 16.7C the coupler used on many radios for random-length antennas. Two variable inductors are used: L_1 resonates the antenna, and L_2 matches the feedpoint impedance to the characteristic impedance of the transmission line. In some

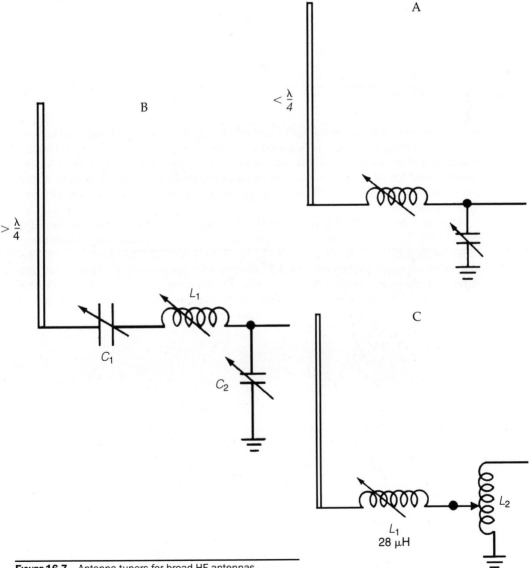

FIGURE 16.7   Antenna tuners for broad HF antennas.

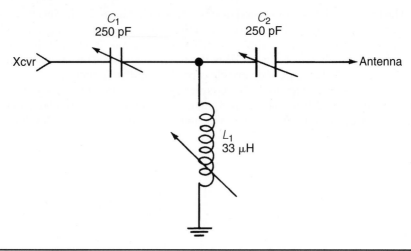

FIGURE 16.8   Line flattener tuner.

designs, the inductors are not actually variable but, rather, use switch-selected taps on the coils. The correct coil taps are selected when the operator selects a channel. This approach is less frequently encountered today, when frequency synthesizers give the owner a selection of channels to use. In those cases, either the antenna must be tuned every time the frequency is changed or an automatic (or motor-driven) preselected tuner is used.

The *line flattener* (Fig. 16.8) is a standard transmatch antenna tuner that reduces the coaxial-cable transmission line VSWR. This type of tuner is especially useful for transmitters with solid-state finals that react badly to high VSWR. Most recent transceiver designs incorporate shutdown (*foldback*) circuits that progressively reduce (and ultimately cut off) power as the VSWR increases. The line flattener tunes out the VSWR at the transmitter. It does nothing to tune the antenna but does facilitate maximum power transfer from the transmitter to the transmission line.

CHAPTER 17

Emergency and Portable Antennas

Some time ago, one of the authors met an interesting character at a convention. This man was a medical doctor, working as a medical missionary at a relief station in Sudan. Because of his unique business address, he was a veritable fount of knowledge regarding the use of mobile and portable antennas for communications from the boondocks. His bona fides for this included the fact that he was licensed to operate not only on the amateur radio bands but as a *land mobile* (or point-to-point) station in the 6.2-MHz band. The desert where he traveled is among the worst in the world, and the path he euphemistically called a "road" was frequently littered with camel corpses because of the brutal environment. The doctor's organization required him to check in twice daily on 6.2 MHz or a backup frequency. If he missed two check-ins in a row, then the search-and-rescue planes would be sent out. As a result of his unique "house calls", he did a lot of mobile and portable operating in the lower-HF region of the spectrum. With that as background, the problem he posed to the author was this: How do you reliably get through the QRM and tropical QRN with only 200 W PEP and a standard loaded mobile antenna?

Another fellow worked in Alaska for a government agency. He faced many of the same problems as the doctor in Sudan, but at temperatures nearly 100 degrees colder. He frequently took his 100W mobile rig into the boondocks with him in a four-wheel-drive vehicle. Again, the same question: With only 100 W into a low-efficiency loaded mobile antenna, how does one reliably cut through the interference to be heard back at the homestead?

Suppose an earthquake or a hurricane strikes your community. Antenna supports collapse, tribanders become tangled masses of aluminum tubing, dipoles are snarled globs of #14 copper wire, and the rig and linear amplifier are smashed under the rubble of one corner of your house. All that remains is the 100W HF rig in your car. How do you reliably establish communications in "kilowatt alley" with a 100W mobile transceiver driving an inductively loaded mobile whip? Of course, you always got through one way or another before, but now communications are not for fun—they are deadly serious. Somehow, the distant problems of a Sudanese missionary doctor and an Alaskan government forester seem much closer to home.

For these operators, the ability to communicate—to "get through"—often means life or death for someone. Given the inefficiency of the loaded whips typically used as mobile antennas in the low-HF region, the low power levels generally found in mobile rigs, and the crowded conditions on the 80-, 75-, and 40-m bands, it becomes a matter of more than academic interest how they might increase the signal strength from their portable (or mobile) emergency station. Anything they can do, easily and cheaply, to improve their signal is like having money in the bank. Fortunately, there are several tricks of the trade that will help.

In cases of emergencies on most highways, we are likely to be in range of some repeater, so we would use a VHF band (probably 2 m) to contact police or other emergency services—perhaps through manual assistance from another ham on the repeater, perhaps through a repeater autopatch. In fact, with the wide availability of repeaters around the country it behooves any amateur backpacking or four-wheeling into remote areas to be familiar with nearby repeater locations and frequencies.

As we saw in the preceding chapter, the HF mobile configuration is inefficient by its nature, and little can be done to improve matters . . . at least while remaining capable of being operated "in motion". But if we recognize that the operator needing emergency or urgent communications is most likely not driving down the road somewhere but is stopped in a remote location and needs to contact someone outside his local area, we can improve the performance of the "mobile" station by employing techniques not available to the mobile operator who is "in motion".

Most amateurs with experience in mobile HF operation will attest to the need to enhance their signal if they are to be effective in providing emergency communications. Setting aside for the moment the issue of output power levels, mobile radiation efficiencies are simply too poor to consistently compete with the larger and better installations found at most home and fixed (or base) stations. There are three primary reasons for this:

- The radiating portion of virtually every mobile HF antenna is substantially less than $\lambda/4$ in length, making its radiation resistance extremely low and difficult to match with low losses.

- Below 2 m, most vehicle bodies make very poor RF grounds, thus further compounding the problem by dissipating a greater portion of the transmitter output power in the resistance of the lossy ground.

- In most emergency situations, the preferred coverage distance is considered "close in" for HF propagation modes, even though it may be beyond the reach of the nearest 2-m repeaters. The use of a conventional mobile antenna mounted on the vehicle (in situ) results in a vertically polarized signal with maximum amplitude at relatively low wave angles, while ionospheric propagation modes for the distances involved overwhelmingly favor an antenna that puts most of its energy into the higher wave angles. This is a perfect match for a low, horizontally polarized radiator.

Let's discuss each of these limitations in more detail, with the intention of identifying simple workarounds for the temporary situations that most emergency and portable operations imply. Keep in mind that throughout the discussion we are assuming that the vehicle and the mobile gear it is carrying are in a fixed location for the duration of the emergency or portable activity.

Improving Radiator Efficiency

If at all possible, the smartest solution to the first component of loss is to replace the mobile antenna with a longer, more efficient, *stowable* antenna that can be brought out and erected when needed. In real-world antenna systems, efficiency of radiation increases dramatically as the length of the radiator approaches $\lambda/4$. One candidate is a surplus military HF whip antenna. Intended for jeeps and communications trucks, these antenna/tuner combinations are collapsible and are as efficient as any on the market. Alternatively, a quarter-wave length of wire (or, with a suitable antenna tuner, anything longer than the mobile whip) can be attached where the mobile whip was located previously. Carrying a ball of twine attached to a baseball with a small-diameter hole drilled through its center might not be a bad idea. It should be possible to get one end of the antenna wire up in the air perhaps 20 or 25 ft, draped over a tree branch, with such a rig. If necessary, arrange to support the far end of the wire a little distance from the rear of the vehicle so that the lack of height is made up for by the added horizontal distance the wire is covering. (Move the vehicle, if necessary.) This creates a sloping vertical, which is not all that bad an antenna, believe it or not.

Reducing Ground Losses

Even for frequencies as high as 6 m the typical vehicle body is not large enough to provide an adequate ground plane. A better solution is to provide a counterpoise or elevated ground plane. As we saw in the chapter on verticals (Chap. 9), the basis for this recommendation is that a vertical monopole, such as a mobile whip, is really only half an antenna. The other half is distributed throughout the ground return for the antenna, and a failure to provide an adequate ground system is equivalent to further compromising the antenna itself.

Figure 17.1A is the schematic representation of a proposed way to enhance the ground return efficiency for a mobile whip used in a portable or emergency (i.e., vehicle not in motion) situation. At its minimum it consists of the addition of three or four radials, each $\lambda/4$ in length on the band(s) of interest and spread as equally as possible around the full 360-degree compass rose. Ideally, such a ground-plane antenna should be mounted $\lambda/8$ or higher with some kind of additional support, but the degradation in performance is gradual as the height becomes progressively lower, so you do the best you can with what you have.

When selecting radial wires, the primary requirement is that they be strong enough to withstand the *mechanical* rigors of being repeatedly installed, taken down, coiled up, and stored. Current-handling capability in radials is seldom an issue, regardless of wire diameter. A number of suppliers of antennas and accessories sell spools of bare #18 wire for radials; a 500-ft spool takes up little space in a car trunk. For many portable and emergency setups, however, insulated wire may be best because there may be a need to drape one or more radials across a metallic object of some sort; clearly, direct metal-to-metal contact with other objects would be undesirable and could cause extra noise in the receiver. For 40 m, four $\lambda/4$ radials total 132 ft. Allowing some extra wire for connections to the base of the whip and for attaching to rope or twine at the other end, 150 ft of #14 wire is all that is needed.

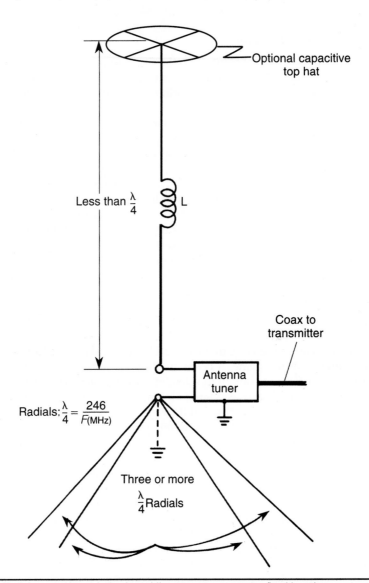

Optional capacitive top hat

Less than $\frac{\lambda}{4}$ L

Coax to transmitter

Antenna tuner

Radials: $\frac{\lambda}{4} = \frac{246}{F(\text{MHz})}$

Three or more
$\frac{\lambda}{4}$ Radials

FIGURE 17.1A   Radials can enhance a basic mobile antenna system at a fixed location.

Figure 17.1B shows a workable system that will further improve the performance of a mobile rig in stationary situations. The mobile antenna uses the normal base mount attached to the rear quarter-panel of the car adjacent to the trunk lid. An all-metal grounding-type binding post is installed through an extra hole drilled in the base insulator (see Fig. 17.1C). Radials for portable operation are attached at this point. The binding post should be sized to easily accommodate three or four #18 radials. In a pinch, additional radials can be attached to the base indirectly by connecting them to the top of one or more of the radials already on the binding post.

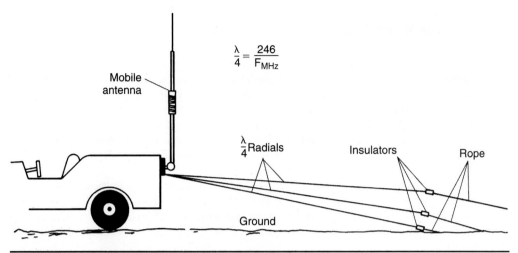

$$\frac{\lambda}{4} = \frac{246}{F_{MHz}}$$

Mobile antenna

$\frac{\lambda}{4}$ Radials

Insulators

Rope

Ground

FIGURE 17.1B Multiple radials are even better.

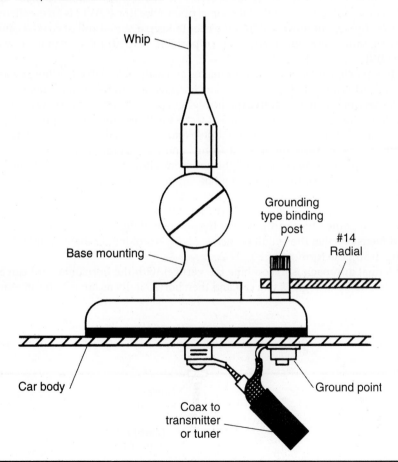

Whip

Grounding type binding post

#14 Radial

Base mounting

Car body

Ground point

Coax to transmitter or tuner

FIGURE 17.1C Connection of radials to mobile antenna.

Selecting the Right Polarization

The final component of improving the HF signal from an emergency or portable station setup is to match antenna polarization to the intended coverage distance(s) and the surrounding ground characteristics. Although there is occasionally a reason (most often for low-band operation by DXpeditions in major amateur radio competitive events) to use vertical polarization for portable operation on the lower HF bands, by far the most likely scenarios dictate the use of a horizontally polarized antenna.

Putting up a dipole is an often overlooked approach when operating portable. In addition to increased radiation at high wave angles, this eliminates the need for ground radials and may cut the required amount of wire in half.

The biggest impediment to erecting a dipole sufficiently high in the air to be useful is the number and spacing of suitable available supports. Trees, flagpoles (but not utility poles), roofs of buildings, and upstairs windows can all be candidates. Depending on the distances to be covered, a height of 30 or 35 ft for an 80-m dipole can be quite useful.

Because of the low power level involved, the dipole can be fed with lightweight RG-58 or RG-8X coaxial cable. Of course, open-wire line (OWL) is an excellent companion to a dipole, but most will find it easier to keep a stored roll of coaxial cable in good working order between emergencies or portable "expeditions" than to stow a spool of true OWL.

In a pinch, *zip cord* or *lamp cord* makes an easily obtainable feedline; measurements on zip cord some years ago indicated a characteristic impedance of about 72. However, the losses can become excessive on the higher bands. In a worst-case scenario, lengths of rotor cable, telephone cable, or Cat 5 or Cat 6 Ethernet cable can be pressed into service!

Figure 17.2 shows the common dipole and the normal equation for determining *approximate* length. Typically it is cut "long" and then trimmed until the VSWR drops to its lowest point, a nicety that might not seem altogether important in an emergency. One or both ends of the dipole must be supported on trees, masts, or some other elevated structure.

If only a single high support is available to use, a sloping dipole or an inverted-vee can be substituted for the dipole. For the antenna of Fig. 17.3, assume as a starting point for dimensions that the length of each leg is 6 percent longer than would be obtained by using the dipole formula of Fig. 17.2.

A final approach is to combine the vertical with the horizontal and run a random-length wire up as high as you can and then horizontally as far as you can. A total of 140

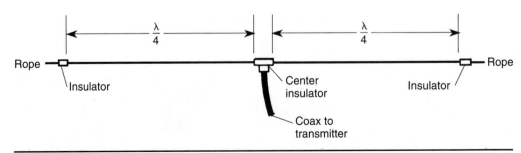

FIGURE 17.2   Basic dipole.

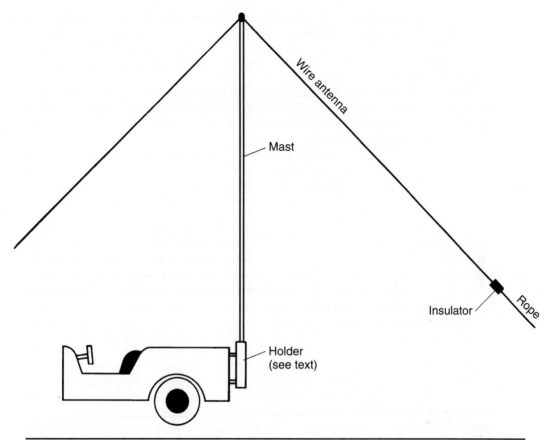

FIGURE 17.3 Portable inverted-vee antenna.

to 160 ft makes a pretty good run for 160 m, and a total length of 70 to 100 ft should work well for 80 m. This is an *inverted-L*, which is a very efficient cousin to a conventional λ/4 monopole; its radiation and matching characteristics are covered in detail in Chap. 18, but its advantage here is that it's very simple to erect, and it provides a mix of high-angle and low-angle radiation from a single antenna. *Note*: For proper operation, the inverted-L *must* have radials, a counterpoise, or a tuned RF ground (see Chap.15) at the feedpoint!

Almost without fail, all of these portable antennas will outperform the basic mobile whip installation. Because the application is both emergency-related and temporary in nature, we can get away with installation and support techniques that would be unthinkable in more permanent installations. But it is absolutely imperative to have an antenna tuner and VSWR monitor to properly set them up.

Temporary Supports

Anyone serious about being prepared for emergency or portable radio operation will very quickly develop an extensive list of "must have" items. The extent of the list will depend on the magnitude of the planned operation and the conveyance mechanisms available. A list of possible components for a functioning radio station for backpackers will be substantially shorter than one for stowing some gear in a vehicle prior to motoring into the backcountry. Size and weight are major constraints in both cases, but the scale is different.

Probably the single biggest antenna item for portable or emergency use is the antenna support, if required. In many parts of the world, there are trees all around but in other areas, sand or fields are all that is visible for miles around, and some form of temporary mast(s) must be employed.

Where trees are available, use of a bow with suitably modified arrows may be the best way to get support lines up high. For nearly 30 years one of the authors used a "portable" bow that could be broken down into small sections to erect not only temporary dipoles but also all his home station wires that were supported by trees. See Chap. 28 for more information on this and similar techniques.

If trees are not plentiful, various mast alternatives, ranging from extremely heavy (military surplus or steel TV antenna sections) to reasonably light (plastic pipe and fiberglass rods) are available. Your choice is determined primarily by whether you expect to have help erecting the mast and by how much room you have for stowing and transporting it. Also, be aware that some masts consist of telescoping sections that get progressively smaller (and more limber) toward the top. Most masts collapse to 6 or 8 ft in length for transport and stowage, but can be pushed up to heights of 18, 25, 30, 40, or 50 ft, depending upon the type selected. Keep in mind, when shopping for these masts, that the larger models are considerably heavier than shorter models, and they require two or more people to install them. Erecting a 40- or 50-ft telescoping mast is not a one-person job!

Masts made of PVC plastic plumbing pipe are somewhat lighter. If mast sections can be carried on top of the vehicle, then lengths up to 10 ft are available and can be joined together at the site with couplings. If the sections are going to be transported *inside* the vehicle, be sure to determine the maximum length that will fit before cutting longer sections into smaller ones. Be very careful not to use PVC pipe that is too small, however; PVC pipe is relatively thin walled and can become quite flexible as it is extended in length. Sizes below 1.5-in diameter will not easily stand alone without guying. Nor is PVC pipe particularly self-supporting. While a single 10-ft section might be, two or more sections together will not support themselves plus the weight of any reasonable antenna. Guying will be necessary—either with ropes or on a temporary basis with synthetic (or other water-repellant) twine.

If you plan to erect temporary or portable antennas that represent a heftier downward or sideward pull on the support mast(s)—an 80-m dipole, for instance—and you have a motor vehicle for transportation, you may wish to employ the heavier but more rigid steel TV antenna masts. Available in 5- and 10-ft lengths, these masts are flared on one end and crimped on the other for joining together to form longer lengths. Make sure you procure the mating guy wire rings, too; although not absolutely necessary, they can make the job of putting the antenna up and keeping it up easier.

Over the years the authors have used military and steel TV mast sections many times—especially during *Field Day,* the annual American Radio Relay League (ARRL) emergency preparedness weekend exercise. The inverted-vee antenna is, of course, a natural for this kind of event, but once one group the author was with mounted a *Hustler* mobile antenna at the top of a 20-ft mast with four radials; it worked surprisingly well given the low output power (30 W) being used.

Rather than reinventing the wheel, often it's productive to look around and see how other groups of people are solving similar problems. Certainly one of the first areas to look to is the RVing public. Recreational vehicles have come a long way in recent years; they sport an amazing variety of creature comforts, many of which require interesting and useful add-ons to the basic RV. Extras such as satellite and over-the-air television, side awnings, and the like require the optional addition of supports and support tubing to these vehicles. As a result, a huge aftermarket industry has developed, and many of these dealers and suppliers are likely to have products that can be adapted easily to specific antenna support needs.

Another source of creativity was pointed out to one of the authors by a CB operator some years ago. Surf fishers on the Outer Banks of North Carolina use four-wheel-drive vehicles to get out onto the beach to the surf, where the big sea bass lurk. Welded to either the front or rear bumper attachments of the 4WD vehicles are steel tubes (see Fig. 17.3) used for mounting the very, very long surf casting rods. This particular CB operator had a 20-ft mast, consisting of two 10-ft TV mast sections, mounted in one of the rod holders. At the upper end of the mast was his 11-m ground-plane vertical. The same method of mounting would also support similar amateur HF antennas, inverted-vee dipoles, or even a VHF Yagi or UHF dish.

Given that the antenna installation will be temporary, normally lasting only a few hours or a few days, we need not worry about long-term integrity or the practicality of the installation. Mounting the mast to the back of a four-wheeler or pickup truck with a pair of U-bolts is not terribly practical if you must move the vehicle, but it works nicely if you plan to camp (or are stranded) for a few days.

For lightweight masts located away from the vehicle—up to about 25 ft, say—the support could be an X-shaped base made of 2- × 4-in lumber (Fig. 17.4A), or even a larger Christmas tree stand. Alternatively, a TV antenna rooftop tripod mount (Fig. 17.4B) is easily adapted for use on the ground. None of these three alternatives can be depended upon to provide a safe self-supporting installation, and all of them must be guyed, even if used for only a short period. Again, because of the temporary nature of the installation, aluminum or wooden tent pegs can be used to anchor the guy wires, depending on the soil

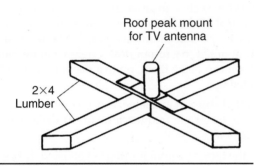

FIGURE 17.4A Wooden antenna stand.

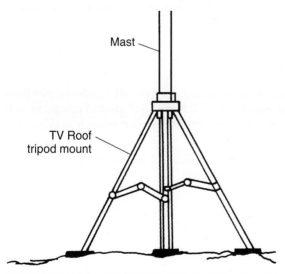

FIGURE 17.4B TV antenna tripod also works in portable situation for HF antenna masts.

characteristics in the area. Although they are clearly not appropriate for long-term installations, they should work fine for the short run.

Summary

The ideas presented in this chapter are meant to help the reader maximize his/her signal under adverse conditions in the field. While methods of getting antennas up in the air under difficult circumstances are substantially different from those techniques we would normally employ at our home or business, the electromagnetic principles that maximize signal transmission and reception do not change. In particular, remember these key points:

- Try to make horizontal (or inverted-vee) antennas as close to an electrical half-wavelength as possible; don't rule out the use of traps to allow easier multiband operation in a limited space.

- Try to make any vertical monopoles as close to an electrical quarter-wavelength as you can; if you can get the base of the vertical off the ground, create a ground-plane antenna with two or three λ/4 radials on each band to be used. If the vertical must remain ground-mounted, try to put down eight or more radials total, with lengths as long as you can make them, up to a maximum of λ/4 on the lowest frequency of intended use.

- If you have only a single elevated support available, consider either a sloping dipole or an inverted-vee on the lower HF frequencies. Spread the legs of the inverted-vee as far apart as possible.

- Select your antenna to match the type of communications you need. If you are primarily interested in contacting others in your own general area (i.e., within a few hundred miles) on HF, use a horizontal wire. If you're on a DXpedition to a remote South Sea island, you'll probably want to use a vertical on the beach.

- Make sure you have a meter for reading both forward and reflected power or, alternatively, VSWR. Most of today's small HF transceivers have that capability built in. At VHF and UHF, however, it may be necessary to bring along a stand-alone wattmeter or SWR bridge.

- Similarly, make sure you have an antenna tuner for the HF bands—either built into the rig or as a separate accessory. For VHF and UHF ground planes, as well as mobile whips, however, all you may need is an Allen wrench to adjust the top section of the vertical whip.

- Be sure to bring rope, twine, insulators, and short lengths of coax and other cables, with the proper connectors on each end.

Antennas for Other Frequencies

CHAPTER 18

Antennas for 160 Meters

One of the most explosive growth areas of amateur radio interest worldwide in the past decade has been the 160-m band. *Topband*, as it is called by its aficionados, presents today's active amateur with extreme technical and operational challenges and opportunities. Part of this comes from the physical dimensions required for efficient antennas, part from the high noise levels often found at these low frequencies, and part from the fact that there are aspects of 160-m propagation that remain an enigma to even the experts.

Efficient transmitting antennas for the medium-frequency band (300 kHZ to 3 MHz) pose certain difficulties for the user. Of course, the first thing that springs to mind is the large size of those antennas. As explained in earlier chapters, the most efficient transmitting antennas for the MF and HF bands approach a half-wavelength (horizontal dipoles) or a quarter-wavelength (vertical monopoles) in their long dimension. For instance, a half-wavelength dipole for 160 m is about 270 ft long, usually exceeding the capabilities of a typical suburban lot. Of course, as the frequency decreases, a half-wavelength increases, so it is not for nothing that towers for the AM broadcast band (530 to 1700 kHz, just below the 160-m band) are usually hundreds of feet tall.

Vertical Monopoles

For the amateur blessed with lots of real estate, a generous budget, and no zoning limitations, the transmit antenna of choice for 160 borrows from the best engineering practices of the AM broadcast industry; it is almost always one or more λ/4 ground-mounted verticals with anywhere from 12 to 60 radials originating at the base of each tower.

But AM broadcast stations are usually concerned with maximizing their *ground-wave* signal strength throughout their own local metropolitan area. Why would amateurs—especially those interested in very long distance communication—adopt broadcast station techniques? Why, for instance, wouldn't a λ/2 dipole for 160 m be a better choice?

There are at least two reasons:

- To radiate well at the low elevation angles that are most useful for long-haul communications, a horizontally polarized antenna should be at least a half-wavelength above ground. On 160 m, this suggests a minimum height of nearly 300 ft for the antenna supports!

- Although a low dipole will normally provide acceptable nighttime coverage on 160 m throughout a radius of a few hundred miles around the antenna, sunlight greatly increases absorption of these near-vertical signals in the lower layers of the ionosphere, making daytime short-haul communication problematic. Further, during periods of very low sunspot activity, the *critical frequency*, f_c, can in fact drop below 1.8 MHz, with the result that nighttime skip entirely disappears on the topband. An efficient vertically polarized antenna can cover the same region with a strong ground wave that is present 24/7 and is not dependent upon conditions in the ionosphere.

What, then, is a "decent" vertical installation? Surprising to many who are wrestling with antenna space requirements for the first time, vertical antennas require nearly as much backyard space (or *footprint*) as $\lambda/2$ dipoles! Although a 40-m vertical (33 ft high) is not an unreasonable mechanical assembly and has modest guying requirements, a full-size quarter-wave vertical for 160 stands 125 to 140 ft tall, and will likely require a guy wire footprint having a diameter of 200 ft or greater. In addition, local building codes may not require any special inspections or permits for the 33-ft antenna but may impose rigid and very exacting requirements on the higher structure. Worse yet, an antenna with overall height substantially taller than the maximum allowed for a principal structure (often 35 or 40 ft in the United States) may not be allowed *at all* by the local zoning ordinance! On suburban or urban lots, a typical 40-m vertical might well be able to fall over without crossing the property line or landing on power lines. Not so the 160-m vertical!

Thus, the biggest problem for most topband DXers, as you can see, is the size of antennas for those frequencies; but there are ways to shorten an antenna—not without penalty, mind you, because the TANSTAAFL* principle still applies—to the point where an antenna for 160 becomes feasible with limited yard space. In return, the user typically gives up a few decibels of distant signal strength for a given power delivered to the antenna and suffers a narrower VSWR bandwidth before retuning of the antenna is necessary.

As discussed in Chap. 16, to get the most (radiated field) from an antenna we must maximize the area under its *current-versus-length* curve while simultaneously minimizing the portion of the RF energy from the transmitter that is wasted in feedline losses and ground system losses. In general, this means that our preferred antennas are those that give maximum exposure to the natural high-current portions of the antenna element(s), exhibit a high radiation resistance relative to other losses in the system, and are a reasonably close match to available feedlines. Further, if the antenna is a monopole (such as a ground-mounted or *ground-plane* vertical) being operated against ground, we must not neglect the design of the RF ground system in the immediate vicinity of the antenna.

$\lambda/4$ Vertical Monopole

By far the conceptually simplest way to give maximize exposure to the high-current portion of the antenna is to make sure that horizontal dipoles are $\lambda\backslash2$ or slightly longer and verticals are $\lambda/4$ or slightly longer. Since the title of this book includes the word "practical", we will concentrate on the $\lambda/4$ ground-mounted vertical instead of $\lambda/2$ dipoles erected at heights of 300 ft or more!

---

*TANSTAAFL = There ain't no such thing as a free lunch!

If very tall trees are available, the simplest physical configuration for a vertical is to hang a 130-ft wire from one, using any of the techniques in Chap. 28. In general, the wire should be insulated and, if high-power operation is contemplated, a catenary or halyard system may be required to prevent arcing from the antenna to tree limbs when transmitting. The height of the antenna should be adjusted so that the bottom end of the wire is within a foot or two of the ground beneath it for direct access to the feedline, a base-mounted antenna tuning unit (ATU), or some other form of base matching network. As always, proper safety precautions must be observed to prevent people and animals from coming in contact with any part of the antenna.

For proper operation as a grounded monopole, it is *imperative* that a system of wire *radials* be installed on or near the surface of the ground directly beneath the vertical. Wires in contact with the earth or within a small fraction of a wavelength from it lose any pretense of a resonant length; better, instead, to think of them as long, skinny capacitors. Nonetheless, a good rule of thumb is to have them extend at least as far out along the earth's surface as the vertical is tall. So a good radial field for a 160-m $\lambda/4$ vertical has wires approximately 120 to 150 ft long coming together electrically and physically at the base of the vertical.

How many radials are sufficient? AM broadcast engineering standards require stations to use at least 120 such wires, equally spaced around the full 360 degrees—i.e., every 3 degrees. But recent review of the work leading to this standard indicates that 60 radials was the point at which no further improvement in ground system performance could be measured and even then the improvement with each added radial was but a small fraction of a decibel. In amateur practice, anything over 20 or so radials is going to be fine as long as the length of the vertical radiating element is sufficient to keep the radiation resistance reasonably high.

As discussed in Chap. 9, a full-sized $\lambda/4$ ground-mounted monopole has an input impedance of 72/2 or 36 Ω at its base. For most of today's transceivers capable of RF outputs up to about 200 W, the resulting SWR at or near the resonant frequency of the wire is not a serious mismatch and it may be possible to feed the antenna directly with 50-Ω coaxial cable. The bottom end of the vertical wire goes to the center conductor of the cable, and the common junction of all the radials that make up the RF ground system beneath the wire attaches to the shield or braid side of the cable.

At higher power levels it may be necessary to add an ATU at this point in order to convert the native impedance of the vertical to that of the actual transmission line used: 50 Ω, 75 Ω, 300 Ω, as appropriate. There are three reasons this might be necessary:

- The power amplifier being used may not tolerate SWRs much greater than 1.0:1 and either shuts down or reduces its output power substantially in response to high SWRs.

- The transmission line may be very near its voltage breakdown rating and any SWR on the line will serve to reduce the maximum power it can handle.

- If the user intends to operate over the entire 160-m band, he or she may need a way to reduce the SWR on the transmission line when operating far from the resonant frequency of the antenna.

Be aware that adding any form of ATU may well increase total loss in the antenna and transmission line system—usually as a consequence of I^2R losses from high circulating currents in any inductors. In a properly designed ATU, however, the added loss within the ATU itself may be less than the reduction in transmission line losses from high SWR, and it may also represent the only way to get full performance and/or legal power from the transceiver or amplifier feeding the line.

Shortened Verticals

The antennas in this section are all variations on the basic $\lambda/4$ vertical of the preceding paragraph. In general, each is a shortened monopole augmented by distributed and/or lumped-constant circuit elements to keep the radiation efficiency as high as possible and to minimize the VSWR on the transmission line back to the transmitter.

Again, these antennas will not work quite as well as a properly installed full-size antenna, but they will serve to get you on the air with a very passable signal from cramped quarters that would otherwise preclude operation on these frequencies.

Several different popular configurations are shown in Fig. 18.1. The basis for all these antennas is a vertical that is physically short as compared to the "standard": a quarter-wavelength ($\lambda/4$) (i.e., 90-degree electrical length) monopole with an ideal ground plane beneath it.

Recall that a vertical monopole too short for its operating frequency (i.e., less than $\lambda/4$) will exhibit capacitive reactance. In order to resonate such an antenna, it is necessary to cancel the capacitive reactance with an equal amount of inductive reactance, such that $|X_L| = |X_C|$. By placing an inductance in series with the radiating element, therefore, we can effectively "lengthen" it electrically, as measured at the feedpoint of the combined antenna element and series inductor.

An antenna that is reactance-compensated is said to be *loaded*; in the preceding example, the antenna is *inductively loaded*. Three generic forms of loading are popular: *discrete loading, continuous loading,* and *linear loading.*

Discrete loading means that there is a discrete, or lumped, component (inductance or capacitance) connected to the antenna radiator. The simplest form of inductive loading inserts a loading coil near the center (Fig. 18.1A) or at the base (Fig. 18.1B) of the radiator element. Because there is very little current near the open circuit end of a vertical, inductive loading at the top does little; instead, a capacitive hat (Fig. 18.1C)—often in conjunction with a center loading coil—is much more effective at increasing the area under the current-versus-length curve.

You may recognize these configurations as being the same as those found on mobile antennas. Indeed, low-band mobile antennas can be used in both mobile and fixed installations. Note, however, that although it is convenient to use mobile antennas for fixed locations (because they are easily available in "store-bought" form), they are far less efficient than other, longer versions of the same concept. The reason is that the mobile antenna for the lower HF bands has historically been based on the standard 96- to 102-in whip antenna used by amateur operators on 10 m or by citizens band operators on 11 m, with the overall height of the antenna limited by safety and mounting considerations. In fixed locations, on the other hand, longer radiator elements (which are more efficient) are usually possible. For example, a 16- to 30-ft-high aluminum radiator element can easily be constructed of readily obtainable materials.

Loaded antennas tend to be rather high Q devices, and their bandwidth is quite narrow. An antenna tuned for the center of a band may present a high VSWR at the ends of the band—especially on 160 m, where the U.S. amateur band extends ±5 percent from the center frequency. One way around this problem is to make the inductor variable, so that slightly different inductance values can be selected for different band seg-

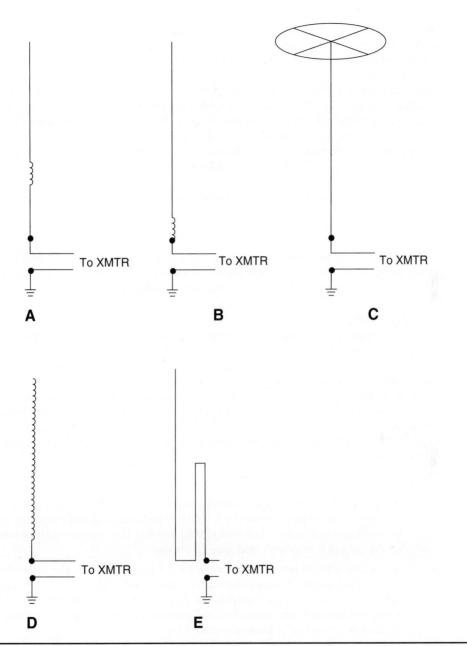

FIGURE 18.1   Examples of loaded vertical antennas. (A) Center. (B) Base. (C) Top. (D) Continuous. (E) Linear.

ments. On 1.8 Mhz this is probably easily accomplished only for the base-loaded version (Fig. 18.1B). For the other configurations, it is far easier to use an adjustable matching network or ATU at the base of the antenna.

A *continuously loaded* antenna has the inductance distributed along the entire length of the radiator (Fig. 18.1D). Typical of these antennas is the helically wound vertical, in which about a half-wavelength of insulated wire is wound over an insulating form (such as a length of PVC pipe or a wooden dowel); the turns of the coil are spread out over the entire length of the insulated support. I^2R losses in the winding are somewhat higher than in a straight vertical member because the conducting element is not only longer but often has a smaller cross-sectional area.

Linear loading (Fig. 18.1E) is an arrangement whereby a section of the antenna is folded back on itself like a stub. Antennas of this sort have been successfully built from the same type of aluminum tubing as regular verticals. For 160-m operation, a length of 60 ft for the radiator represents 41 degrees, while a normal λ/4 vertical corresponds to 90 degrees. The difference between 90 degrees and 41 degrees of electrical length is made up by the "hairpin" structure at the base. A potential disadvantage of the linear loading approach is the requirement for an additional insulator somewhere along the monopole's physical support structure if the antenna is supported from the bottom (as a tower or rigid mast might be), rather than suspended from above (as a wire would be).

The relative cost and mechanical complexity of continuously loaded and linearly loaded verticals for 160 m generally render them inferior to other approaches—especially the top hat of Fig. 18.1C—and we shall not consider them further in this chapter.

Loading the Tower

Many amateur radio installations include one or more rotatable Yagi antennas on top of an appropriate support structure, such as a ground-mounted guyed triangular lattice tower. By far the preponderance of amateur tower heights is in the 40- to 90-ft range—often considered too short to be an efficient vertical monopole radiator on 160 m. But with proper attention to installation details and a modest set of radials, most such tower and antenna combinations can become excellent transmitting and receiving antennas for 160 m.

There are two critical components to making a "short" tower an efficient radiator on 160 (or 80) m:

- The apparent electrical length must be increased to approximately λ/4 or slightly greater. This brings the natural high-current part of the antenna out into the open so as to increase the $I \times h$ (or $I \times l$) product, and coincidentally raises the radiation resistance, thus beneficially altering the resistive divider formed by the radiation resistance and ground losses.

- A low-loss ground return path for the RF fields surrounding the vertical must be provided. In the absence of saltwater at the base of the tower, 20 to 60 copper radial wires—each approximately as long as the tower is tall—should be connected (through an appropriate anticorrosion metal) to the grounded side of the transmission line or base-mounted ATU.

As implied at the beginning of this topic, an excellent way to turn a short grounded tower into an efficient 160-m antenna is to utilize one or more HF Yagi antennas at the top to provide top loading, but there are at least two other ways to do the same thing:

- If the tower is guyed, allow the top set of guy wires to extend out and down from the top of the tower by a calculated amount before inserting a set of insulators.

- Add three or four dedicated *top hat* wires equally spaced around the horizon at or near the top of the tower, below any rotatable antennas, and extend them out as close to horizontal as possible.

Here are some drawbacks, or compromises, associated with some or all of these approaches:

- Using a portion of each guy wire in the top set of guys is inefficient (compared to the horizontal boom and elements of a Yagi) because the top guy wires usually follow a fairly steep slope back down alongside the tower itself.

- Use of the top set of guys as a top hat means you should select an appropriate length for the uninsulated sections of guy wire attached to the top of the tower *prior* to installing the tower—a task probably best done with the help of an antenna modeling program, such as those discussed in Chap. 25.

- Horizontal or sloped lengths of wire at the top of the tower—whether constructed from the top set of guy wires or entirely separate wires—may detune any rotatable antenna(s) on the mast if there is but a few feet of separation between the lowest rotating antenna and the wire attachment level.

- In most installations it will not be possible to find supporting termination points for the far ends of any added top hat wires that allow any less of a slope than that encountered when using the guy wires themselves. Having said that, however, the advantage of using additional wires is that these wires can be added, removed, shortened, or lengthened at any time, if necessary. Since they are not guy wires, the only mechanical requirement on them is that they be strong enough to support their own weight plus the weight of any rope or twine used to tie them off and hold them away from the tower. When used with a tower surrounded by tall trees, it may be possible to string the top hat wires almost horizontally by using halyards from high in the trees.

- As a practical matter, the use of Yagis for top loading is limited to HF beams or very long boom VHF/UHF beams, since smaller beams will most likely not provide adequate capacitive loading. Further, since Yagi dimensions are "fixed" (with the exception of the SteppIR beams), the equivalent electrical length of your tower plus beams is not totally within your control.

- Many Yagi beams do not have their elements directly connected to their booms; in such cases, only the boom fully contributes to the top loading effect.

- Insulators and/or matching networks for some Yagi beams may not tolerate high voltages across their terminals at frequencies far from their intended band(s) of operation.

As a first approximation to the length of top hat wires or uninsulated guy wire segments at the top of the tower, subtract the tower height from 140 ft. (If you have a choice, err on the side of longer total length, rather than shorter.) If you are relying on one or more Yagis at the top of the tower, get the turning radius of the lowest beam from the manufacturer's specifications or calculate an equivalent top hat radius r from

$$r = \sqrt{\left(b^2 + e^2\right)} \qquad (18.1)$$

where *b* is the distance (usually half the total boom length) from the mast to where the elements farthest from the mast are attached, and *e* is half the average length of elements nearest the ends of the boom. Only elements directly grounded to the boom should be considered as fully contributing to the top hat effect.

NOTE *If it is to ever be used as a radiating monopole, any guyed tower must have at least one insulator in each and every guy wire attached to the tower, including guy wires attached at guy stations farther down the tower. With the exception of any initial lengths established in the top set of guy wires for top hat purposes, the distance from the tower attachment point to the uppermost insulator in each guy wire should not exceed 3 ft or so. Even though the tower itself may be grounded at its base, failure to insulate each and every guy wire near the point where it attaches to the tower completely changes the electrical characteristics of the structure, complicates calculations and assumptions, and may well destroy the overall efficacy of the tower as a radiator. Even if you believe you have no intention of ever using your towers as low-frequency antennas, interests and plans can change—insulate your guy wires when you erect your towers—it's a whole lot easier to do it then than it is to do it later!*

As discussed in the chapter on verticals (Chap. 9), if a grounded vertical monopole is increased in length somewhat beyond $\lambda/4$, its elevation pattern is "squashed"; that is, radiation (and reception) at low vertical angles is enhanced at the expense of the higher angles. For most intended uses of an efficient transmitting vertical on 160 or 80 m, this is desirable—hence, the suggestion that it's okay to let the total equivalent electrical length exceed $\lambda/4$. When this is the case, the feedpoint impedance at the base of the antenna will increase and the reactive component will appear as a series inductance, which is quite easily canceled with a transmitting air variable capacitor in series with the feedpoint.

When the base of a vertical is insulated from ground so that it can be fed directly from the transmission line, the technique is called *series feed,* and it is most often employed in conjunction with wire verticals. But for mechanical simplicity and low cost, towers and rigid 160-m verticals in most amateur installations are grounded at the base. For these, *shunt feed* must be used. Here's how it works.

In general, the impedance at any given point on a grounded vertical monopole increases with height up to $\lambda/2$. Numerous ways of connecting a feedline to such a structure have been employed over the years. Figure 18.2 shows one of the simplest: an unbalanced output ATU feeds a wire connected to the side of the tower at an appropriate point. Electrically, this is half of the *delta match* often employed to provide a balanced feed to the center of dipoles and Yagi driven elements. The "correct" height for the tap on the tower depends on the design and tuning range of the ATU; the objective is to find a height where both the resistive and the reactive parts of the impedance seen at the lower end of the wire by the ATU are within its tuning range.

Similar to the sloping wire or half-delta feed of Fig. 18.2 is the shunt feed of Fig. 18.3. This is known as a *gamma match*; it is basically the unbalanced form of the *T-match* often used to feed HF and VHF Yagi driven elements. Physically, the only difference between this and the delta match is that the gamma wire or rod is held a constant distance from the side of the tower, allowing the ATU and *gamma rod* to be placed very close to the tower. The gamma match is adjusted to provide minimum SWR on the transmission line at the center of the operating frequency range by alternately changing the tap point on the vertical and the spacing of the gamma rod from the radiating element while retuning the series capacitor.

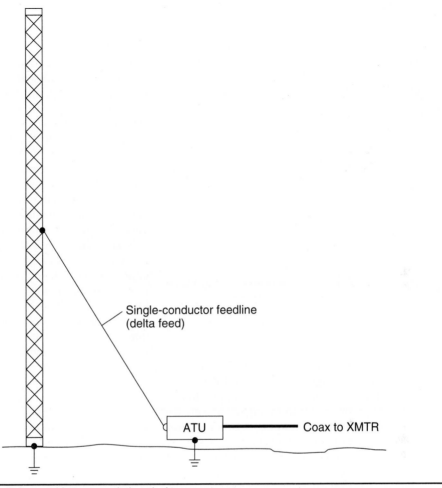

FIGURE 18.2 Delta-matching a grounded vertical tower antenna.

Electrically, the shunt feed acts much like an autotransformer, converting the (higher) impedance of the tap on the tower to the (lower) impedance of the feedline. If dimensioned properly, the need for a separate ATU can be eliminated completely or the ATU can be replaced with a single air variable capacitor having adequate spacing between its plates for the transmitter power levels involved. The capacitor is placed in series between the bottom end of the shunt rod and the hot side of the coaxial cable or unbalanced transmission line feeding the antenna.

Antenna modeling software (Chap. 25) is an excellent way to experiment with tower heights and top loading schemes. But its greatest value here is in determining the "proper" length and spacing for the gamma rod or wire. Properly used, antenna modeling software can eliminate literally dozens of trips up and down the tower!

Here's a true story involving a 90-ft guyed tower built with Rohn 45 triangular sections and topped off with a 10-ft heavy-duty mast (Fig. 18.4). At the 92-ft point is a four-

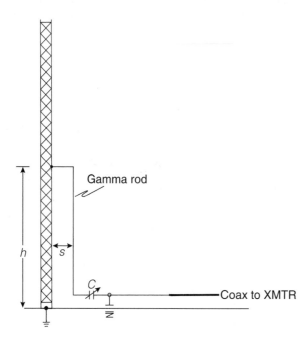

FIGURE 18.3   Gamma-matching a grounded vertical tower antenna.

element 20-m monobander with a 32-ft boom and full-size elements approximately 36 ft long. Above that, at the 98-ft level, is a two-element 40-m "shorty 40" monobander with a 20-ft boom and shortened (linear-loaded) elements about 42 ft long. All elements of both beams are directly grounded to their respective booms. The tower is well grounded at its base, and approximately 16 radials of varying length extend outward in all compass directions. Of course, all guy wires have their uppermost insulators located 3 ft or less from their respective attachment points to the tower. Since a 98-ft tower is a substantial percentage (70 percent) of a quarter-wave on 160 m, the capacitive top loading provided by the two beams should easily bring the total electrical height to more than 90 degrees—that is, to an equivalent height of more than 135 ft at 1.8 MHz.

Initial experiments with attempting to find the "right" gamma rod tap point, using a fixed spacing of 2 ft from the tower, were frustrating; dozens of trips up the tower by the owner to heights ranging between 25 and 65 ft to adjust the gamma rod tap point had resulted only in establishing a very sharp resonance (narrow bandwidth) that was very difficult to reliably tune with the transmitting air variables on hand and that could never quite be tuned to a perfect 1.0:1 VSWR with either a simple gamma capacitor or the more complicated *omega match.*

The owner then turned to *EZNEC* antenna modeling software from W7EL to model the combined effect of his tower, gamma rod, and HF monoband beams on 160 m. The most time-consuming part of this was in creating accurate models of the two beams, although simplified models would probably have been quite adequate for determining their capacitive loading effect on 160 m.

The results were eye-opening! Instead of using a gamma rod spaced 2 ft from the nearest edge of the triangular tower, the model suggested that a spacing of between 6 and 8 ft in combination with a tap point of about 57 ft would optimize bandwidth at the desired center frequency while developing a feedpoint impedance very close to 50 Ω resistive with a small amount of inductive reactance deliberately introduced to allow fine-tuning of the VSWR on the feedline with a single air-variable transmitting capacitor at the base.

Here is what the owner finally installed:

- An 8-ft length of sturdy aluminum angle stock from a local hardware store supports the gamma rod (which is made from lengths of copper plumbing

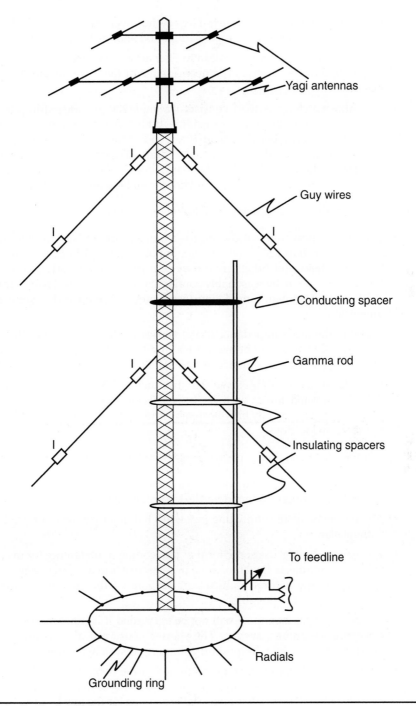

Yagi antennas

Guy wires

Conducting spacer

Gamma rod

Insulating spacers

To feedline

Radials

Grounding ring

FIGURE 18.4   Top-loaded 90-ft tower.

tubing and couplings) 7 ft from the side of the tower and provides the horizontal electrical connection between the gamma rod and the tap point on the tower.

- Two 10-ft lengths of scrap 1- × 2-in lumber act as insulated spacers holding the gamma rod 6 ft away from the nearest edge of the tower at the 20- and 40-ft heights.

- The entire matching network consists of a single 200-pF transmitting air variable capacitor atop a cinder block inside a lidded plastic storage container and connected in series between the bottom end of the gamma rod and the center (signal) conductor of the 52-Ω coaxial feedline. (The cinder block is an important element of the overall design: Not only does it insulate the capacitor from ground—it prevents the plastic container from blowing away in the wind!)

After installation, the owner observed the following results:

- The antenna now tunes easily and has a much broader SWR bandwidth (under 2.0:1 from 1.800 to 1.877 MHz, with a minimum of 1.43:1 at 1.827 MHz). Since the radial field and other ground system components and connections are unchanged, we can be reasonably confident that the increased bandwidth is the result of using antenna modeling techniques to optimize the gamma match dimensions.

- A plot of the input impedance of the vertical with a VNA shows the resistive part to be 32 Ω or greater between 1.800 and 1.860 MHz, with a maximum value of 36 Ω at the design frequency of 1.830 MHz.

- In on-the-air use in competitive activities such as DXing and contests, this top-loaded shunt-fed tower with its 16 radials of various lengths is a superior performer that is usually only beaten by arrays of multiple verticals and ocean-front station locations.

Some "glittering generalities" to take away from this section:

- When cost, strength, and mechanical complexity push strongly toward an uninsulated ground-mounted vertical, consider the use of a gamma match.

- Always break up all conducting guy wires holding up a tower or rigid vertical with insulators.

- Whenever possible, use some form of top loading in preference (or in addition) to bottom, middle, or linear loading; the objective is to maximize signal strength by bringing the natural high-current portions of the antenna up out of the feedline as much as possible.

- The best vertical monopole will not be very good if it is not married to a low-resistance RF ground system. The shorter your vertical is, relative to a full quarter-wave, the more important your ground system becomes.

- Consider whether elements are grounded to the boom when selecting beams for the higher HF bands if there is any chance the tower they are going to be mounted on will be used as a vertical monopole for the low bands.

- Inexpensive or free antenna modeling software (available for both Windows and Mac OS X environments) can help you avoid much unnecessary climbing.

Insulated-Base Loaded Verticals

In some situations, insulating the tower base or feedpoint from ground may be necessary or desirable. One prime example is when an *array* of vertical monopoles is going to be used to provide greater forward gain and/or *front-to-side* (F/S) or *front-to-back* (F/B) ratios than can be obtained from a single (omnidirectional) radiating element. As discussed in Chap. 11 ("Vertical Arrays"), getting full performance from a multielement array requires strict control of both the magnitude and the phase of the feedpoint current in each element of the array. As a general rule, this control is easier to maintain when the feedpoint is not grounded—i.e., when series feed is used.

A loaded insulated-base vertical is essentially the same thing as one side of a shortened dipole or one side of the driven element of a beam that uses loading coils or some other technique to shorten the elements. As before, the performance of such a vertical will be no better than the conductivity of the ground system beneath it. Unless the vertical is mounted in seawater, radials are a *must*.

As with the grounded vertical, loading of an insulated vertical can take many forms; the simplest and most efficient are capacitive top hats with or without a small amount of base loading inductance. However, keep in mind:

- Radial wires lying on the ground or in very close proximity to ground are not resonant wires; there is no one specific length that is optimum. In general, 20 or more radials between $\lambda/4$ and $\lambda/8$ in free-space length will provide an excellent ground for a vertical monopole, but there is nothing magic about either the length or the number. "The more the merrier", and if they have to be shorter than normal to fit a backyard, so be it.

- Up to an effective electrical height of $5/8\lambda$ (or 225 degrees), increased height results in a modest increase in the low-angle radiation of a vertical monopole, obtained at the expense of high-angle radiation. This additional gain reaches its maximum at $5/8\lambda$, beyond which the low-angle gain gradually decreases. Thus, a monopole having an equivalent electrical length of $\lambda/4$ or 90 degrees on 160 can do well on 80 (although matching its high impedance there can be tricky), but will suffer reduced performance on 40 unless a phasing section is added such that the physical structure acts like a collinear array.

- In an array of multiple vertical elements, each element should have its own radial field and the fields should be as similar as possible. If radials from different elements overlap, it may be useful to connect them together, but it's not clear from the available research or modeling that the possible improvement in ground conductivity is worth the extra effort. Radials that overlap should *not* be allowed to intermittently touch or develop oxidized connections, however! Either connect them together firmly or insulate them from each other.

Figure 18.5 shows a situation in which a 90-ft tower is insulated from ground and operated on 160, 80, and 40 m. Here it is fed by a simple L-section or reverse L-section ATU. Note the approximate values of the inductors and capacitors in each configuration, and also their relationships. On 40 and 75/80 m, the feedpoint impedance of the tower is inductive, so a series capacitance and shunt inductance are used for the ATU (i.e., reverse L-section coupler). On 160, the feedpoint impedance is capacitive, so the inductor and capacitor are reversed.

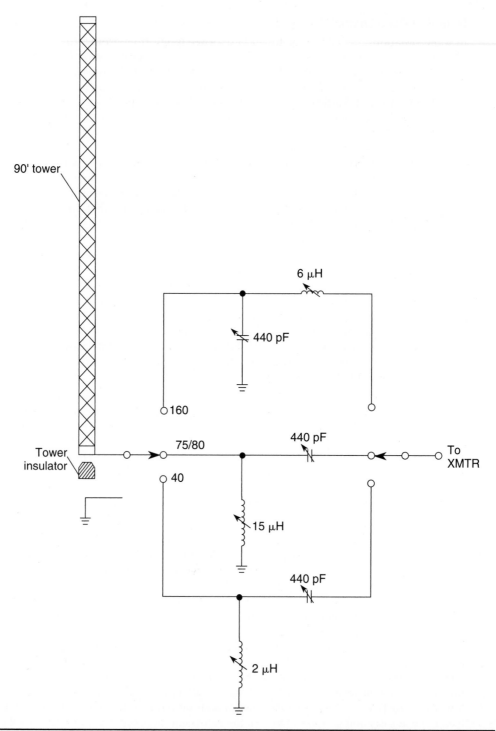

FIGURE 18.5 Methods for matching a series-fed tower.

If the decision to insulate the tower from ground has been made, the insulator section is not limited to being located at ground level. A number of DXers and contesters are using *elevated-base ground-plane antennas* on 160 m. One advantage is that fewer radials are needed; if the base of the ground plane is $\lambda/8$ or higher above earth ground, somewhere between four and eight elevated radials will provide ground efficiencies comparable to 20 or more radials lying on the ground for a vertical monopole whose (insulated) base is at ground level. The big disadvantage, of course, is that, all other things being equal, the overall height of the tower increases by as much as $\lambda/8$, or 65 ft! But see Chap. 30 for a clever alternative.

Finally, it's worth explicitly noting when considering any form of insulated-base vertical:

- Insulated-base towers are more expensive.

- Lightning protection is simpler with a grounded-base tower.

Phased Verticals

Almost all serious *topband* DXers employ arrays of phased verticals to add transmit gain in switch-selectable compass directions. To maximize radiation efficiency and bandwidth, most try to make the array elements $\lambda/4$ tall or higher, but arrays of shortened verticals have merit, as well, as long as the user takes pains to minimize losses in ground systems and matching units.

One favorite form of array on 160 m (and 80 m, too) is the *four-square* (also written *4-square*). In general, these arrays are base-insulated and series-fed; because of this, some amateurs install them as much as $\lambda/8$ above ground and use elevated radials. Whether elevated or ground-mounted, a 160-m four-square of $\lambda/4$ elements is a marvel to behold!

Typically, feedpoint currents are phased as 0 degrees in the rear element, –90 degrees in the two side elements, and –180 degrees in the front element (i.e., the element nearest the desired direction for maximum gain), but many variations in both element phasing and interelement spacing to optimize one or another parameter exist. The definitive compendium of the many forms these arrays can take is "ON4UN's Low-Band DXing" by John Devoldere, available from ARRL.

Four-squares are *all-driven* arrays, but parasitic arrays are equally useful and often quite a bit less expensive or time-intensive to install. For most of us, *any* array of full-size $\lambda/4$ elements does not fall in our definition of "practical", so we won't devote any more space to them. However, some very worthy—and *practical*— alternatives exist. Chief among those—whether employed in an *all-driven* or a *parasitic* array—is the *inverted-L*, described later in this chapter.

Of course, the use of directional arrays for transmitting antennas enhances their utility as receiving arrays. Some operators have run *A/B tests* comparing full-size four-squares with dedicated receive antennas, such as 1λ to 2λ Beverages and concluded that these transmitting arrays are usually as good as or better than the specialized receiving antennas. But then again, as the saying goes, "You can never have too many antennas!"

Horizontal Antennas

Although subject to the limitations discussed at the beginning of this chapter, a horizontal dipole can be an inexpensive solution to the problem of putting out a decent nighttime signal on 160—especially if tall trees are available. Of course, the user has the choice of at least four variations on a dipole: fully horizontal, sloping, inverted-vee, or bent. Having enjoyed the use of three fully horizontal 160-m dipoles at 100+ ft for more than a decade some years ago, the author feels qualified to observe that even though the antennas were low (in terms of wavelength at 1.8 MHz), they were excellent for distances out to 1500 mi or so, beyond which they generally took a backseat to a single top-loaded vertical. And with the appropriate ATUs capable of handling high feedpoint impedances when operating a dipole on its second harmonic, the antennas were superb performers on 80 m, as well, where they exhibited about 2 dB more gain in the main broadside lobe than a conventional λ/2 dipole.

For limited spaces, the author prefers bent dipoles to inverted-vees, for reasons described in detail in Chap. 6. However, the former antenna requires two high supports, while the latter needs but one. With only one high support but no particular space limitations, the sloping dipole is often a better choice than the inverted-vee.

Shortened Horizontal Antennas

The strategies that work for loading short vertical antennas also work for horizontal antennas, although in the horizontal case we are simulating a half-wavelength (180-degree) balanced antenna rather than a λ/4 unbalanced antenna. Because of the large dimensions involved, probably the only practical form of shortened horizontal antenna is one with inductive loading in the middle of each leg, either with loading coils or with traps that allow use of the same antenna on 80 m.

Inverted-L Antennas

One of the simplest wire antennas, but one that can produce excellent results, is the λ/4 *inverted-L* (Fig. 18.6). Typically considered when there is no chance of having a high enough support to hang a full λ/4 wire vertically over its entire length, this antenna is simplicity itself. Starting as close to the antenna coupler (ATU) or transmitter as is reasonable, run as much of a 170-ft piece of wire vertically as high as you can. Then run the remainder of the wire horizontally until the full 170 ft have been deployed. As an example, this might result in a 65-ft vertical section and a 105-ft horizontal section, but wide variations from these lengths are often found—even vertical sections as short as 30 ft have merit. The recommendation to make the total wire length somewhat longer than λ/4 on 160 is partly to make sure the natural high-current portion of the standing wave is out on the vertical portion of the wire, not trapped inside the feedline, and partly to give a little boost to the radiation resistance. It also has the effect of forcing the input impedance to be slightly inductive so that the feedpoint reactance can be tuned out easily with a series air variable transmitting capacitor. Fed as described, the antenna should exhibit a reasonable match to 52- or 75-Ω coaxial cable.

Basically, the inverted-L is a cousin to a quarter-wave vertical monopole. Some prefer to look at it as a *bent* vertical, while others see it as a *short* vertical with asymmetrical top loading. If constructed as shown in Fig. 18.6, a small portion of the transmitter power will be radiated from the horizontal half of the wire, providing some high-angle signal useful for working stations within a few hundred miles. If you don't have a tower you can shunt-feed, but you need a good antenna for 160 that will acquit itself well on DX yet still allow you to work stations "in close", you will be hard-pressed to beat the inverted-L.

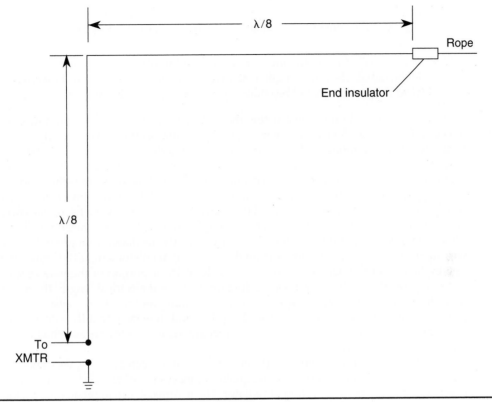

FIGURE 18.6   The inverted-L.

Because the horizontal portion of the wire extends in a single direction from the vertical portion, it causes a very slight distortion in the azimuthal radiation pattern of the antenna.

Far more important, however, is this: Because it is a member of the family of vertical monopoles, the inverted-L will not work well without an adequate ground system under it. Chapter 30 and the beginning of this chapter cover RF grounding in more detail, but here's a summary of what's important for an inverted-L:

- Run 12 to 50 radials of 50 to 150 ft, depending on the available space. A good rule of thumb is to put down as many equally spaced radials as necessary to have their far tips be no more than about 0.05λ apart. On 160, that's about 8-ft spacing between tips. As explained in Chap. 9, the longer you can make your radials, the more you should use, and the higher your overall antenna and ground system efficiency will be.

- The center of the radial circle should be at the feedpoint of the inverted-L. In other words, tie the inner end of all the radials to the shield side of the coaxial transmission line, using a chassis-mount SO-239 UHF receptacle or whatever best matches the connector on your feedline. Tape the joint with electrical tape and spray it with clear acrylic spray. Cover the entire joint with a small plastic shelter, such as a coffee can or a child's beach bucket. If you are using a matching capacitor, consider investing in a plastic storage container to enclose everything.

- Don't worry if property lines, driveways, or buildings keep you from making all the radials the same length. Just do the best you can. Radials in close proximity to the ground are not resonant, so their exact length is immaterial.

One popular method of constructing the inverted-L employs an existing tower for the vertical section and a length of wire for the horizontal section. If you already have a 60-ft tower to accommodate a beam antenna used on higher frequencies, it is relatively easy to turn it into an inverted-L for 160 m. If the base of the tower and any guy wires are insulated, all that is necessary is to connect the shield of the coaxial feedline and the tips of the radials together at the bottom end of the insulator and connect the center conductor of the feedline (or one end of the matching capacitor) to one leg of the tower above the insulator. (But never connect copper wire directly to a galvanized tower. See Chap. 29 for additional information.) If the tower is not insulated from ground, run a wire up the side of it, using aluminum angle-stock with insulators, rigid PVC pipe, or a two-by-four from the hardware store to hold the vertical portion of the wire at least a few feet away from the tower. Depending on its natural electrical length, the tower may still affect the tuning of the inverted-L—becoming an *incidental reradiator* in the process—but that's not necessarily bad. Keep in mind, however, that the effect of the tower may vary as feedlines to HF Yagis or other antennas at its top are switched in and out at their far end in the radio room.

Like the little girl in the nursery rhyme, the inverted-L can be "very, very good"—but its success is far more related to the quality of its system of radials than it is to the exact height of the vertical section of radiator. The inverted-L can approach the performance of a full-size $\lambda/4$ monopole, but only if these two factors are optimized:

- The radial field must be extensive, so as to minimize losses in the earth beneath it. This is more important than for a full-size $\lambda/4$ or taller vertical radiator because the radiation resistance of an inverted-L is typically lower (20 to 25 Ω) than that of the taller (37 Ω) vertical, Hence, it takes a lower ground resistance to dissipate no more transmitter power in ground losses, than in the case of the full-length $\lambda/4$ radiator.

- The total length of wire in the inverted-L should be at least 150 ft in order to get the high-current portion of the standing current out of the feedline and onto the open vertical portion of the antenna.

Dual-Band Inverted-L

Once you've gone to the trouble of erecting a wire antenna for 160 m, the natural question is "Can I use this on 80 m?" The answer is "Yes, but only with a different way of feeding it". The problem with trying to use the 160 inverted-L on its second harmonic is that the impedance at the feedpoint, near the ground, is very high (high voltage, low current) because it is roughly $\lambda/2$ from the far end of the wire. Instead of being close to that of the coaxial cable, the feedpoint impedance is apt to be a few thousand ohms. This can be matched with a good parallel-tuned ATU or a step-up transformer, but the system is likely to tune more sharply.

The second problem with using the antenna on its second harmonic is this: Much more of the transmitter power (roughly half, in fact) is going to be radiated from the horizontal section of wire. (The total wire is a half-wavelength on 80, and the current maximum is at the midpoint of the wire, at the point where the vertical and horizontal sections come together.)

A work-around that minimizes both problems is to insert an 80-m trap at the top of the vertical section—or even a slight distance out on the horizontal section if the total vertical run is less than 60 ft or so. The trap will also permit some shortening of the horizontal top section, allowing the total wire length to be less than 150 ft.

Phased Inverted-L Antennas

Because it is inexpensive and normally series-fed, the inverted-L is an excellent candidate to be the basic element or building block of an *all-driven* 160-m phased array. All the considerations of Chaps. 5 and 11 for ensuring proper and stable phasing pertain to the inverted-L, as well.

But inverted-Ls make good building blocks for *parasitic* arrays, too. Consider, for instance, an inline array of three inverted-Ls as being equivalent to one side of a vertically polarized three-element Yagi, the other side of each pseudo-Yagi element being replaced by ground. Using interelement spacings of 0.15λ results in a total baseline (or boom equivalent) length of about 170 ft. Forward gain, F/B, etc., are then optimized by adjusting the *vertical* portions of the Reflector and Director, preferably with the aid of an antenna modeling program. Unlike the horizontal Yagi this array is derived from, a good radial system is required under each element for maximum efficiency and stability of parameters.

Flat-Top or T Antenna

If you are primarily interested in DXing, then you will want to minimize the amount of signal that you put straight up in the air. One way to do that, if you have the backyard space, is to convert the "inverted-L" into a "T" by extending another horizontal wire, exactly the same length as the existing horizontal segment, in the opposite direction (Fig. 18.7). Now whatever standing wave of current exists in the first horizontal wire will be balanced by an equal current in the *opposite* direction in the new wire.* At distant points nearly straight up (meaning the ionosphere directly above you), the total signal from the antenna will approach zero. The effect of this is to re-form the radiation pattern of the antenna such that nearly all the transmitter RF emanates from the vertical portion of the wire, thus increasing its low-angle performance—and coincidentally eliminating some of the azimuthal pattern skew introduced by the single horizontal wire. The flat-

---

*You can see this by considering an instance when the current in the vertical portion of the antenna is flowing in the *downward* direction. Think of the two horizontal sections as having been created by splitting the original horizontal section of wire in half lengthwise. Since the two horizontal sections are part of the same quarter-wave monopole as the vertical section, the current in both of them must be flowing *into* the three-way junction at the top of the vertical section. For the current in both horizontal sections to be flowing into the junction, the current in the left-hand section must flow to the *right*, and the current in the right-hand section to the *left*. These two segments of wire constitute a two-element horizontally polarized array, and at any point in space directly above the antenna (or nearly so), the fields from the two wires cancel because they are of opposite polarity.

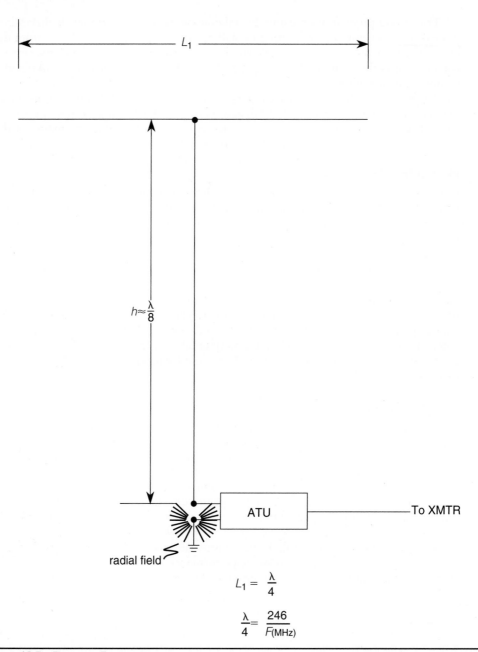

FIGURE 18.7   Flat-top or T antenna.

top thus reduces high-angle radiation on both 160 and 80 m. The radial system can be the same as that for the inverted-L.

In the early days of AM broadcasting, many stations in the United States and elsewhere successfully employed the T antenna as a way of putting most of their radiated signal into the groundwave; only in time did the T come to be supplanted by the purely vertical monopole.

The downside to the flat-top is that it requires twice as much span for the horizontal wires than the inverted-L does. However, if total available space in the long dimension is an issue, the two sides of the flat-top portion of the T can be bent or run in zigzag fashion. To minimize high-angle radiation, such bending should be applied to both sections of wire as symmetrically as possible, but don't lose sleep over it.

80-Meter Dipole on 160

Although not optimum, an 80-m dipole can be used on 160 m with some success. In fact, there are two different ways to do this.

The first approach requires an ATU capable of tuning out a wide range of reactance on 160 m. Assuming you have such a unit, simply match your antenna to your transmitter (presumably designed to match 50 Ω) on 160 m by tuning the ATU for minimum SWR at the chosen 160-m operating frequency. Even though the dipole is too short to fully expose the high-current portions of the natural radiating standing wave, there's enough total wire length to get signal out of the feedline and on its way. There are two disadvantages to this scheme:

- A suitable wide-range ATU typically costs more than most units commonly seen in use, and may be harder to locate. The old Dentron *Super Super Tuner* (no longer manufactured) is one that can match short antennas on 160, but not without some internal losses.

- The antenna feedpoint impedance is highly reactive and a terrible match to any available transmission line. Consequently, the line will experience very high standing waves with concomitant voltage and current peaks and increased I^2R losses that are characteristic of high SWR. Open-wire line is probably the best feedline for this application because it has less added loss in the presence of high SWR than coaxial cable or twin-lead does, and it is less likely to break down because of the high voltages encountered. On the other hand, if you're running only a few hundred watts or less, just about any RG-8 or equivalent cable should be able to handle the voltages.

Another approach is to use your existing 80-m dipole and its feedline as a pseudo-flat-top or T-antenna on 160. This is particularly attractive if you have a "reasonable" feedline length (say, more than 60 ft) and if the center of your dipole is substantially higher than your radio room. If so, it's clear that your feedline goes not only horizontally (to get from indoors to outdoors) but *up* as well. In other words, there's a vertical component to the feedline path that can serve as the vertical section of a "poor man's" inverted-L.

In this case, simply disconnect the transmitter end of the feedline from its normal connection, tie the bare ends of the two feedline conductors together, and reconnect the joined ends to the hot (center) terminal of the SO-239 coaxial socket on your ATU or transceiver (Fig. 18.8). Tune the ATU for minimum SWR, but it shouldn't be excessive

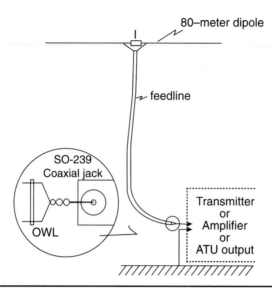

FIGURE 18.8   Pseudo-T from an 80-m dipole and feedline.

to begin with. If you are using twin-lead or open-wire line, simply twist the indoor ends of the two sides of the feedline together and stuff them back into the center (or "hot") receptacle of the SO-239. If you are using coaxial cable, you may find it best to build up a small adapter consisting of an SO-239 socket and either a wire pigtail (to push into the SO-239 socket) or a PL-259 plug attached via a short wire to the added SO-239. Some important considerations, however:

- A really good RF ground system is an absolute necessity if this antenna is to be effective. Especially if the radio equipment room is above the first floor of the building, this can be a challenge. Without a ground system, not only will the antenna fail to perform but the radio equipment will be "hot" with transmitted RF and capable of causing very painful burns to your skin. At the very least, a single $\lambda/4$ counterpoise wire must be attached to the chassis at the point where the new feedline is connected. Use insulated wire (except where it connects to the chassis, of course!) and try to get it outdoors and close to earth in as short a run as possible.

- One negative of this approach is that you have to disconnect and reconnect the 80-m feedline whenever switching from 80 to 160, or vice versa.

- This technique works only if neither side of the feedline is grounded between the transmitter SO-239 and the antenna. If, for instance, all your coaxial feedlines go through a remote switch or lightning arrestor connected to ground, your signal will be shorted to ground before it gets to the antenna.

- As described in the preceding T-antenna paragraphs, radiation from the two sides of the dipole mostly cancels at high elevation angles. However, because the feedline in this example travels both horizontally and vertically on its way to the dipole center, it generates both vertically and horizontally polarized radiation. As a result, it's not possible to use this scheme to completely eliminate high-angle radiation.

- Keep in mind that the primary radiation from this antenna comes from the feedline, not the 80-m dipole! The purpose of the dipole is to provide a total length (feedline + one side of the dipole) that is close to $\lambda/4$ on 160 m. (And to hold the far end of the feedline up in the air!)

Years ago, before the power limitations and interference imposed by LORAN on 1.8-MHz amateur operation had been removed, the author—whose station was located in upstate New York—made his very first 160-m amateur radio contact with Hawai'i by using exactly this antenna and a 30W transmitter (but without the "really good RF ground system" recommended here). Of course, primary credit for the contact should really go to the operator in Hawai'i who picked the author's weak signal out of the interference and tropical noise masking it!

Receiving on the Transmit Antenna

If your station location enjoys a low background noise level on 160 m, your transmit antenna may well be your best receiving antenna, as well. In some respects this is surprising because most 160-m transmit antennas in use today are nondirectional around the 360-degree compass, just like the antennas previously described.

For most of us, who are not fortunate enough to have transmitting arrays on 160 m or who are not blessed with extremely low noise levels, other antennas for improving the detectability of weak signals may be necessary. Chapter 14 describes a variety of receiving antennas—including Beverages and small loops—that can often perform miracles. Most are very low efficiency (that's why they're not also transmitting antennas), but any decent receiver of the past half-century has more than enough gain to compensate for the low output levels of these antennas.

Other than to discuss (in Chap. 23) direction-finding antennas that can be helpful in locating specific noise sources, and to note the existence of *noise-canceling* accessories available from manufacturers and suppliers listed in App. B, noise elimination is beyond the scope of this book.

CHAPTER 19

VHF and UHF Antennas

The VHF spectrum is defined as extending from 30 MHz to 300 MHz, and the UHF spectrum from 300 MHz to 3 GHz. In common parlance, however, the UHF upper limit is sometimes taken to be 900 MHz, with frequencies above 900 MHz loosely defined as *microwave*.

When designing, constructing, and using antennas for VHF and UHF, keep the following points in mind:

- Shorter wavelengths translate to smaller element dimensions. Thus, VHF and UHF antennas are often less expensive and simpler to construct than their HF counterparts, and their supports don't need to be as strong.

- Shorter wavelengths may allow the use of *free-space* design rules. A 2-m antenna atop a 70-ft tower is 10 wavelengths above ground, far enough away to model the behavior of the antenna as if it were in free space.

- Yagis and other antenna types that use aluminum tubing or rods for elements tend to deliver performance over a wider percentage bandwidth than at HF because the element diameter is much larger relative to wavelength.

- Directive antennas with more gain, narrower beamwidths, and better rejection of signals off the sides and rear are relatively easy to implement. A 14-element Yagi for 144 MHz is a lot easier to put up and keep up than even a 3-element Yagi for 14 MHz.

That's the "good" news. Now keep these points in mind, too:

- Smaller element dimensions mean that greater precision and accuracy in assembly are required. Small hardware items, such as element-to-boom mounting brackets, become a critical part of each element's electrical characteristics and can have a huge effect on how the assembled antenna performs compared with predictions based on models or calculated values.

- Some of the propagation modes require aiming highly directive antennas above the horizon, so two rotors (called the *el-az* mount) may be required.

- The higher the forward gain of the antenna, the narrower its main beam and the more precise your ability to aim it must be.

- In contrast to the ionospheric skip of the HF bands, most VHF and UHF propagation modes demand that your antennas be as high as possible, to attain true line-of-sight paths extending as far from your site as possible before encountering the first obstacle. Amateurs dedicated to VHF and UHF DXing live on hilltops or, better yet, mountaintops. HF DXers don't need to.

CAUTION *A 14-element 2-mYagi is so lightweight it can be carried up the tower in one hand by one person. But long-boom beams, even at VHF/UHF, have a relatively high "windsail area", and even relatively light winds can apply a lot of force or torque to them. One of the authors once witnessed a muscular technician blown off a ladder by a sudden puff of breeze acting on a modest, "suburban"-sized TV antenna he was holding. **It can happen to you, too. Always install antennas with a helper, and use hoists and other tools to actually lift and support the array while it is being attached to the mast or rotor.***

To speak of a "VHF" or "UHF" antenna is somewhat misleading because virtually all forms of antenna used on the HF or MW bands can also be used at VHF and UHF. The main factors that distinguish supposedly VHF/UHF designs from others are mechanical: Some things are simply much easier to accomplish with small antennas. For example, consider the delta match. At 80 m, the delta-match dimensions are approximately 36×43 ft, and at 2 m they are 9.5×12 *in.* Clearly, delta matching is a bit more practical at VHF!

Although VHF/UHF antennas are not substantially different from HF antennas, for various practical reasons there are several forms that are especially well suited to the VHF/UHF bands. In this section we will take a look at some of them.

Ground-Plane Vertical

At VHF and above, the *ground-plane antenna* is a vertical radiator situated immediately above an artificial RF ground consisting of quarter-wavelength radiators or a sheet of highly conductive material such as copper. Most ground-plane antennas are either ¼-wavelength or ⅝-wavelength (although for the latter case impedance matching is needed—see the example later in this chapter).

Figure 19.1 shows how to construct an extremely simple ground-plane antenna for 2 m and above. The base of the antenna is a single SO-239 chassis-type coaxial connector. Be sure to use the type that requires four small machine screws to hold it to the chassis, not the single-nut variety.

The radiating element is a piece of ³⁄₁₆-in or 4-mm brass tubing usually obtained at hobby stores that sell model airplanes and modeling supplies. The sizes quoted just happen to fit over the center pin of an SO-239 with only a slight tap from a lightweight hammer. If the inside of the tubing and the connector pin are pretinned with solder, sweat-soldering the joint will make a good electrical connection that is resistant to weathering. Cover the joint with clear lacquer spray for added protection.

Tubing is also used for the radials. Alternatively, solid rods can also be used for this purpose. At least three radials equally spaced around a circle are needed for a proper antenna. (Only two are shown in Fig. 19.1.) For this method of construction, four might be a better number, with one radial attached to each of the four SO-239 mounting holes. Gently flatten one end of the radial with the same small hammer, and drill a small hole in the center of the flattened area. Mount the radial to the SO-239 using hardware of the appropriate size (typically 4-40 screws and nuts for this particular SO-239).

The SO-239 can be attached to a metal L-bracket that is mounted to a length of 2- × 2-in lumber that serves as the mast. While it is easy to fabricate such a bracket from aluminum stock, it is also possible to buy suitable brackets in any well-equipped hardware store.

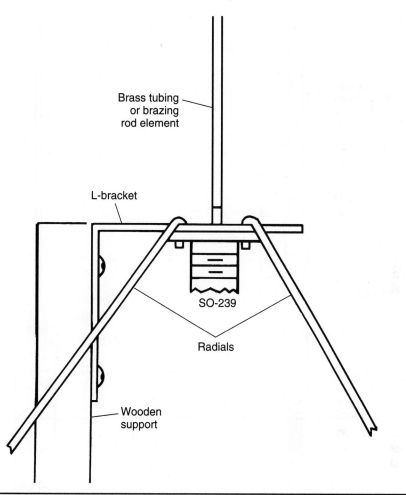

Brass tubing
or brazing
rod element

L-bracket

SO-239

Radials

Wooden
support

FIGURE 19.1   Construction of VHF ground-plane antenna.

Coaxial Vertical

The coaxial vertical is a quarter-wavelength vertically polarized antenna that is popular on VHF/UHF. It is really a $\lambda/2$ dipole, however, since the exposed quarter-wave center conductor is working against an equal length of exposed shield or braid in line with it, rather than in a ground-plane configuration. The only thing unusual about this antenna is that its method of construction forces the feedline to come away from the antenna inside the braid—an asymmetrical arrangement that has no particular merit and which probably would benefit from the addition of common mode chokes at the bottom of the antenna.

Two varieties of this antenna are usually encountered, depending on how permanent it is supposed to be. In Fig. 19.2A we see the coaxial antenna made with coaxial cable. Although not suitable for long-term installations, it is very useful for short-term, portable, or emergency applications. The length of either the exposed insulation or the exposed shield is found from Eq. (19.1):

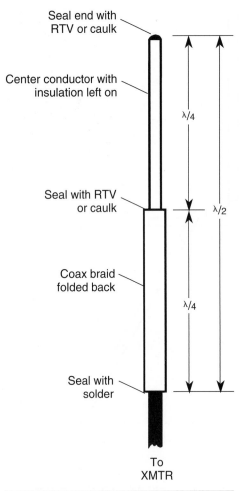

Seal end with RTV or caulk

Center conductor with insulation left on

$\lambda/4$

Seal with RTV or caulk

$\lambda/2$

Coax braid folded back

$\lambda/4$

Seal with solder

To XMTR

FIGURE 19.2A Coaxial vertical based on coaxial cable.

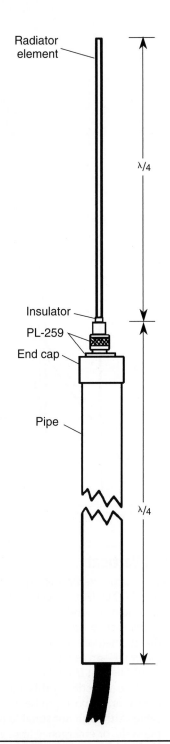

Radiator element

$\lambda/4$

Insulator

PL-259

End cap

Pipe

$\lambda/4$

FIGURE 19.2B Tubing coaxial vertical.

$$L \text{ (inches)} = \frac{2952}{F \text{ (MHz)}} \qquad (19.1)$$

The antenna can be suspended from above with a short piece of string, twine, or fishing line. From a practical point of view, the major problem with this implementation is that the coaxial cable begins to deteriorate after a few rainstorms because the center of the dipole, where the braid and the center conductor go in opposite directions, is exposed to the elements. This effect can be slowed (but not prevented) by sealing the end and the break between the sleeve and the radiator with either silicone RTV or bathtub caulk.

A more weather-resistant implementation is shown in Fig. 19.2B. The sleeve is a piece of copper or brass tubing (pipe) about 1 in in diameter. An end cap is fitted over the end and sweat-soldered into place to prevent the weather from destroying the electrical contact between the two pieces. An SO-239 coaxial connector is mounted on the end cap. The coaxial cable is connected to the SO-239 inside the pipe, which requires making the connection before mounting the end cap.

The radiator element is a small piece of tubing (or brazing rod) soldered to the center conductor of a PL-259 coaxial connector. An insulator is used to prevent the rod from shorting to the outer shell of the PL-259. (Note: An insulator salvaged from the smaller variety of banana plug can be shaved a small amount with a fine file and made to fit inside the PL-259. It allows enough center clearance for $\frac{1}{8}$-in or $\frac{3}{16}$-in brass tubing.)

Alternatively, the radiator element can be soldered to a banana plug. A standard banana plug nicely fits into the female center conductor of an SO-239.

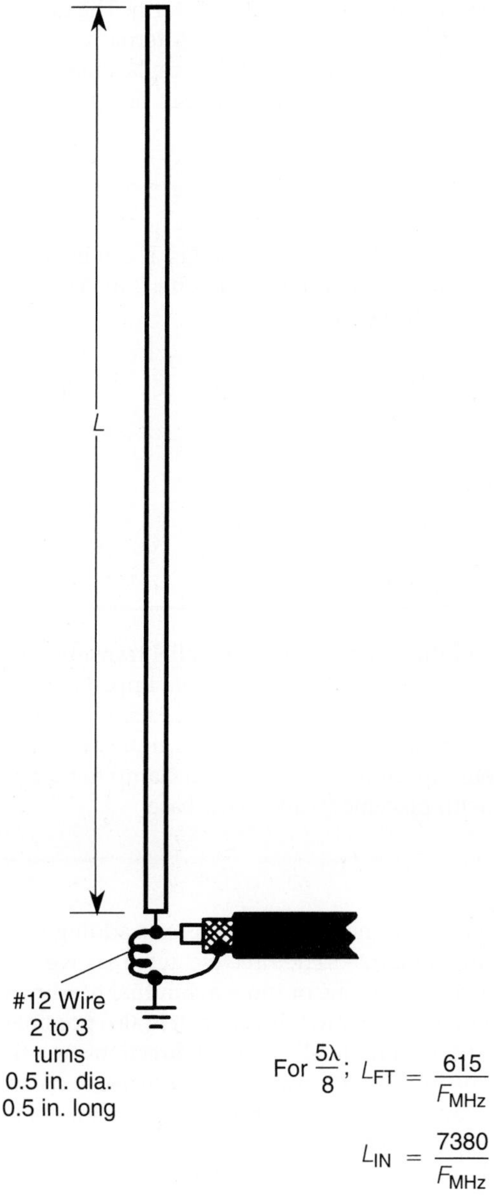

⅝-Wavelength 2-Meter Antenna

The ⅝-wavelength antenna (Fig. 19.3) is popular on 2 m for mobile operation because it is easy to construct, and at low elevation angles it provides a small amount of gain relative to the standard $\lambda/4$ ground plane. The radiator element is ⅝-wavelength, so its physical length is found from:

#12 Wire
2 to 3
turns
0.5 in. dia.
0.5 in. long

For $\frac{5\lambda}{8}$; $L_{FT} = \dfrac{615}{F_{MHz}}$

$$L_{IN} = \frac{7380}{F_{MHz}}$$

Figure 19.3   ⅝-wavelength 2-m antenna.

$$L \text{ (inches)} = \frac{7380}{F \text{ (MHz)}} \qquad (19.2)$$

The ⅝-wavelength antenna is not a good match to any of the common forms of coaxial cables. Either a transmission line matching section or an inductor match is normally used. Figure 19.3 shows an inductor match, using a matching coil of two to three turns of #12 wire wound over a ½-in-long ½-in outside diameter (OD) form. The radiator element can be tubing, brazing rod, or a length of heavy *piano wire*. Alternatively, for low-power systems, it can be a telescoping antenna that is sold as a replacement for portable radios or television sets. These antennas have the advantage of being adjustable to resonance without the need for cutting.

J-Pole Antennas

The *J-pole antenna* is another popular form of vertical on the VHF bands. It can be used at almost any frequency, although the example shown in Fig. 19.4 is for 2 m. The antenna radiator is ¾-wavelength long, so its length is found from:

$$L(\text{inches}) = \frac{8838}{F(\text{MHz})} \qquad (19.3)$$

and the quarter-wavelength matching section length from:

$$L(\text{inches}) = \frac{2952}{F(\text{MHz})} \qquad (19.4)$$

Taken together, the matching section and the radiator form a parallel transmission line with a characteristic impedance that is four times the coaxial cable impedance. If 50-Ω coax is used, and the elements are made from 0.5-in OD pipe, a spacing of 1.5 in will yield an impedance of about 200 Ω. Impedance matching is accomplished by a gamma match consisting of a 25-pF variable capacitor, connected by a clamp to the radiator and tapped about 6 in (experiment with placement) above the base.

Collinear Vertical

Gain in antennas is obtained by "stealing" power from some directions and adding it to other directions. While many of the vertically polarized HF directive antennas we examined in previous chapters have taken energy from one or more azimuthal headings and used it to augment the power radiated in the "forward" direction, it is also possible to obtain gain over a limited range of elevation angles in all compass directions by utilizing energy that might have otherwise gone into elevation angles of little use to us. In this manner it is possible to have a vertical antenna that exhibits gain—at least for certain elevation angles—around the entire 360-degree range of azimuths.

Figure 19.5 is an example of just such an antenna. Through the use of collinear elements for the two sides of a vertical dipole, it develops gain at low elevation angles

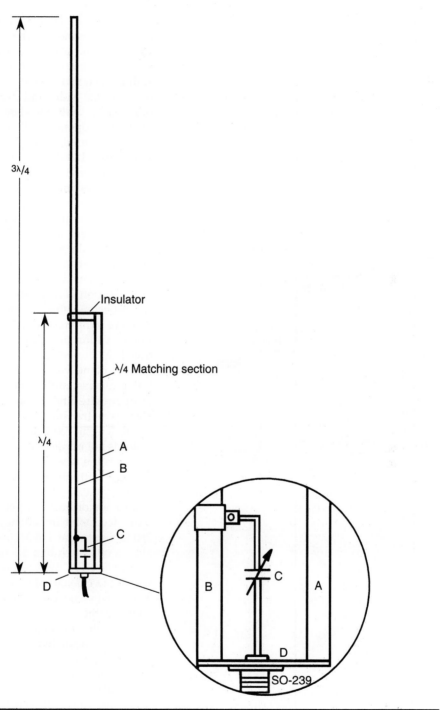

3λ/4

Insulator

λ/4 Matching section

λ/4

A

B

C

D

B

C

A

D

SO-239

FIGURE 19.4   J-pole antenna.

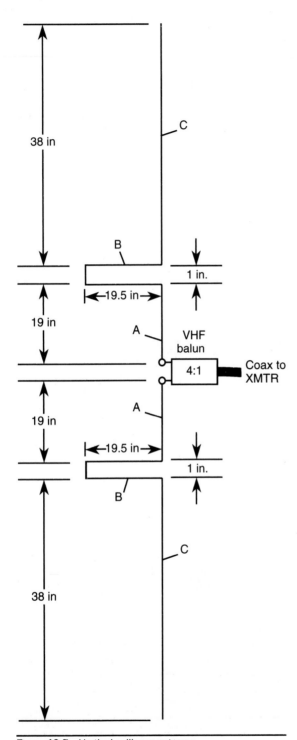

FIGURE 19.5   Vertical collinear antenna.

in an omnidirectional pattern. Each array consists of a quarter-wavelength section A and a half-wavelength section C separated by a phase reversing stub B. (The total length of the stub—out on one side and back on the other—is a half-wavelength, but if made from open-wire line, only a λ/4 line is required. The phase reversal stub adds a 180-degree phase shift between the inner λ/4 section and the outer λ/2 section. Because either outboard half-wavelength section is λ/2 or 180 degrees above or below the center half-wave section, the in-phase currents in the sections cancel in the vertical direction. In any horizontal direction, however, the current in each outer section is in phase with that of its respective inner section, and the antenna exhibits about 3 dB gain at low elevation angles over a conventional λ/2 vertical dipole or λ/4 ground-plane vertical. Note that the phase reversal stub, even if made out of open-wire line, is not being operated as a transmission line and the usual length correction for velocity factor is not applied.

As is the case with an ordinary dipole, the feedpoint is at the midpoint of the array (i.e., between the A sections). Unlike an ordinary dipole, the free-space feedpoint impedance is around 280 Ω resistive at 146 MHz—a reasonably good match to a 4:1 balun transformer (see Fig. 19.6), especially if fed with 72-Ω coaxial cable. Alternatively, 300-Ω twin-lead can be connected directly to the two sides of the dipole center. If the transmitter lacks the balanced output needed to feed twin-lead, use an appropriate balun at the input end of the twin-lead (i.e., right at the transmitter). For proper performance, make sure the feedline is kept at right angles to the dipole axis near the dipole.

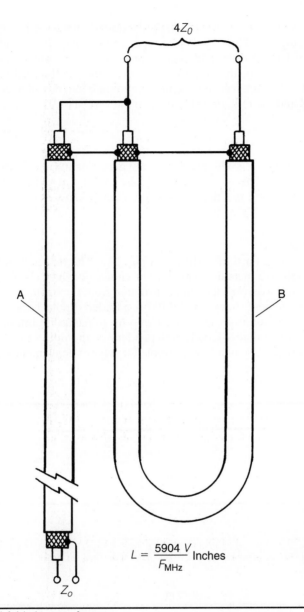

$$L = \frac{5904 \; V}{F_{MHz}} \; \text{Inches}$$

FIGURE 19.6   Coaxial 4:1 balun transformer.

Grounds for Ground-Plane Verticals

With the exception of the sleeve antenna and the collinear dipole, all the preceding verticals are *monopoles*—in effect, one side of a dipole. All will be poor performers if a good return path simulating the other half of the dipole is not provided. When the base of these antennas is elevated more than a small fraction of a wavelength above ground, an artificial ground plane is an absolute necessity. The simplest way to provide this is with three or four radials cut to $\lambda/4$ at the design frequency of the vertical and spaced equally around the compass (120 or 90 degrees apart, respectively). The radials should slope down from the base at about a 45-degree angle. Some experimentation with their exact length and slope may be in order. In contrast to radials laid directly on the earth's surface, elevated radials are resonant elements, and their length will have a strong effect on both the SWR and the radiation efficiency of the vertical.

Yagi Antennas

The Yagi beam antenna is a highly directional gain antenna and is used in both HF and VHF/UHF systems. The antenna can be easier to build at VHF/UHF than at HF. The basic Yagi was covered in Chap. 12, so we will only show examples of practical VHF devices.

Six-Meter Yagi

A four-element Yagi for 6 m is shown in Fig. 19.7. The reflector and directors can be mounted directly to a metallic boom, but the choice of driven element and its feed system require that the driven element be insulated from the boom.

The driven element shown in Fig. 19.7 is a folded dipole; because both conductors are the same diameter, it provides a 4:1 step-up in feedpoint impedance relative to a conventional dipole that is largely independent of the ratio of conductor spacing to conductor diameter. Use of a folded dipole for the driven element is common practice

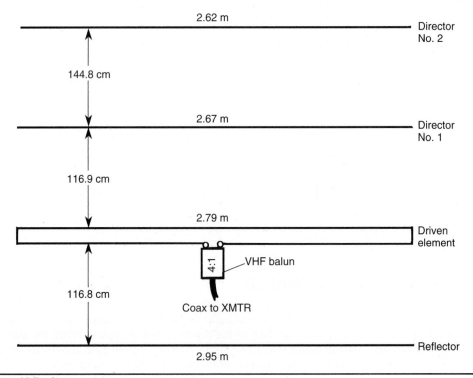

FIGURE 19.7   Six-meter beam antenna.

at VHF because it tends to broaden the SWR bandwidth of the antenna's input impedance and because the native feedpoint impedance of Yagis is often very low. For this construction project, two ¾-inch aluminum tubes, spaced 6.5 in apart, were used.

The antenna can be directly fed with balanced 300-Ω line or from 50-Ω or 75-Ω coax through a 4:1 balun. If broadbanding is not important, the folded dipole of the driven element can be replaced with a conventional dipole and the antenna fed directly from a 52- or 75-Ω feedline.

Two-Meter Yagi

Figure 19.8 shows the construction details for a six-element 2-m Yagi beam antenna. This antenna is built using a 2- × 2-in wooden boom and elements made of either brass or copper rod. Threaded brass rod simplifies assembly, but is not strictly necessary. The job of securing the elements (other than the driven element) is easier with threaded rod

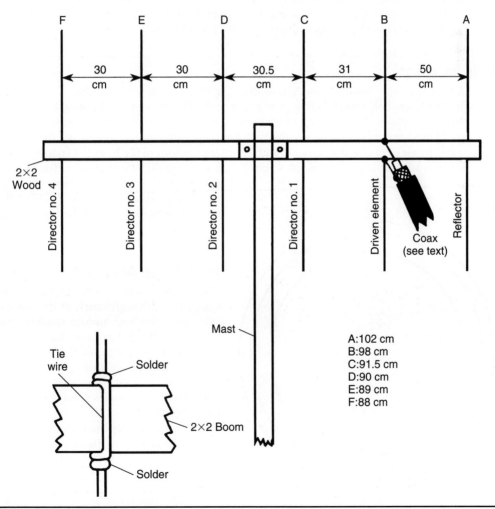

FIGURE 19.8 Two-meter vertical beam.

because it allows securing of each parasitic element with a pair of hex nuts, one on either side of the 2- × 2-in boom. Unthreaded elements can be secured with RTV to seal a press-fit. Alternatively, secure the rods with tie wires (see inset to Fig. 19.8): drill a hole through the 2 × 2 to admit the rod or tubing, secure the element by wrapping a tie wire around the rod on either side of the 2 × 2, and solder it in place. Use #14 to #10 solid bare or tinned wire for the tie wires.

Mount the mast of the antenna to the boom with an appropriate boom-mast bracket or clamp. One alternative is to use an end-flange clamp, such as is sometimes used to support pole lamps and the like. The mast should be attached to the boom at the *center of gravity*, also known as the *balance point*. If you try to balance the antenna on one hand unsupported, there is one (and only one) point at which it is balanced (and won't tilt end down). Attach the mast hardware at, or near, this point in order to prevent normal gravitational torques from tearing the mounting apart.

The antenna is fed with coaxial cable at the center of the driven element. Often, either a matching section of coax or a gamma match will be needed because the usual effect of parasitic elements on a Yagi's feedpoint impedance is to make it too low to be a good direct match to standard coaxial cable impedances.

Halo Antennas

One of the more "saintly" antennas used on the VHF boards is the *halo* (Fig. 19.9), formed by bending a half-wavelength dipole into a circle. The ends of the dipole are connected to the two plates of a capacitor. In some cases, a transmitting-type mica "button" capacitor is used, but more commonly the VHF halo capacitor consists of two 3-in disks separated by air or plastic dielectric. In some implementations of the halo, the spacing of the disks is adjustable via a threaded screw. While air is a good (and perhaps better) dielectric, the use of a plastic spacer provides mechanical rigidity to help stabilize the exact spacing and, hence, the value of the capacitor.

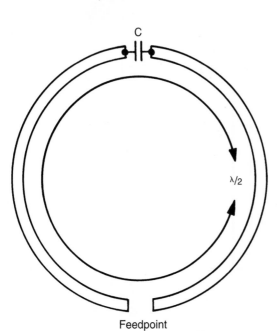

Feedpoint

Quad Beam Antennas

The quad antenna was introduced in Chap. 13. Although much of the discussion there revolved around implementations for HF, it has, nonetheless, also emerged as a very good VHF/UHF antenna. It should go without saying that the antenna is a lot easier to construct at VHF/UHF frequencies than it is at HF frequencies!

Figure 19.10 depicts only one of several methods for building a quad for VHF or UHF. The radiator element can be any

FIGURE 19.9   Halo antenna.

Overall lengths:

$$\text{Reflector: } L = \frac{12{,}562}{F_{\text{MHz}}} \text{ in}$$

$$\text{Director: } L = \frac{11{,}248}{F_{\text{MHz}}} \text{ in}$$

$$\text{Driven element: } L = \frac{11{,}826}{F_{\text{MHz}}} \text{ in}$$

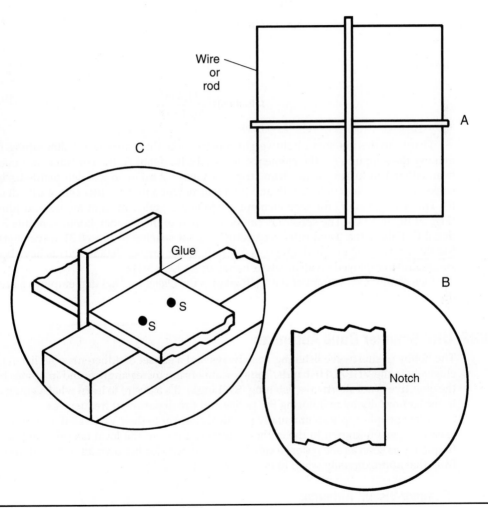

FIGURE 19.10   Quad loop construction on VHF.

of several materials, including heavy solid wire (#8 to #12), tubing, or metal rods. The overall lengths of the elements are given by:

Driven element:

$$L(\text{inches}) = \frac{11,826}{F(\text{MHz})} \qquad (19.5)$$

Reflector:

$$L(\text{inches}) = \frac{12,562}{F(\text{MHz})} \qquad (19.6)$$

Director:

$$L(\text{inches}) = \frac{11,248}{F(\text{MHz})} \qquad (19.7)$$

Thanks to this antenna's lightweight construction, there are several alternatives for making the supports for the elements. In Fig. 19.10, detail A, the spreaders are made from either 1-in furring strips, trim strips, or (above 2 m) even wooden paint-stirring sticks. The sticks are cut to length and then half-notched in the center (Fig. 19.10, detail B). The two spreaders for each element are joined together at right angles and glued (Fig. 19.10, detail C). The spreaders can be fastened to the wooden boom at points S in detail C. Follow the usual rules regarding element spacing (0.15 to 0.31 wavelength). See the information on quad antennas in Chap. 13 for further details. Quads have been successfully employed on all amateur bands up to 1296 MHz.

One commercial variant of the VHF quad is the *Quagi*—a Yagi that employs a quad driven element.

VHF/UHF Scanner Band Antennas

The hobby of shortwave listening has always had a subset of adherents who listen exclusively to the VHF/UHF bands. Today, scanners are increasingly found in homes but the objective is not shortwave listening (SWLing)—it's a desire to learn what's going on in the community by monitoring the police and fire department frequencies.

A few people apply an unusually practical element to their VHF/UHF listening. At least one person known to the author routinely tunes in the local taxicab company's frequency as soon as she orders a cab. She then listens for her own address and knows from that approximately when to expect her cab!

"Scanner-Vision" Antennas

The antennas used by scanner listeners are widely varied and (in some cases) overpriced. Although it is arguable whether a total-coverage VHF/UHF antenna is worth the money, there are other possibilities that should be considered.

First, don't overlook the use of television antennas for scanner monitoring! The television bands (about 80 channels from 54 MHz to around 800 MHz) encompass most of the commonly used scanner frequencies. Although antenna performance is not optimized for the scanner frequencies, it is also not terrible on those frequencies. If you already have an "all channel" TV antenna installed, then it is a simple matter to connect the antenna to the scanner receiver with a 2:1 splitter and (possibly) a 4:1 balun transformer that accepts 300 Ω in and produces 75 Ω out. These transformers are readily available at RadioShack and anywhere that TV and video accessories are sold.

The directional characteristic of the TV antenna can be boon or bane to the scanner user. If the antenna has a rotator, then there is no problem. Just rotate the antenna to the direction of interest (unless someone else in the family happens to be watching TV at the time!). But much of the time it won't even be necessary to rotate the TV/FM antenna— partly because no antenna completely rejects signals arriving from off the sides or the back, and partly because most of the local scanner repeaters will be quite strong.

An excellent alternative to hijacking the family's primary TV antenna is to locate an indoor set-top TV antenna at a garage sale or flea market (or maybe your own attic). Some of these have a simple phasing circuit built in with a front panel knob that allows the user to move the peak of the antenna pattern around some. With the possible addition of a wideband TV antenna preamplifier (available at RadioShack, among other places), such a setup may be perfectly adequate for scanner monitoring.

Scanner Skyhooks

Some popular scanners even cover much of the HF spectrum. For true SWLing on those bands, a random-length wire antenna ought to turn in decent performance, and it may even be able to do double duty as a VHF/UHF longwire antenna. Typically, this antenna is simply a 30- to 150-ft length of #14 wire attached to a distant support.

Additional gain, about +3 dB, can be achieved by stacking VHF/UHF antennas together. Figure 19.11 shows a typical arrangement in which two half-wavelength dipole antennas are connected together through a quarter-wavelength harness of RG-59/U coaxial cable. This harness is physically shorter than an electrical quarter-wavelength by the velocity factor of the coaxial cable:

$$L \text{ (inches)} = \frac{2832 v_F}{F \text{(MHz)}} \tag{19.8}$$

where L = length, in inches
v_F = velocity factor (typically 0.66 or 0.80 for common coax)
F = frequency, in megahertz

The antennas can be oriented in the same direction to increase gain, or orthogonally (as shown in Fig. 19.11) to obtain a more omnidirectional cloverleaf pattern.

Because the impedance of two identical dipoles, fed in parallel, is one-half that of a single dipole, it is necessary to have an impedance-matching section made of RG-58/U coaxial cable. This cable is then fed with RG-59/U coax from the receiver.

There is nothing magical about scanner antennas that is significantly different from other VHF/UHF antennas except, perhaps, the need to cover multiple frequency ranges. Although the designs might be optimized for VHF or UHF, these antennas are basically the same as others shown in this book. As a matter of fact, almost any antennas, from any chapter, can be used over at least part of the scanner spectrum.

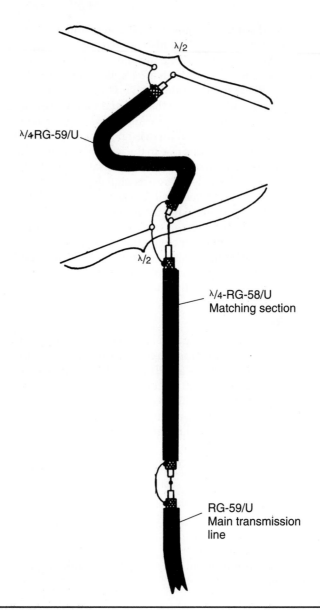

Figure 19.11   Stacking VHF antennas.

VHF/UHF Antenna Impedance Matching

VHF/UHF antennas are no different from their HF brethren in their need for impedance matching to the feedline. However, some methods (coax baluns, delta match, etc.)
are easier at the higher frequencies, while others become difficult or impossible. An
example of the latter case is the tuned *LC* impedance-matching network. At 6 m, and
even to some extent at 2 m, lumped-component *LC* networks can be used. But at 2 m

and above, other methods are often easier to implement. For example, we can replace the *LC* tuner with stripline tank circuits.

The *balun* transformer provides an impedance transformation between *balanced* and *unbalanced* circuit configurations. Although both 1:1 and 4:1 impedance ratios are possible, the 4:1 ratio is most commonly used for VHF/UHF antenna work. At lower frequencies it is easy to build broadband transformer baluns, but these become more of a problem at VHF and above.

Figure 19.6 is a 4:1 impedance ratio coaxial balun often used on VHF/UHF frequencies. Two sections of identical coaxial cable are needed. One section (A) can be any convenient length needed to reach from the antenna to the transmitter. Its characteristic impedance is Z_0. The other section (B) is cut to be a half-wavelength long at the center of the frequency range of interest. The physical length is found from

$$L(\text{inches}) = \frac{5904 v_F}{F(\text{MHz})}$$ (19.9)

where L = cable length, in inches
   F = operating frequency, in megahertz
   v_F = velocity factor of coaxial cable

The velocity factors of common coaxial cables are shown in Table 19.1.

Regular polyethylene	0.66
Polyethylene foam	0.80
Teflon	0.72

TABLE 19.1 Coaxial Cable Velocity Factors

Hard line velocity factors typically lie in the 0.8 to 0.9 range, but it is wise to check the manufacturer's specifications for the exact number. Alternatively, a better approach for either coaxial cable or hard line is to measure the velocity factor of the specific piece of cable you intend to use. See Chap. 27 ("Instruments for Testing and Troubleshooting") for more information.

Example 19.1 Calculate the physical length required of a 146-MHz 4:1 balun made of polyethylene foam coaxial cable.

Solution

$$L = \frac{5904 v_F}{F}$$

$$= \frac{(5904)(0.08)}{146}$$

$$= \frac{4723.2}{146} = 32.4 \text{ in}$$

◆

One approach to neatly joining the coaxial cables is shown in Fig. 19.12. In this example, three SO-239 coaxial receptacles are mounted on a metal plate. This arrangement has the effect of shorting together the shields of the three ends of coaxial cable while the center conductors are connected in the manner shown. This method is sometimes provided as part of a commercial antenna; the author's commercial 6-m Yagi has just such a bracket on it, the only difference being the use of BNC connectors.

Because the balun of Figs. 19.6 and 19.12 is mounted at the antenna feedpoint and, hence, is almost always located outdoors, special care must be taken to weatherproof the ends of the transmission line—especially the long run back to the transmitter.

The *delta match* gets its name from the fact that the physical layout of the matching network used when tapping out on the two sides of a driven element or dipole a distance corresponding to the transmission line impedance has the shape of the Greek letter delta. Figure 19.13A shows the basic delta match scheme. The delta match connections to the driven element are made symmetrically on both sides of center, and can be made from brass, copper, or aluminum tubing, or a bronze brazing rod bolted to the main radiator element.

The width (A) of the delta match is given by

$$A(\text{inches}) = \frac{1416}{F(\text{MHz})} \tag{19.10}$$

while the length (B) of the matching section is

$$B(\text{inches}) = \frac{1776}{F(\text{MHz})} \tag{19.11}$$

The transmission line feeding the delta match is balanced line, such as parallel transmission line or twin-lead. The exact impedance is not terribly critical because the

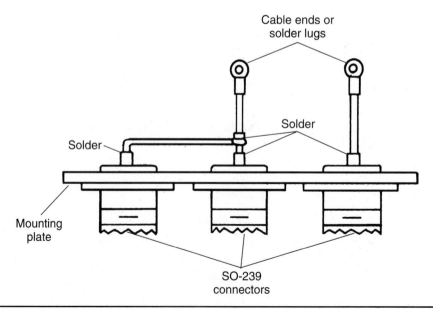

FIGURE 19.12 Practical implementation of 4:1 balun using connectors.

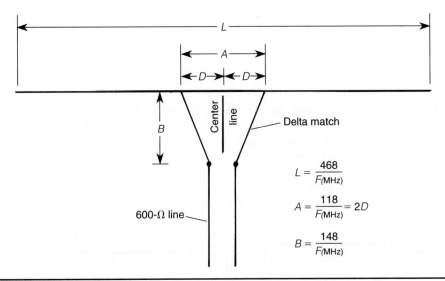

FIGURE 19.13A Delta feed matching system.

dimensions (especially A) can be adjusted for a good match. Although 450-Ω line is commonly used, 300-Ω or 600-Ω lines are equally good alternatives. Figure 19.13B shows a method for using coaxial cable or hard line with the delta match. The 50- or 75-Ω coax or hardline impedance is transformed up to the delta match impedance with a 4:1 balun transformer such as the one in Fig. 19.6.

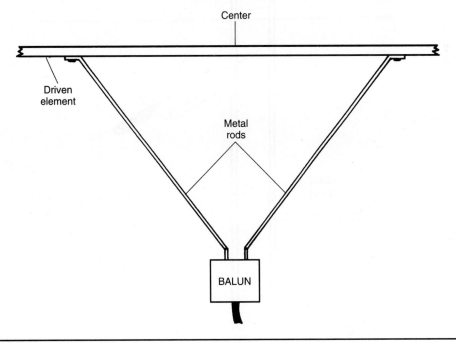

FIGURE 19.13B Practical VHF delta match.

A stub-matching system is shown in Fig. 19.14. The exact impedance of the line is not very critical and is found from

$$Z_0 = 276 \log_{10} \frac{2S}{d} \qquad (19.12)$$

where S is the space between the two sides of the line and d is the diameter of each conductor.

The matching stub section is made from metal elements such as tubing, wire, or rods, since all three are practical at VHF/UHF frequencies. For a $\frac{3}{16}$-in rod, the spacing is approximately 2.56 in to make a 450-Ω transmission line. A sliding short circuit sets the electrical length of the half-wave stub. The stub is tapped at a distance from the antenna feedpoint that matches the impedance of the transmission line. In the example

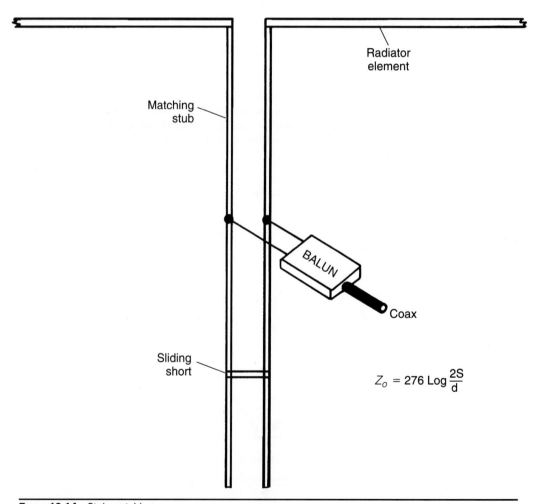

FIGURE 19.14   Stub matching.

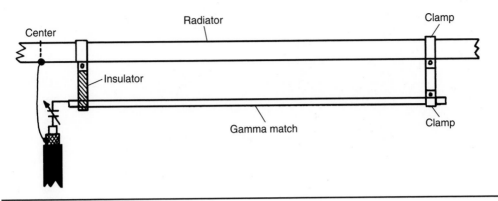

Figure 19.15 Gamma matching.

shown, the transmission line is coaxial cable, so a 4:1 balun transformer is used between the stub and the transmission line. The two adjustments to make in this system are: (1) the distance of the short from the feedpoint and (2) the distance of the transmission line tap point from the feedpoint. The two are alternately adjusted for minimum VSWR.

The gamma match is basically a half-delta match and operates according to similar principles (Fig. 19.15). The shield (outer conductor) of the coaxial cable is connected to the exact center of the driven element. The center conductor of the coaxial cable series feeds the gamma element through a variable capacitor.

CHAPTER 20

Microwave Waveguides and Antennas

CAUTION *Microwave RF energy is dangerous to your health! Anything that can cook meat can also cook you! The U.S. government sets the safety limit for microwave exposure at 10 mW/ cm² averaged over 6 min; some other countries use a level $\frac{1}{10}$ of the U.S. standard. The principal problem is tissue heating, and eyes seem especially sensitive to microwave energy. Some authorities believe that cataracts form from prolonged exposure. Some authorities also believe that genetic damage to offspring is possible as well as other long-term effects as a result of cumulative exposure.*

Because of their relatively high gain, microwave antennas can produce hazardous field strengths in close proximity—even at relatively low RF input power levels. *At least one technician in a TV satellite earth station suffered abdominal adhesions, solid matter in his urine, and genital dysfunction after servicing a 45-m-diameter 3.5-GHz antenna with RF power applied.*

Be very careful around microwave antennas. Do not service a radiating antenna. When servicing nonradiating antennas, be sure to stow them in a position that prevents the inadvertent exposure of humans should power accidentally be applied. A Radiation Hazard *sign should be prominently displayed on the antenna. Good design practice requires an interlock system that prevents radiation in such situations. "Hot" transmitter service should be performed with a shielded dummy load replacing the antenna.*

The microwave portion of the radio spectrum covers frequencies from about 900 MHz to 300 GHz, with wavelengths in free space ranging from 33 cm down to 1 mm. At these wavelengths we find that conventional antennas derived from the basic wire dipole are not the only efficient radiating structures available to the designer or user. Similarly, conventional transmission lines—which exhibit a variety of losses that increase with frequency—become impractical above about 5 GHz, except for very short runs. Instead, at microwave frequencies a number of alternative devices, including *waveguides, horns,* and other components reminiscent of a plumber's parts bin, dominate antenna system design.

Waveguides

Conventional transmission lines exhibit three kinds of losses: *ohmic, dielectric,* and *radiation*. Ohmic losses are the result of skin effect, which increases a conductor's apparent resistance with increasing frequency. Dielectric losses are caused by the electric field acting on the molecules of the insulator, thereby causing heating through molecular agitation, and tend to be largest when the wavelength is comparable to the interatomic resonances of the dielectric. Radiation losses result from incomplete shielding (in a co-axial line) or from imbalance and incomplete cancellation of RF in the two sides of the line—usually because at some frequency the spacing between the wires becomes too large a percentage of the RF signal wavelength. All these losses are proportional to both frequency and transmission line length and serve to limit both the maximum power that can be applied at the transmitter end of the line and the maximum frequency for a given line length.

To circumvent these problems, *waveguides* are used. What is a waveguide? Probably the simplest response is to say that a waveguide is to radio waves what an optical fiber is to light waves. Each is a hollow tube or tunnel through which an electromagnetic wave is "encouraged" by the tube's characteristics to pass. Consider the light pipe analogy depicted in Fig. 20.1. A flashlight serves as our RF source, which (given that light is also an electromagnetic wave) is not altogether unreasonable. In Fig. 20.1A the source radiates into free space and spreads out as a function of distance. The intensity per unit area falls off as a function of distance (D) according to the *inverse square law* ($1/D^2$).

But now consider the transmission scheme in Fig. 20.1B. The light wave still propagates over distance D but is now confined to the interior of a mirrored pipe. Almost all of the energy coupled to the input end is delivered to the output end, where the intensity is practically undiminished. Although not perfect, the light pipe analogy illuminates (pardon the pun . . . again) the value of microwave waveguides. Similarly, fiber-optic technology is waveguide-like at optical (IR and visible) wavelengths. In fact, the analogy between fiber optics and waveguide is a more rigorous comparison than the simplistic light pipe analogy.

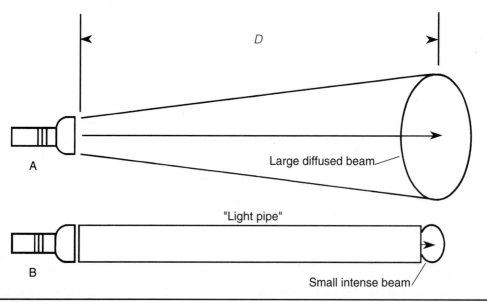

FIGURE 20.1 Waveguide analogy to light pipe.

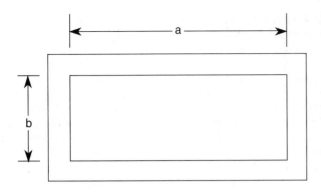

FIGURE 20.2 Rectangular waveguide (end view).

The internal walls of a waveguide are not mirrored surfaces, as in our optical analogy, but are, rather, electrical conductors—typically aluminum, brass, or copper. In order to further reduce ohmic losses, the internal surfaces of some waveguides are electroplated with either gold or silver, both of which have lower resistivities than the other metals mentioned.

Waveguides are hollow metal pipes and can have either circular or rectangular cross sections (although the rectangular are, by far, the more common). Figure 20.2 is a cross-sectional view of a typical rectangular waveguide; *a* is the wider *interior* dimension, and *b* is the narrower. These letters are considered the standard form of notation for waveguide dimensions and are used throughout this chapter.

Rectangular Waveguide Operation

One way of visualizing how a waveguide works is to develop the theory of waveguides from the theory of elementary parallel-wire transmission lines. Figure 20.3A shows the basic parallel transmission line that was introduced in Chap. 4. The line consists of two parallel conductors separated by an air dielectric. Because air won't support the conductors, spaced ceramic or other rigid insulators are used as supports.

There are several reasons why the parallel transmission line per se is not used at microwave frequencies:

- Skin effect increases ohmic losses to an unacceptable level.

- Insulators supporting the two conductors are significantly more lossy at microwave frequencies than at lower frequencies.

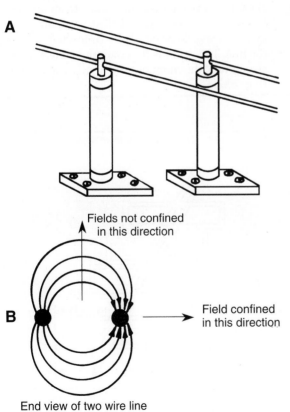

End view of two wire line

FIGURE 20.3 Parallel transmission line and fields.

- Radiation losses increase dramatically because the spacing between the wires is a greater fraction of a wavelength. Figure 20.3B shows the electric fields surrounding the conductors. The fields add algebraically (either constructively or destructively), resulting in pinching of the resultant field along one axis and reinforcement along the other.

Now consider a quarter-wavelength shorted stub. As we saw in Chap. 4, a short circuit is transformed to an open circuit by a quarter-wavelength of any transmission line. Thus, the "looking-in" impedance of such a stub is infinite, so when it is connected in parallel across a transmission line (Fig. 20.4A) the stub has no electrical effect on passing waves for which the stub length is $\lambda/4$. In other words, at its resonant frequency, the stub is a *metallic insulator* and can be used to physically support the transmission line.

Similarly, we can connect a second $\lambda/4$ stub in parallel with the first across the same points on the transmission line (Fig. 20.4B) without loading down the line. This arrangement effectively forms a half-wavelength pair. The impedance is still infinite, so no harm is done. But if we can do this at one point along the transmission line, we can do it at *many* (Fig. 20.4C). Ultimately, if we do it everywhere, we have totally surrounded the original parallel-wire transmission line with a metal skin! The waveguide is analogous to an infinite number of center-fed "half-wave pairs" of quarter-wave shorted stubs connected across the line. The result is the continuous metal pipe structure of the common rectangular waveguide (Fig. 20.2).

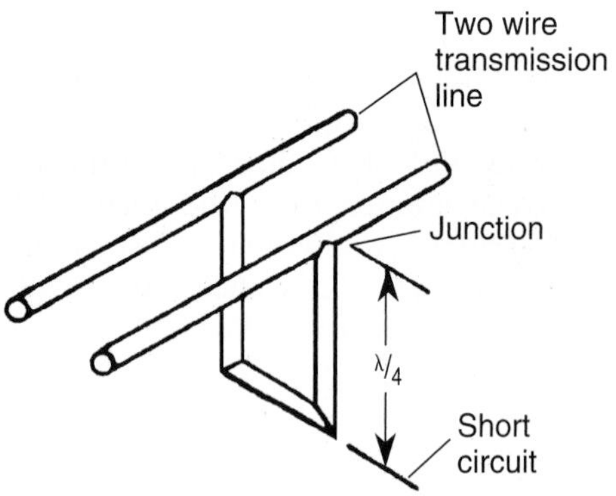

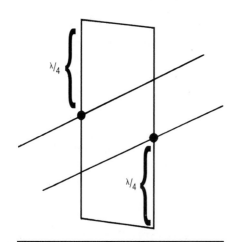

Figure 20.4A   Quarter-wave stub analogy.

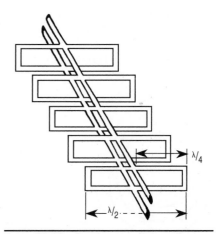

Figure 20.4B   Quarter-wave stub analogy extended.

Figure 20.4C   Quarter-wave stub analogy extended even further.

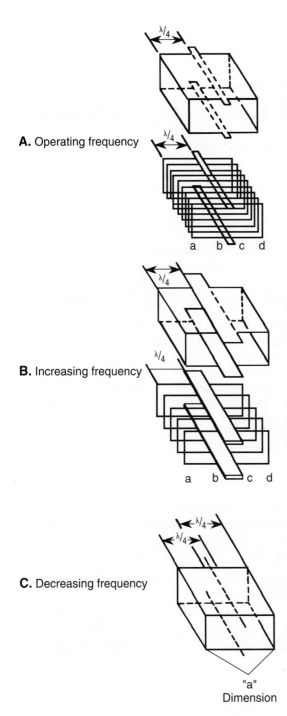

A. Operating frequency

B. Increasing frequency

C. Decreasing frequency

"a"
Dimension

Figure 20.5 Changing frequency does not affect the analogy.

On first glance, an explanation based on λ/4 shorted stubs would seem to be valid only at one frequency. It turns out, however, that the analogy also holds up at other frequencies so long as the frequency is higher than a certain *minimum cutoff frequency*. The waveguide thus acts like a high-pass filter. Waveguides also have an upper frequency limit; between the upper and lower limits, waveguides support a bandwidth of 30 to 40 percent of the cutoff frequency. As shown in Fig. 20.5, between segments the centerline (which represents the conductors in the parallel line analogy) of the waveguide becomes a "shorting bar" that widens as the operating frequency gets higher and the λ/4 stub on each side gets shorter or narrows as the operating frequency becomes lower and the λ/4 stub lengthens.

At the cutoff frequency, the fraction of the *a* dimension that acts as though it is a transmission line conductor is a minimum; in other words, the back-to-back λ/4 stubs virtually touch at the midpoint of *a*. At this frequency, $a = \lambda/2$. Above that frequency, the effective or apparent width of the conductor increases and $\lambda/2 < a$. Below the cutoff frequency, the chamber ceases to function as a waveguide; instead, it acts like a conventional parallel transmission line with a pure reactance connected across the two conductors. Thus, the (low-frequency) cutoff frequency is defined as the frequency at which the *a* dimension is less than λ/2.

Propagation Modes in Waveguides

Whether in a conventional transmission line or a microwave waveguide, the signal propagates as an electromagnetic wave, not as a longitudinal current. The transmission line supports a *transverse electromagnetic* (TEM) field. As explained in Chap. 3, the word *transverse* indicates that the wave is propagating at right angles to both the electric and the magnetic fields. In addition to the word *transverse* the

E- and H-fields are said to be "normal" or "orthogonal" to the direction of travel—three different ways of saying the same thing: right-angledness.

Boundary Conditions

In contrast, a TEM wave will not propagate in a waveguide because different constraints, called *boundary conditions,* apply. Although the wave in the waveguide propagates through the air (or inert gas dielectric) in a manner similar to free-space propagation, the phenomenon is bounded by the walls of the waveguide—clearly a far different situation than for a TEM wave in free space! The boundary conditions for waveguides are these:

- The electric field must be orthogonal to the conductor in order to exist at the surface of that conductor.

- The magnetic field must not be orthogonal to the surface of the waveguide.

In order to satisfy these boundary conditions the waveguide supports two types of propagation modes: *transverse electric mode* (TE mode) and *transverse magnetic mode* (TM mode). The TEM mode for radio waves propagating through free space violates the boundary conditions because the magnetic field is not parallel to the surface and so does not occur in waveguides.

The transverse electric field requirement means that the E-field must be perpendicular to the conductor wall of the waveguide. This requirement is met by use of a proper coupling scheme at the input end of the waveguide. A vertically polarized coupling radiator will provide the necessary transverse field.

One boundary condition requires that the magnetic (H) field must not be orthogonal to the conductor surface. An H-field that is at right angles to the E-field (which *is* orthogonal to the conductor surface) meets this requirement (see Fig. 20.6). The planes formed by the magnetic field are parallel to both the direction of propagation and the wide surface dimension of the waveguide.

As the wave propagates away from the input radiator, it resolves into two components that are not along the axis of propagation and are not orthogonal to the walls. The component along the waveguide axis violates the boundary conditions, so it is rapidly attenuated. For the sake of simplicity, only one component is shown in Fig. 20.7. Three cases are shown in Fig. 20.7A, B, and C, respectively: high, medium, and low frequency.

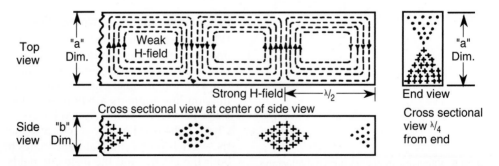

Figure 20.6 Magnetic fields in waveguide.

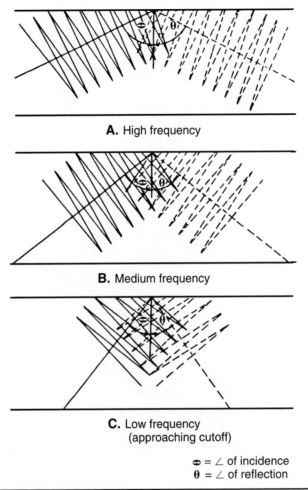

A. High frequency

B. Medium frequency

C. Low frequency
(approaching cutoff)

⊕ = ∠ of incidence
θ = ∠ of reflection

Figure 20.7 Frequency effect on propagating wave.

Note that the angle of incidence with the waveguide wall increases as frequency drops. The angle rises toward 90 degrees as the cutoff frequency is approached from above. Below the cutoff frequency the angle is 90 degrees, so the wave bounces back and forth between the walls without propagating.

Coordinate System and Dominant Mode in Waveguides

Figure 20.8 shows the coordinate system used to denote dimensions and directions in microwave discussions. The *a* and *b* dimensions of the waveguide correspond to the *x* and *y* axes of a cartesian coordinate system, and the *z* axis (into or out of the page) is the direction of wave propagation.

Shorthand notation for describing the various modes of propagation is as follows:

$$Tx_{mn}$$

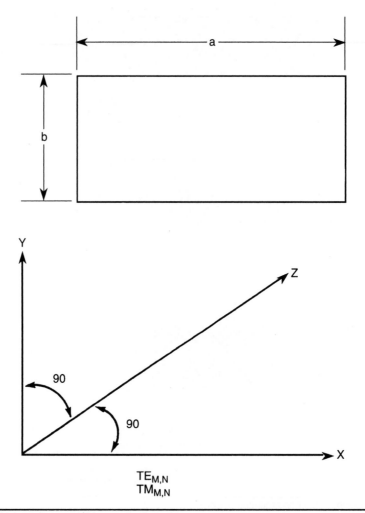

FIGURE 20.8 Rectangular waveguide coordinate system.

where $x = E$ for transverse electric mode and M for transverse magnetic mode
m = number of half-wavelengths along x axis (a dimension)
n = number of half-wavelengths along y axis (b dimension)

The TE_{10} mode is called the *dominant mode and is the best mode for low-attenuation propagation along the z axis.* The nomenclature TE_{10} indicates that there is one half-wavelength in the *a* dimension and zero half-wavelengths in the *b* dimension. The dominant mode exists at the lowest frequency at which the waveguide is a half-wavelength.

Velocity and Wavelength in Waveguides

Figures 20.9A and 20.9B show the geometry for two wave components, simplified for the sake of illustration. There are three different wave velocities to consider with respect to waveguides:

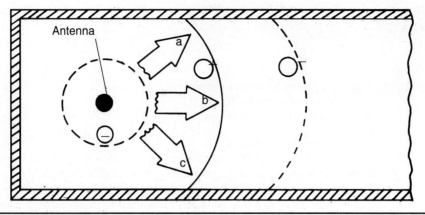

FIGURE 20.9A   Antenna radiator in capped waveguide.

- Free-space velocity c
- Group velocity v_g
- Phase velocity v_p

Of course, c is the velocity of propagation in unbounded free space, 3×10^8 m/s.

The *group velocity* is the straight-line velocity of propagation of the wave down the centerline (z axis) of the waveguides. The value of V_g is always less than c, because the actual path length taken, as the wave bounces back and forth, is longer than the straight-

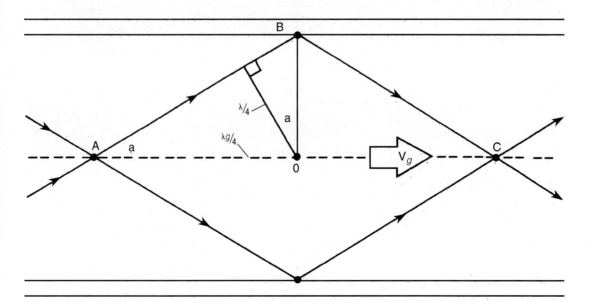

FIGURE 20.9B   Wave propagation in waveguide.

line path (i.e., path ABC in Fig. 20.9B is longer than path AC). The relationship between c and V_g is

$$v_g = c \sin a \qquad (20.1)$$

where v_g = group velocity, in meters per second
  c = free-space velocity (3×10^8 m/s)
  a = angle of incidence in waveguide

Clearly, v_g can never be greater than c.

The *phase velocity* is the velocity of propagation of the spot on the waveguide wall where the wave impinges (e.g., point B in Fig. 20.9B). Depending upon the angle of incidence, this velocity can actually be faster than both the group velocity and the speed of light! How can that be? Let's look at an analogy. Consider an ocean beach, on which the waves arrive from offshore at an angle other than 90 degrees—in other words, the arriving wavefronts are not parallel to the shore. The arriving waves have a group velocity v_g. But as a wave hits the shore, it will strike a point far down the beach first. Then the "point of strike" races up the beach at a much faster phase velocity v_p, which is even faster than the group velocity. Similarly, in a microwave waveguide, the phase velocity can be greater than c, as can be seen from Eq. (20.2):

$$v_p = \frac{c}{\sin a} \qquad (20.2)$$

Example 20.1   Calculate the group and phase velocities for an angle of incidence of 33 degrees.

Solution

(a) Group velocity

$$v_g = c \sin a$$

$$= (3 \times 10^8)(\sin 33°)$$

$$= (3 \times 10^8)(0.5446) = 1.6 \times 10^8 \text{ m/s}$$

(b) Phase velocity

$$v_p = c / \sin a$$

$$= (3 \times 10^8 \text{ m/s}) / \sin 33°$$

$$= (3 \times 10^8 \text{ m/s}) / (0.5446)$$

$$= 5.51 \times 10^8 \text{ m/s}$$

For this problem the solutions are

$$c = 3 \times 10^8 \text{ m/s}$$
$$v_p = 5.51 \times 10^8 \text{ m/s}$$
$$v_g = 1.6 \times 10^8 \text{ m/s}$$

◆

We can also write a relationship between all three velocities by combining Eqs. (20.1) and (20.2), resulting in

$$c = \sqrt{v_p v_g} \tag{20.3}$$

In any wave the product of *frequency* and *wavelength* is the *velocity*. Thus, for a TEM wave in unbounded free space we know that

$$c = f\lambda_{fs} \tag{20.4}$$

Because the frequency f is fixed by the generator and the waveguide is a linear medium, only the wavelength can change when the velocity changes. In a microwave waveguide we can relate phase velocity to wavelength as the wave is propagated in the waveguide:

$$v_p = \frac{\lambda \times c}{\lambda_{fs}} \tag{20.5}$$

where v_p = phase velocity, in meters per second
c = free-space velocity (3×10^8 m/s)
λ = wavelength in waveguide, in meters
λ_{fs} = wavelength in free space (c/f), in meters (see Eq. [20.4])

Equation (20.5) can be rearranged to find the wavelength in the waveguide:

$$\lambda = \frac{v_p \lambda_{fs}}{c} \tag{20.6}$$

Example 20.2 A 5.6-GHz microwave signal is propagated in a waveguide. Assume that the internal angle of incidence to the waveguide surfaces is 42 degrees. Calculate phase velocity, wavelength in unbounded free space, and wavelength in the waveguide.

Solution

(a) Phase velocity

$$v_p = \frac{c}{\sin a}$$

$$= \frac{3 \times 10^8 \text{ m/s}}{\sin 42°}$$

$$= \frac{3 \times 10^8 \text{ m/s}}{0.6991}$$

$$= 4.5 \times 10^8 \text{ m/s}$$

(b) Wavelength in free space

$$\lambda_{fs} = c/f$$

$$= (3 \times 10^8 \text{ m/s})/(5.6 \times 10^9 \text{ Hz})$$

$$= 0.054 \text{ m}$$

(c) Wavelength in waveguide

$$\lambda = \frac{v_p \lambda_o}{c}$$

$$= \frac{(4.5 \times 10^8 \text{ m/s})(0.054 \text{ m})}{3 \times 10^8 \text{ m/s}} = 0.08 \text{ m}$$

Comparing, we find that the free-space wavelength is 0.054 m, while the wavelength inside of the waveguide has increased to 0.08 m.

◆

Cutoff Frequency (f_c)

The propagation of signals in a waveguide is a function of the frequency of the applied signal. As the frequency drops, the angle of incidence increases toward 90 degrees. Indeed, both phase and group velocities are functions of the angle of incidence. When the frequency drops to a point where the angle of incidence is 90 degrees, then group velocity is meaningless and propagation ceases. This occurs at the low frequency cutoff.

We can define a general mode equation based on our system of notation:

$$\frac{1}{(\lambda_c)^2} = \left(\frac{m}{2a}\right)^2 + \left(\frac{n}{2b}\right)^2 \tag{20.7}$$

where λ_c = longest wavelength that will propagate
a, b = waveguide dimensions (see Fig. 20.2)
m, n = integers defining number of half-wavelengths that will fit in a and b dimensions, respectively

Evaluating Eq. (20.7) reveals that the longest TE-mode signal that will propagate in the dominant mode (TE$_{10}$) is given by

$$\lambda_c = 2a \tag{20.8}$$

from which we can write an expression for the cutoff frequency:

$$f_c = \frac{c}{2a} \tag{20.9}$$

where f_c = lowest frequency that will propagate, in hertz
c = speed of light (3×10^8 m/s)
a = wider of the two waveguide dimensions

Example 20.3 A rectangular waveguide has dimensions of 3×5 cm. Calculate the TE$_{10}$ mode cutoff frequency.

Solution

$$f_c = \frac{c}{2a}$$

$$= \frac{(3 \times 10^8 \text{ m/s}}{(2)\left(5 \text{ cm} \times \dfrac{1 \text{ m}}{100 \text{ cm}}\right)}$$

$$= \frac{3 \times 10^8 \text{ m/s}}{(2)(0.05 \text{ m})}$$

$$= 3 \text{ GHz}$$

◆

Equation (20.7) assumes that the dielectric inside the waveguide is air. A more generalized form, which can accommodate other dielectrics, is

$$f_c = \frac{1}{2\sqrt{\mu\varepsilon}} \sqrt{\left(\frac{m}{a}\right)^2 + \left(\frac{n}{b}\right)^2} \tag{20.10}$$

where ε = dielectric constant
μ = permeability constant

For air dielectrics, $\mu = \mu_0$ and $\varepsilon = \varepsilon_0$, from which

$$c = \frac{1}{\sqrt{\mu_0\varepsilon_0}} \tag{20.11}$$

To determine the cutoff wavelength, we can rearrange Eq. (20.10) to the form:

$$\lambda_C = \frac{2}{\sqrt{\left(\frac{m}{a}\right)^2 + \left(\frac{n}{b}\right)^2}} \tag{20.12}$$

One further expression for air-filled waveguide calculates the actual wavelength in the waveguide from a knowledge of the free-space wavelength and actual operating frequency:

$$\lambda_g = \frac{\lambda_0}{\sqrt{1 - \left(\frac{f_c}{f}\right)^2}} \tag{20.13}$$

where λ_g = wavelength in waveguide
λ_0 = wavelength in free space
f_c = waveguide cutoff frequency
f = operating frequency

Example 20.4   A waveguide with a 4.5-GHz cutoff frequency is excited with a 6.7-GHz signal. Find (*a*) the wavelength in free space and (*b*) the wavelength in the waveguide.

Solution

(*a*)

$$\lambda_o = c / f$$

$$= \frac{3\times10^8 \text{ m / s}}{6.7 \text{ GHz} \times \dfrac{10^9 \text{ Hz}}{1 \text{ GHz}}}$$

$$= \frac{3 \times 10^8 \text{ m/s}}{6.7 \times 10^9 \text{ Hz}}$$

$$= 0.0448 \text{ m}$$

(b)

$$\lambda_g = \frac{\lambda_0}{\sqrt{1 - \left(\dfrac{f_c}{f}\right)^2}}$$

$$= \frac{0.0448 \text{ m}}{\sqrt{1 - \left(\dfrac{4.5 \text{ GHz}}{6.7 \text{ GHz}}\right)^2}}$$

$$= \frac{0.0448 \text{ m}}{1 - 0.67}$$

$$= \frac{0.0448}{0.33}$$

$$= 0.136 \text{ m}$$

Transverse magnetic modes also propagate in waveguides, but the base TM_{10} mode is excluded by the boundary conditions. Thus, the TM_{11} mode is the lowest magnetic mode that will propagate.

Waveguide Impedance

All forms of transmission line, including the waveguide, exhibit a characteristic impedance, although in the case of waveguide it is a little difficult to pin down conceptually. The characteristic impedance of ordinary two-conductor transmission lines was developed in Chap. 4. For a waveguide, the characteristic impedance is approximately equal to the ratio of the electric and magnetic fields (*E/H*) and converges (as a function of frequency) to the intrinsic impedance of the dielectric (Fig. 20.10). The impedance of the waveguide is a function of waveguide characteristic impedance (Z_0) and the wavelength in the waveguide:

$$Z = \frac{Z_0 \lambda_g}{\lambda_0} \tag{20.14}$$

Or, for rectangular waveguide, with constants taken into consideration:

$$Z = \frac{120 \pi \lambda_g}{\lambda_0} \tag{20.15}$$

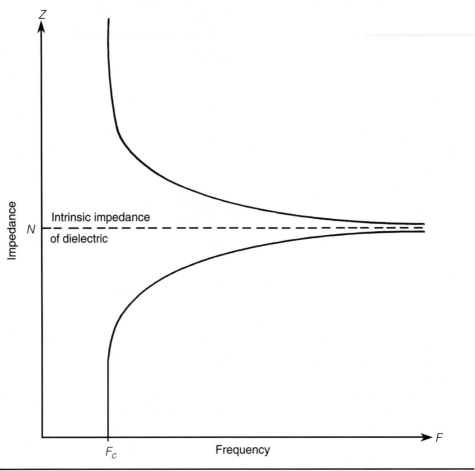

FIGURE 20.10 Impedance versus frequency.

The *propagation constant* β for rectangular waveguide is a function of both cutoff frequency and operating frequency:

$$\beta = \omega\sqrt{\mu\varepsilon}\,\sqrt{1-\left(\frac{f_c}{f}\right)^2} \qquad (20.16)$$

from which we can express the TE-mode impedance:

$$Z_{\text{TE}} = \frac{\sqrt{\mu\varepsilon}}{\sqrt{1-\left(\frac{f_c}{f}\right)^2}} \qquad (20.17)$$

and the TM-mode impedance:

$$Z_{TM} = 377 \sqrt{1 - \left(\frac{f_c}{f}\right)^2}$$ (20.18)

Waveguide Terminations

When an electromagnetic wave propagates along a waveguide, it must eventually reach the end of the guide. If the end is open, then the wave will propagate into free space. The horn radiator is an example of an unterminated waveguide. If the waveguide terminates in a metallic wall, then the wave reflects back down the waveguide, from whence it came. The interference between incident and reflected waves forms standing waves (see Chap. 4). Such waves are stationary in space but vary in the time domain.

In order to prevent standing waves or, more properly, the reflections that give rise to standing waves, the waveguide must be *terminated* in a matching impedance. When a properly designed antenna is used to terminate the waveguide, it forms the matched load required to prevent reflections. Otherwise, a *dummy load* must be provided. Figure 20.11 shows several types of dummy load.

The classic termination is shown in Fig. 20.11A. The "resistor" making up the dummy load is a mixture of sand and graphite. When the fields of the propagated wave enter the load, they cause currents to flow, which in turn cause heating. Thus, the RF power dissipates in the sand-graphite mixture rather than being reflected back down the waveguide.

A second type of dummy load is shown in Fig. 20.11B. The resistor element is a carbonized rod critically placed at the center of the electric field. The E-field causes currents to flow, resulting in I^2R losses that dissipate the power.

Bulk loads, similar to the graphite-sand chamber, are shown in Fig. 20.11C, D, and E. Using bulk material such as graphite or a carbonized synthetic material, these loads are used in much the same way as the sand load (i.e., currents set up, and I^2R losses dissipate the power).

The resistive vane load is shown in Fig. 20.11F. The plane of the element is orthogonal to the magnetic lines of force. When the magnetic lines cut across the vane, currents are induced, which gives rise to the I^2R losses. Very little RF energy reaches the metallic end of the waveguide, so there is little reflected energy and a low VSWR.

There are situations where it isn't desirable to terminate the waveguide in a dummy load. Several reflective terminations are shown in Fig. 20.12. Perhaps the simplest form is the permanent end plate shown in Fig. 20.12A. The metal cover must be welded or otherwise affixed through a very low resistance joint. At the substantial power levels typically handled in transmitter waveguides, even small resistances can be important.

The end plate (shown in Fig. 20.12B) uses a quarter-wavelength cup to reduce the effect of joint resistances. The cup places the contact joint at a point that is a quarter-wavelength from the end. This point is a minimum-current node, so I^2R losses in the contact resistance become less important.

The adjustable short circuit is shown in Fig. 20.12C. The walls of the waveguide and the surface of the plunger form a half-wavelength channel. Because the metallic end of the channel is a short circuit, the impedance reflected back to the front of the plunger is

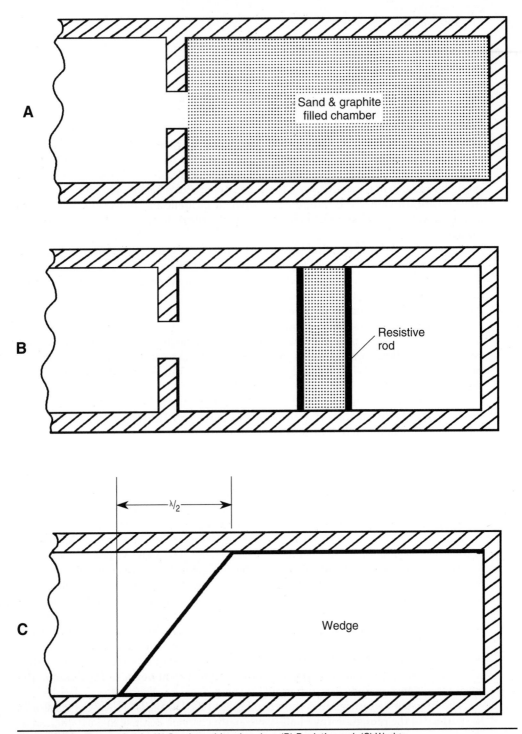

FIGURE 20.11 Dummy loads. (A) Sand-graphite chamber. (B) Resistive rod. (C) Wedge.

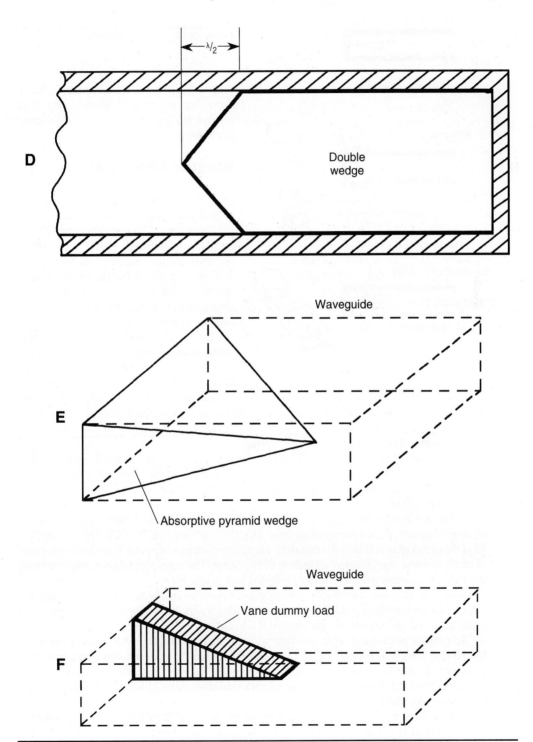

FIGURE 20.11   Dummy loads. (D) Double wedge. (E) Pyramid wedge. (F) Vane dummy load.

A

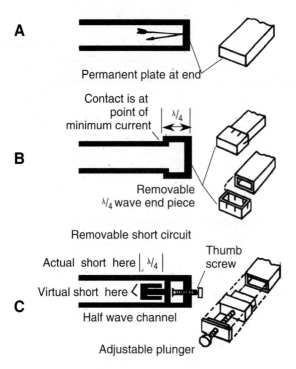

Permanent plate at end

Contact is at point of minimum current $\left|\overset{\lambda/4}{\longleftrightarrow}\right|$

B

Removable $\lambda/4$ wave end piece

Removable short circuit

Actual short here $\left|\overset{\lambda/4}{}\right|$

Virtual short here

C

Thumb screw

Half wave channel

Adjustable plunger

FIGURE 20.12 End terminations.

0 Ω, or nearly so. Thus, a virtual short exists at the points shown. By this means, the contact (or joint) resistance problem is overcome.

Waveguide Joints and Bends

Joints and bends in any form of transmission line or waveguide are seen as impedance discontinuities, and so are points at which disruptions occur. Thus, improperly formed bends and joints are substantial contributors to a poor VSWR. In general, bends, twists, joints, or abrupt changes in waveguide dimension can deteriorate the VSWR by giving rise to reflections.

Extensive runs of waveguide are sometimes difficult to make in a straight line. Although some installations do permit a straight waveguide, many others require directional change. This possibility is especially likely on shipboard installations. Figure 20.13A shows the proper ways to bend a waveguide around a corner. In each case, the radius of the bend must be at least two wavelengths at the lowest frequency that will be propagated in the system.

The "twist" shown in Fig. 20.13B is used to rotate the polarity of the E- and H-fields by 90 degrees. This type of section is sometimes used in antenna arrays for phasing the elements. As in the case of the bend, the twist must be made over a distance of at least two wavelengths.

When an abrupt 90-degree transition is needed, it is better to use two successive 45-degree bends spaced one quarter-wavelength apart (see Fig. 20.13C). The theory (behind this kind of bend) is to deliberately cause interference between the direct reflection of one bend and the inverted reflection of the other. The resultant relationship between the fields is reconstructed as if no reflections had taken place.

Joints are necessary in practical waveguides because it simply isn't possible to construct a single length of guide that is practical for all situations. Three types of common joints are used: *permanent*, *semipermanent*, and *rotating*.

To make a permanent joint, the two waveguide ends must be machined extremely flat so that they can be butt-fitted together. A welded or brazed seam bonds the two sections together. Because such a surface represents a tremendous discontinuity, reflections and VSWR will result unless the interior surfaces are milled flat and then polished to a mirrorlike finish.

A *semipermanent joint* allows the joint to be disassembled for repair and maintenance, as well as allowing easier on-site assembly. The most common example of this class is the *choke joint* shown in Fig. 20.14.

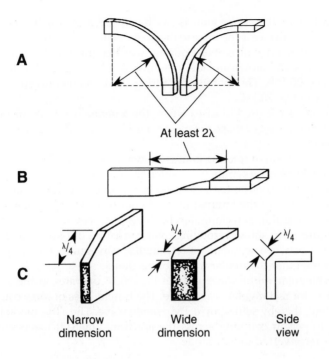

At least 2λ

Narrow
dimension

Wide
dimension

Side
view

FIGURE 20.13   Bends in waveguide must be gentle.

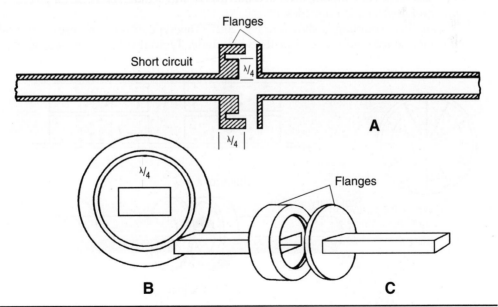

Flanges

Short circuit

$\lambda/4$

$\lambda/4$

A

$\lambda/4$

Flanges

B

C

FIGURE 20.14   Choke joint.

One surface of the choke joint is machined flat and is a simple butt-end planar flange. The other surface is the mate to the planar flange, but it has a quarter-wavelength circular slot cut at a distance of one quarter-wavelength from the waveguide aperture. The two flanges are shown in side view in Fig. 20.14A, and the slotted end view is shown in Fig. 20.14B. The method for fitting the two ends together is shown in the oblique view in Fig. 20.14C.

Rotating joints are used in cases where the antenna has to point in different directions at different times. Perhaps the most common example of such an application is the radar antenna.

The simplest form of rotating joint is shown in Fig. 20.15. The key to its operation is that the selected mode is symmetrical about the rotating axis. For this reason, a circular waveguide operating in the TM_{01} mode is chosen. In this rotating choke joint, the actual waveguide rotates but the internal fields do not (thereby minimizing reflections). Because most waveguide is rectangular, however, a somewhat more complex system is needed. Figure 20.16 shows a rotating joint consisting of two circular waveguide sections inserted between segments of rectangular waveguide. On each end of the joint, there is a rectangular-to-circular transition section.

The rectangular input waveguide of Fig. 20.16 operates in the TE_{10} mode that is most efficient for rectangular waveguide. The E-field lines of force couple with the circular segment, thereby setting up a TM_{01} mode wave. The TM_{01} mode has the required symmetry to permit coupling across the junction, where it meets another transition zone and is reconverted to TE_{10} mode.

Waveguide Coupling Methods

Except possibly for the case where an oscillator exists inside a waveguide, it is necessary to have some form of input or output coupling in a waveguide system. There are three basic types of coupling used in a microwave waveguide: *capacitive* (or *probe*), *inductive* (or *loop*), and *aperture* (or *slot*).

Capacitive coupling is shown in Fig. 20.17. This type of coupling uses a vertical radiator inserted into one end of the waveguide. Typically, the probe is a quarter-

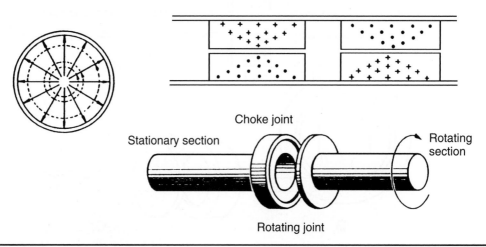

Choke joint

Stationary section

Rotating section

Rotating joint

FIGURE 20.15 Basic rotating joint.

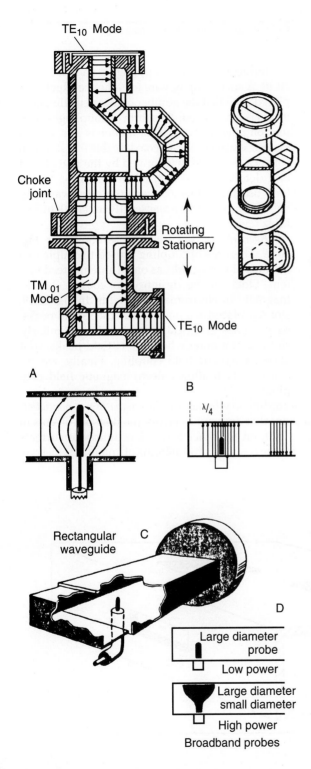

A

B

Rectangular
waveguide   C

D

Large diameter
probe

Low power

Large diameter
small diameter

High power

Broadband probes

FIGURE 20.16   Representative practical
rotating joint.

wavelength in a fixed-frequency system. The probe is analogous to the vertical monopoles employed at lower frequencies. A characteristic of this type of radiator is that the E-field is parallel to the waveguide top and bottom surfaces. This arrangement satisfies the first boundary condition for the dominant TE_{10} mode.

The radiator is placed at a point that is a quarter-wavelength from the rear wall (Fig. 20.17B). By traversing the quarter-wave distance (90-degree phase shift), reflecting off the rear wall (180-degree phase shift), and then retraversing the quarter-wavelength distance (another 90-degree phase shift), the wave undergoes a total phase shift of one complete cycle, or 360 degrees. Thus, the reflected wave arrives back at the radiator in phase to reinforce the outgoing wave, and none of the excitation energy is lost.

Some waveguides have an adjustable end cap (Fig. 20.17C) to accommodate multiple frequencies. The end cap position is varied as required for signals of different wavelengths.

Figure 20.17D shows high- and low-power broadband probes that are typically not a quarter-wavelength except at one particular frequency. Broadbanding is accomplished by attention to the diameter-to-length ratio. The *degree of coupling* can be varied in any of several ways: the *length* of the probe can be varied, the *position* of the probe in the E-field can be changed, or *shielding* can be used to partially shade the radiator element.

FIGURE 20.17   Probe (capacitive) coupling.

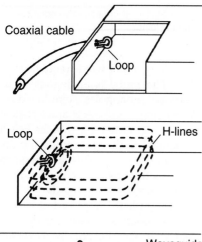

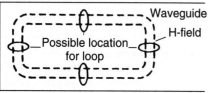

FIGURE 20.18 Loop (inductive) coupling.

Inductive, or loop coupling, is shown in Fig. 20.18. A small loop of wire (or other conductor) is placed such that the number of magnetic flux lines it cuts is maximized. This form of coupling is popular on microwave receiver antennas as a way of making a waveguide-to-coaxial cable transition. In some cases, the loop is formed by the pigtail lead of a detector diode that, when combined with a local oscillator, downconverts the microwave signal to an *intermediate frequency* (IF) in the 30- to 300-MHz region.

Aperture, or *slot*, *coupling* is shown in Fig. 20.19. This type of coupling is used to couple two sections of waveguide, as on an antenna feed system. Slots can be designed to couple electric, magnetic, or electromagnetic fields. In Fig. 20.19, slot A is placed at a point where the E-field peaks, so it allows electrical field coupling. Similarly, slot B is at a point where the H-field peaks, so it allows magnetic field coupling. Finally, we see slot C, which allows electromagnetic field coupling.

Slots can also be characterized according to whether they are *radiating* or *nonradiating*. A nonradiating slot is cut at a point that does not interrupt the flow of currents in the waveguide walls. The radiating slot, on the other hand, does interrupt currents flowing in the walls. A radiating slot is the basis for several forms of antenna, which are discussed at the end of this chapter.

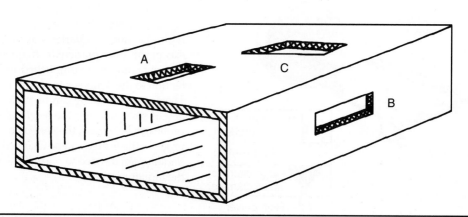

FIGURE 20.19 Slot coupling.

Microwave Antennas

Antennas are used in communications and radar systems over a phenomenally wide range of radio frequencies. In both theory and practice, antennas are used until operating frequencies reach infrared and visible light, at which point optical techniques take over. Microwaves are the transition region between ordinary "radio waves" and "optical waves", so (as might be expected) microwave technology makes use of techniques from both worlds. For example, both dipoles and parabolic reflectors are used in microwave systems.

The purpose of an antenna is to act as a *transducer*, converting signals propagating on the two conductors of a conventional transmission line or "bouncing off the walls" inside a waveguide to an electromagnetic wave propagating in free space. In the process, the antenna also acts as an *impedance matcher* between the waveguide or transmission line impedance and the impedance of free space.

Antennas can be used equally well for both receiving and transmitting signals because they obey the *law of reciprocity*. That is, the same antenna can be used to receive and transmit with equal success. Although there might be practical or mechanical reasons to prefer specific antennas for one or the other mode, electrically they are the same.

In the transmit mode, the antenna must radiate electromagnetic energy. For this job, the important property is *gain G*. In the receive mode, the job of the antenna is to gather energy from impinging electromagnetic waves in free space. The important property for receiving antennas is the *effective aperture A_e*, which is a function of the antenna's physical area. Reciprocity suggests that large gain goes hand in hand with a large effective aperture. *Effective aperture* is defined as the area of the impinging radio wavefront that contains the same power as is delivered to a matched resistive load across the feedpoint terminals.

The Isotropic "Antenna"

Antenna definitions and specifications can become useless unless a means is provided for putting everything on a common footing. Although a variety of systems exist for describing antenna behavior, the most common system compares a specific antenna with a theoretical construct, called the *isotropic radiator*, which we first encountered in Chap. 3.

Since an isotropic radiator is a spherical point source that radiates equally well in all directions, the directivity of the isotropic antenna is unity (1) by definition, and all other antenna gains are measured against this standard. From spherical geometry, we can calculate isotropic power density at any distance R from the point source:

$$P_d = \frac{P}{4\pi r^2} \tag{20.19}$$

where P_d = power density, in watts per square meter
   P = power in watts input to the isotropic radiator
   r = radius in meters at which point power density is measured

Example 20.5   Calculate the power density at a distance of 1 km (1000 m) from a 1000W isotropic source.

Solution

$$P_d = \frac{P}{4\pi r^2}$$

$$= \frac{(1000\,\text{W})}{4\pi\,(1000\,\text{m})^2}$$

$$= 7.95 \times 10^{-5}\ \text{W}/\text{m}^2$$

◆

The rest of this chapter expresses antenna gains and directivities relative to isotropic radiators.

Near Field and Far Field

Antennas are defined in terms of *gain* and *directivity*, both of which are measured by examining the radiated field of the antenna. Published antenna patterns usually report only far-field performance. In free space, the far field for most antennas falls off according to the *inverse square law*. That is, the intensity falls off according to the square of the distance $(1/r^2)$, as in Eq. (20.19).

The *near field* of the antenna contains more energy than the far field because of its proximity to the antenna radiator element, but it diminishes very rapidly with increasing distance according to a $1/r^4$ function. The minimum distance to the edge of the near field is a function of both the wavelength of the radiated signals and the antenna dimensions:

$$r_{\text{min}} = \frac{2d^2}{\lambda} \tag{20.20}$$

where r_{min} = near-field distance
    d = largest antenna dimension
    λ = wavelength of radiated signal (all factors in same units)

Example 20.6   An antenna with a length of 6 cm radiates a 12-cm wavelength signal. Calculate the near-field distance.

Solution

$$r_{\text{min}} = \frac{2d^2}{\lambda}$$

$$= \frac{(2)\,(6\,\text{cm})^2}{12\,\text{cm}}$$

$$= \frac{72}{12}$$

$$= 6\,\text{cm}$$

◆

Antenna Impedance

Impedance is a measure of device or system opposition to the flow of alternating current (e.g., RF); in the general case it is *complex*; that is, it includes both resistive and reactive components. The reactive components can be either capacitive or inductive, or a combination of both. Impedance can be expressed in either of two notations.

The magnitude can be obtained from

$$Z = \sqrt{R^2 + (X_L - X_c)^2} \tag{20.21}$$

Alternatively, in complex plane notation, where pure reactances lie along the imaginary axis,

$$Z = R \pm jX \tag{20.22}$$

Of these, Eq. (20.22) is more useful in RF applications because it provides both *magnitude* and *phase* information. This is especially important when dealing with the antenna's near-field performance because energy transfer in the near field is highly reactive, consisting predominantly of large circulating amounts of electric and magnetic energy (close to the radiator) being exchanged with the radiating antenna throughout the course of each and every cycle.

As discussed in earlier chapters dealing with antennas for lower frequencies, the resistive part of an antenna's impedance consists of two elements: *ohmic losses* R_Ω and *radiation resistance* R_{RAD}. The ohmic losses are due to heating of the antenna conductor elements by RF current passing through, as when current passes through any conductor. Efficiency ξ is then:

$$\xi = \frac{R_{RAD}}{R_{RAD} + R_\Omega} \tag{20.23}$$

A major goal of the antenna designer—for transmitting antennas, at least—is to maximize ξ by minimizing R_Ω and by developing an antenna design that results in a value for R_{RAD} that is as large as is practical.

Dipole Antenna Elements

The *dipole* can be modeled as either a single radiator fed at the center (Fig. 20.20A) or as a pair of radiators fed back to back (Fig. 20.20B). As discussed in Chaps. 3 and 6 in particular, by definition the polarization of an electromagnetic field is the direction of the electrical field vector. Since the dipole's far E-field is parallel to the radiating element, a horizontal dipole produces a horizontally polarized signal, while a vertical element produces a vertically polarized signal.

A microwave dipole is shown in Fig. 20.21. The antenna radiating element consists of a short conductor at the end of a section of waveguide. Although most low-frequency dipoles are a half-wavelength, microwave dipoles might be exactly a half-wavelength, less than a half-wavelength, or greater than a half-wavelength, depending upon the application. For example, because most microwave dipoles are used to illuminate a reflector of some sort, the length of the dipole depends upon the exact illumination function required for proper operation of the reflector. Most, however, will be a half-wavelength.

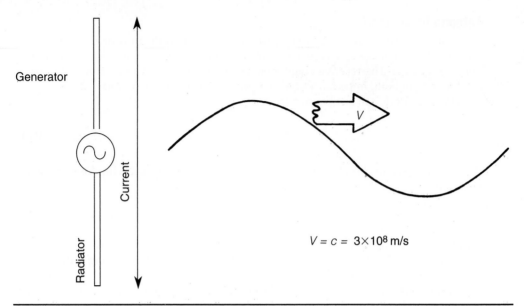

FIGURE 20.20A Basic dipole antenna showing propagation.

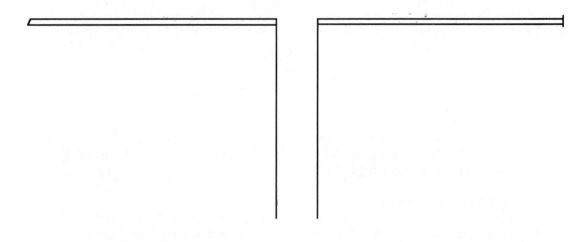

FIGURE 20.20B Basic dipole antenna.

Directivity

The directivity of an antenna is a measure of its ability to concentrate RF energy in a subset of directions rather than in all (spherical) directions equally. Two methods for quantifying unidirectional antenna patterns are shown in Fig. 20.22. Figure 20.22A is a polar plot viewed from above, oriented such that the main lobe is centered on 0 degrees. The plot of Fig. 20.22B is a rectangular method for displaying the same information. Typical field patterns follow a $(\sin x)/x$ function or, for power, $[(\sin x)/x]^2$.

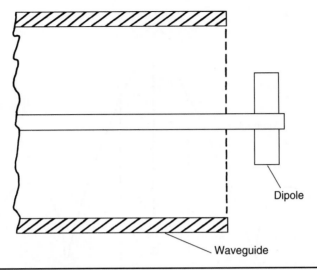

FIGURE 20.21 Microwave dipole radiator.

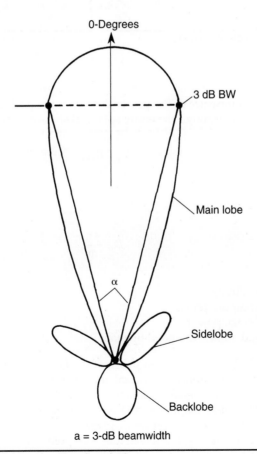

FIGURE 20.22A Directional antenna pattern (top view).

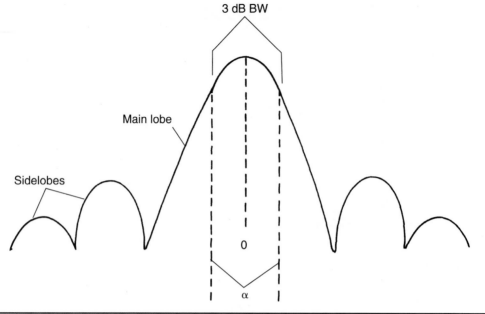

Figure 20.22B Graphically presented pattern.

Directivity D is a measure of the power density at the peak of the main lobe to the average power density through the entire spherical surface surrounding the antenna:

$$D = \frac{P_{MAX}}{P_{AVG}} \tag{20.24}$$

or, referred to isotropic,

$$D = \frac{4\pi}{\Phi} \tag{20.25}$$

where D = directivity
P_{MAX} = maximum power
P_{AVG} = average power
Φ = solid angle subtended by main lobe

The term Φ is a solid angle, which emphasizes the fact that antenna patterns must be examined in at least two extents: horizontal and vertical.

A common method for specifying antenna directivity is *beamwidth* (BW). The definition of BW is the angular displacement between points on the main lobe (see Figs. 20.22A and 20.22B) where the power density drops to one-half (–3 dB) of its maximum at the peak of the lobe. The angle between the –3 dB points is α, as shown in Fig. 20.22A.

In an ideal antenna system, 100 percent of the radiated power is in the main lobe, and there are no other lobes. But in real antennas, certain design and installation anomalies cause additional minor lobes, such as the *sidelobes* and *backlobe* shown in Fig. 20.22A. Several problems derive from the existence of minor lobes:

- *Loss of useable power for the (desired) main lobe.* For a given power density required at a distant receiver site, the transmitter must supply additional power to make up for losses in the minor lobe(s).

- *Intersystem interference.* A major benefit of directional antennas is the reduction in mutual interference between neighboring *co-channel* stations. In radar systems, large sidelobes translate to errors in detecting target bearings. If, for example, a sidelobe is strong enough to detect a target, then the radar display will show this off-axis target as though it were in the main lobe of the antenna. The result is an azimuth error that could be important in terms of marine and aeronautical navigation.

Gain

Antenna *gain G* is a measure of the apparent power radiated in the peak of the main lobe relative to the power delivered to the antenna feedpoint from the transmitter. Thus, G incorporates the effect of ohmic and other losses in the antenna. Specifically:

$$G = \xi D \tag{20.26}$$

Antenna/transmitter systems are often rated in terms of *effective radiated power* (ERP). ERP is the product of the transmitter power and the antenna gain:

$$ERP = G \, P_{DLVD} \tag{20.27}$$

where P_{DLVD} is the actual transmitter RF power delivered to antenna terminals. If an antenna has a gain of +3 dB, the ERP is twice the transmitter output power. In other words, a 100W output transmitter connected to a +3-dB antenna will produce a power density at a distant receiver equal to a 200W transmitter feeding an isotropic radiator.

There are two interrelated gains to be considered: *directivity gain G_d* and *power gain G_p.*

Directivity gain is defined as the quotient of the maximum radiation intensity divided by the average radiation intensity. (Note the similarity to the directivity definition.) This measure of gain is based on the shape of the antenna radiation pattern and can be calculated with respect to an isotropic radiator ($D = 1$) from

$$G_D = \frac{4\pi \, P_\alpha}{P_{RAD}} \tag{20.28}$$

where G_D = directivity gain
P_α = maximum power radiated per unit of solid angle
P_{RAD} = total power radiated by antenna

Power gain is slightly different from directivity gain because it includes dissipative losses in the antenna. Not included in the power gain are losses caused by cross-polarization or impedance mismatch between the waveguide (or transmission line) and the antenna. There are two commonly used means for determining power gain:

$$G_P = \frac{4\pi P_\alpha}{P_{NET}} \qquad (20.29)$$

and

$$G_P = \frac{P_{AI}}{P_I} \qquad (20.30)$$

where P_α = maximum radiated power per unit solid angle
P_{NET} = net power accepted by antenna (less mismatch losses)
P_{AI} = average intensity at a distant point
P_I = intensity at same point from isotropic radiator fed same RF power level as antenna

(Equations assume equal power delivered to antenna and comparison isotropic source.)
Provided that ohmic losses are kept negligible, the relationship between directivity gain and power gain is given by

$$G_P = \frac{P_{RAD}G_D}{P_{NET}} \qquad (20.31)$$

where all terms are as previously defined.

Aperture

Antennas obey the law of reciprocity, which means that any given antenna will work as well receiving as it does transmitting. The function of the receive antenna is to gather energy from the electromagnetic field radiated by one or more transmit antennas. The aperture is related to, and often closely approximates, the physical area of the antenna. But in some designs the effective aperture A_e is less than the physical area A, so there is an effectiveness factor η that must be applied. In general, however, a high-gain transmitter antenna also exhibits a high receiving aperture, and the relationship can be expressed as

$$G = \frac{4\pi A_e \eta}{\lambda^2} \qquad (20.32)$$

where A_e = effective aperture
η = aperture effectiveness ($\eta = 1$ for a perfect, lossless antenna)
λ = wavelength of signal

Horn Antenna Radiators

The horn radiator is a tapered termination of a length of waveguide (see Fig. 20.23A–C) that provides the impedance transformation between the waveguide and free space. Horn radiators are used both as antennas in their own right and as illuminators for reflector antennas. Horn antennas are not a perfect match to the waveguide, although standing wave ratios of 1.5:1 or less are achievable. The gain of a horn radiator is proportional to the area A of the flared open flange ($A = ab$ in Fig. 20.23B), and inversely proportional to the square of the wavelength:

$$G = \frac{10A}{\lambda^2} \tag{20.33}$$

where A = flange area
λ = wavelength (both in same units)

The −3-dB beamwidth for vertical and horizontal extents can be approximated from:

(a) Vertical

$$\Phi_V = \frac{51\lambda}{b} \quad \text{degrees} \tag{20.34}$$

(b) Horizontal

$$\Phi_H = \frac{70\lambda}{a} \quad \text{degrees} \tag{20.35}$$

where Φ_V = vertical beamwidth, in degrees
Φ_H = horizontal beamwidth, in degrees
a, b = dimensions of flared flange
λ = wavelength

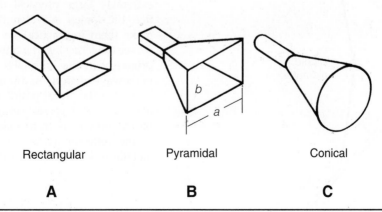

Rectangular   Pyramidal   Conical

A   B   C

FIGURE 20.23 Horn radiators.

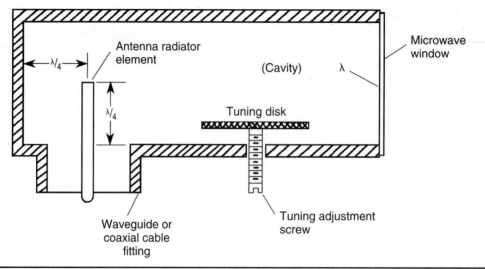

Figure 20.24   Cavity antenna.

Related to the horn is the cavity antenna of Fig. 20.24. Here a quarter-wavelength radiating element extends from the waveguide (or transmission line connector) into a resonant cavity. The radiator element is placed a quarter-wavelength into a resonant cavity, spaced a quarter-wavelength from the rear wall of the cavity. A tuning disk provides a limited tuning range for the antenna by altering cavity dimensions. Gains to about 6 dB are possible with this arrangement.

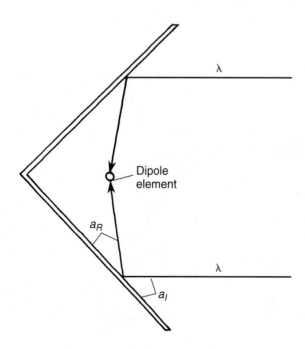

Reflector Antennas

At microwave frequencies, it becomes possible to use *reflector antennas* because of the short wavelengths involved. Reflectors are theoretically possible at lower frequencies, but the extremely large physical dimensions that the longer wavelengths require make them impractical. In Fig. 20.25 we see the *corner reflector* antenna, used primarily in the high-UHF and low-microwave region. A dipole element located at the "focal point" of the corner reflector receives (in phase) reflected *wavefronts* from many regions of the reflector surface. Either solid metallic reflector surfaces or wire mesh

Figure 20.25   Corner reflector.

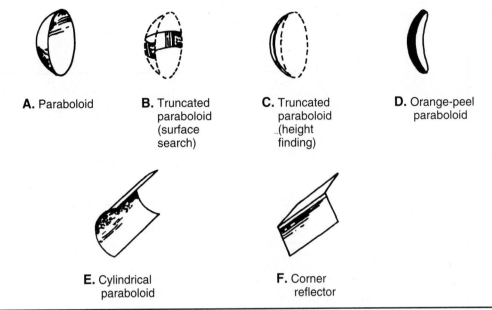

A. Paraboloid **B.** Truncated paraboloid (surface search) **C.** Truncated paraboloid (height finding) **D.** Orange-peel paraboloid

E. Cylindrical paraboloid **F.** Corner reflector

FIGURE 20.26 Reflector antennas.

may be used. When mesh is used, however, the holes in the mesh must be $\lambda/12$ or smaller.

An assortment of several other reflector shapes, most of which are found in radar applications, can be seen in Fig. 20.26.

Parabolic "Dish" Antennas

The parabolic reflector antenna is one of the most widespread of microwave antennas, and one that normally comes to mind when thinking of microwave systems. It derives its operation from the field of optics—possible in part because microwaves are in a transition region between ordinary radio waves and infrared/visible light.

The dish antenna has a paraboloid shape, as defined by Fig. 20.27. In this figure, the dish surface is positioned such that the center is at the origin (0,0) of an x-y coordinate system. For purposes of defining the surface, we place a second vertical axis called the *directrix* (y') a distance behind the surface equal to the focal length (u). The paraboloid surface follows the function $y^2 = 4uX$, and has the property that a line from the focal point F to any point on the surface is the same length as a line from that same point to the directrix. (In other words, $MN = MF$.)

If a radiator element is placed at the focal point F, it will illuminate the reflector surface, causing wavefronts to be propagated away from the surface in phase. Similarly, wavefronts intercepted by the reflector surface are reflected to the focal point.

The gain of a parabolic antenna is a function of several factors, including dish diameter, feed illumination, and surface accuracy. The dish diameter D should be large compared with its depth. Surface accuracy refers to the degree of surface irregularities. For commercial antennas, $1/8$-wavelength surface accuracy is usually sufficient, although on certain radar antennas the surface accuracy specification must be tighter.

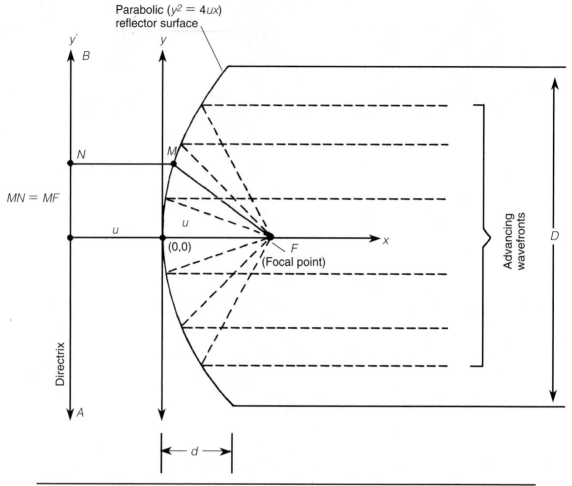

Figure 20.27 Ray tracing shows operation of parabolic antenna.

Feed illumination refers to how evenly the feed element radiates to the reflector surface. For circular parabolic dishes, a circular waveguide feed produces optimum illumination, and rectangular waveguides are not as good a match. The TE_{11} mode is preferred. For best performance, the illumination should drop off evenly from the center to the edge, with the edge being 210 dB down from the center. The diameter, length, and beamwidth of the radiator element (or horn) must be optimized for the specific F/d ratio of the dish. The cutoff frequency is approximated from

$$f_{cutoff} = \frac{175,698}{d} \qquad (20.36)$$

where f_{cutoff} = cutoff frequency in MHz
$\quad\quad d$ = inside diameter of circular feedhorn in mm

The gain of the parabolic dish antenna is found from

$$G = \frac{\xi\,(\pi D)^2}{\lambda^2} \tag{20.37}$$

where G = gain over isotropic
D = diameter
λ = wavelength (same units as D)
ξ = reflection efficiency (0.4 to 0.7, with 0.55 being most common)

The –3-dB beamwidth of the parabolic dish antenna is approximated by

$$\mathrm{BW} = \frac{70\lambda}{D} \tag{20.38}$$

and the focal length by

$$F = \frac{D^2}{16d} \tag{20.39}$$

For receiving applications, the effective aperture is the relevant specification and is found from

$$A_e = \xi\pi(D\,/\,2)^2 \tag{20.40}$$

The antenna pattern radiated by the antenna is similar to Fig. 20.22B. With horn illumination, the sidelobes tend to be 23 to 28 dB below the main lobe, or 10 to 15 dB below isotropic. Of the energy radiated by the parabolic dish, 50 percent is within the –3-dB beamwidth, and 90 percent is between the first nulls bracketing the main lobe.

If a dipole element is used for the feed device, then a *splash plate* is placed a quarter-wavelength behind the dipole in order to improve illumination. The splash plate must be several wavelengths in diameter and is used to reflect the backlobe back toward the reflector surface. When added to the half-wave phase reversal inherent in the reflection process, the two-way quarter-wavelength adds another half-wavelength and thereby permits the backwave to move out in phase with the front lobe wave.

Parabolic Dish Feed Geometries

Figure 20.28 shows two methods for feeding parabolic dish antennas, independent of the choice of radiator (horn, dipole, etc.). In Fig. 20.28A the radiator element is placed at the focal point, and a waveguide (or transmission line) is routed to it. This method is used in low-cost installations such as home satellite TV receive-only (TVRO) antennas.

Figure 20.28B shows the *Cassegrain feed* system modeled after the Cassegrain optical telescope. The radiator element is placed at an opening at the center of the dish. A hyperbolic subreflector is placed at the focal point, and it is used to reflect the wavefronts to the radiator element. The Cassegrain system results in lower-noise operation for several reasons:

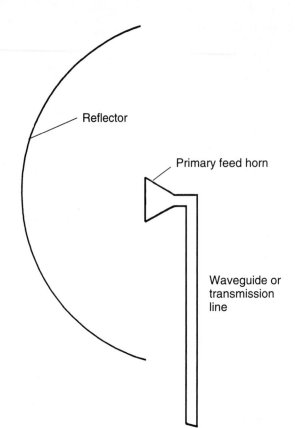

Figure 20.28A   Parabolic antenna conventional feed.

- Shorter transmission line length
- Lower sidelobes
- The open horn sees sky instead of earth (hence, a lower temperature)

On the negative side, galactic and solar noise might be slightly higher on a Cassegrain dish, depending on where it's aimed.

Figure 20.29A shows the *monopulse* feed geometry. In this system, two radiator elements placed at the focal point are fed to a power splitter network that produces both *sum* and *difference* signals. When these are combined, the resultant beam shape (Fig. 20.29B) has an improved –3-dB beamwidth as a result of the algebraic summation of the two.

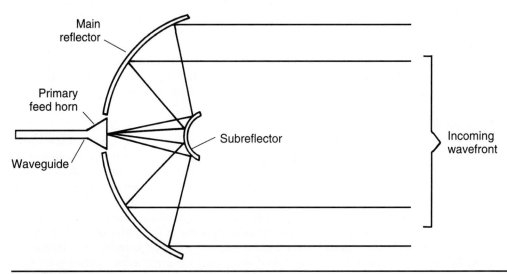

Figure 20.28B   Parabolic antenna Cassegrain feed.

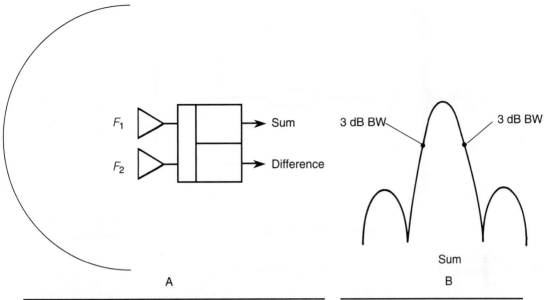

FIGURE 20.29A Monopulse feed.

FIGURE 20.29B Monopulse patterns.

Array Antennas

Microwave arrays can be formed from dipole elements, as in the broadside array of Fig. 20.30 (which is used in the UHF region), or a series of slots, horns, or other radiators. The overall gain of an array antenna is proportional to the number of elements and the details of their spacing. These and other antennas require a method of *phase shifting* for proper beam formation by the array. In Fig. 20.30, the necessary phase shifts are created by the crossed feeding of the elements, but in more modern arrays other forms of phase shifter are used.

Two other methods of feeding an array are shown in Fig. 20.31. The *corporate feed* method connects all elements and their phase shifters in parallel with the source. The *branch feed* method breaks the waveguide network into two (or more) separate paths.

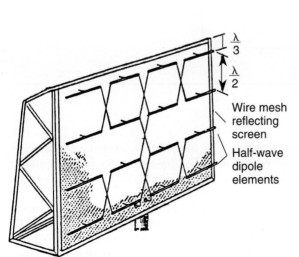

FIGURE 20.30 Reflector array antenna.

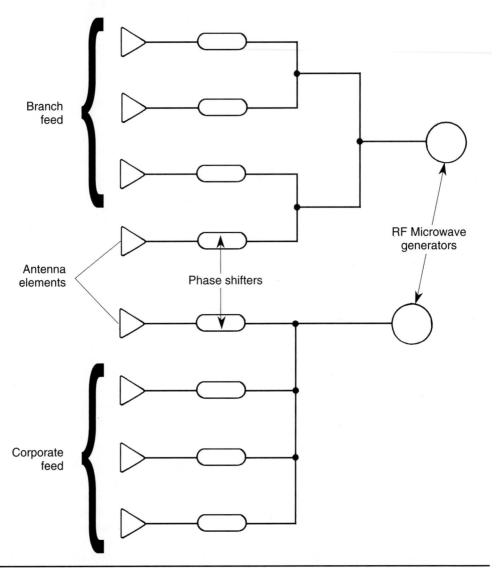

FIGURE 20.31   Branch feed and corporate feed.

Solid-State Arrays

Some modern radar sets use solid-state arrays consisting of a large number of elements, each of which is capable of independently shifting the phase of a microwave input signal. In practice, both *passive* (Fig. 20.32A) and *active* (Fig. 20.32B) phase shifters are found.

In the passive implementation, a ferrite (or PIN diode) phase shifter is placed in the transmission path between the RF input and the radiator element (usually a slot). By changing the phase of the RF signal selectively, it is possible to form and steer the beam at will. A 3-bit phase shifter is capable of eight distinct states and thus allows the phase

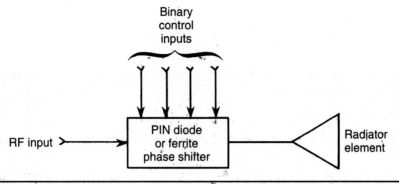

FIGURE **20.32A**   Phase shifter.

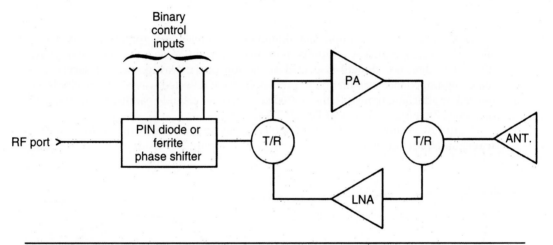

FIGURE **20.32B**   Phase-shift T/R element.

to shift in 45-degree increments, while a 4-bit phase shifter (with sixteen states) allows 22.5-degree increments of phase shift.

In addition to a phase shifter, the active element of Fig. 20.32B contains a transmit power amplifier (1 or 2 W) and a low-noise amplifier (LNA) for receiving. A transmit/receive (T/R) switch isolates the LNA input from the high-power RF emitted by the power amplifier, while a second T/R switch simultaneously selects whether the phase shifter is performing beamforming operations on the transmitted or the received RF. The total output power of this antenna is the sum of all output powers from all elements in the array. For example, an array of one thousand 2W elements forms a 2000W system.

Slot Array Antennas

A resonant slot (cut into a wall of a section of waveguide) is somewhat analogous, if not identical, to a dipole. Slot arrays are used for marine navigation radars, telemetry systems, and the reception of microwave television signals in the Multipoint Distribution Service (MDS) on 2.145 GHz.

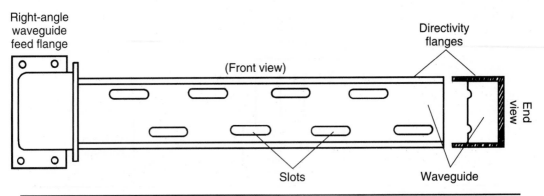

Right-angle
waveguide
feed flange

Directivity
flanges

(Front view)

End
view

Slots

Waveguide

FIGURE 20.33   Slot antenna (front view).

Figure 20.33 shows a simple slot antenna used in telemetry applications. A slotted section of rectangular waveguide is mounted to a right-angle waveguide flange. An internal wedge (not shown) is placed at the top of the waveguide and serves as a matching-impedance termination to prevent internal reflected waves. Directivity is enhanced by attaching flanges to the slotted section of waveguide parallel to the direction of propagation (see end view of Fig. 20.33).

Figure 20.34 shows two forms of *flatplate array* antennas constructed from slotted waveguide radiator elements (shown as insets). Figure 20.34A depicts a rectangular

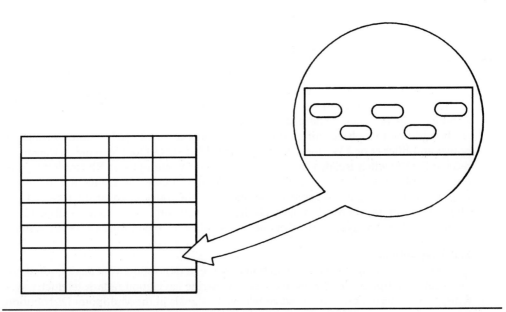

FIGURE 20.34A   Flatplate slot array.

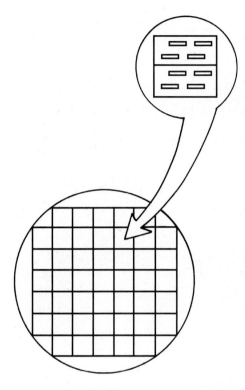

FIGURE 20.34B Flatplate antenna.

array and Fig. 20.34B shows the circular array. These flatplate arrays are used extensively in microwave communications and radar applications.

The feed structure for a flatplate array is shown in Fig. 20.34C. A distribution waveguide is physically mated with the element, and a coupling slot is provided between the two waveguides so that energy propagating in the distribution system waveguide can couple into the antenna radiator element. In some cases metallic or dielectric phase-shifting stubs are also used to fine-tune the antenna radiation pattern.

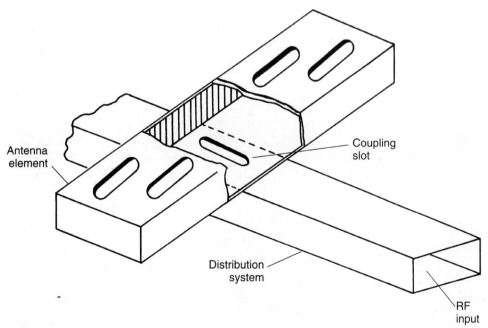

FIGURE 20.34C Flatplate antenna feed coupling.

CHAPTER 21

Antenna Noise Temperature

Radio reception is essentially a matter of *signal-to-noise ratio* (SNR). Signals must be at or above some amplitude relative to the noise floor of the system in order to be detected properly for their intended use. All electronic systems (receivers and antennas included) generate noise internally, even if there is no power flowing in them. While we normally think in terms of the antenna designer maximizing signal strength to overcome noise, often a second goal is to minimize the noise from some or all of the various sources.

In general, at HF and below, the external signals and noise sources detected by the antennas described in this book (including even the very low efficiency antennas of Chap. 14) are strong enough to swamp any noise generated within a well-designed receiver. Thus, we typically have to take special measures to override internally generated receiver noise only at VHF and above, and the balance of this chapter is primarily for the benefit of those who are concerned with weak signal reception above 30 MHz or so.

One of the basic forms of noise seen in systems is *thermal noise*. Even if the amplifiers in the receiver add no additional noise (they will!), there is thermal noise at the input. In fact, if we replace the antenna attached to the receiver input with a totally shielded resistor matched to the system impedance, some noise will still be present. This noise is produced by the random motion of electrons inside the resistor. At all temperatures above absolute zero (about −273.16°C), the electrons in the resistor material are in random motion. In the absence of an external bias voltage creating a uniform field acting on the resistor body, the short-term random motions of the electrons cancel each other out to the extent that no discernable current can be observed.

Thermal noise in a resistor can be modeled as a voltage source, $\bar{V}_N$, in series with a noise-free resistor R, where $\bar{V}_N$ is the root mean square (rms) value of the fluctuating, thermally generated noise voltage. If R—in ohms—is constant over the frequency range of interest, $\bar{V}_N$ is proportional to $\sqrt{kTBR}$, whose components are defined below. When the resistor is connected across a matched load, the noise power transferred to that load is given by Eq. (21.1). Note that the noise power delivered to a matched load is independent of the value of the resistor.

$$P_N = kTB \qquad (21.1)$$

where P_N = noise power, in watts
k = Boltzmann's constant (1.38×10^{-23} joules/K)
T = temperature, in degrees kelvin (K)
B = bandwidth, in hertz

Degrees kelvin (K) is the international way of defining all temperatures relative to absolute zero. (No degree symbol is used with K.) To express temperature in degrees kelvin (K) we add 273.16 to a temperature expressed in Celsius. The formula is

$$T(\text{K}) = T(^\circ\text{C}) + 273.16 \qquad (21.2)$$

Of course, temperatures expressed in Fahrenheit (°F) are related to Celsius by

$$T(\text{Fahrenheit}) = \frac{9}{5}T(\text{Celsius}) + 32 \qquad (21.3)$$

so we also have

$$T(\text{K}) = \frac{5}{9}T(^\circ\text{F}) + 255.38 \qquad (21.4)$$

Thus, water turns to ice at 32°F, 0°C, or 273.16 K. Water boils at 212°F, 100°C, or 373.15 K. Absolute zero corresponds to –459.67°F, –273.16°C, or 0 K.

Another important point on the various temperature scales is *room temperature,* typically taken to be 27°C (or about 80°F) by some scientific and engineering specialties. This may be higher than what you or I would want our room to be, but it has the advantage of corresponding to 300 K—a nice round number for doing calculations!

Finally, by international agreement, *T* for terrestrial components and system elements is assumed to be 290 K (about 17°C, or 62°F) unless otherwise stated. To improve weak-signal detection at microwave frequencies and above, however, receiver input stages are often cooled by liquid nitrogen. The markedly lower *T* of those stages results in their contributing a substantially reduced noise power to the overall system noise figure.

Example 21.1   A terrestrial receiver with a 1-MHz bandwidth and a 50-Ω input impedance is connected to a 50-Ω resistor. The noise power delivered to the receiver input stage is $(1.38 \times 10^{-23}\,\text{J/K}) \times (290\,\text{K}) \times (1{,}000{,}000\,\text{Hz}) = 4 \times 10^{-15}\,\text{W}$. This noise is called *thermal noise, thermal agitation noise,* or *Johnson noise.*

Noise Factor, Noise Figure, and Noise Temperature

The noise performance of a receiving system can be defined in three different, but related, ways: *noise factor F_n, noise figure (NF),* and *equivalent noise temperature T_e;* these properties are definable as a simple ratio, decibel ratio, or kelvin temperature, respectively.

Noise Factor (F_n)

For components such as resistors, the noise factor is the ratio of the noise produced by a real resistor to the simple thermal noise of an ideal resistor. The noise factor of a radio receiver (or any system) is the ratio of output noise power P_{no} to input noise power P_{ni}:

$$F_n = \left.\frac{P_{no}}{P_{ni}}\right|_{T=290\,K} \tag{21.5}$$

In order to make comparisons easier, the noise factor is usually measured at the standard temperature (T_0) of 290 K, although in some countries 299 K or 300 K is commonly used (the differences are generally negligible).

It is also possible to define noise factor F_n in terms of input and output signal-to-noise ratios:

$$F_n = \frac{S_{ni}}{S_{no}} \tag{21.6}$$

where S_{ni} = input signal-to-noise ratio
$\quad\quad S_{no}$ = output signal-to-noise ratio

Noise Figure (NF)

The noise figure is a frequently used measure of a receiver's "goodness", or its departure from "idealness". Thus, it is a *figure of merit*. The noise figure is the noise factor converted to decibel notation:

$$NF = 10 \log F_n \tag{21.7}$$

where NF = noise figure, in decibels
$\quad\quad F_n$ = noise factor

"log" refers to the system of base-10 logarithms. (See App. A for an explanation of logarithms.)

Noise Temperature (T_e)

Noise temperature is a means for specifying noise in terms of an equivalent temperature. Examination of Eq. (21.1) shows that the noise power is directly proportional to temperature in kelvins, and also that noise power collapses to zero at the temperature of absolute zero (0 K).

NOTE *The equivalent noise temperature T_e is* not *the physical temperature of the amplifier but, rather, a theoretical construct that is an* equivalent *temperature that would produce the same amount of noise power in a resistor at that temperature.*

Noise temperature is related to the noise factor by

$$T_e = (F_n - 1)\, T_0 \tag{21.8}$$

and to noise figure by

$$T_e = KT_0 \log^{-1}\left(\frac{NF}{10}\right) - 1 \qquad (21.9)$$

Noise temperature is often specified for receivers and amplifiers in combination with, or in lieu of, the noise figure. Applied to antennas, the noise temperature concept relates the amount of thermal noise generated to the resistive loss components of the antenna feedpoint impedance.

The antenna/receiver system will be afflicted by three different noise sources external to the receiver:

- The *thermal noise temperature* (T_R) of the resistive loss portion of the feedpoint impedance
- A sky noise temperature (T_{SKY}) that depends on where the antenna main lobe is pointed
- A *ground noise temperature* (T_{GND}) that consists of components reflected from the sky as well as components caused by internal thermal agitation of the ground

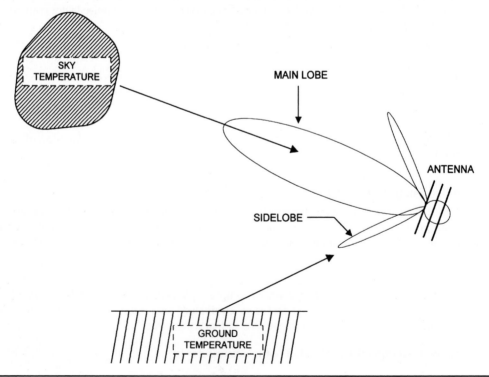

FIGURE 21.1   Contributors to antenna noise temperature.

In a typical system (Fig. 21.1) the main lobe will be pointed toward the sky noise source, while the sidelobes will pick up noise from the ground. The total noise temperature of the antenna is thus

$$T_{ANT} = (M \times T_{SKY}) + \xi(1-M) T_{GND} + T_R \tag{21.10}$$

where T_{ANT} = equivalent noise temperature of antenna
T_{SKY} = noise temperature of sky
T_{GND} = noise temperature of ground
T_R = feedpoint loss resistance noise temperature
M = fraction of total energy that enters main lobe
ξ = fraction of sidelobes that are viewing the ground (only one of several sidelobes is shown in Fig. 21.1)

CHAPTER 22

Radio Astronomy Antennas

For centuries astronomers have scanned the heavens with optical telescopes. But today, astronomers have many more tools in their bag, and one of them is *radio astronomy*. The field of radio astronomy emerged in the 1930s and 1940s through the work of Grote Reber and Carl Jansky. Even during World War II, progress was made as many tens of thousands of operators were listening to frequencies from "dc to daylight" (well, actually, the low end of the microwave bands). British radar operators noted during the Battle of Britain that the distance at which they could detect German aircraft dropped when the Milky Way was above the horizon.

Although there is a lot of amateur radio astronomy being done, most of it requires microwave equipment with low-noise front ends and is beyond the scope of this book. For example, most *deep space* (beyond our solar system) radio sources will require the use of high-gain configurations—such as *long baseline arrays*—that are far beyond any one individual's ability to implement. However, there are several things that almost anyone can do as an introduction to this "hobby within a hobby".

Radio astronomy antennas can assume nearly all forms. It is common to see Yagis, ring Yagis, cubical quads, and other antennas for lower-frequency use (18 to 1200 MHz). Microwave gain antennas, such as parabolic reflectors, can be used for higher frequencies. Indeed, many amateur radio astronomers employ TV receive-only (TVRO) satellite dish antennas for astronomy work. In this chapter we limit our coverage to some antennas that are not discussed in other chapters—at least, not in this present context.

General Considerations

The strongest extraterrestrial radio sources we can typically detect are from our own sun. These emissions are very broadband and so our receivers typically detect only a small portion of the total radiated spectrum at any one time. When we listen on a receiver, using conventional heterodyne reception feeding a loudspeaker, what we hear is noise that varies in amplitude—sometimes slowly, sometimes in bursts.

As a rule, we are restricted to monitoring the sun and other astronomical bodies at frequencies higher than those blocked by earth's ionosphere. The low end of the useable frequency range varies with the sunspot cycle, time of day, etc., but frequencies above 18 MHz, as mentioned, should work virtually all the time. Many radio astronomers, both amateur and professional, concentrate on the region around 20.1 MHz, but the author has heard the sun on 28 MHz many times with a simple three-element Yagi. Of course, if these bands are open for HF skip propagation, or if many local groundwave signals are present, other frequencies may contain fewer interfering terrestrial signals.

So what do you need to chase DX throughout our solar system? All it takes is a receiver that works well over the range from 18 to 30 MHz—most modern communications receivers are fine for the purpose—and a relatively simple antenna tuned to some portion of that frequency span. Any *automatic gain control* (AGC) or *automatic volume control* (AVC) should be turned off, the audio volume control set to a comfortable level, and the *RF gain* control advanced at least to the point where external noise overrides the internal noise of the receiver. Set your receiver to the *single sideband* (SSB) mode and select the widest filter bandwidth(s) available.

Today, of course, inexpensive PCs and software applications allow us to "listen" with our eyes, as well. *Spectrograms* and other displays of the received signals allow us to see and print out evidence of these astronomical noise bursts. A simple Internet search should return a host of how-to articles, blogs, and discussion groups.

Listening to "Ol' Sol"

Even a dipole has directivity, so it's helpful to orient even the simplest of antennas with the peak response of its pattern in the direction of the source. Since the sun's path for many of us covers such a broad range of both azimuth and elevation angles, it's probably smartest to zero in on its location for those hours of the day that we're most apt to be able to listen for it.

Of course, for dipoles at 18 MHz and above, rotating them is a relatively simple task, whether done manually or with an antenna rotator.

At some point, you may wish to add greater gain and directivity to improve your ability to pull extraterrestrial signals out of your background noise environment, especially if you are in an urban neighborhood. Even then, rotating a two- or three-element Yagi at 20 MHz and higher is not an insurmountable task. Best yet, antenna height is not a factor—especially if you stick to listening periods when the sun is well above the horizon.

Signals from Jupiter

Second only to the sun, Jupiter is a strong radio source. It produces noiselike signals from VLF to 40 MHz, with peaks between 18 and 24 MHz. One theory attributes the radio signals to massive storms on the largest planet's surface, apparently triggered by the transit of the Jovian moons through the planet's magnetic field. The signals are plainly audible on the upper HF bands any time Jupiter is above the horizon, day or night. However, in order to eliminate the possibility of both local and terrestrial skip signals from interfering, Jupiter DXers prefer to listen only when the maximum useable frequency (MUF) drops significantly below 18 MHz—typically the darkness hours. In preparation, listen to the amateur 17- or 15-m bands; if you hear no skip-distance activity, then it's a good bet that the MUF has dropped enough to make listening worthwhile. Even during the day, however, it is possible to hear Jovian signals, but differentiating them from other signals and solar noise can be difficult.

Jupiter emits two distinctly different types of radio noises. Listening in SSB mode, one form can be heard as "swooshing" noises that rise and fall in amplitude over a relatively long interval. The second type of noise from the planet is heard as a more rapid-fire "popping" sound.

Unlike the sun, Jupiter does not often rise high in our sky. Its transit, as viewed at midlatitudes in the northern hemisphere, is typically close to the horizon—rising in the southeast and setting in the southwest. In that respect, Jupiter is an excellent target for a typical HF Yagi or cubical quad whose height has been optimized for low elevation angles. Also, Jupiter is visible (and audible) at times totally unrelated to our day and night periods. As a result, there may be periods when Jupiter's emissions may be difficult to hear because the sun is also "in your face". Search the Internet for detailed calendars of Jupiter's transits.

For monitoring Jupiter, a good beginning antenna can be a simple dipole cut for the middle of the 18- to 24-MHz band, which happens to coincide with the 15-m amateur radio band. The antenna should be installed in the normal manner for any dipole, except that if it is fixed in one position the wire should run east-west in order to maximize pickup from this southerly rising planet.

Figure 22.1 shows a broadband dipole that covers the entire frequency span of interest (18 to 24 MHz) by paralleling three different dipoles: one cut for 18 MHz, one cut for 21 MHz, and one cut for 24 MHz. The dimensions are

A	19.5 ft	24 MHz
B	22.3 ft	21 MHz
C	26 ft	18 MHz

As discussed in the chapter on multiband wires (Chap. 8), there are several approaches to making this type of antenna. One is to use three-conductor wire and cut the wires to the lengths indicated here. Another is to use a homemade spacer to spread the wires apart.

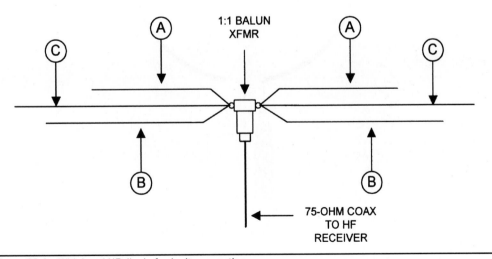

FIGURE 22.1 Wideband HF dipole for Jupiter reception.

Ring Antenna

Another popular Jovian radio antenna is the ring radiator, two versions of which are shown in Fig. 22.2. This antenna is made of a 5-ft-diameter loop of ½-in-diameter soft-drawn copper plumbing pipe found in any local hardware or plumbing supply store. The single-ended version is shown in Fig. 22.2A. In this form the far end of the loop is unterminated (i.e., open-circuited). The center conductor of the coaxial cable feedline is connected to the near end of the ring radiator, while the coax shield is connected to the chicken wire ground plane. The balanced version (Fig. 22.2B) has an RF transformer (T_1) at the feedpoint.

The ring radiator antenna should have a bandpass-filtered preamplifier—needed because of the antenna's low output levels. The preamp should be mounted as close as possible to the antenna. The intent is to minimize the likelihood of strong terrestrial signals in the adjacent bands *desensing* the preamp or receiver. Even a 5W CB transmitter a few blocks away can drive the preamplifier into saturation, so it's wise to eliminate the undesired signals before they get into the preamplifier. In the case of the single-ended amplifier, a single-ended preamplifier is used. But for the balanced version (Fig. 22.2B) a differential preamplifier is appropriate.

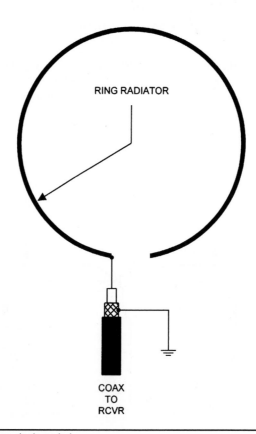

Figure 22.2A   Ring radiator: single-ended.

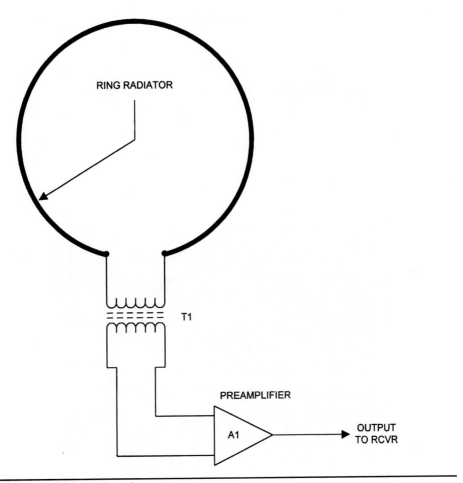

RING RADIATOR

T1

PREAMPLIFIER

A1

OUTPUT
TO RCVR

FIGURE 22.2B   Ring radiator: balanced.

Either version of the loop or ring is mounted about 7 or 8 in above a ground plane made of chicken wire, metal window screen, copper sheeting, or copper foil. (Copper sheeting or foil is best but costs a lot of money and turns ugly green after a couple weeks in the elements.) If you use screening, make sure that it is metallic. Some window and porch screening material is made of synthetic materials that are insulators.

Figure 22.3A is a mechanical side view of the ring radiator antenna, while Fig. 22.3B is a top view. The antenna is mounted above the screen with insulators. These can be made of wood, plastic, or any other insulating material. The frame holding the ground-plane screen (Fig. 22.3B) can be made from 1- × 2-in lumber. Note that the frame is strengthened by interior crosspieces that also support the antenna. The larger outer perimeter is needed because the screen ground plane should extend beyond the diameter of the radiator element by at least 10 to 15 percent of that dimension.

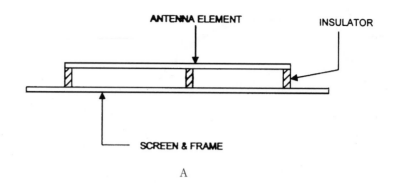

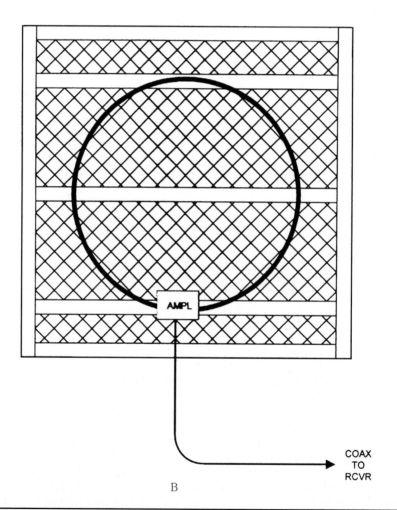

FIGURE 22.3 Ring radiator mounted over a ground screen.

DDRR

The directional discontinuity ring radiator (DDRR) antenna is shown in Fig. 22.4A, while a side view showing the mounting scheme is shown in Fig. 22.4B. The DDRR is typically mounted about 1 ft off the ground (H = 12 in). It consists of two sections, one vertical and one horizontal. The short vertical section has a length equal to the height H of the antenna above the ground, and the lower end of the vertical segment is grounded.

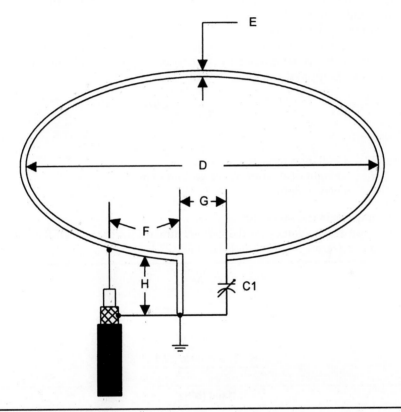

FIGURE 22.4A   DDRR antenna.

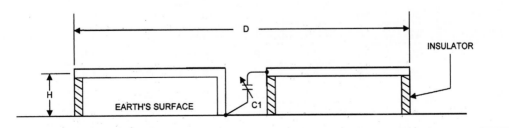

FIGURE 22.4B   DDRR antenna: side view.

The horizontal section is an open-circuited loop of diameter D, which varies with frequency.

The optimum conductor diameter E for the DDRR is at least 0.5 in at 28 MHz, increasing to 4 in at 4 MHz. Thus, for monitoring Jupiter, a diameter of 1 in or slightly less is appropriate.

Because of the loop, some people call this the *hula hoop* antenna. One author recommends using a 2-in automobile exhaust pipe bent into the correct shape by a cooperative automobile muffler mechanic. The far end of the loop is connected to ground through a small-value tuning capacitor C_1. The actual value of C_1, which is used to resonate the antenna at a particular operating frequency, is found experimentally.

The feedline of the DDRR antenna is coaxial cable; its shield is grounded at the bottom end of the vertical section. The center conductor is connected to the ring radiator a distance F from the vertical section in a circular variant of the *autotransformer* or *gamma match*. The length F is determined by the impedance that must be matched. The radiation resistance is approximated by

$$R_{RAD} = \frac{2620 H^2}{\lambda^2} \tag{22.1}$$

where R_{RAD} = radiation resistance, in ohms
       H = height of antenna above ground plane
       λ = wavelength

H and λ must be in the same units.

The approximate values for the various dimensions of the DDRR are given as follows in general terms, with examples in Table 22.1:

D	0.078λ
H	0.11D
F	0.25H
E	0.5 to 4 in
G	See Table 22.1

Dimension	Band (MHz)							
	1.8	**4**	**7.5**	**15**	**22**	**30**	**50**	**150**
G (in)	16	7	5	3	2.5	2	1.5	1
C_1 (pF)	150	100	75	35	15	12	10	6
F (in)	12	6	6	1.5	1.5	1	1	0.5
H (in)	48	24	11	6	4.75	3	1.5	1
D (ft)	36	18	9	4.5	3.33	2.33	1.4	6
E (in)	5	4	2	1	0.75	0.75	0.5	0.25

TABLE 22.1   Examples of Dimensions for DDRR

The construction details of the DDRR are so similar to those of the ring radiator that the same diagram can be used (see Fig. 22.3 again).

The normal *attitude* of the DDRR for communications is horizontal. However, for maximum signal when monitoring Jupiter, the entire antenna assembly, including the ground-plane screen, can be tilted to track Jupiter's position in the sky.

Helical Antennas

The helical antenna (Fig. 22.5) provides moderately wide bandwidth and *circular polarization*. Thanks to the latter, some find the helical antenna particularly well suited to radio astronomy reception. The antenna (of diameter D) will have a circumference C of 0.75λ to 1.3λ. The pitch of the helix (S) is the axial length of a single turn, while the overall length $L = NS$ (where N is the number of turns). The ratio S/C should be 0.22λ to 0.28λ. At least three turns are needed to produce axial-mode main lobe maxima.

The diameter or edge of the ground plane G should be on the order of 0.8λ to 1.1λ if the conducting surface is circular or square, respectively. The offset between the ground plane and the first turn of the helix is 0.12λ.

The approximate gain of the helical antenna is found from

$$\text{Gain} = 11.8 + 10 \log (C^2 NS) \quad \text{dBi} \tag{22.2}$$

The pitch angle ϕ and turn length γ for the helical antenna are given by

$$\phi = \tan^{-1}\left(\frac{\pi D}{S}\right) \tag{22.3}$$

and

$$\gamma = \sqrt{(\pi D)^2 + S^2} \tag{22.4}$$

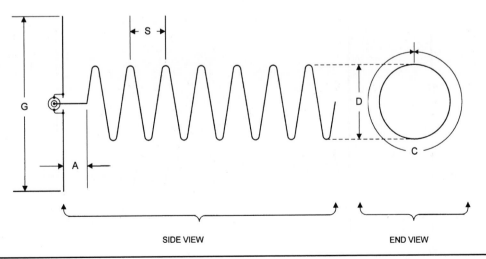

SIDE VIEW END VIEW

FIGURE 22.5 Helical antenna.

The beamwidth of the helical antenna is

$$\theta = \frac{K}{\sqrt{NS}}$$

(22.5)

where θ = beamwidth, in degrees
N = number of turns
S = pitch, in wavelengths
C = circumference, in wavelengths

K is 52 for the –3-dB beamwidth and 115 for the beamwidth to the first null in the pattern.

The short section between the helix and the ground plane is terminated in a coaxial connector, allowing the antenna to be fed from the rear of the ground plane. The feed-point impedance is approximately 140 Ω.

Multiple Helical Antennas

Stacking helical antennas allows a radiation pattern that is much cleaner than the normal one-antenna radiation pattern. It also provides a good way to obtain high gain with only a few turns in each helix. If two helixes are stacked, then the gain will be the same as for an antenna that is twice the length of each element, while for four stacked antennas the gain is the same as for a single antenna four times as long. Figure 22.6 shows a side view of the stacked helixes.

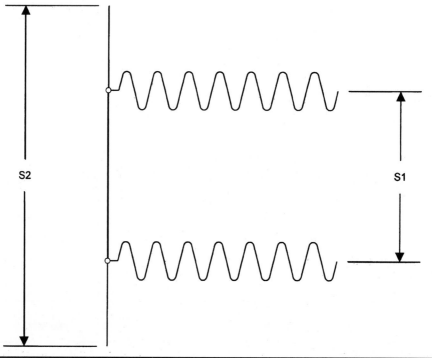

FIGURE 22.6 Dual/quad helical antenna.

The feed system for stacked helixes is more complicated than for a single helix. Figure 22.7A shows an end view of a set of four stacked helical antennas. Tapered lines (TL) are used to carry signal from each element and the coaxial connector (B). In this case, the coaxial connector is a feed-through *barrel* or back-to-back SO-239 device at the center of the ground plane (B). A side view of the tapered line system is shown in Fig. 22.7B. The length of the tapered lines is 1.06λ, while the center-to-center spacing between the helical elements is 1.5λ. The length of each side of the ground plane is 2.5λ. In the case of Fig. 22.7, the antenna is fed from the front of the ground plane.

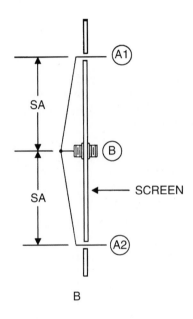

B

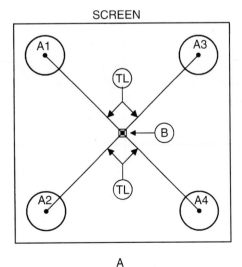

A

FIGURE 22.7   Front view of quad helical antenna.

Interferometer Antennas

The resolution of an antenna is a function of its dimensions relative to the wavelength λ of the received signal. Better resolution can be achieved by increasing the size of the antenna, but that is not always the best solution and beyond a certain point is totally impractical. Figure 22.8 shows a *summation interferometer array*. Two antennas, each with aperture a, are spaced S wavelengths apart. The radiation pattern is a fringe pattern (Fig. 22.9), consisting of a series of maxima and minima. The resolution angle ϕ to the first null is

$$\phi = \frac{S + a}{57.3} \tag{22.6}$$

The interferometer can be improved by adding antennas to the array. Professional radio astronomers use very long or wide baseline antennas. With modern communications it is possible to link radio telescopes on different continents to make the widest or longest possible terrestrial baseline.

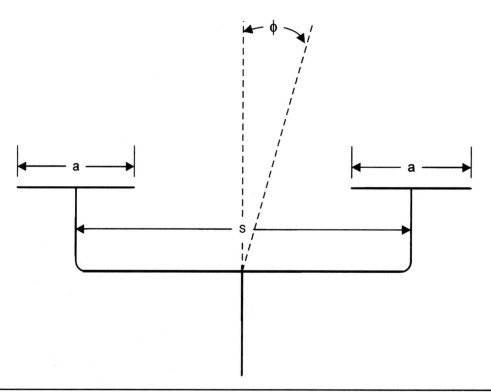

FIGURE 22.8   Interferometer array.

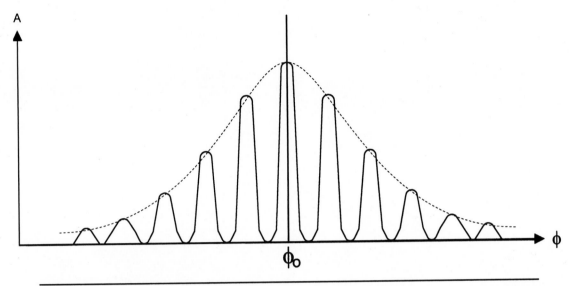

FIGURE 22.9   Interferometer pattern.

CHAPTER 23

Radio Direction-Finding (RDF) Antennas

Radio direction finding is the art and practice of using antenna directivity to pinpoint either your own location or that of a distant source of RF energy. When the Federal Communications Commission (FCC) or other telecommunications authorities want to locate an illegal station that is transmitting, one of their primary tools is the use of radio direction finders to triangulate the bootleg station's position. In the same way that the *global positioning system* (GPS) receiver in your navigation system or smart phone determines where *you* are located, monitors can determine the location of *another* station from the intersection of bearing lines obtained at two or more receiving sites. In contrast to the bearing accuracy available from GPS satellites, there is a fair degree of ambiguity in bearings obtained for *terrestrial* targets—especially on the MF and HF bands—because multiple propagation modes, paths, and anomalies can easily obscure the true bearing of the signal source. As a result, radio direction finders typically use three or more sites. Each receiving site that can find a bearing to the station reduces the overall error.

At one time, aviators and seafarers relied on radio direction finding. It is said that the Japanese air fleet that attacked Pearl Harbor, Hawaii, on December 7, 1941, homed in on a Honolulu AM broadcast station. During the 1950s and early 1960s, AM radio dials in the United States were marked with two little circled triangles at 640 and 1240 kHz. These were CONELRAD frequencies that people were encouraged to tune to in case of a nuclear attack. In each area of the country multiple selected radio stations would occupy a CONELRAD frequency, taking turns transmitting in a carefully orchestrated (and secret) sequence. At the same time, all radio stations except assigned CONELRAD stations would have to cease transmitting. Thus, local broadcast news and information services could be maintained at the same time the enemy was being prevented from using any AM broadcast signals for RDF because no one CONELRAD station was on the air long enough to allow a fix.

Radio direction-finding equipment utilizing the AM broadcast band (BCB) looks much like Fig. 23.1. A receiver with an *S-meter* to measure signal strength is equipped with a rotatable ferrite loopstick antenna to form the RDF unit. A 360-degree scale around the perimeter of the antenna base can be rotated to line up with true north so that compass bearing can be read off directly.

Loopstick antennas exhibit a figure eight reception pattern (Fig. 23.2A) with the maxima (lobes) parallel to the loopstick rod and the minima (nulls) off the ends. When

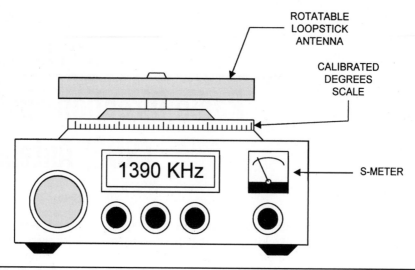

Figure 23.1   Radio direction-finding (RDF) receiver with loopstick antenna.

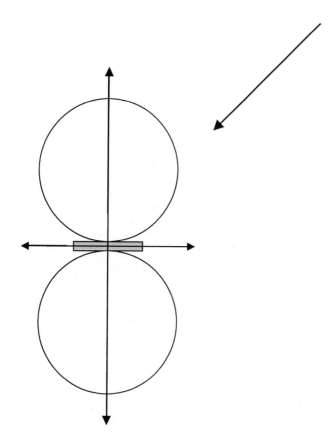

the antenna is broadside to the signal source, received signal strength is maximum. Unfortunately, the maxima are so broad that it is virtually impossible to find the true point on the compass dial where the signal peaks. Fortunately, the minima are very sharp, and a good fix on the source can be obtained by nulling the signal, instead of peaking it. The null is found by rotating the antenna until the audio disappears or the S-meter dips to its minimum observed value (Fig. 23.2B).

The loopstick is a really neat way to do RDF—except for one little problem: The darn thing is bidirectional! There are two minima because the antenna pattern is symmetrical about the anten-

Figure 23.2A   Pattern of loopstick antenna: random orientation of source.

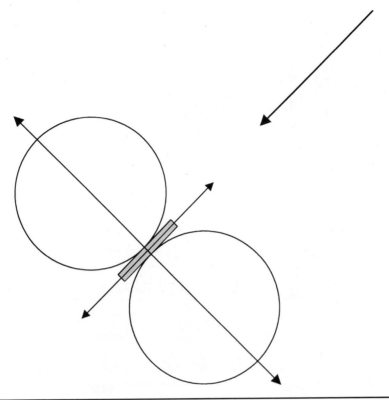

FIGURE 23.2B   Pattern of loopstick antenna: nulling the source.

na's axis. You will get exactly the same response from aiming either end of the stick in the direction of the station. As a result, the unassisted loopstick can only show you a line along which the radio station is located but can't tell you which half of the line to focus on. Sometimes this doesn't matter; if you know the station is in a certain city and that you are generally south of the city, and can distinguish the general direction from other clues, then the half of the null line that is closer to the north will provide the answer.

One solution to the ambiguity problem is to add a sense antenna to the loopstick (Fig. 23.3). The sense antenna is an omnidirectional vertical whip, and its signal is combined with that of the loopstick in an *RC* phasing circuit. When signals from the two antennas have roughly equal amplitudes, the resultant pattern will resemble Fig. 23.4. This pattern is called a *cardioid* because of its heart shape. Because it has only one null, it resolves the ambiguity of the loopstick pattern.

Note, however, that a single RDF receiver can provide only bearing information. Receivers at two or more sites, widely separated (relative to bearing angle of the target), are necessary to also obtain distance information and pinpoint the location of the target.

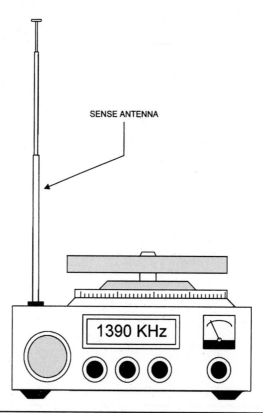

FIGURE 23.3   Addition of sense antenna overcomes directional ambiguity of the loopstick.

FIGURE 23.4   Cardioid pattern of sense-plus-loopstick antennas.

Field Improvisation

Suppose you are out in the woods trekking around the habitat of lions, tigers, and bears (plus a rattlesnake or two for good measure). Normally you find your way with a compass, a U.S. Geological Survey 7.5-min *topological map,* and a GPS receiver. Those little GPS marvels can usually give you excellent latitude and longitude indication when you can see sky above you, but often fail to deliver reliable position information in woodlands and dense forests. In such terrain, the answer to your direction-finding problem might be the little portable AM BCB radio (Fig. 23.5) that you brought along for company.

Open the back of the radio and find the loopstick antenna, because you will need to know its orientation. In the radio shown in Fig. 23.5 the loopstick lies across the top of the radio, from left to right, but in other radios it is vertical, from top to bottom. Once you know its position, you can tune in a known AM station and orient the radio until you find a null. Next, point your compass in the same direction to obtain the numerical bearing (0–360°) of the signal. If you know the approximate location of the broadcast station, subtract the bearing you determined from 180 degrees (or vice versa, depending on which method gives you a positive value equal to 360 or less) and draw a line on the topo map at that angle, originating at the known AM broadcast station. Of course, it's still a bidirectional indication, so all you know is the line along which you are lost.

Now tune in a different station in a different city (or one at least far enough from the first station to make a difference) and take another reading. Again subtract (or add) 180 degrees to obtain a positive number between 0 and 360. Draw a line at this bearing originating from the second broadcast station location. These two lines should intersect at your approximate location. If you know the locations of additional broadcast stations, take a third, fourth, and fifth reading and you will generally be able to pinpoint your location pretty tightly. If you were smart enough to plan ahead, you will have se-

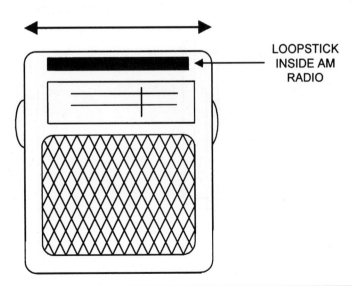

LOOPSTICK
INSIDE AM
RADIO

FIGURE 23.5   AM portable radio. Dark bar represents position and direction of internal loopstick.

lected candidate stations in advance and located the latitude and longitude coordinates of each. Alternatively, you would have brought topo maps premarked with their locations as well as where you plan to hike, and from those maps you can find the latitudes and longitudes of the distant stations.

With the appropriate receiver, the same technique works as well with FM stations, although it's important not to let multipath propagation modes (see Chap. 4) confuse you.

Loop Antennas for RDF

Conventional wire loop antennas (Fig. 23.6) are also used for radio direction finding. In fact, in some cases the regular loop is preferred over the loopstick. The regular loop antenna may be square (as shown), circular, or any other equi-sided "*n*-gon" (e.g., hexagon), although the square is generally easiest to build. Only a few turns are needed to accumulate adequate inductance for resonating in the AM broadcast band with a standard 365-pF "broadcast" variable capacitor. For a square loop 24 in on a side (dimension A in Fig. 23.6), about 10 turns of wire spaced over a 1-in width (B in Fig. 23.6) should be sufficient. For portable direction finding, a small loop can be mounted to the top of a rigid support mast such as a length of 2- × 4-in lumber, as shown in Fig. 23.7. The important design constraints are that the loop is high enough to have some separation from your vehicle (if you need to stand close to it in order for the feedline to reach the receiver) and that the loop is never so high that it can touch the lowest power lines in the vicinity.

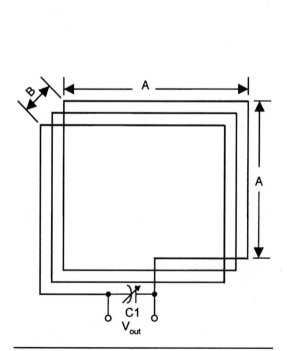

FIGURE 23.6 Wire loop antenna.

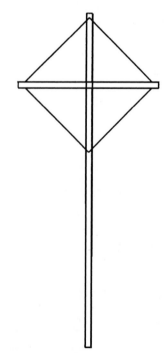

FIGURE 23.7 Simple portable RDF loop antenna.

Be aware that the small loop antenna has a figure eight pattern like the loopstick, but the maxima and minima are oriented 90 degrees from those of the loopstick. In the regular loop the nulls are perpendicular to the plane of the loop, while the maxima are off the sides. Thus, in Fig. 23.6 the nulls are in and out of the page, while the main lobes are left and right (or top and bottom).

Sense Antenna Circuit

Figure 23.8 shows a method for summing together the signals from an RDF antenna (such as a loop) and a sense antenna. The two terminals of the loop are connected to the primary winding of an RF transformer. This primary (L_{1A}) is centertapped and the centertap is grounded. The secondary of the transformer (L_{1B}) is resonated by a variable capacitor C_1. The dots on the transformer coils indicate the 90-degree phase points.

The top of L_{1B} is connected to both the sense circuit and the receiver antenna input. Potentiometer R_1 is a phasing control. The value of this pot is usually 10 to 100 kΩ, with 25 kΩ common. Switch S_1 is used to disconnect the sense antenna from the circuit. Because the nulls of the loop or loopstick are typically a lot deeper than the null on the cardioid pattern, the null is first located with the switch open. If the receiver S-meter reading drops when the switch is closed, the correct null has been selected. If not, reverse the direction of the antenna and repeat the process.

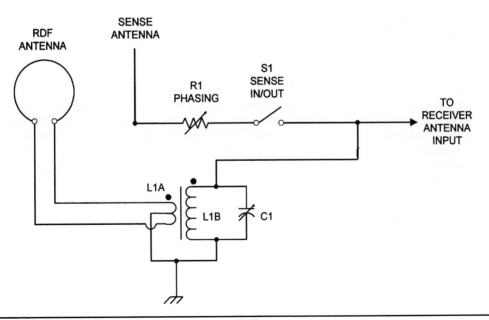

FIGURE 23.8 Sense circuit for a loop/sense antenna RDF.

RDF at VHF and UHF

It should be apparent from the preceding paragraphs that obtaining an easily portable antenna with a unidirectional pattern at HF requires some mechanical and electronic complexities. At VHF and above, such patterns are far easier to attain. Even a four-element 6-m Yagi is reasonably easy to manipulate by hand, and the task gets simpler as one goes higher in frequency. At even higher frequencies, the antenna of choice is the *parabolic reflector*.

Perhaps a more difficult task at VHF and above is finding a way to get signal strength information from the receiver being used. In today's marketplace, very few VHF/UHF FM mobile rigs include any visual metering, even when the AM aircraft band is present. A rare rig may have a dc voltage proportional to received signal level available on the rear panel, but the most likely transceivers to have S-meters (or their modern equivalent, the LED "bar" meters) are the relatively expensive "multimode" rigs intended primarily for base station use.

Fox Hunting

An activity popular with ham radio operators for many decades is *fox hunting*. A fiendishly clever ham (the "fox") would hide with a mobile or portable transmitter (usually on either 10 or 75 m). The "hunters" would then RDF the fox's brief transmissions and try to locate the transmitter. If you were first to locate the antenna, you won—and got to be the fox the following month!

Today, fox hunting is as likely to be held on VHF and UHF as on the lower frequencies. Because the propagation modes can be so different around a geographic region, both types provide many lessons.

Tracking Down Electrical Noise

A practical use for fox hunting skills is tracking down sources of electrical noise that can mask weak signals. Most electrical noise comes from close by: every electrical appliance that plugs into the wall (and even a few that don't!), every wall adapter, every computer or fax machine or wireless modem, every power line and power pole, every high-efficiency heating or cooling system, every on-demand hot water heater, every electric blanket, every touch lamp, and every streetlight or yard light is a suspect until proven "innocent". In most cases, a good technique is to start your "fox hunting" activities on the frequencies where you first noticed the noise and then gradually move upward in frequency as you get closer to the source. There are at least three reasons for that:

- Noise energy in the VHF and UHF ranges is not likely to travel via as many ambiguous propagation paths that can lead to contradictory results.

- Noise energy in the VHF and UHF ranges from most *unintentional radiators* is much weaker than at the lower frequencies; if you can hear it at all at VHF and UHF, you're probably very close to the source.

- It's far, far easier to hold and rotate a high-gain, sharp beam VHF or UHF antenna by hand than it is to get the same directivity on HF!

Here are some additional hints for locating that stubborn noise that's wiping out your weak-signal reception on the HF bands:

- Power line and power pole noises are particularly difficult because the noise can propagate for miles along the lines and go through many maxima and minimum before tapering off. Don't let one of those maxima cause you to jump to conclusions.

- Do your initial "hunting" quickly and easily while mobile by tuning the AM broadcast radio in your car to an unused frequency near the top of the broadcast band (somewhere in the 1600- to 1700-kHz range). The author found an offending municipal streetlight near his home in this manner.

- Keep your portable receiver in AM mode if at all possible. When it's time to switch to VHF antennas, amateur 2-m FM handhelds or mobile rigs that include coverage of the AM aircraft band between 108 and 130 MHz are especially useful for zeroing in on a specific power pole or building.

- Keep a log of dates, times, day/night status, and weather conditions each time you observe the noise. Some noises are the result of arcing when insulating surfaces get wet; others are the exact opposite. Some noises result from outdoor lighting powered by day/night sensors, and so on.

- To determine if the noise source is in your own home, turn off your house power at the main breaker box at a time when the noise is evident in your receiver. Continue to run your equipment on a 12-V car battery or an *uninterruptible power supply* (UPS). However, don't forget that UPSs and other battery-operated devices may continue to run for hours on their own and are fully capable of creating RFI long after primary power has been removed.

- Be careful, courteous, and circumspect when checking out noises potentially emanating from a neighbor's house; otherwise you may end up in front of the local judge, charged with being a "peeping Tom"!

CAUTION *Never, under any circumstances, "body-slam" a power pole or hit it with a sledgehammer or your vehicle's bumper to see if the offending noise is coming from it! Besides being highly illegal, this could be sure death if the trouble is, in fact, with any of the equipment on the pole. If you have reason to suspect a specific section of utility wiring, go back home and report your concern directly to your utility.*

The usual lack of an S-meter on mobile or portable VHF/UHF receivers is as much a problem for tracking down noise sources as it is for RDFing on distant stations. However, one possible work-around while listening to the AM aircraft band— if the background noise environment is favorable—is to employ the audio noise output of the speaker in a mobile VHF/UHF rig in conjunction with an audio signal level monitor app for a smart phone. Examples of apps currently available for the Apple *iPhone* include *Sound Level* (and *Sound Level Pro*). As with the S-meter approach, two operators are typically required—one to maneuver the antenna and another to watch the meter and call out "Warmer!" or "Colder!" as the signal level goes up or down.

Shortwave and AM BCB "Skip" RDF

Radio direction finding is most accurate over relatively short distances. If you have the choice, it's better to use the ground wave (which is what you use during daylight hours for nearly all AM BCB stations). Skip rolls in on the AM BCB after local sundown, so you can hear all manner of stations up and down the dial. However, with a little practice, you can even RDF distant stations.

Unfortunately, there are some practical difficulties associated with skip RDFing. When we look at propagation drawings of skip in textbooks we usually see what's happening in a single plane at a time. Typically, a single imagined "pencil beam" radio signal is shown traveling at some angle up the ionosphere, where it is "reflected" (actually, it's a refraction phenomenon but looks like reflection to an observer on the surface) back to earth. We can tell from the drawing that the angle of incidence equals the angle of reflection, just like they told us in high school science classes. Unfortunately, the real world is not so neat and crisp.

In the real world the skywave signal we hear is an amalgam of *many* such "pencil beams", coming in from a near-infinite number of directions and elevation angles. The strongest component of such a signal aggregate is always the path that has the least total loss from start to end, but for skip distances that path is always shifting because the ionosphere that makes skip reception possible is a continuously shifting medium. A radio wave refracting through the ionosphere must feel like you and I do when taking a ride through the house of warped mirrors at an amusement park. It's entirely possible, therefore, that at any given instant the preferred path to your antenna for a signal source located to the northwest of you might be a so-called skew path, coming in from as far around the compass rose as the southwest!

Terrestrial reflections also cause problems, especially when RDFing a station in the high end of the HF band or the VHF/UHF bands. Radio waves will reflect from geological features such as mountains and from man-made structures (e.g., buildings). If the reflection is strong enough, it might appear to be the real signal and cause a severe error in RDFing.

Bottom line: Be wary of RDF results on HF when the "skip is in". Similarly, don't let the canyons of New York City or the mountains of upstate fool you on VHF and UHF.

Adcock Antennas

The Adcock antenna (Fig. 23.9) has been around since 1919, when it was patented by British Royal Engineer and Lieutenant Frank Adcock as a way of overcoming the tendency of RDF loops to lose directional accuracy at nighttime. Adcock surmised that the horizontally polarized component of incoming skywave signals induced loop output voltages indicating certain spurious directions when the output voltage due to the vertically polarized ground wave was near zero. To minimize this error, Adcock designed a number of different configurations based on the use of vertical elements, both monopole and dipole.

The Adcock antenna shown here once found use in the United States by the FCC at their old monitoring sites. It consists of two center-fed, nonresonant (but identical) vertical radiators. Each side of each element is at least 0.1λ long, but need not be reso-

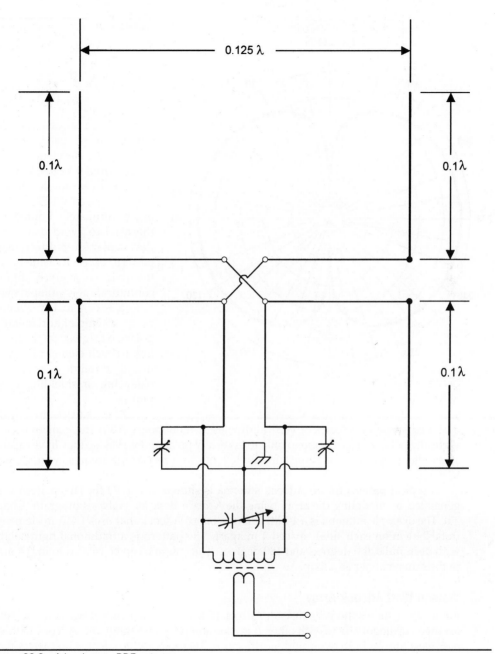

FIGURE 23.9 Adcock array RDF antenna.

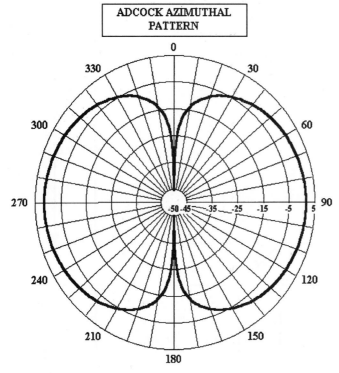

ADCOCK AZIMUTHAL
PATTERN

FIGURE 23.10   Adcock pattern.

nant (which means the antenna can be used over a wide band). The elements are spaced from 0.1λ to 0.75λ, although the example shown here is spaced 0.125λ. Spacing is ultimately limited by the need to avoid distorting the circular (omnidirectional) pattern of each element in the pair, so a given dipole configuration is suitable over only a 3- or 4-to-1 frequency range. Pickup of horizontally polarized signals on the feedline to each element is minimized through the use of balancing or shielding techniques.

Other Adcock arrays are designed around *two* pairs of vertical dipoles or monopoles. All Adcock antennas are vertically polarized and respond only to vertically polarized waves, so they have earned an excellent reputation as being superior to loops for precise high-frequency shortwave RDFing.

A typical pattern for an Adcock antenna is shown in Fig. 23.10. This pattern was generated by modeling the array using the *NecWin Basic for Windows* program (Chap. 25). The example antenna is a 10-MHz (30-m band) Adcock that uses 1.455-m elements (total 2.91 m on each side), spaced 4 m apart. The pattern is a traditional figure eight with deep nulls at 0 degrees and 180 degrees. The antenna can be rotated to find a null in the same manner as a loop.

Watson-Watt Adcock Array

Figure 23.11 shows the Watson-Watt Adcock RDF array, consisting of two Adcock pairs arranged orthogonally to each other. It is common practice to align one Adcock with an east-west line and the other with north-south. These are fed to identical receivers that are controlled by a common *local oscillator* (LO). The outputs of the receivers are balanced and are used to drive the vertical and horizontal plates of a cathode-ray oscilloscope (CRO). Figure 23.12 shows the patterns that result when signals of various phases arrive at the Watson-Watt array. The patterns of Figs. 23.12A and 23.12B are from signals 180 degrees out of phase, while the pattern of Fig. 23.12C is the result of a 90-degree phase difference.

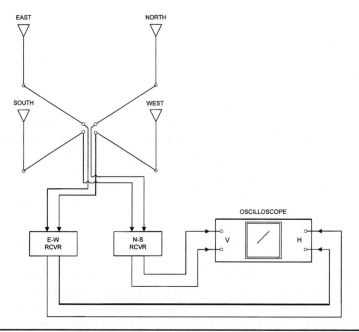

FIGURE 23.11 Watson-Watt Adcock array.

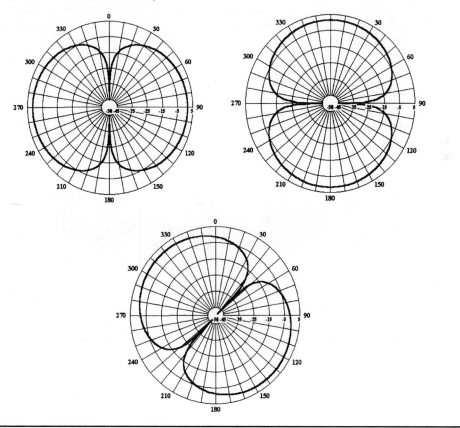

FIGURE 23.12 Patterns for different signals on the Watson-Watt array.

Doppler RDF Antennas

Figure 23.13 shows the basic concept of an RDF antenna based on the *Doppler effect*, which was discovered in the nineteenth century. A practical example of the Doppler effect is seen when an ambulance with wailing siren first approaches a stationary observer, then passes beyond. The pitch of the siren heard by the motionless observer is initially at its highest (and higher than the pitch heard at the source by the driver of the ambulance). It continuously drops as the source approaches and then passes by. Although useful, the acoustic analogy is not perfect, since the Doppler mechanism for EM waves in a vacuum is a little bit different than that for the subsonic acoustic waves of the ambulance.

In a radio system, when distance between a receiving antenna and a signal source is changing, a Doppler shift in the received signal frequency is detected; the amount of shift is proportional to the differential *radial* velocity between the two. (No Doppler shift occurs as a result of any change in motion at right angles to the line connecting the source and the receiver.) Thus, when a hobbyist listens to the telemetry signals from a passing satellite or the international Morse code (CW) of a radio amateur making contacts through an *AMSAT* repeater satellite (visit www.amsat.org), the received frequency continuously slides lower as the satellite comes into radio "view", passes nearby, and finally disappears until its next orbit. At the point of the satellite's closest approach to the receiving station (whether directly overhead or off to the side), the received frequency equals the actual transmitted frequency because neither antenna has any radial velocity relative to the other for that brief moment. Before then, the received frequency is higher; afterward, it is lower.

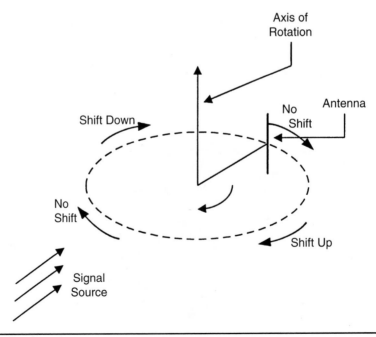

FIGURE 23.13 Doppler antenna.

The Doppler RDF antenna of Fig. 23.13 rotates at a constant angular velocity. The signal wavefront approaches from a single direction (the lower left corner of the illustration in this example), so there will be a predictable Doppler shift at any point on the circular path of the antenna. The magnitude of the frequency shift is maximum when the antenna is moving directly toward the source or away from it—i.e., when it is on either side of the figure. When the antenna is closest to the source or farthest from it, there is very little Doppler shift. The maximum shift at the two sides can be found to be

$$S = \frac{R\omega F_c}{c} \tag{23.1}$$

where S = Doppler shift, in hertz
R = radius of rotation, in meters
ω = angular velocity of antenna, in radians per second
F_c = carrier frequency of incoming signal, in hertz
c = velocity of light (3×10^8 m/s)

In theory this antenna works nicely, but in practice there are problems, one of which is getting a Doppler shift large enough to easily measure. Unfortunately, the mechanical rotational speed required of the antenna is very high—too high for practical use. However, the effect can be simulated by sequentially (i.e., electronically) scanning a number of antennas arranged in a circle. The result is a piecewise approximation of the effect seen when the antenna is rotated at high speed. *Byonics* (www.byonics.com) offers a kit by Daniel F. Welch, W6DFW, that incorporates the N0GSG Doppler system with *digital signal processing* (DSP) originally described in the November 2002 issue of *QST*. In this system, four verticals are electronically switched at high speed to create the Doppler shift that allows pinpointing of the source to a 22.5 percent segment of the compass rose. The user is responsible for constructing the four verticals. Googling the words "Doppler RDF antenna" will provide numerous hits on the topic.

Wullenweber Array

One of the problems associated with small RDF antennas is that relatively large distortions of their pattern result from even small anomalies because they have such a small aperture. If you build a *wide-aperture direction finder* (WADF), however, you can average the signals from a large number of antenna elements distributed over a large-circumference circle. The Wullenweber array (Fig. 23.14) is such an antenna. It consists of a circle of *many* vertical elements. For HF the circle can be 500 to 2000 ft in diameter!

A *goniometer* rotor spins inside the ring to produce an output that will indicate the direction of arrival of the signal as a function of the position of the goniometer. The theoretical resolution of the Wullenweber array is on the order of 0.1 degree, although practical resolutions of less than 3 degrees are commonly seen.

Time Difference of Arrival (TDOA) Array

If two antennas are erected at a distance d apart, arriving signals can be detected by examining the time-of-arrival difference. Figure 23.15A shows a wavefront advancing

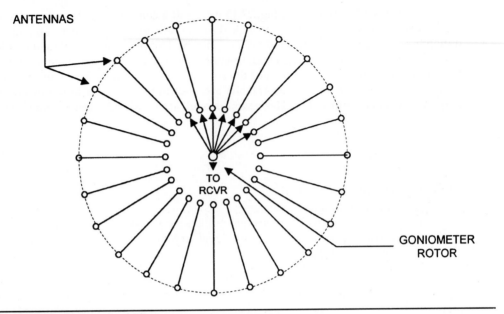

FIGURE 23.14 Wullenweber RDF array.

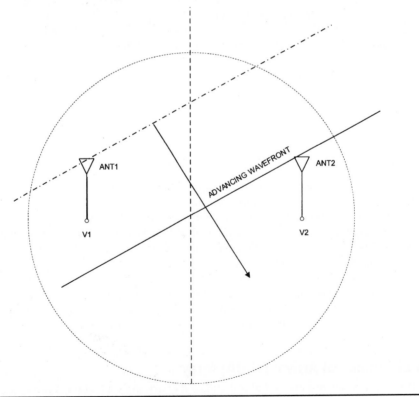

FIGURE 23.15A Time difference of arrival (TDOA) RDF array.

at an arbitrary angle to the line between two antennas. If the wavefront is parallel to the antenna axis, it will arrive at both antennas at the same time. But if the signal arrives at an angle, it will arrive at one antenna some time before it reaches the second. From the difference in the arrival times at the two antennas we can calculate the arrival angle.

There is an ambiguity in the basic TDOA array in that the combined output will be the same for conjugate angles, i.e. at the same angle on opposite sides of a line bisecting the array. This problem can be resolved by adding the electronic system of Fig. 23.15B. Signals from ANT1 and ANT2 (designated V_1 and V_2, respectively) are detected by separate receivers (RCVR1 and RCVR2) and then threshold detected in order to prevent signal-to-noise problems from interfering with the operation. The outputs of the threshold detectors are used to trigger a sawtooth generator that controls the horizontal sweep on an oscilloscope. The two signals are then delayed, and one is inverted so that the operator can distinguish between their traces on the CRT screen, thus eliminating the ambiguity.

Example: If the antennas in Fig. 23.15A are arrayed east and west, the line perpendicular to the line drawn between them runs north and south. If we designate north as 0 degrees, the signal shown arrives at an angle of 330 degrees, and it will arrive at ANT1 before it arrives at ANT2. A signal arriving from a bearing of 30 degrees will produce the same output signal, even though it arrives at ANT2 before ANT1. All signals of bearing $0 \leq \theta < 180°$ will arrive at ANT2 first, while all signals $180 \leq \theta < 360°$ will arrive at ANT1 first. Yet both will produce the same blip on the oscilloscope screen. The solution to discerning which of the two conjugate angles is the true arrival angle is to invert the ANT1 signal. When this is done, the ANT1 signal falls below the baseline on the CRT screen, while the ANT2 signal is above the baseline. By noting the time difference between the pulses and their relative positions, we can determine the bearing of the arriving signal.

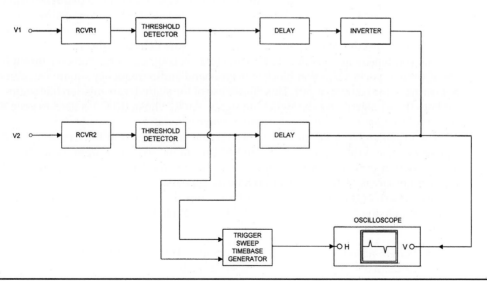

FIGURE 23.15B   Block diagram of typical TDOA circuit.

Switched-Pattern RDF Antennas

If an advancing wavefront arrives at two identical antennas from a direction perpendicular to the line connecting the antennas, the phase of the wavefront will be the same at both antennas. If signals from the two antennas are "chopped" at an audio rate with a square wave and alternately fed to a simple receiver, the resulting phase modulation will appear as an audio tone at the receiver output. Rotation of the antenna array until the tone is nulled out, which occurs when there is no phase difference between the signals at the two antennas, provides an accurate bidirectional "fix" on the source of the wavefront.

Double-Ducky Direction Finder (DDDF)

Figure 23.16 shows in simplified form a system originally presented to amateur radio operators by David T. Geiser, W5IXM, for use with "rubber ducky" VHF antennas; he called it the *double-ducky direction finder* (DDDF). Because of the sharpness of the null, Geiser chose to use these short, loaded, and rubber-damped antennas rather than full $\lambda/4$ or $5/8\lambda$ whips because vibration and wind can cause enough relative motion between elements of an array built with full-length whips to make it difficult to get a clean null in normal outdoor use.

Spacing of the two antennas is not critical; 0.25λ to 1λ apart over a good ground plane (such as the roof of a car or truck) should work well. As with all vertical monopoles, if no ground plane exists, a sheet metal ground plane should be provided.

In Fig. 23.16 the antennas are fed from a common transmission line to the receiver. In order to keep them electrically the same distance apart, identical half-wavelength sections of transmission line are used to connect the antennas to the shared section of line.

Switching is accomplished by using a bipolar square wave and PIN diodes. The bipolar square wave (see inset to Fig. 23.16) has equal positive and negative peak voltages on opposite halves of the cycle. The PIN diodes (D_1 and D_2) are connected in opposition such that diode D_1 conducts on negative excursions of the square wave and D_2 conducts on positive excursions. In any given half-cycle of the audio frequency square wave, one antenna is connected to the receiver when its diode is conducting and the other diode is back-biased, creating a high impedance between its antenna and the shared line. Thus, the active antenna alternates rapidly between ANT1 and ANT2.

The combined signal from the two antennas is coupled to the receiver through a small-value capacitor (C_1) that blocks the baseband audio frequency square wave from appearing at the receiver input. This allows use of the shared transmission line segment for both the RF signal and the switching signal. An RF choke (RFC_3) is used to keep RF from the antenna from entering the square-wave generator.

The DDDF antenna phase-modulates the incoming signal at the frequency of the square wave, and the resulting tone can be heard in the receiver output. When the signal's direction of arrival is perpendicular to the line between the two antennas, the phase difference is zero and the audio tone disappears.

The pattern of the DDDF antenna is bidirectional, so the ambiguity problem found with loop antennas exists here, as well. In this system the ambiguity can be resolved by either of two methods:

- Place a reflector $\lambda/4$ behind each antenna. This is an attractive solution, but it tends to distort the antenna pattern a little bit.
- Rotate the antenna through 90 degrees (or walk an L-shaped path).

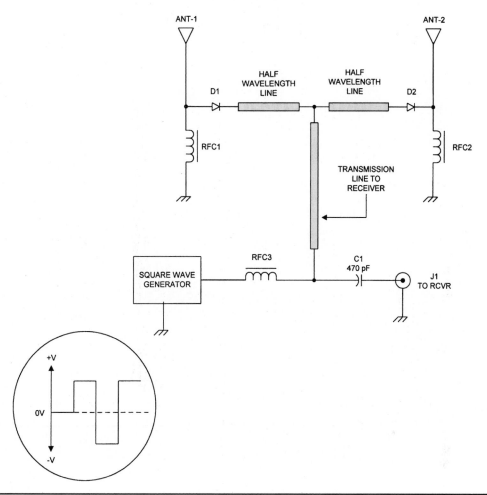

Figure 23.16   Double-ducky switched-pattern RDF system.

Summary

This chapter has touched on the basics of radio direction finding. Much more depth can be found on the Internet and in newsletters from clubs and others specializing in RDF technology and techniques. RDF can be very useful for locating RF sources, both good and bad; as such, it is an important tool for locating illegal stations or undesired noise sources. RDF is a great way to build club spirit and teach a wide variety of technical concepts through fox-hunting training exercises and group construction projects. It can also be used to determine your own location if you can get bearings on at least two RF sources at known positions. Try it . . . you'll like it!

PART VII

Tuning, Troubleshooting, and Design Aids

CHAPTER 24

Antenna Tuners (ATUs)

The primary task of an impedance-matching or antenna tuning unit (ATU) located at the *transmitter* end of the transmission line is to present the transmitter output stage with an apparent antenna feedpoint impedance (after transformation through an arbitrary-length transmission line) equal to the output impedance of the transmitter. This results in maximum power transfer from the power amplifier stage of the transmitter to the feedline and antenna. So located, a properly designed ATU provides the station owner with multiple benefits:

- It minimizes the standing wave ratio (SWR) that the transmitter or power amplifier—whether vacuum tube or solid-state—sees. RF power amplifiers are not tolerant of high SWR; when operating into a mismatched load, expensive tubes, transistor(s), and other components can be destroyed instantly.

- It minimizes the occurrence of reduced output power triggered by power amplifier "fold-back" protection circuitry when a high VSWR is sensed.

- It facilitates operation of the transmitter final amplifier stage at maximum efficiency.

- It provides additional attenuation of out-of-band emissions, including harmonics generated in, or amplified by, the amplifier.

- If left in the line to the receiver during "key up" periods, it helps protect the receiver from overload caused by strong out-of-band signals.

If, on the other hand, the ATU is located at the *antenna* end of the transmission line, it not only accomplishes all those benefits listed here but it also transforms the feedpoint impedance of the antenna to the *system impedance* or principal station transmission line impedance—totally eliminating the reactive part of the feedpoint impedance and presenting the transmission line with an R_{RAD} that is identical to Z_0. Thus, an ATU located at the antenna provides the user with the following additional benefits:

- It minimizes the occurrence and amplitude of high-voltage standing waves along the line that can destroy the transmission line and any associated switches, relays, or other components when high power is employed.

- It minimizes additional, or *mismatch,* signal loss on long transmission lines relative to the loss that is present with line SWRs greater than 1.0:1.

- It can include additional protection for station electronics (receivers, transmitters, accessories, etc.) from lightning-induced events, serving as a first line of defense in preventing voltage or current surges from entering the radio room.

As the second set of bullets makes clear, it is generally preferable to locate the ATU at the antenna. However, at low frequencies, low power levels, and/or for relatively short distances between the transmitter output and the antenna, an ATU at the transmitter can provide acceptable results (except with respect to keeping lightning out of the radio shack). Today most commercially available transceivers sport an internal ATU—either standard or as an option. When operated "barefoot" (i.e., with no amplifier) and with short feedlines (such as mobile or portable installations) on HF, this configuration is compact and provides nearly as efficient power transfer to the antenna as a separate box located at the antenna would.

In many stations, however, the approach is to incorporate matching devices at *both* ends of the system transmission line—that is, to locate fixed-tuned impedance-matching networks at the antenna feedpoint to get the SWR on the system transmission line down to a manageable level and a tunable ATU back at the transmitter and receiver end of the main transmission line to complete the task of presenting the transmitter with a matched load, especially when the operating frequency is apt to be varied by a modest amount—typically no more than a few percent. Often, the impedance-matching network at the antenna is a simple *balun* or a *distributed* network, such as the transmission line transformers described in the second half of Chap. 4.

Occasionally fully tunable ATUs are located at or near the antenna end of the system transmission line. In that case, they are often capable of being tuned remotely, or the intended operating range is over such a small percentage bandwidth that adjustment of the ATU is required infrequently or never. An example of such a configuration is found in Fig. 8.2; open-wire line (which is the best choice when the possibility of high SWR on the line exists) connects the center of a multiband dipole to a tunable ATU directly beneath it. The ATU is adjusted to provide minimum SWR on the coaxial transmission line that comes from the station equipment. It may also incorporate RF chokes to drain static buildup from the antenna and OWL, or remotely actuated mechanical contactors to directly ground the OWL and dipole when they're not in use. An even more compelling case for a remote ATU is the bobtail curtain of Chap. 10. Since the feedpoint for this array is at a high-impedance point, the SWR on any common coaxial cable back to the transmitter would put the cable at risk of destruction when substantial power levels are used. Any attempt to run open-wire line over that distance could materially degrade the antenna pattern. Locating a tunable ATU at the base of the antenna, as shown in Fig. 10.11, is by far the best possible approach.

ATU Circuit Configurations

While it is tempting to treat the ATU as a *black box,* in truth there are many different kinds of impedance-matching circuit configurations, each with its own list of advantages and disadvantages. At any given operating frequency, the antenna feedpoint impedance, Z_{ANT}, can be represented as a resistive part, R_{ANT}, in series with a reactive part, X_{ANT}:

$$Z_{ANT} = R_{ANT} + jX_{ANT} \qquad (24.1)$$

where X_{ANT} can be either positive or negative. If we could count on R_{ANT} always equaling the system transmission line impedance, Z_0, we could use a single variable capacitor in series to tune out a positive X_{ANT} and a single variable inductor in series to tune out a negative X_{ANT}. Unfortunately, as the operating frequency changes, not only does X_{ANT} vary—often between positive and negative values—but R_{ANT} does so, as well. As explained in Chap. 3, neither R_{ANT} nor X_{ANT} corresponds to a specific resistor, capacitor, or inductor except in the very simplest electronic circuits. The magnitude and frequency dependence of either R_{ANT} or X_{ANT}—and most likely *both*—are, in general, complicated functions of many components in a lumped-element circuit and of many geometrical interrelationships in a distributed circuit such as an antenna. Thus, the task of the successful antenna matching unit or ATU is to be flexible enough to present the system transmission line with an *apparent* antenna feedpoint impedance of $Z_{ANT} = Z_0 = R_0 + j0$ over a reasonable range of operating frequencies and a practical range of complex antenna impedances. In the process, the ATU should dissipate as little of the transmitter RF power output in internal losses as possible. As you might expect, various network topologies can do this—but with varying degrees of success.

L-Section Network

Judging by the number of times it has appeared in print, the L-section network is one of the most used antenna matching networks in existence, rivaling even the pi network. A typical circuit for one form of L-section network is shown in Fig. 24.1A. Values for L and C can be found that will allow this circuit to match all possible load impedances having R_1, the resistive part of the source impedance, less than R_2, the resistive part of the load impedance. The circuit can also match *some* load impedances for the range $R_1 > R_2$, depending on the magnitude of X_2, the reactive part of the series-representation load im-

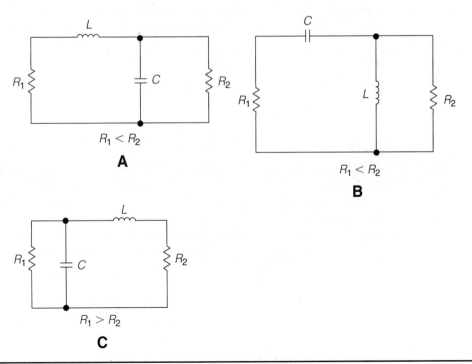

FIGURE 24.1   (A) L-section network. (B) Reverse L section. (C) Inverted L-section network.

pedance. For the special case when both the source and the load impedances are purely resistive, the design equations are:

$$Q = \sqrt{\frac{R_2}{R_1} - 1} \qquad\qquad (24.2)$$

$$X_L = Q \times R_1 \qquad\qquad (24.3)$$

$$X_C = \frac{R_2}{Q} \qquad\qquad (24.4)$$

where, of course,

$$X_C = \frac{1}{2\pi f C} \qquad\qquad (24.5)$$

and

$$X_L = 2\pi f L \qquad\qquad (24.6)$$

The L-section network of Fig. 24.1B differs from the previous circuit in that the coil and capacitor locations are swapped. As before, values for L and C can be found that will allow this circuit to match all possible load impedances having $R_1 < R_2$, and *some* load impedances for the range $R_1 > R_2$, depending on the magnitude of X_2. For the special case when both the source and the load impedances are purely resistive, the design equations are:

$$Q = \sqrt{\frac{R_2}{R_1} - 1} \qquad\qquad (24.7)$$

$$X_L = \frac{R_2}{Q} \qquad\qquad (24.8)$$

$$X_C = R_1 Q \qquad\qquad (24.9)$$

Figure 24.1C is similar to Fig. 24.1A with the exception that the capacitor is at the input rather than the output of the network. Values for L and C can be found that will allow this topology to match all possible load impedances having $R_1 > R_2$, and *some* load impedances for the range $R_1 < R_2$. The equations governing this network are:

$$Q = \sqrt{\frac{R_1}{R_2} - 1} \qquad\qquad (24.10)$$

$$X_L = R_2 Q \qquad\qquad (24.11)$$

$$X_C = \frac{R_1}{Q} \qquad\qquad (24.12)$$

Because of the limitations on the range of load impedances it can match when either the source or the load has a reactive component, an L network located at the transmitter end of a feedline should be configurable to accommodate the different topologies shown in Fig. 24.1. Alternatively, an L network located at the antenna feedpoint can be a fixed topology once the nature of the feedpoint impedance of the antenna is known.

Pi Networks

The pi network (or *pi-section network*) shown in Fig. 24.2 is used to match a high source impedance to a low load impedance. These circuits are typically used in vacuum tube RF power amplifiers that need to match output impedances of a few thousand ohms to much lower system transmission line impedances—typically 50 or 75 Ω. The name of the circuit comes from its resemblance to the Greek letter pi (π). The equations for the pi network are:

$$R_1 > R_2 \text{ and } 5 < Q < 15 \tag{24.13}$$

$$Q > \sqrt{\frac{R_1}{R_2} - 1} \tag{24.14}$$

$$X_{C2} = \frac{R_2}{\sqrt{\left(\frac{R_2}{R_1}\right)(1+Q^2) - 1}} \tag{24.15}$$

$$X_{C1} = \frac{R_1}{Q} \tag{24.16}$$

$$X_L = \frac{R_1\{(Q + (R_2 / X_{C2})\}}{Q^2 + 1} \tag{24.17}$$

Of course, if $R_1 < R_2$, the pi network can be flipped left to right; that is, R_1 can become the load and R_2 the source. Because the pi network typically has a higher Q than an L-section network, its bandwidth will generally be narrower. If bandwidth is an important issue, a pair of cascaded L sections may be a better solution to a particular matching task.

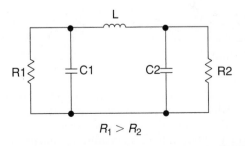

$$R_1 > R_2$$

Figure 24.2 Pi network.

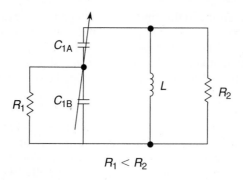

Figure 24.3   Split-capacitor network.

Split-Capacitor Network

The split-capacitor network shown in Fig. 24.3 is used to transform a source impedance that is less than the load impedance. In addition to matching antennas, this circuit is also used for interstage impedance matching inside communications equipment. The equations for design are:

$$R_1 < R_2 \tag{24.18}$$

$$Q > \sqrt{\frac{R_2}{R_1} - 1} \tag{24.19}$$

$$X_L = \frac{R_2}{Q} \tag{24.20}$$

$$X_{C1B} = \sqrt{\frac{R_1 (Q_2 + 1)^2}{R_2} - 1} \tag{24.21}$$

$$X_{C1A} = \frac{R_2 Q}{Q^2 + 1} \left(1 - \frac{R_1}{Q X_{C1B}} \right) \tag{24.22}$$

Transmatch Circuit

One version of the *transmatch* is shown in Fig. 24.4. This circuit is basically a combination of the split-capacitor network and an output tuning capacitor (C_2). For the HF bands, the capacitors are on the order of 150 pF per section for C_1, and 250 pF for C_2. The tapped or roller inductor should have a maximum value of 28 μH. A popular use of the transmatch is as a coax-to-coax impedance matcher.

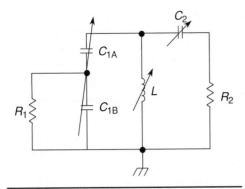

FIGURE 24.4 Split-capacitor transmatch network.

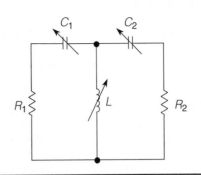

FIGURE 24.5 Tee-network transmatch.

The long-popular E. F. Johnson *Matchbox* is a fundamentally balanced variant of this circuit, designed primarily for use with balanced feedlines such as open-wire line, twinlead, et al. In the *Matchbox*, coupling to the transmitter or receiver is accomplished via an inductive link (an isolated secondary winding on *L*), thus eliminating the connection of R_1 across half of C_1. C_2 becomes a split capacitor connected across the full length of tapped inductor *L*, and the junctions of the two halves of C_1 and C_2 are tied together and grounded to the chassis. Each half of the new C_2 is itself split again, and the two sides of the balanced transmission line are connected at the midpoint of the split-capacitor half of C_2 on each side of ground. The center conductor of an SO-239 chassis-mounted co-axial receptacle for feeding unbalanced transmission lines is hard-wired to the midpoint of one side of C_2.

Perhaps the most common form of transmatch circuit is the tee network shown in Fig. 24.5. Like the *reverse L-network* of Fig. 24.1B, it is basically a *high-pass filter* and thus does nothing for transmitter harmonic attenuation.

An alternative network, called the *SPC transmatch*, is shown in Fig. 24.6. This version of the circuit offers some harmonic attenuation.

Lumped-component matching networks (or ATUs) for MF and HF are relatively easy to build, although the cost of inductors and variable capacitors capable of withstanding the voltages and currents associated with transmitter power levels is eye-opening. (Even when purchased used at flea markets or via the Internet, good high-power components are not cheap!) Probably the most important thing to remember when building your own ATU is to leave lots of space between any metallic enclosure and the internal components. Failure to do so can alter the circuit *Q* and reduce overall efficiency of the unit.

In the early days of radio, very few home-built transmitters or ATUs were shielded. The drive to shield RF-generating circuits gained strength in the 1940s with the spread of television broadcasting and the concomitant potential for interference to the neighbors' TV reception. Today, all commercially produced transmitters, transceivers, and amplifiers are thoroughly shielded and filtered

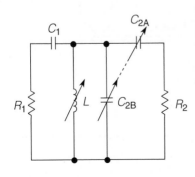

FIGURE 24.6 Improved transmatch offers harmonic attenuation.

against the unwanted emission of harmonics and other spurious products of the signal generation and amplification process, so shielding of the ATU may not be seen as an imperative. However, every tuned circuit in the path between transmitter and antenna helps reduce these undesired emissions across some frequency ranges, so total shielding of the ATU circuitry is always the best practice.

Baluns

A *balun* is a transformer that matches a *bal*anced load (such as a horizontal dipole or Yagi antenna) to an *un*balanced resistive source (such as a transmitter output or a coaxial feedline). The baluns discussed in this section are lumped-circuit implementations typically wound on ferrite cores, but baluns can also be formed from sections of transmission line properly interconnected. The latter type is included in the review of distributed matching networks near the end of Chap. 4.

Toroid Impedance-Matching Transformers

The toroidal transformer is capable of providing a broadband match between antenna and transmission line, or between transmission line and transmitter or receiver. The other matching methods (shown thus far) are frequency-sensitive and must be readjusted whenever the operating frequency is changed by even a small percentage. Although this problem is of no great concern to fixed-frequency radio stations, it is of critical importance to stations that operate on a variety of frequencies or widely separated bands of frequencies.

Figure 24.7A shows a trifilar transformer that provides a 1:1 impedance ratio, but it will transform an unbalanced transmission line (e.g., coaxial cable) to a balanced signal required to feed a dipole antenna. Although it *provides no impedance transformation*, it does tend to balance the feed currents in the two halves of the antenna. This fact makes it possible to obtain a more symmetrical figure eight dipole radiation pattern in the horizontal plane. Many station owners make it standard practice to use a balun at the antenna feedpoint when using coaxial (unbalanced) transmission line, even if the feedpoint impedance is close to the Z_0 of the cable.

The balun shown in Fig. 24.7B is designed to provide unbalanced to balanced transformation in conjunction with a 4:1 impedance transformation. Thus, a 300-Ω folded

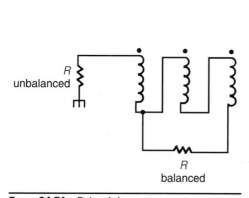

FIGURE 24.7A Balun 1:1.

FIGURE 24.7B Balun 1:4.

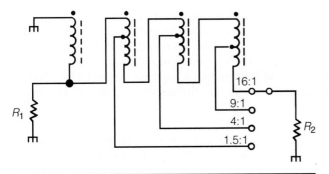

FIGURE 24.7D Commercial tapped transformer.

FIGURE 24.7C Tapped multiple-impedance BALUN.

dipole feedpoint impedance will be transformed to 75-Ω unbalanced. This type of balun is often included inside ATUs. L_1 and L_2 have equal turns.

Another multiple-impedance transformer is shown in Fig. 24.7C. In this case, the operator can select impedance transformation ratios of 1.5:1, 4:1, 9:1 or 16:1. A commercial version of this type of transformer, shown in Fig. 24.7D, is manufactured by Palomar Engineers; it's intended for feeding vertical HF antennas. It will, however, work well in other impedance-transformation applications, too.

Figures 24.7B and 24.7C are specific examples of what can be accomplished using a toroidal core of the right material and power-handling capability in conjunction with an *autoformer* winding, where both input and output loads are referenced to a common ground. If ground isolation is important, there is no reason completely separate windings can't be used, in a configuration analogous to a link-coupled coil.

Ferrite Core Inductors

The word *ferrite* refers to any of several examples of a class of materials that behave similarly to powdered iron compounds and are used in radio equipment as the cores for inductors and transformers. Although the materials originally employed were of powdered iron (and indeed the name *ferrite* still implies iron), many modern materials are composed of other compounds. According to literature from Amidon Associates, ferrites with a *relative permeability*, or μ_r, of 800 to 5000 are generally of the manganese-zinc type of material, and cores with relative permeabilities of 20 to 800 are of nickel-zinc. The latter are useful in the 0.5- to 100-MHz range.

Toroid Cores

In electronics parlance, a *toroid* is a doughnut-shaped object made from a ferrite material and used as the form for winding an inductor or transformer. Many different core material formulations are available for the designer to choose from, based on the expected frequency range and power levels of the application.

Unlike just about any other lumped component available to the experimenter, most toroids carry little or no information about themselves on their surface. In general, blindly grabbing an arbitrary toroid from the junk box will lead to a failed project. Toroids must be selected carefully—with the use of a test jig, if necessary.

Charts of product characteristics for toroids are provided by manufacturers such as Amidon and Fair-Rite, and by their distributors, and are best obtained from an Internet search of their sites. In addition, some of these data are also available in *The ARRL Handbook for the Radio Amateur*. As an example, a T-50-2 core is useful from 1 to 30 MHz, has a permeability of 10, is painted red, and has the following dimensions: OD = 0.500 in (1.27 cm), ID = 0.281 in (0.714 cm), and a height (i.e., thickness) of 0.188 in (0.448 cm).

Toroidal Transformers

Windings on a toroid are generally wound simultaneously in a "multifilar" manner. In this approach, multiple wires of equal length are slightly twisted together first, then wrapped around and through the doughnut form as if they were a single winding. The preferred number of wires in parallel is a function of the required *step-up* or *step-down* ratio and whether an autotransformer or isolated primary and secondary windings are required. Figure 24.8A is the schematic of a 4:1 balun transformer with separate primary and secondary windings. The dot at the top of each winding indicates the polarity or phase of the winding; for proper operation as an autoformer, all three wires must wrap around the doughnut in the same direction, and the builder must mark the dotted end of all three—conceptually, if not literally—when soldering ends together or bringing them out to connectors. This particular circuit is said to have a *trifilar wound* core.

Figure 24.8B exposes detail of the trifilar winding. For the sake of clarity, each of the three wires has been drawn with its own simulated insulation "pattern" so that you can more easily see how the winding is formed. Most small construction projects (e.g., Beverage antenna transformers) use #26, #28, or #30 enameled wire to wind coils; consider stocking three colors of each size, so that each winding can be a unique color. For higher-power transmitting antenna baluns, use #16, #14, #12, or #10 wire. Enamel and Teflon are common insulation layers. To help identify wires of these diameters, label their ends with adhesive labels.

Figure 24.9 shows two accepted methods for winding a multifilar coil on a toroidal core. Figure 24.9A is an actual photograph of toroid wound with the method shown pictorially in Fig. 24.8B. Here the wires are laid down parallel to each other, as shown previously. The method of Fig. 24.9B uses twisted wires. Prior to winding the toroid, the three wires are "chucked up" in a drill and twisted together before

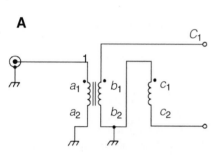

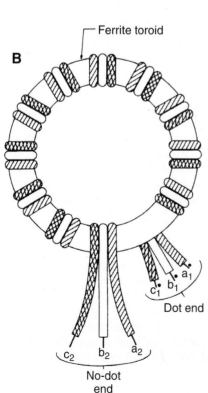

FIGURE 24.8   Broadband RF transformer.

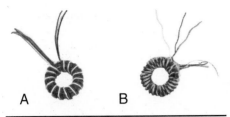

FIGURE 24.9   Winding a toroidal transformer. (A) Parallel wound. (B) Twist wound.

being wound on the core. With one end of the three wires secured in the drill chuck, anchor their far ends in something that will hold them taut. Some people use a bench vise for this purpose. Run the drill at slow speed and allow the wires to twist together until the desired pitch is achieved. Remember to wear eye protection in case a wire breaks or gets loose from its mooring at the far end.

To actually wind the bundle of wires on the toroid, pass the bundle through the doughnut hole until the toroid is about in the middle of the length of wire. Loop the wire over and around the outside surface of the toroid and pass it through the hole again. Repeat this process until the correct number of turns is wound onto the core. Be sure to press the wire against the toroid form, and keep it taut as you wind the coils.

To minimize the chance of a chipped core breaking through the wire insulation, wrap the core with a slightly overlapped layer of fiberglass packing tape before starting to wind. To prevent windings (especially those made with very fine wire) from unwinding, secure the ends of the wires with a tiny dab of rubber cement or RTV silicone sealer.

Mounting Toroid Cores

To mount a properly wound toroidal inductor or transformer, consider these options:

- If the wire is heavy enough, just use the wire connections to the circuit board or terminal strip to support the component—especially if the connections are directly above or below the toroid.

- Smaller, lighter toroids can be laid flat on the circuit board and cemented in place with silicone seal or rubber cement.

- Drill a hole in the wiring board and use a screw and nut to secure the toroid. Do not use metallic hardware for mounting the toroid! Metallic fasteners will alter the inductance of the component and possibly render it unuseable. Use nylon hardware for mounting the toroidal component.

How Many Turns?

Three factors must be taken into consideration when making toroid transformers or inductors: *toroid size, core material,* and *number of turns of wire.* Toroid size is a function of power-handling capability and handling/installation convenience. Core material depends on the frequency range of the circuit and the application (balun, common mode choke, etc.).

Equation (24.23) provides a rough rule of thumb for the turns count. Based on the availability of a parameter called the A_L factor, the number of turns needed to obtain a specified inductance is given by

$$N = 100\sqrt{\frac{L}{A_L}} \qquad (24.23)$$

where N = number of turns on winding
   L = inductance, in microhenrys
   A_L = core factor, in microhenrys per 100 turns

Example 24.1   Calculate the number of turns required to make a 5-µH inductor on a T-50-6 core. The A_L factor for the core is 40.

Solution

$$N = 100\sqrt{\frac{L}{A_L}}$$

$$= 100\sqrt{\frac{5}{40}}$$

$$= 100\sqrt{0.125}$$

$$= (100)(0.35)$$

$$= 35$$

◆

Don't take the value obtained from the equation too seriously, however, because a wide tolerance exists on amateur-grade ferrite cores. Although it isn't too much of a problem when building baluns and other transformers, it can be a concern when making fixed inductors for a tuned circuit. If the tuned circuit requires considerably more (or less) capacitance than called for in the standard equation, and all of the stray capacitance is properly taken into consideration, the actual A_L value of your particular core may be different from the table value.

Ferrite Rods

Another form of ferrite core available on the market is the rod, shown in Fig. 24.10. Often used as a high-current RF choke for the vacuum tube filaments of grounded-grid linear amplifiers, the rod can be pressed into service as a balun, as well. Primary and secondary windings are wound in a bifilar manner over the ferrite rod.

Ferrite rods are also used in receiving antennas—especially when high inductance in a small space is required. Although the amateur use is not extensive, the ferrite rod antenna (or *loopstick*) is popular in small receivers for MF and below, and in portable radio direction-finding equipment. Some amateurs use sharp nulls of loopstick receiving antennas to null out interfering signals on the crowded HF bands. Of course, you would not want to use a small loopstick rod in a transmitting application.

Ferrite rods are light enough to be suspended from their own wires or to be glued or cemented to a panel or printed circuit board inside the receiver. In place of the simple nylon screws that hold toroids in place, we can use insulating cable clamps to secure the ends of the rod to the board.

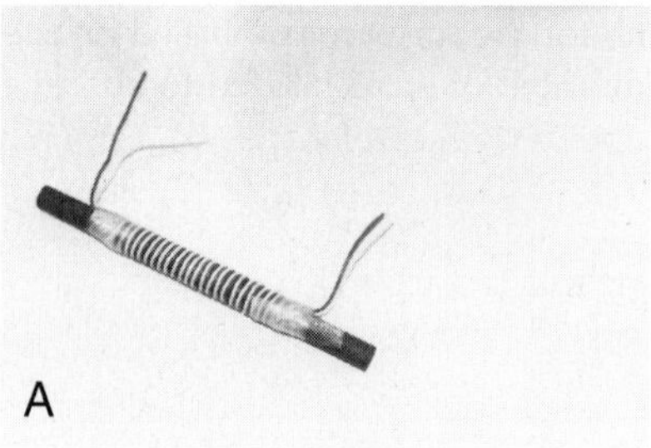

A

FIGURE 24.10   Ferrite rod inductor construction.

CHAPTER 25

Antenna Modeling Software

O ne of the significant contributions of computer technology to the field of antennas is the availability of modeling and simulation tools that eliminate most of the drudgery associated with the complex calculations involved in analyzing and optimizing antenna designs. For most readers of this book, the best part of this is the extension of those tools to personal computers at very modest prices and even for free! Today, any individual interested in experimenting with novel antenna configurations can do so, and in the process spend little or nothing on modeling software.

Modeling and simulation are used in a wide variety of applications, including management, science, and engineering. We can now model just about any process, any device, or any circuit that can be described mathematically. The purpose of modeling, at least in engineering, is to validate the design quickly and cheaply on the computer before "bending metal". Modeling and simulation make it possible to look at alternatives and gauge the effect of a proposed design change *before* the change is actually implemented. The old "cut and try" method works, to be sure, but it is costly in both time and money—two resources perpetually in short supply. If performance issues and problems can be solved on a computer, then we are time and money ahead of the game.

Nowhere is this more true than in the field of antennas, where each design iteration historically involved lowering the antenna from its support, adjusting its dimensions, reinstalling it many feet up in the air, and conducting a new round of test measurements. Alternatively, towers with sizeable nonconducting work platforms were utilized, and all the physical adjustments and fine-tuning accomplished "up topside"—in situ, as it were. A half-century ago or less, antenna test ranges with these capabilities were found only at government research centers and commercial antenna manufacturers' facilities.

Under the Hood

Virtually all commonly available antenna modeling software for personal computers is based on a numerical analysis technique known as the "method of moments", in which each wire or antenna element in the system is broken into many short segments, and the current in each segment is calculated based on the rules of electromagnetism and the *boundary conditions* seen by each segment. The fields generated by the antenna are determined by summing the contributions of all wire segments throughout *model space*, varying both azimuth and elevation angle to compute, tabulate, and graphically plot relative field strength compared to an *isotropic radiator* as a function of angular position from the antenna for a constant RF drive power applied to the antenna feedpoint(s). Ancillary data, including plots of voltage standing wave ratio (VSWR) versus frequency, calculation of feedpoint voltage and current, currents in all wire segments, etc., are also available.

The core for most of the programs in use today was originally developed for the U.S. Navy. Called the *Numerical Electromagnetic Computation* (NEC) *program*, it has had a number of updates over the years; the most recent that is available to users and software authors is NEC-4. The government still charges for an NEC-4 license, and it cannot be exported without a permit from the U.S. government, but an earlier version, NEC-2, is available free of charge and is currently the version most commonly used by hobbyists and as the core of free or inexpensive NEC-based applications. When PCs became popular, a smaller version known as miniNEC was made available. Initially it ran under BASIC, but faster FORTRAN versions became available later. You can still download a number of miniNEC variants from several Internet sites.

Despite its computational power, the core NEC software is not particularly user-friendly, with an input/output format analogous to the old IBM punched card decks of a half-century or more ago. In particular, it lacks any pretense of a GUI (*graphical user interface*), such as those enjoyed by modern-day PC and Macintosh users. However, a few vendors offer compound products consisting of the core miniNEC program wrapped in shells that provide decent GUIs for the user. Four excellent examples at the time of this writing are *EZ-NEC, NEC-Win Plus,* and *4nec2*—all for the PC—and *cocoaNEC* for the Mac. The latter two are free. (Vendor information for these and other modeling packages can be found in App. B.) Some vendors have more than one NEC-based product, with pricing and performance set by the maximum complexity of the antenna system that can be analyzed and whether or not it employs NEC-4 (with that core's added requirement for separately purchasing a government license). In addition, there are other free products—often with somewhat less versatility or capability—that may well be perfectly suitable for a specific antenna design, analysis, or optimization task.

NEC-based programs model wire antennas using the standard three-dimensional cartesian coordinate system, as shown in Fig. 25.1. In normal practice, the z axis is vertical, and the x and y axes are at right angles (*orthogonal*) to it and to each other. Φ and θ are the azimuth and elevation angles, respectively. The user "places" the antenna to be examined in this space by identifying the x, y, z coordinates of both ends of each straight wire element that is part of the antenna, along with similar representation of any nearby conducting surfaces likely to affect the performance of the antenna being modeled. (No model of an attic antenna would be complete or yield accurate results, for instance, without including any proximate house wiring in the attic or between the attic floor and the ceiling of the story below.) If the antenna is located in free space, the orientation of the antenna with respect to this x, y, z coordinate system is not particularly important, but if the antenna is near a conducting or dielectric medium (such as ground, seawater, etc.) having characteristics different from those of free space, orientation of the antenna properly with respect to the z-axis becomes important, since most of the inexpensive or free programs available assume a flat surface (earth, a car top, the ocean, etc.) spanning the xy-plane that corresponds to $z = 0$.

Figure 25.2 depicts a single-wire antenna in free space. It has been laid out in *model space* centered on the origin (0,0,0) and extending from $x = x_1$ (expressed as a positive number) to $x = x_2$ (expressed as a negative number). In the more common case of an antenna close (in wavelengths) to an underlying ground such as earth, z at all points along the wire would be a positive, nonzero number if the wire represented a horizontal

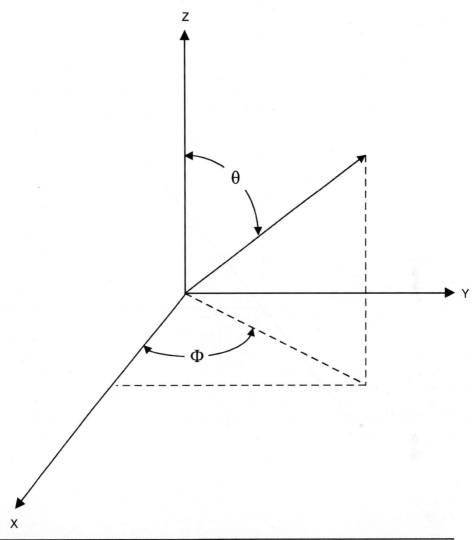

FIGURE 25.1 *x,y,z* coordinate system for *NecWin* modeling software.

antenna. The wire in Fig. 25.2 has been placed where it is, relative to the three axes, for user convenience. There is no computational reason it couldn't have also been placed along the *y* axis or at any angle in between the *x* and *y* axes. In fact, in free space, it could be at any angle and offset with respect to all three axes!

In addition to its dimensions and location in model space, the antenna designer must specify how many *segments* each wire element is to be divided into and where any sources of RF current or voltage are to be located. Thus, the wire in Fig. 25.2 can serve as a center-fed dipole or an end-fed dipole or long wire, depending on where the feed-point, or source, is located and the choice of measurement frequency. With practice, the

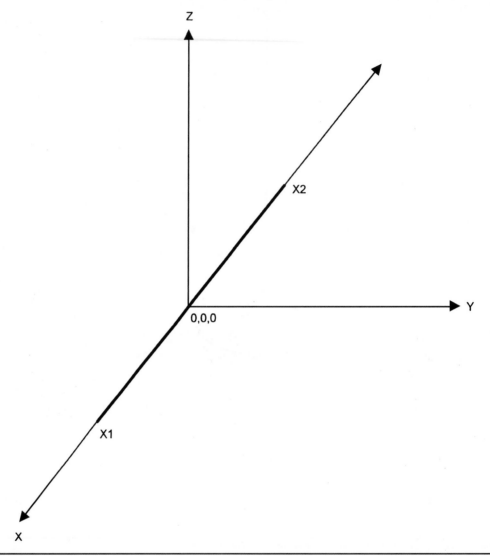

FIGURE 25.2   Example of a single wire laid out on *x,y,z* coordinate system.

user will learn how to quickly specify wires and their positions in model space, select an appropriate number of segments for each wire, and position sources within segments.

TIP *When modeling antenna elements, such as dipoles, that are fed in the center, use an odd number of segments for the wire. This allows placement of the source or feedpoint in the exact center of the middle segment, thus duplicating the actual antenna very closely.*

A Typical NEC-2 GUI

Roy Lewallen, W7EL, produces the *EZNEC* family of software for Windows PCs. The software extends Roy's earlier work on DOS versions of *EZNEC* but is still based on the NEC-2 engine. (A professional version based on the NEC-4 engine is available for those who obtain the required government license.)

Figure 25.3A shows the opening screen in *EZNEC+ ver 5.0* for a typical antenna project (in this case, an 80-m bent dipole). Visible in this window are file name, frequency, free-space wavelength (in meters), number of wires and segments, number of sources, number of loads, transmission lines, ground type, wire loss, the units employed (meters, feet, etc.), plot type (azimuth, elevation), elevation angle, step size, a reference level, and an alternate Z_0 for the SWR plot (to conveniently handle both 50- and 75-Ω system impedances).

Command buttons to the left of the main window include Open, Save As, Ant[enna] Notes, Currents, Src [Source] Dat[a], Load Dat[a], F[ar]Field] Tab[le], N[ear]F[ield] Tab[le], SWR, View Ant[enna], and F[ar]F[ield] Plot. The "FF Plot" button graphs the far

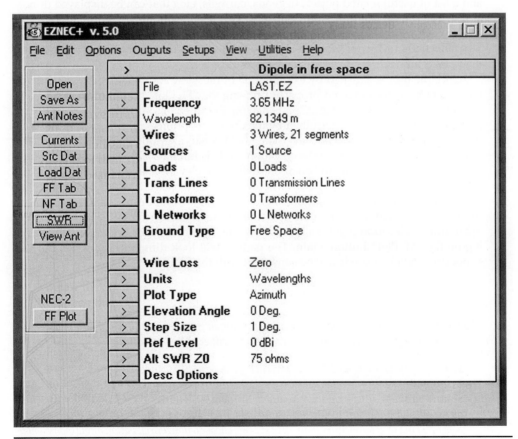

FIGURE 25.3A *EZNEC+ ver. 5 opening screen.*

field of the antenna as a function of azimuth (compass heading) for a given radiation (elevation) angle, or as a function of radiation angle for a user-selected azimuth (heading). Alternatively, the user can select a 3D presentation covering all elevations and azimuths in a single plot. Clicking on the "SWR" button brings up a new window for the user to define a frequency range and sample frequency intervals prior to running an SWR plot.

The wire and source tables used to create this antenna are shown in Fig. 25.3B. Note that the number of segments chosen for each of the three wires that make up the antenna keep the segment lengths as similar as possible. The horizontal wire in this example is divided into an odd number of segments to allow placement of a source at the exact midpoint of the center segment and, hence, at the midpoint of the antenna, in case it is to be a center-fed dipole. As can be seen from the rightmost two columns of the wire table, this antenna is being modeled with insulated wire.

The "View Antenna" window (Fig. 25.4) displays the x,y,z coordinate system and places all the wires specified in the Wires table according to coordinates entered there for each wire end. This window includes its own set of tools, including scroll bars to the left of the main window for zooming and repositioning the image within the window. Other options available from the drop-down menus include printer selection and setup, and a list of useful related objects (legends, currents, etc.) that can be displayed or not, as the user chooses. Once a set of calculations has been run, this window graphically shows (in a separate color) the magnitude of the current at all points on the antenna. A separate tabular listing of the current and its phase in each segment is also available for viewing, printing, or saving to a stand-alone file.

Once the user is satisfied by examination of the View Antenna display that all antenna data have been entered properly, pressing the "FF Plot" button in the main window opens a new window (Fig. 25.5) displaying the antenna's far-field radiation pattern in the format previously determined by the choice of options in the main window. The plot shown here happens to be the radiated field versus azimuth (compass heading) for a bent dipole in free space (so the elevation angle is immaterial). Data entries beneath the plot display supporting information such as the elevation angle used for the calculation, the gain of the strongest lobe (wherever the pattern touches the 0-dB outer circle) over that of an isotropic radiator, front-to-back (F/B) ratio, beamwidth, and the azimuthal angle (listed as "Cursor Az") for the strongest lobe. Alternatively, the user can opt to run an Elevation plot for a specific heading by making those selections and clicking on the "FF Plot" button again. The pattern will look different, of course, but the supporting data beneath it will be consistent with the first run.

NOTE *Unless the antenna model is specified as being in free space, azimuth plots for a 0-degree elevation angle will return an error message for either of two reasons:*

1. As discussed in Chap. 3, the presence of a perfectly conducting ground beneath a horizontally polarized antenna will result in zero radiated field at zero elevation angle because a voltage difference (an E-field, in other words) cannot exist between any two points on a lossless ground plane. When running azimuthal patterns for a horizontal dipole or Yagi over any kind of ground other than "free space", choose a nonzero elevation angle (somewhere between 10 and 20 degrees, say) to see the low-angle azimuthal radiation pattern of the

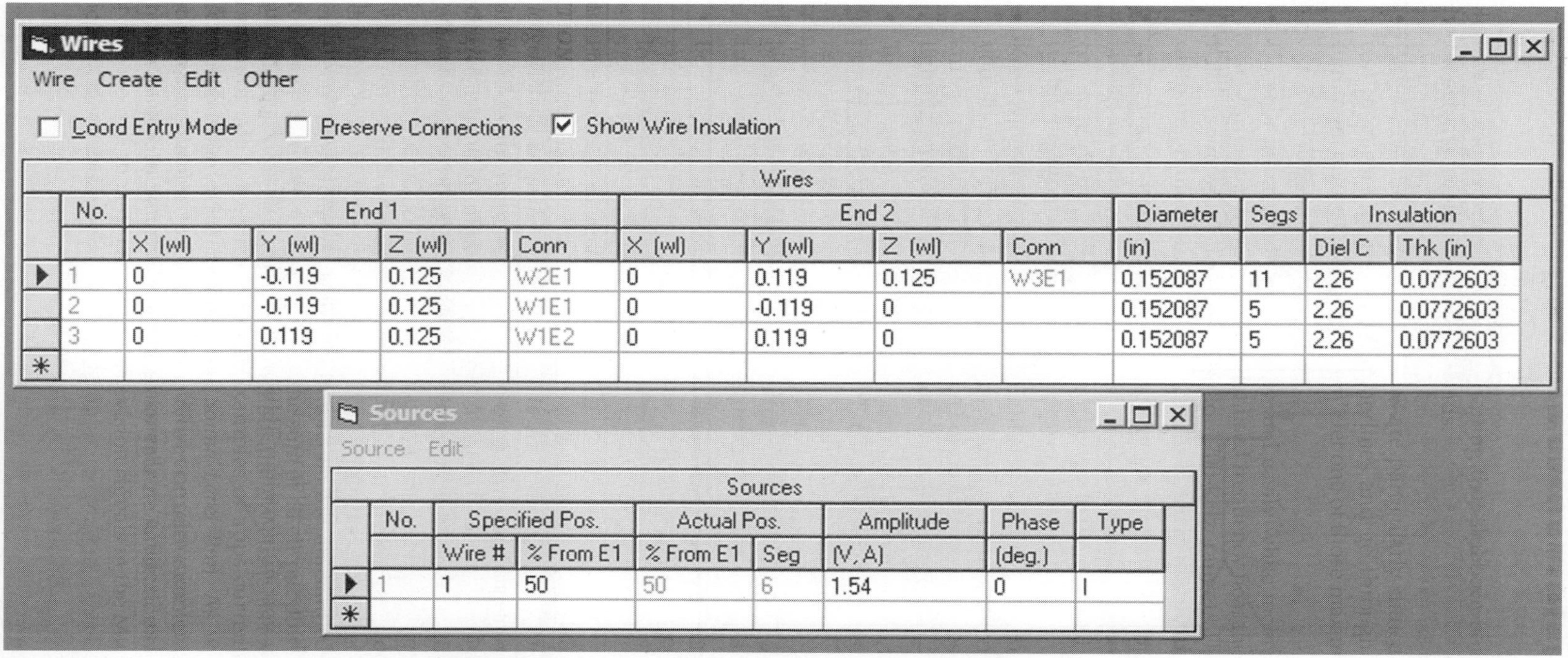

Wires

Wire Create Edit Other

☐ Coord Entry Mode   ☐ Preserve Connections   ☑ Show Wire Insulation

Wires

| No. | End 1 | | | | End 2 | | | | Diameter | Segs | Insulation | |
	X [wl]	Y [wl]	Z [wl]	Conn	X [wl]	Y [wl]	Z [wl]	Conn	(in)		Diel C	Thk (in)
▲ 1	0	-0.119	0.125	W2E1	0	0.119	0.125	W3E1	0.152087	11	2.26	0.0772603
2	0	-0.119	0.125	W1E1	0	-0.119	0		0.152087	5	2.26	0.0772603
3	0	0.119	0.125	W1E2	0	0.119	0		0.152087	5	2.26	0.0772603
✱												

Sources

Source Edit

Sources

| No. | Specified Pos. | Actual Pos. | | Amplitude | Phase | Type |
	Wire #	% From E1	% From E1	Seg	(V, A)	(deg.)	
▲ 1	1	50	50	6	1.54	0	I
✱							

FIGURE 25.3B   Bent dipole wire and source tables.

551

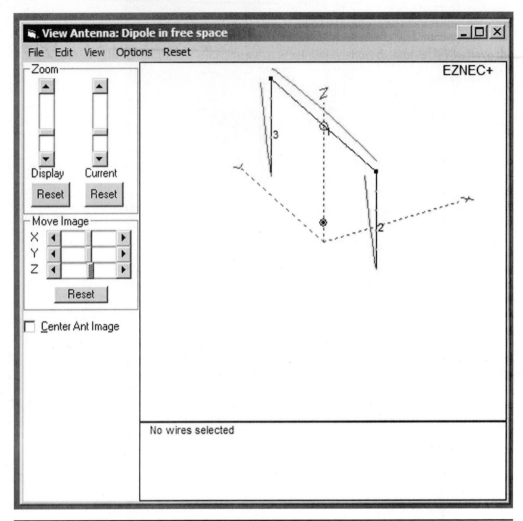

FIGURE 25.4   Bent dipole antenna view.

antenna. Cubical quads and loops fed in the center of their top or bottom sides should be handled the same way.

2. *The resistive component of a "real" ground having finite conductivity will sap power from an advancing wavefront in contact with it, causing the zero elevation angle component of the wave to become nonexistent long before reaching a distant receiver. Again, when modeling any antenna type over one of NEC's "real" ground options, choose a nonzero elevation angle to view the low-angle azimuthal radiation pattern of the antenna.*

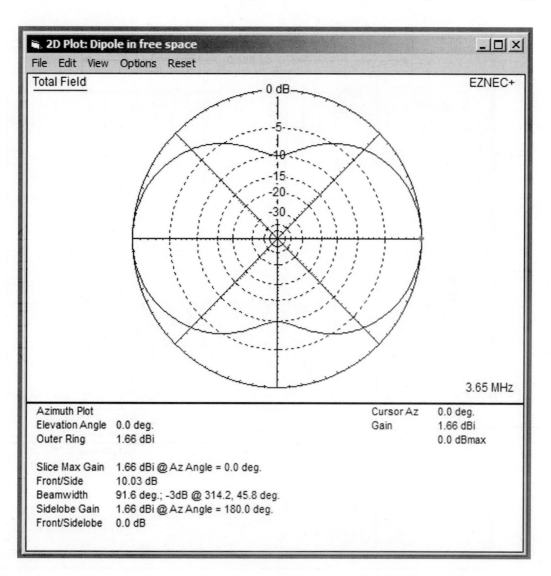

FIGURE 25.5 Bent dipole far field pattern.

Recent versions of *EZNEC* include SWR plotting capability previously available only in the professional edition. Figure 25.6 shows the 50-Ω plot for the dipole under consideration. The displayed data include SWR, complex feedpoint impedance (Z), reflection coefficient, and return loss—all specified at the starting frequency or at any other stepped frequency within the plot range. Simply place your mouse pointer on the new frequency of interest and read off the (new) data below the graph.

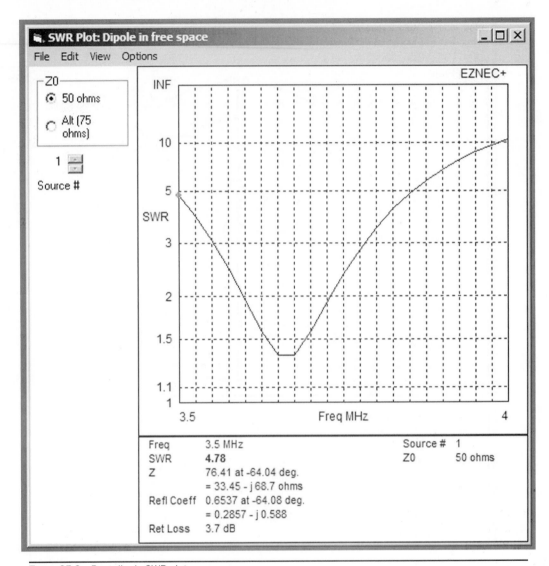

FIGURE 25.6 Bent dipole SWR plot.

Recently Added Features

Historically, the standard *method of moments* algorithms in NEC-based software for personal computers have not done a particularly good job with elements whose diameter changed at specified distances along the element, such as is almost always the case with HF beams. In recent years, however, special *taper algorithms* have been incorporated to correct that shortcoming. One example is an algorithm originally developed by David B. Leeson, W6ML (ex-W6QHS), in the process of mechanically strengthening a popular commercial HF Yagi. Recent versions of some user shells incorporate the Leeson taper algorithm, accessible as a menu option. To avoid confusion with a modeling technique called *tapered segment lengths,* the *EZNEC* menus, tables, and help files refer to the Leeson algorithm as *stepped-diameter correction* even though it has been a long-standing practice to call these *tapered elements.*

Another recent addition to many of the available programs is the ability to model a transmission line attached to the antenna. In some packages, commonly used "standard" transmission line models are already included and can be incorporated quickly into an otherwise custom antenna model. In general, the user must specify where on the modeled antenna each conductor at the antenna end of the transmission line connects, the line length and characteristic impedance (Z_0), its velocity factor, and the loss per 100 ft at a user- or catalog-specified frequency. One benefit of this recent capability is that comparisons of parameters such as feedpoint impedance and SWR versus frequency between a model and its corresponding real antenna system can be performed from the comfort of the radio room.

An interesting feature of *EZNEC 5* is the ability to optionally model *two* adjacent grounds of differing characteristics; the two grounds can even be at different heights! To access this capability, either of the "Real" types of ground ("High Accuracy" or "MiniNEC") must first be selected. The user can then choose to have a single ground extending infinitely far in all directions or split the surface between two grounds in either of two ways: by a straight line dividing the grounds at a user-specified x coordinate or by a circle having a user-specified radius "R" where the ground characteristics change. The first option requires the user to rotate model space around the x and y axes until the model's wires are oriented correctly with respect to the boundary between the two grounds, since this boundary is always parallel to the y axis. Both options provide an extremely useful feature for certain users . . . especially those who live near a body of water! The author's antennas, for instance, are on a point of land approximately 10 ft higher than the freshwater lake that surrounds his property on three sides, so the circular boundary allows much better approximation of ground effects on his own antennas than a single ground does.

Other Considerations

One of the factors that serve to form practical limits on the capabilities of a given antenna modeling package is the number of wire segments needed to represent the antenna and neighboring conductors. For complex models (a Yagi beam in close proximity to other beams and wire antennas, for instance), the user may require an upgraded version of the software. Alternatively, judicious reduction in the number of segments used to model each wire may allow the model to run. As a general rule, however, it is good practice to think in terms of using 10 or more segments for each half-wavelength of wire length plus a few added segments wherever two wires meet at an angle or where two wires of different diameter are joined. The user can then fairly quickly add up the number of segments likely to be required for a multielement Yagi or quad.

Of course, the more segments in the model, the greater the number of computations that must be performed by the computer, although virtually any PC or Mac purchased within the past few years has more than enough "horsepower" for the inexpensive, segment-limited software packages that most of us are apt to purchase.

Although an "ideal" model of an antenna and the conductors around it would take into account *all* conducting structures in the vicinity of the target antenna, some common sense is required. For horizontal HF antennas there is likely to be little effect if vertical masts and lattice towers having horizontal dimensions less than $\lambda/20$ at the modeling frequencies of interest are not included. On the other hand, modeling an 80-m grounded vertical or a 40-m elevated ground-plane antenna within a wavelength of a 15-m or taller guyed tower supporting a triband Yagi might not yield accurate predictions unless the tower, Yagi, and metallic guy wires are added to the model.

Many "gotchas" are hidden in the inner workings of the NEC engine. So many, in fact, that antenna modeling should not be attempted without a thorough reading of the user manual that comes with each program. Even better is to take a course in the subject. ARRL, for instance, offers for a small fee a multisession online course that comes with an online tutor for the duration of the course and a substantial textbook for reference long after the course is completed. Examples and exercises in the ARRL book are split between *EZ-NEC* and *NEC-Win Plus*. The user is responsible for procuring his or her own copy of an appropriate modeling program prior to starting the course.

Perhaps the biggest "gotcha" for most users of these programs is the likelihood of serious inaccuracies in modeled performance of antennas that rely for their performance on one or more wires in contact with, or very close to, the ground. Over the years, each new version of the NEC core has improved in its ability to accurately describe the effects of nearby grounds and dielectrics on antenna performance. Current offerings based on the NEC-2 engine allow the user to select from as many as three different approaches to modeling the ground, and they contain an assortment of representative ground parameters for multiple soil types, freshwater lakes and ponds, seawater, and even custom permeability and dielectric constants defined by the user. Nonetheless, except for the more expensive packages built around the NEC-4 computing engine, all have a prohibition against modeling any wires in direct contact with the ground. As a result, grounded MF and HF verticals with extensive radial systems must be approximated by modeling the radials at a height of at least a few inches above ground. Model results for antennas based on the use of counterpoises (low wires running in close proximity to ground directly underneath the antenna wire) are similarly suspect, and the user should proceed with caution, reinforced with copious reading.

Once you have put together a model for your particular antenna, you should *test* the model for *convergence*—usually by making small changes to such things as the number of segments in each wire. A good model will yield results (maximum forward gain or feedpoint impedance, to name two examples) that vary by 1 percent or less with small changes in segment counts, wire diameters, etc. Discussion of all the techniques for confirming the quality of your model would be far more extensive than our limited treatment of antenna modeling has space for.

Terrain and Propagation Modeling Software

When modeling antennas, never lose sight of the fundamental objective: to maximize the ability to communicate via radio waves. Thus, while development of an optimized antenna is an important step, putting that antenna in the right location and/or at the right height can be just as important in sending those radio waves on their way.

While not strictly antenna modeling tools, two other categories of modeling software are especially germane to the process of optimizing signal paths between two stations for maximum received signal strength:

- *Terrain analysis* examines the effect of nearby (i.e., within a few miles) topographic features using principles of optical reflection and diffraction. In some countries, government or private topographical databases can be downloaded for automatic use by the software. If this is not the case in your country, manual entry is always possible. Terrain analysis is of great benefit when utilized prior to determining the height or location of antennas and their supports on a specific parcel of land, but it is probably of most benefit when used even earlier to compare the relative merit of multiple candidate sites for an antenna facility.

- Ionospheric propagation programs carry the ball the rest of the way, providing statistical predictions of *optimum working frequencies*, the probabilities and times for communicating between two points at any given time of the day, and the propagation modes providing the most efficient path between those points. In addition, some will predict received signal levels at distant receiving locations anywhere on the globe for a given transmitter power level.

See Chap. 2 ("Radio-Wave Propagation") for additional information regarding these software tools.

Summary

Despite a plethora of caveats, "gotchas", rules, and guidelines for users, today's accurate yet inexpensive modeling tools are a boon to antenna experimenters everywhere. Hours or even days that might previously have been spent trudging through swamps, prickly brush, or deep snowdrifts to erect yet another configuration of wires atop the trees can now be put to better use, thanks to the power of the modern personal computer and the considerable efforts and accomplishments of a dedicated group of scientists, engineers, hobbyists, and entrepreneurs.

Many examples of antenna modeling results from both *EZNEC* (for the PC) and *cocoaNEC* (created by Kok Chen, W7AY, for the Mac) can be found in the chapters of this book dedicated to specific families of antennas. Any errors in the results presented are the fault of this author, not the software packages or their authors.

See App. B for a list of antenna modeling software vendors and the link for obtaining an NEC-4 license (U.S. users only).

CHAPTER 26

The Smith Chart

The mathematics of transmission lines can become cumbersome at times, especially when dealing with complex impedances and nonstandard situations. In 1939, Phillip H. Smith published a graphical device for solving these problems, and an improved version of the chart followed in 1945. That graphic aid, somewhat modified over time, is still in constant use in microwave electronics and other fields where complex impedances and transmission line problems are found. The *Smith chart* is indeed a powerful tool for the RF designer.

Smith Chart Components

The modern Smith chart is shown in Fig. 26.1. It consists of a series of overlapping *orthogonal* circles (i.e., circles that intersect each other at right angles). This chapter will dissect the Smith chart, so that the origin and use of these circles is apparent. You may find it useful in going through the tutorial and examples of this chapter to download and print out any of the free Smith chart templates and document files that can be found on the Internet. Simply search on the phrase "smith chart".

Pure Resistance Line

Figure 26.2 highlights (in bold) the horizontal axis that bisects the Smith chart outer circle. This line is called the *pure resistance line,* and it forms the reference for measurements made on the chart. Recall that a complex impedance contains both resistance and reactance, and can be expressed in the form:

$$Z = R \pm jX \qquad (26.1)$$

where Z = complex impedance
$\quad\quad R$ = resistive component of impedance
$\quad\quad X$ = reactive component of impedance*

Points along the pure resistance line represent all possible impedances where $X = 0$. To make the Smith chart universal, the printed values of impedance along the pure resistance line are *normalized* with respect to a *system impedance* (usually chosen to be equal to the Z_0 of the predominant transmission line or the output impedance of the transmitter). For most microwave RF systems the system impedance is standardized at 50 Ω; for

---

*According to convention, inductive reactance (X_L) is positive (+) and capacitive reactance (X_C) is negative (–). The term X in Eq. (26.1) is the difference between the two reactances ($X = X_L - X_C$).

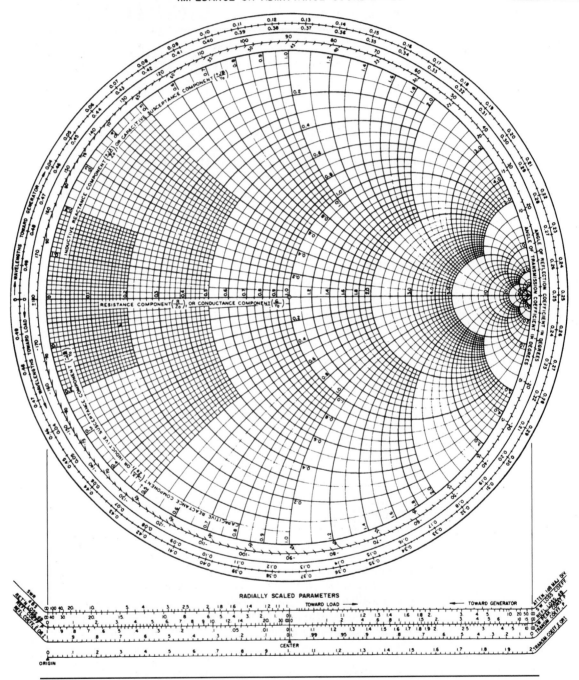

NAME	TITLE	DWG. NO.
SMITH CHART FORM 82-BSPR (9-66)	KAY ELECTRIC COMPANY, PINE BROOK, N.J. © 1966. PRINTED IN U.S.A.	DATE

IMPEDANCE OR ADMITTANCE COORDINATES

RADIALLY SCALED PARAMETERS

FIGURE 26.1 Smith chart. (*Courtesy of Kay Elementrics*)

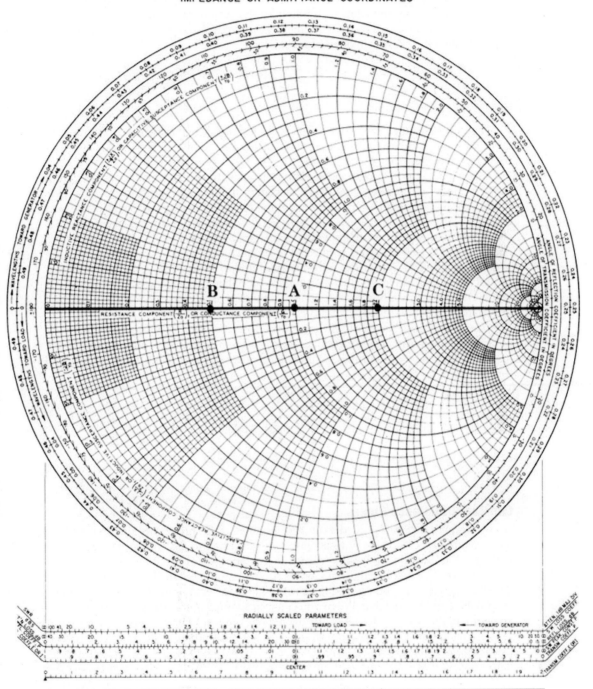

IMPEDANCE OR ADMITTANCE COORDINATES

RADIALLY SCALED PARAMETERS

FIGURE 26.2 Normalized impedance line. (*Courtesy of Kay Elementrics*)

CATV systems, it is 75 Ω. Most amateur applications use one of those impedances, but open-wire lines provide additional choices—most frequently 300 Ω, 450 Ω, or 600 Ω.

To normalize any impedance, divide it by the system impedance. For example, if the load impedance of a transmission line is Z_L and the characteristic impedance of the line is Z_0, then $Z' = Z_L/Z_0$. In other words,

$$Z' = \frac{R \pm jX}{Z_0} \tag{26.2}$$

The pure resistance line is structured such that the system impedance is in the center of the chart and has a normalized value of 1.0 (see point A in Fig. 26.2). This value results from the fact that $Z' = Z_0/Z_0 = 1.0$.

To the left of the 1.0 point on the pure resistance line are decimal fractional values used to denote impedances less than the system impedance. For example, in a 50-Ω transmission line system with a 25-Ω load impedance, the normalized value of the load impedance is 25 Ω/50 Ω, or 0.50 (point B in Fig. 26.2). Similarly, points to the right of 1.0 are greater than 1 and denote impedances that are higher than the system impedance. For example, in a 50-Ω system connected to a 100-Ω resistive load, the normalized impedance at the load end of the transmission line is 100 Ω/50 Ω, or 2.0; this value is shown as point C in Fig. 26.2. By employing normalized impedances, you can use the Smith chart for virtually any practical combination of system, load and source, and impedances, whether resistive, reactive, or complex.

Conversion of the normalized impedance to actual impedance values is done by multiplying the normalized impedance by the system impedance. For example, if the resistive component of a normalized impedance in a 50-Ω system is 0.45, then the actual impedance is

$$Z = (Z')(Z_0) \tag{26.3}$$

$$= (0.45)(50\ \Omega)$$

$$= 22.5\ \Omega$$

Constant Resistance Circles

An *isoresistance circle*, also called a *constant resistance circle*, represents all possible chart locations of a specific value of resistance, corresponding to the family of complex impedances that include all possible values of reactance in combination with that single resistance value. Several of these circles are shown highlighted in Fig. 26.3. These circles are all tangent to the point $R = \infty$ at the right-hand extreme of the pure resistance line and are bisected by that line. When you construct complex impedances (for which X is nonzero) on the Smith chart, all points on any one of these circles will have an identical resistive component. Circle A, for example, passes through the center of the chart, so it has a normalized constant resistance of 1.0. Note that impedances that are pure resistances (i.e., $Z' = R' + j0$) will fall at the intersection of a constant resistance circle and the pure resistance line, and complex impedances (i.e., $X \neq 0$) will appear at any other points on that same constant resistance circle. In Fig. 26.3, circle A passes through the

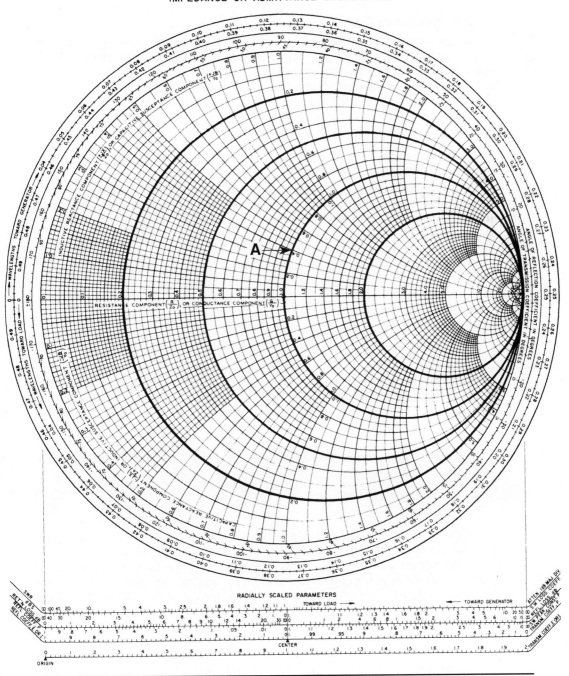

FIGURE 26.3 Constant resistance circles. (*Courtesy of Kay Elementrics*)

center of the chart, so it represents all points on the chart with a normalized resistance of 1.0. This particular circle is sometimes called the *unity resistance circle*.

Constant Reactance Circles

Constant reactance circles are highlighted in Fig. 26.4. Circles (or circle segments) *above* the pure resistance line (Fig. 26.4A) represent *inductive reactance* (+X), while circles (or segments) *below* the pure resistance line (Fig. 26.4B) represent *capacitive reactance* (–X). In both graphs, circle A corresponds to a normalized reactance of 0.80.

One of the outer circles (i.e., circle A in Fig. 26.4C) is called the *pure reactance circle*. Points along circle A represent reactance only; in other words, $R = 0$ everywhere on this outer circle, and all impedances along it are of the form $Z' = 0 \pm jX$.

Plotting Impedance and Admittance

Let's use Fig. 26.4D to see how to plot any impedance and admittance on the Smith chart. Consider an example in which system impedance Z_0 is 50 Ω, and the load impedance is $Z_L = 95 + j55$ Ω. This load impedance normalizes to

$$Z_L' = \frac{Z_L}{Z_0}$$

$$= \frac{95 + j55 \ \Omega}{50 \ \Omega} \tag{26.4}$$

$$= 1.9 + j1.1$$

An *impedance radius* is constructed by drawing a line from the point represented by the normalized load impedance, $1.9 + j1.1$, to the point represented by the normalized system impedance (1.0, 0.0) in the center of the chart. A circle, called the *VSWR circle*, is constructed by drawing a circle with this radius, centered on the location (1.0,0.0).

Admittance is the reciprocal of impedance, so in normalized form it is found from

$$Y' = \frac{1}{Z'} \tag{26.5}$$

On the Smith chart, any admittance value is located exactly opposite, or 180 degrees from, its corresponding impedance. That is, if the radius line connecting the center of the Smith chart to the impedance point is doubled to become a *diameter*, the coordinates of the point at the opposite end of the line from the original impedance point describe the normalized *admittance* corresponding to the reciprocal of the original normalized impedance. In other words, by extending a simple radius to a diameter, we have solved Eq. (26.5) with *graphical* techniques, rather than with mathematical ones!

Example 26.1   To show the value of the Smith chart in easily solving real problems, let's first find the *complex admittance* manually: Obtain the reciprocal of the complex impedance by multiplying the simple reciprocal by the *complex conjugate* (see App. A) of the

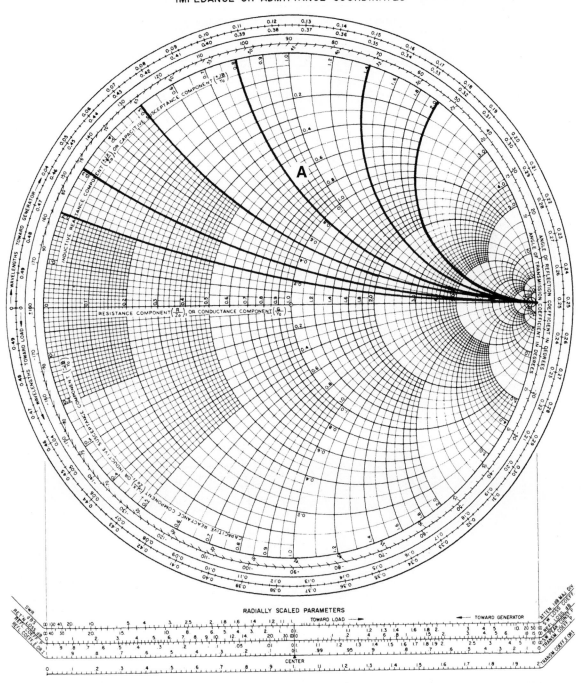

IMPEDANCE OR ADMITTANCE COORDINATES

A

RADIALLY SCALED PARAMETERS

TOWARD LOAD → ← TOWARD GENERATOR

CENTER

FIGURE 26.4A Constant inductive reactance lines. (*Courtesy of Kay Elementrics*)

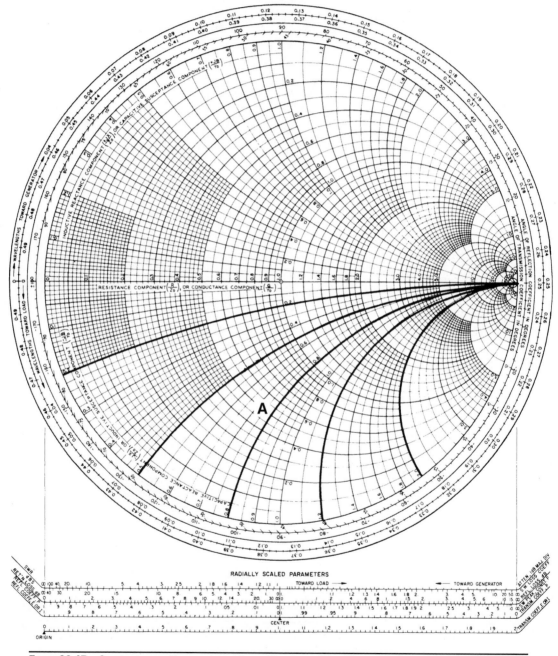

NAME	TITLE		DWG. NO	
SMITH CHART FORM 82-BSPR(9-66)	KAY ELECTRIC COMPANY, PINE BROOK, N.J., © 1966. PRINTED IN U.S.A.		DATE	

IMPEDANCE OR ADMITTANCE COORDINATES

FIGURE 26.4B   Constant capacitive reactance lines. (*Courtesy of Kay Elementrics*)

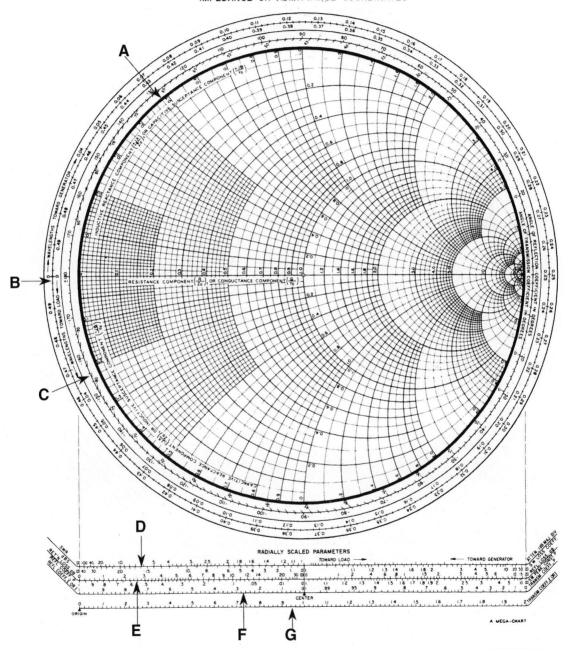

IMPEDANCE OR ADMITTANCE COORDINATES

FIGURE 26.4C Angle of transmission coefficient circle. (*Courtesy of Kay Elementrics*)

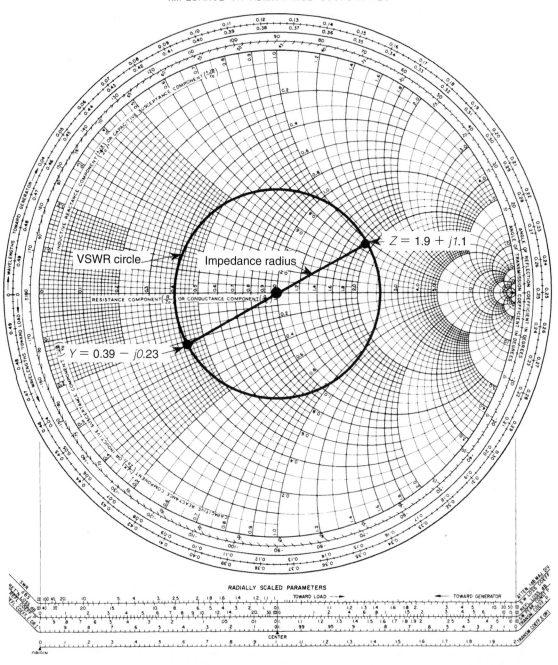

FIGURE 26.4D   VSWR circles. *(Courtesy of Kay Elementrics)*

impedance. For example, using the normalized impedance of Eq. (26.4), the normalized admittance will be:

$$Y' = \frac{1}{Z'}$$

$$= \frac{1}{1.9 + j1.1} \times \frac{1.9 - j1.1}{1.9 - j1.1}$$

$$= \frac{1.9 - j1.1}{3.6 + 1.2}$$

$$= \frac{1.9 - j1.1}{4.8} = 0.39 - j0.23$$

Now compare your answer with the coordinates of the left-hand end of the line in Fig. 26.4D. Simply extend the impedance radius through the 1.0,0.0 center point until it intersects the VSWR circle again. This point of intersection represents the same normalized load as that found in Eq. (26.4); the only difference is that it is now specified as a normalized admittance. As you can see, one of the delights of the Smith chart is that calculations like these are reduced to a quick graphical construction!

◆

Outer Circle Parameters

The standard Smith chart contains three concentric calibrated circles on the outer perimeter of the chart. Circle A has already been covered in Fig. 26.4C; it is the *pure reactance circle*. The next larger circle B defines the distance in wavelengths as you travel around the circle. Clockwise motion is toward the generator or transmitter, and counterclockwise is toward the load or antenna. Circle B is important when transforming an impedance at one point on a transmission line to the resulting impedance at another point some distance away.

There are two scales on the *wavelength* circle (B in Fig 26.4C), and both have their zero origin at the left-hand extreme of the pure resistance line. On both scales one complete revolution represents a distance of *one half-wavelength* along the line. The scales are calibrated linearly from 0 through 0.50 such that these two points are identical with each other on the circle. In other words, starting at the zero point and traveling 360 degrees around the circle brings one back to zero, which represents one half-wavelength, or 0.5λ.

Although both wavelength scales span the same range (0 to 0.50λ), they are traversed in opposite directions. The outer scale is calibrated clockwise and it represents *wavelengths toward the generator*; the inner scale is calibrated counterclockwise and represents *wavelengths toward the load*. These two scales are complementary at all points. Thus, 0.12 on the outer scale corresponds to (0.50 minus 0.12) or 0.38 on the inner scale.

The *angle of transmission coefficient* and *angle of reflection coefficient* scales are shown in circle C in Fig. 26.4C. These scales report the relative phase angle between reflected and incident waves at each point on the line resulting from a given load impedance. Recall from transmission line theory (see Chap. 4) that a short circuit ($R = 0$) at the load end of the line results in the reflected voltage headed back toward the generator being 180 degrees out of phase with the incident voltage at the reflection point, and an open line (i.e., infinite impedance) results in the reflected voltage headed back toward the generator being in phase (i.e., 0 degrees) with the incident voltage at the load. On the Smith chart, a short circuit at the load corresponds to the left-hand end of the pure resistance line, where 180 degrees is printed on circle C. Similarly, an open circuit at the load corresponds to the right-hand end of the pure resistance line, where 0 degrees is printed. Note that the upper half-circle is calibrated 0 to +180 degrees, and the bottom half-circle is calibrated 0 to –180 degrees, reflecting how the load impedance is transformed into either inductive or capacitive reactance, respectively, depending on how far back from the load the observation point has moved along the transmission line.

Radially Scaled Parameters

There are six scales laid out on four lines (D through G in Fig. 26.4C and in expanded form in Fig. 26.5) at the bottom of the Smith chart. These scales are called the *radially scaled parameters*—and they are not only very important but often overlooked. With these scales, we can determine such factors as VSWR (both as a ratio and in decibels), *return loss* in decibels, voltage or current *reflection coefficient,* and the power reflection coefficient.

As discussed in detail in Chap. 4, the reflection coefficient Γ is defined as the ratio of the reflected signal to the incident (or forward) signal. For voltage or current:

$$\Gamma = \frac{V_{REF}}{V_{FWD}} \qquad (26.6)$$

and

$$\Gamma = \frac{I_{REF}}{I_{FWD}} \qquad (26.7)$$

Power is proportional to the square of voltage or current, so:

$$\Gamma_{PWR} = \Gamma^2 \qquad (26.8)$$

or

$$\Gamma_{PWR} = \frac{P_{REF}}{P_{FWD}} \qquad (26.9)$$

Example 26.2   10W of microwave RF power is applied to a lossless transmission line; 2.8W of the 10W total incident power is reflected from the mismatched load. Calculate the reflection coefficient.

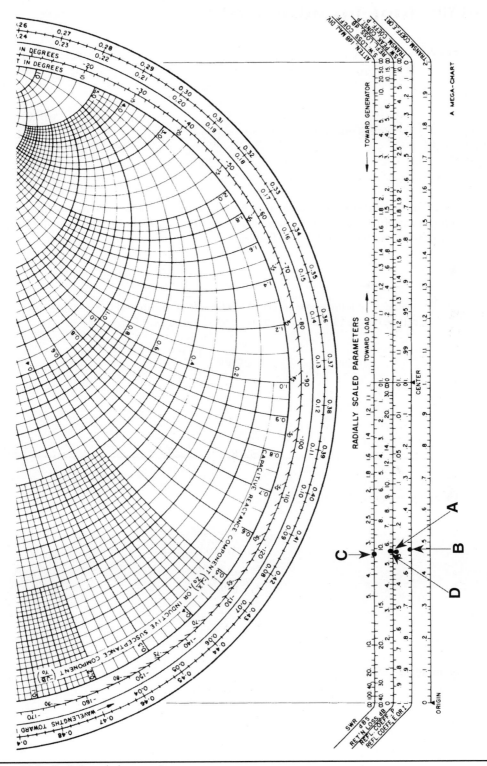

Figure 26.5 Radially scaled parameters.

Solution

$$\Gamma_{\text{PWR}} = \frac{P_{\text{REF}}}{P_{\text{FWD}}}$$

$$= \frac{2.8 \text{ W}}{10 \text{ W}}$$

$$= 0.28$$

The voltage reflection coefficient Γ is found by taking the square root of the power reflection coefficient, so in this example it is equal to 0.529. These points are plotted at A (for power) and B (for voltage) in Fig. 26.5.

The *voltage standing wave ration* (VSWR), often shortened to *standing wave ratio* (SWR), can be defined in terms of the reflection coefficient, Γ:

$$\text{VSWR} = \frac{1+\Gamma}{1-\Gamma} \tag{26.10}$$

or

$$\text{VSWR} = \frac{1+\sqrt{\Gamma_{\text{PWR}}}}{1-\sqrt{\Gamma_{\text{PWR}}}} \tag{26.11}$$

Using the results of Example 26.2,

$$\text{VSWR} = \frac{1+\sqrt{0.28}}{1-\sqrt{0.28}}$$

$$= \frac{1+0.529}{1-0.529}$$

$$= \frac{1.529}{0.471} = 3.25:1$$

or, in decibel form,

$$\text{VSWR}_{\text{dB}} = 20 \log (\text{VSWR})$$

$$= 20 \log (3.25)$$

$$= (20)(0.510) \tag{26.12}$$

$$= 10.2 \text{ dB}$$

These points are plotted at C in Fig. 26.5. Shortly, we will work an example to show how these factors are determined graphically in a transmission line problem when we are given the complex load impedance.

Transmission loss is a measure of the one-way loss of power in a transmission line because of reflection from the load. *Return loss* represents the two-way loss, so it is exactly twice the transmission loss. Return loss is found from

$$\text{Loss}_{\text{RET}} = 10 \log (\Gamma_{\text{PWR}}) \tag{26.13}$$

and, for our example in which $\Gamma_{\text{PWR}} = 0.28$,

$$\text{Loss}_{\text{RET}} = 10 \log (0.28)$$

$$= (10)(-0.553)$$

$$= -5.53\,\text{dB}$$

This point is shown as D in Fig. 26.5.

The *transmission loss coefficient* (TLC) can be calculated from

$$\text{TLC} = \frac{1 + \Gamma_{\text{PWR}}}{1 - \Gamma_{\text{PWR}}} \tag{26.14}$$

or, for our example,

$$\text{TLC} = \frac{1 + (0.28)}{1 - (0.28)}$$

$$= \frac{1.28}{0.72}$$

$$= 1.78$$

The TLC is a correction factor that is used to calculate the attenuation caused by mismatched impedance in a lossy, as opposed to an ideal or "lossless", line. The TLC is found from laying out the impedance radius on the *loss coefficient* scale of the radially scaled parameters at the bottom of the chart.

Smith Chart Applications

One of the best ways to demonstrate the usefulness of the Smith chart is by practical example. The following sections look at two general cases: *transmission line problems* and *stub matching systems*.

Transmission Line Calculations

Figure 26.6 shows a 50-Ω transmission line connected to a complex load impedance Z_L of $36 + j40$ Ω. The transmission line has a velocity factor v_F of 0.80, which means that the wave propagates along the line at $^8/_{10}$ the speed of light, c. ($c = 300,000,000$ m/s.) The length of the transmission line is 28 cm. The generator (a voltage source in this example) V_{in} is operated at a frequency of 4.5 GHz and produces a power output of 1.5 W. Let's see what we can glean from the Smith chart (Fig. 26.7).

First, normalize the load impedance by dividing it by the system impedance (in this case, $Z_0 = 50\ \Omega$):

$$Z' = \frac{36\ \Omega + j40\ \Omega}{50\ \Omega}$$

$$= 0.72 + j0.8$$

Now locate the circle of constant resistive component that goes through the point (0.72,0.0) in Fig. 26.7 and follow it until it intersects the $+j0.8$ constant reactance circle. This point (identified with an arrow in Fig. 26.7) graphically represents the normalized load impedance $Z' = 0.72 + j0.80$.

Next, construct a VSWR circle centered on the point (1.0,0.0) and having an impedance radius equal to the line connecting 1.0,0.0 (at the center of the chart) and the 0.72 + j0.8 point.

At a frequency of 4.5 GHz, the length of a wave propagating in the transmission line, assuming a velocity factor of 0.80, is

$$\lambda_{LINE} = \frac{c\ \upsilon_F}{F(Hz)}$$

$$= \frac{(3 \times 10^8\text{m/s})\ (0.80)}{4.5 \times 10^9\ \text{Hz}}$$

$$= \frac{2.4 \times 10^8\text{m/s}}{4.5 \times 10^9\ \text{Hz}} \qquad (26.15)$$

$$= 0.053\ \text{m} \times \frac{100\ \text{cm}}{\text{m}}$$

$$= 5.3\ \text{cm}$$

One wavelength *in the transmission line* is 5.3 cm, so a half-wavelength is 5.3/2, or 2.65 cm. The 28-cm line is 28 cm/5.3 cm, or 5.28 wavelengths long. With a straightedge, draw a line from the center (1.0,0.0) through the load impedance and extend it to the outermost circle; it should intersect that circle at 0.1325 (or 0.3675). Because one complete revolution around this circle represents one half-wavelength, 5.28 wavelengths from this point represents 10 revolutions plus 0.28 wavelengths (not revolutions) more. The residual 0.28 wavelength figure is added to 0.1325 to form a value of (0.1325 + 0.28) or 0.413. Now locate the point 0.413 on the circle, making sure to use the same markings that the original line intersected at 0.1325, and place a mark there. Draw a line from 0.413 to the center of the circle, and note that it intersects the VSWR circle at 0.49 − j0.49, which represents the input impedance Z'_{in} looking into the line.

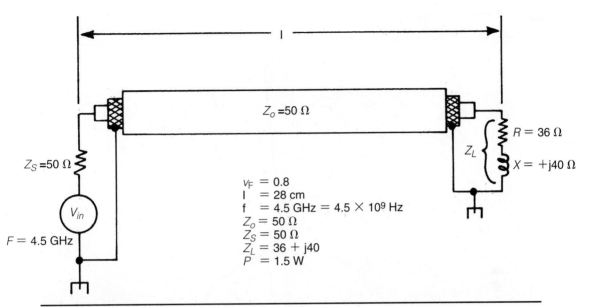

FIGURE 26.6 Transmission line and load circuit.

Finally, to find the actual impedance represented by the normalized input impedance we have just read off the chart, "denormalize" the Smith chart impedance by multiplying the result by Z_0:

$$Z_{in} = (0.49 - j0.49)\ (50\ \Omega)$$

$$= 24.5 - j24.5\ \Omega$$

This is the impedance that must be matched at the generator end of the transmission line by a conjugate matching network.

The admittance of the load is the reciprocal of the load impedance and is found by extending the impedance radius through the center of the VSWR circle until it intersects the circle again on the other side (as an *impedance diameter*). This point is found to be $Y' = 0.62 - j0.69$; it is the *normalized input admittance*. Confirming the solution mathematically:

$$Y' = \frac{1}{Z'}$$

$$= \frac{1}{0.72 + j0.80} \times \frac{0.72 - j0.80}{0.72 - j0.80}$$

$$= \frac{0.72 - j0.80}{1.16}$$

$$= 0.62 - j0.69$$

The VSWR at the generator end of the line is found by transferring the "impedance radius" of the VSWR circle to the radial scales below the circle chart. At the leftmost

NAME	TITLE	DWG. NO
SMITH CHART FORM 82-BSPR (9-66)	KAY ELECTRIC COMPANY, PINE BROOK, N.J. © 1966 PRINTED IN U.S.A.	DATE

IMPEDANCE OR ADMITTANCE COORDINATES

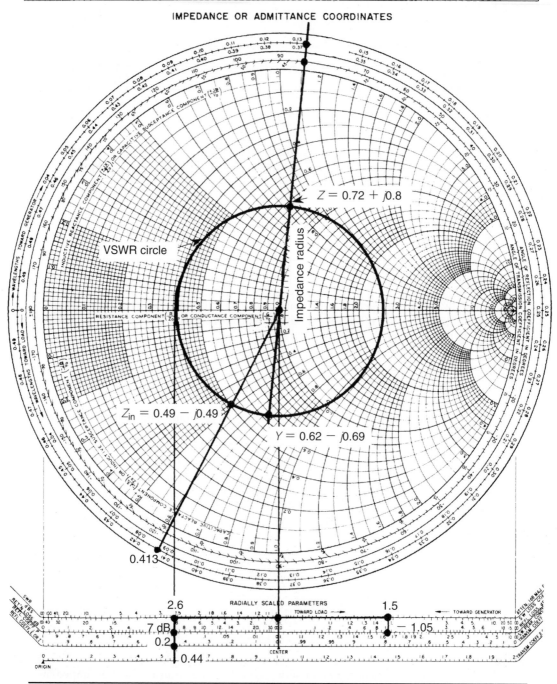

FIGURE 26.7 Smith chart for Example 26.2 (transmission line parameters). (*Courtesy of Kay Elementrics*)

extension of the radius circle, drop a line vertically to the topmost of the radially scaled parameters at the bottom of the chart. For this example, we find an input VSWR of approximately 2.6:1. In decibel form, the VSWR is 8.3 dB (next scale down from VSWR), and this is confirmed by

$$VSWR_{dB} = 20 \log \; (VSWR)$$

$$= (20) \log \; (2.7)$$

$$= (20) \; (0.431)$$

$$= 8.3 \; dB$$

$$(26.16)$$

The *transmission loss coefficient* is found in a similar manner, using the radially scaled parameter scales. In practice, once we have found the VSWR we need only continue the vertical line from the 2.6:1 VSWR line across the other scales below. In this case, the line intersects the voltage reflection coefficient scale at 0.44 and the power reflection coefficient line at 0.20, which is, of course, $(0.44)^2$.

The *return loss* can found by reading the "RET'N LOSS, dB" line where the same vertical line intersects it; the value is found to be approximately 7 dB, which is confirmed with the use of Eq. (26.13):

$$Loss_{RET} = 10 \log \; (\Gamma_{PWR}) \; dB$$

$$= 10 \log \; (0.21) \; dB$$

$$= (10) \; (-0.677) \; dB$$

$$= -6.77 \; dB$$

The *angle of reflection coefficient* is found from the outer circles of the Smith chart. The line connecting the center to the load impedance ($Z' = 0.72 + j0.8$) is extended to the *angle of reflection coefficient in degrees* circle, and intersects it at approximately 84 degrees. The total *magnitude + phase* description of the reflection coefficient is therefore 0.44/84 degrees.

To find the transmission loss coefficient (TLC) from the radially scaled parameter scales, drop a vertical line from the rightmost extension of the impedance radius to the uppermost radially scaled parameter scale, the *loss coefficient* scale, where it is found to intersect 1.5. This value is confirmed with Eq. (26.14):

$$TLC = \frac{1 + \Gamma_{PWR}}{1 - \Gamma_{PWR}}$$

$$= \frac{1 + (0.20)}{1 - (0.20)}$$

$$= \frac{1.20}{0.80}$$

$$= 1.5$$

The *reflection loss* is the amount of RF power reflected back down the transmission line from the load. The difference between incident power supplied by the generator (1.5 W in this example) and the reflected power is the *absorbed power*:

$$P_{ABS} = P_{FWD} - P_{REF} \tag{26.17}$$

If the load is an antenna, P_{ABS} is split between radiated power and losses, such as ground losses. The reflection loss is found graphically by continuing the righthand perpendicular below the TLC point (or by laying out the impedance radius on the "REFL. LOSS, dB" scale), and in this example (Fig. 26.7) it is –1.05 dB. Let's check the calculations.

The return loss is –7 dB, so

$$-7\,dB = 10\log\left(\frac{P_{REF}}{P_{FWD}}\right)$$

$$-7 = 10\log\left(\frac{P_{REF}}{1.5\ W}\right)$$

$$\frac{-7}{10} = \log\left(\frac{P_{REF}}{1.5\ W}\right)$$

$$10^{(-7/10)} = \frac{P_{REF}}{1.5\ W}$$

$$0.2 = \frac{P_{REF}}{1.5\ W}$$

$$(0.2)\,(1.5\ W) = P_{REF}$$

$$0.3\ W = P_{REF}$$

The power absorbed by the load (P_L) is the difference between incident power P_{FWD} and the reflected power P_{REF}. If 0.3 W is reflected, then that means the absorbed power is (1.5 – 0.3), or 1.2 W.

The reflection loss is –1.05 dB and can be checked from:

$$-1.05\ dB = 10\log\left(\frac{P_L}{P_{FWD}}\right)$$

$$\frac{-1.05}{10} = \log\left(\frac{P_L}{1.5\ W}\right)$$

$$10^{(-1.05/10)} = \frac{P_L}{1.5\ W}$$

$$0.785 = \frac{P_L}{1.5\ W}$$

$$(1.5W) \times (0.785) = P_L$$

$$1.2\ W = P_L$$

Now let's summarize everything we've gleaned using the Smith chart. Recall that 1.5 W of 4.5-GHz microwave RF signal was applied to the input of a 50-Ω transmission line that was 28 cm long. The load connected to the far end of the transmission line had an impedance of 36 + j40. From the Smith chart:

Admittance (load)	0.62 – j0.69
VSWR	2.6:1
VSWR (dB):	8.3 dB
Reflection coefficient (V)	0.44
Reflection coefficient (P)	0.2
Reflection coefficient angle	84 degrees
Return loss	–7 dB
Reflection loss	–1.05 dB
Transmission loss coefficient	1.5

Note that in all cases the numerical calculation agrees with the graphical solution of the problem within the limits of the graphical method.

Stub Matching

A properly designed matching system will provide a conjugate match to a complex impedance. Some sort of matching system or network is likely to be needed any time the load impedance Z_L differs significantly from the characteristic impedance Z_0 of the transmission line. In a transmission line system, a *shorted stub* is often connected in parallel with the line, at a specific distance back toward the generator from the mismatched load, in order to effect a match.

As shown schematically in Fig. 26.8, a stub is simply a short section of transmission line whose two conductors are shorted together at one of its ends and individually attached to the two conductors of the system transmission line at its other end. A lossless shorted line exhibits a pure reactance that can vary from $-\infty$ to $+\infty$, depending upon its length (and repeating every $\lambda/2$), so a line of critical length L_2 attached across a load impedance can cancel any possible reactive component. However, because the stub must also transform the resistive part of the load impedance up or down to match the system impedance (Z_0), another adjustable stub parameter is required. This turns out to be the stub attachment point L_1—i.e., the distance from the load back toward the transmitter (or generator).

Because the stub is connected in parallel with the line, it is generally easier to work with admittance parameters, rather than impedances. When we do that, at a certain stage of the process values on the Smith chart represent admittances, consisting of conductances and susceptances, rather than impedances. The *pure resistance line* becomes the *pure conductance line,* and circles of constant *resistance* become circles of constant *conductance.*

Example 26.3 A parallel-wire transmission line is terminated by an antenna with an impedance Z_L = 100 + j60, as shown in Fig. 26.8. Find the closest location (to the load) and the length of a shorted stub that will reduce the VSWR on the transmission line to the left of the stub to 1.0:1.

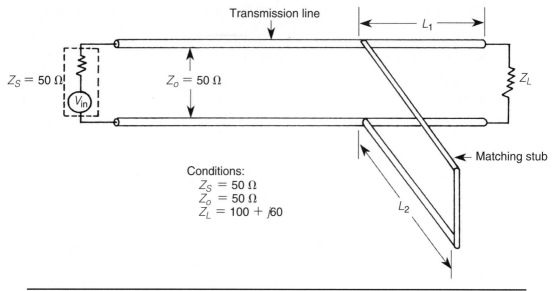

FIGURE 26.8   Stub matching example.

Solution   The load impedance is $Z_L = 100 + j60$, which we normalize to $Z' = 2.0 + j1.2$. After plotting this impedance on the Smith chart in Fig. 26.9, we construct a VSWR circle and draw a diameter across it to find the load admittance, $Y' = 0.37 - j0.22$, corresponding to a shunt conductance $G' = 0.37$ in parallel with a shunt capacitance having susceptance $B' = -0.22$.

In order to properly design the matching stub, we need to find two dimensions: L_1 in Fig. 26.8 is the length (in wavelengths) from the load toward the generator and L_2 is the length of the stub itself.

The first step in finding a solution to the problem is to locate any points where the unit conductance circle that passes through $G' = 1.0, 0.0$ at the chart center intersects the VSWR circle. At these points (and there are two of them in every half-wavelength of transmission line), the transformed load impedance Y''_L has a conductance exactly equal to that of the system transmission line along with a nonzero susceptance (i.e., $Y''_L = G''_0 + jB''_L$, where the double-prime superscripts indicate a load admittance and its components *transformed* to new values that depend on the exact location along the system transmission line). There are two such points shown in Fig. 26.9: $1.0 + j1.1$ and $1.0 - j1.1$. If we choose one of these points and extend a line from the center (1.0,0.0) of the chart through the selected point to the outer circle ("WAVELENGTHS TOWARD GENERATOR"), we will be able to calculate the distance from the load, back along the system transmission line, where we should locate the stub. Then our only task will be to determine the length of stub that exactly cancels the susceptance of the transformed load admittance at that point.

Suppose, therefore, we choose the point $1.0 + j1.1$ and extend a line through it from the center of the chart, all the way out to the outer ring; it intercepts the ring at 0.165λ (i.e., at a distance 0.165 wavelengths toward the generator from a reference point at the left edge of the Smith chart).

IMPEDANCE OR ADMITTANCE COORDINATES

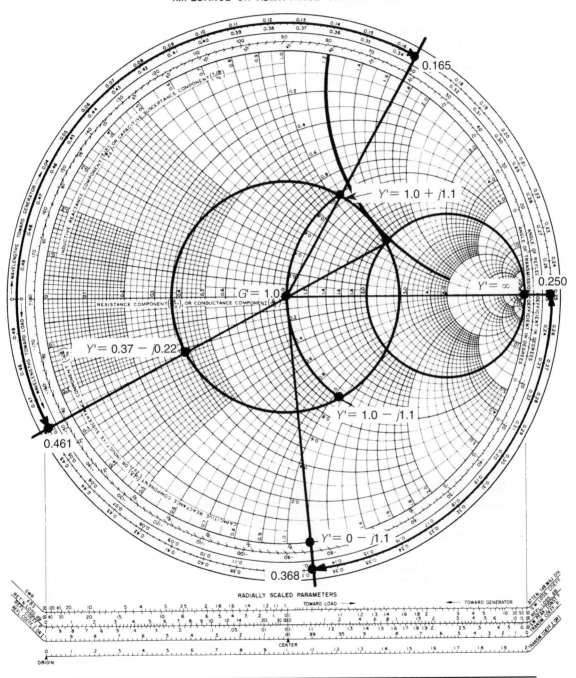

FIGURE 26.9 Smith chart for Example 26.3 (stub matching). (*Courtesy of Kay Elementrics*)

581

Next we extend the $Z' \rightarrow Y'$ diameter we initially drew to the outer circle; its path is from the chart center through the admittance point $0.37 - j0.22$ to where the outer ring is marked 0.461λ. This corresponds to determining the position of the load relative to the same arbitrary reference point on the outer circle at the left edge of the chart. Now we have two lines intersecting the outer circle at the points 0.165λ and 0.461λ, respectively, but they lie on opposite sides of the reference point of 0λ. The shorter distance between those two points on the circle is the sum of 0.165λ and 0.039λ, or 0.204λ. (0.039λ is 0.500λ minus 0.461λ.) Thus,

$$L_1 = 0.204\lambda$$

The final step is to determine the length of the stub required. This is done by finding two additional points on the Smith chart. First, locate the point where admittance is infinite (far right side of the pure conductance line, where $G' = \infty$); this corresponds to the short-circuited far end of the stub. Second, locate the point where the admittance is $0 - j1.1$, which is the pure susceptance we need to place in parallel with the susceptance of the transformed load to perfectly cancel out the latter. (Note that the susceptance portion is the same as that found where the unit conductance circle crossed the VSWR circle.) Because the conductance component of this new point is 0, the point will lie on the $-j1.1$ circle at the intersection with the outer circle. Now draw lines from the center of the chart through each of these points to the outer circle. These lines intersect the outer circle at 0.368 and 0.250λ. The length of the stub is found from the shorter distance between them:

$$L_2 = (0.368 - 0.250)\lambda$$

$$= 0.118\lambda$$

Summary: Using nothing but the Smith chart we have matched a load of impedance $Z_L = 100 + j60$ to a 50-Ω line by attaching a shorted stub of length 0.118λ to the transmission line a distance 0.204λ back from the load.

Incorporating Loss

Thus far, we have dealt with situations in which loss is either zero (i.e., ideal transmission lines) or so small as to be negligible. In the presence of appreciable loss in the circuit or transmission line, however, the VSWR circle is actually a spiral. Let's look at one such case.

Figure 26.10 summarizes a typical situation directly on the Smith chart. Assume a lossy transmission line 0.60λ long, connected to a normalized load impedance of $Z'_L = 1.2 + j1.2$. First, an "ideal" (i.e., lossless) VSWR circle is constructed on the impedance radius represented by $1.2 + j1.2$. Next, a line (A) is dropped perpendicularly from the righthand point where this circle intersects the pure resistance baseline (B) to the "ATTEN, 1 dB/MAJ. DIV." line on the radially scaled parameters. A distance representing a total line loss (3 dB) is stepped off on this scale back toward the center. A second perpendicular line (C) is drawn, from the new 3-dB mark back up to the pure resistance line (D).

The distance from the center of the Smith chart to point D, where line C intersects the pure resistance line becomes the radius for a new circle that contains the actual

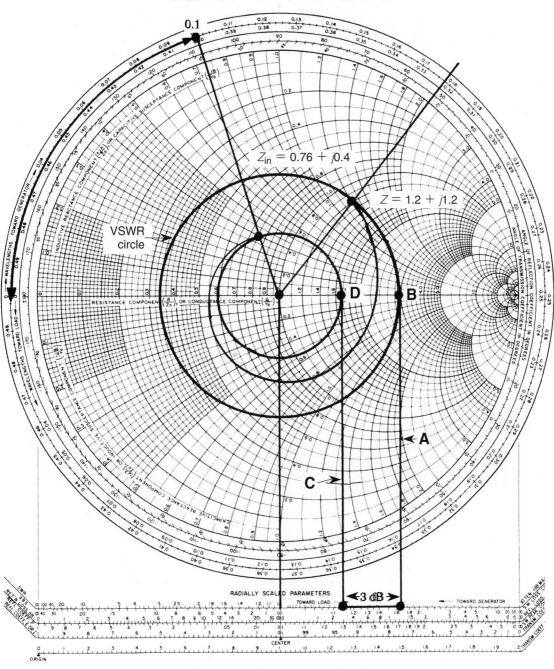

FIGURE 26.10 Smith chart analysis on a lossy transmission line. (*Courtesy of Kay Elementrics*)

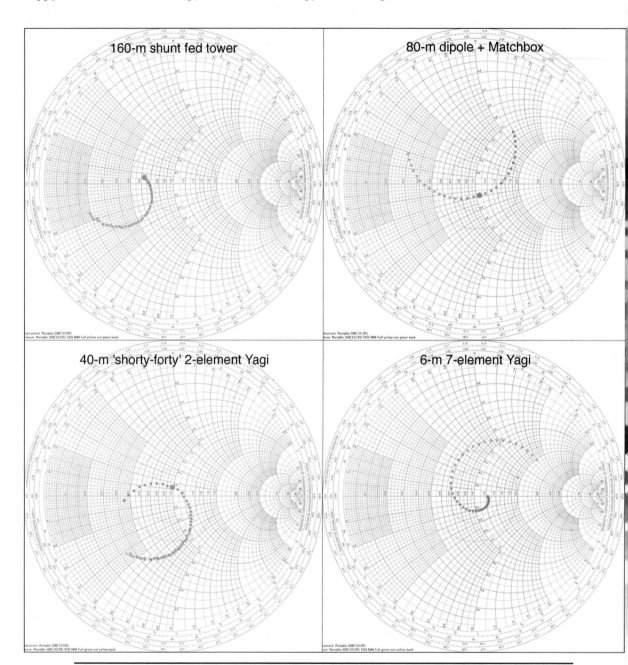

Figure 26.11   Smith chart presentation of Z_{in} for various antennas.

input impedance of the line. The length of the line is 0.60λ, so we must step back (0.60 – 0.50) λ or 0.1λ as marked off on the "WAVELENGTHS TOWARD GENERATOR" outer circle. A radius is drawn from this point to the 1.0, 0.0 center point of the circular chart. The new radius intersects the smaller VSWR circle at 0.76 + j0.4, which is the normalized input impedance (Z'_{IN}) corresponding to an actual input impedance of 38 + j20 Ω.

This example exaggerates the loss per unit length to clearly show the inward spiral of a lossy transmission line's VSWR curve. In fact, if a lossy line is long enough, the VSWR will appear to be 1.0:1 *regardless* of the nature of the load attached to it!

Frequency Response Plots

A complex passive network (such as a filter or an antenna) may contain resistors, inductors, and capacitors—either lumped or distributed. Both the resistive and the reactive components of the input impedance of such a network are generally functions of frequency; we say the network response is *frequency-sensitive*. For antennas, in particular, the Smith chart becomes a valuable tool for plotting (and recording for future reference) Z_{IN} versus frequency.

Figure 26.11 shows Z_{IN} for four different antennas currently in use by the author. The curves were obtained with a VNA. The exact values on these curves are unimportant, but the general shapes of the curves, which span varying percentage bandwidths, are typical of what various practical monoband antennas exhibit for Z_{IN}. The same curves can be created for your antennas with even simpler equipment. All that is necessary is an antenna analyzer that provides a readout of both the resistive and the reactive parts of Z_{IN} at each frequency, as well as the sign of the reactance.

CHAPTER 27
Testing and Troubleshooting

This chapter examines some of the instruments and techniques for testing antenna systems—whether brand-new installations or when trouble crops up on existing antennas. Figure 27.1 shows in block diagram format a basic radio system capable of two-way communication. We have specifically limited the diagram's level of detail so the only interconnections shown are those that carry RF signals and/or transmitter power between the various station components: transmitter, receiver (or transceiver), amplifier, T/R switch, wattmeter/SWR bridge, low-pass filter, impedance-matching unit, and the antenna itself. All of these interconnections, whether long or short, are lengths of transmission line. Today, in almost all active radio systems below the microwaves, the transmission lines are coaxial cable, with the frequent exception on the lower HF bands of the line between the impedance-matching unit (also known as the ATU, or *antenna tuner unit*) and the antenna.

In many installations, one or more of the boxes in the block diagram of Fig. 27.1 may be combined with (or part of) another box. For instance, modern HF transceiv-

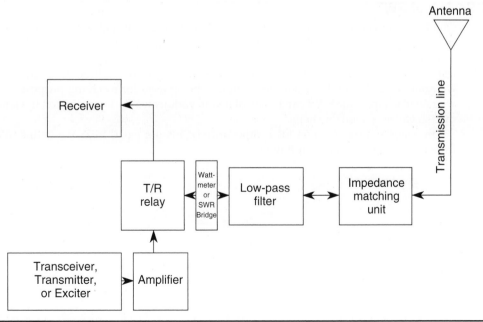

FIGURE 27.1 Basic communications radio station setup.

ers almost always include a wattmeter, or *directional coupler,* that is capable of indicating *forward* (*incident*) power and either *reflected* power or SWR. If no external power amplifier is present, the transceiver wattmeter should be perfectly adequate for all measurements of antenna and transmission line performance that can normally be made at the transmitter end of the line. Similarly, T/R switching—whether at low or high power levels—is now found inside all transceivers and even a few amplifiers. Nonetheless, whether performed internally or provided in individual external boxes, all the functions shown are typically found in a modern-day antenna system.

In this chapter we will explore some of the more commonly used instruments to help us test and evaluate the "health" of our *antenna system,* which we define as including not only the antenna itself but the entire RF path between the antenna and the transmitter or receiver. Because the feedpoints of most antennas—especially on the MF, HF, and VHF bands—are up in the air, far from our reach, many of the devices described here are intended to give us some idea of conditions "up topside" while we are at ground level.

When measuring, monitoring, or troubleshooting an antenna system, we are typically interested in certain key attributes of the system components:

- Signal strength at a distance
- Impedance matching throughout the system
- RF attenuation (loss) throughout the system
- Useable bandwidth of the system

A Review of SWR

One of the most important characteristics of an antenna system we can measure is the *standing wave ratio* (SWR or VSWR) at the input to the system. In our definition, the input is the end of the system *farthest* from the antenna or radiating load when we are transmitting RF energy to the antenna. (The system functions a little differently, and we examine it differently, when the antenna is being used for receiving purposes.) As we saw in Chap. 4, the SWR can be calculated in various ways, depending on the information we have available to us.

In general, if the antenna load impedance (Z_L) is not equal to Z_0 we define SWR from one or the other of the following:

If Z_0 is greater than Z_L:

$$\text{VSWR} = \frac{Z_0}{Z_L} \tag{27.1}$$

If Z_0 is less than Z_L:

$$\text{VSWR} = \frac{Z_L}{Z_0} \tag{27.2}$$

We can also measure the forward and reflected power and calculate the VSWR from those readings:

$$\text{VSWR} = \frac{1 + \sqrt{(P_{\text{REV}} / P_{\text{FWD}})}}{1 - \sqrt{(P_{\text{REV}} / P_{\text{FWD}})}} \qquad (27.3)$$

where VSWR = voltage standing wave ratio
P_{REV} = reflected power
P_{FWD} = forward power

If we can measure either the voltage maximum and minimum or the current maximum and minimum, we can calculate SWR:

$$\text{VSWR} = \frac{V_{\text{MAX}}}{V_{\text{MIN}}} \qquad (27.4)$$

Finally, if the forward and reflected voltage components at any given point on the transmission line can be measured, then we can calculate the VSWR from

$$\text{VSWR} = \frac{V_{\text{FWD}} + V_{\text{REV}}}{V_{\text{FWD}} - V_{\text{REV}}} \qquad (27.5)$$

where V_{FWD} = forward voltage component
V_{REV} = reflected voltage

The last equation, based on the forward and reflected voltages, is the basis for many modern VSWR bridges and RF power meters.

If the feedpoint impedance of an antenna has both resistance and reactance, it is called *complex*. In mathematicians' terms, it has *real* and *imaginary* parts, and the resulting voltages at the feedpoint and all along the transmission line are similarly complex. Summarizing some of the important findings of Chap. 4, we note:

- Only the real part of the impedance of any part of an antenna system dissipates power. At MF and HF, these losses are most often found in the antenna radiation resistance (desirable) and in the I^R losses (undesirable) of the transmission line(s) feeding the antenna, other nearby conductors, and (especially if the radiator is a grounded monopole) the earth beneath the antenna.

- In an ideal system the antenna feedpoint resistance, the transmission line characteristic impedance, and the transmitter output impedance are identical and there are no reactive loads and no resistive losses, so SWR = 1.0:1 everywhere throughout the system. However, that is not always possible, so we often have to employ *impedance-matching devices*, such as ATUs, baluns, transmission line transformers, and the like, to minimize SWR and undesirable losses.

- Even if the real, or resistive, part of the antenna impedance perfectly matches the characteristic impedance of the transmission line at their attachment point, any reactive component of the antenna feedpoint impedance will cause the SWR to deviate from 1.0:1 just as though $R_L \neq Z_0$, and even the resistive part of Z_{IN} will vary along the line.

- All "real" antennas exhibit SWR bandwidth. Bandwidth can be too narrow (typical for antennas that are small with respect to a wavelength) or too wide (typical for antennas having resistive or dissipative losses comparable to the radiation resistance of the antenna itself).

- High SWR on a transmission line causes greater losses along the line than when the line is matched to the load impedance. A large part of this is because the high SWR causes high currents near current maxima along the line, increasing the line's I^2R dissipation.

In the following sections of this chapter, we look at a variety of useful instruments for measuring the RF characteristics of antenna systems and for diagnosing problems in those systems:

- Impedance bridges

- RF noise bridges

- Dip oscillators

- Field strength meters

- SWR bridges

- SWR analyzers

- RF wattmeters

- Vector network analyzers

- Dummy loads

Impedance Bridges

We can make antenna impedance measurements using a variant of the old-fashioned Wheatstone bridge. Figure 27.2A shows the basic configuration of the bridge in its most generalized form. The current flowing in the meter will be zero when $(Z_1/Z_2) = (Z_3/Z_4)$. If one arm of the bridge is the antenna impedance and a second arm is the reference (R_0), we ineratively adjust the other two arms until we obtain a null at the measurement frequency. A typical example is shown in Fig. 27.2B. An antenna connected at J_2 constitutes one arm of the bridge, while R_2 is a second. The value of R_2 should be 50 Ω or 75 Ω, depending upon the value of the expected antenna impedance, but 68 Ω is an acceptable compromise value for simple meters that must be used with both system impedance levels. The other two arms of the bridge are the reactances of C_{1A} and C_{1B}, which are two halves of a single differential capacitor.

In operation, a very low level RF signal is applied at J_1 and C_1 is adjusted for minimum meter reading. The antenna resistance at that frequency is then read from the previously calibrated dial.

Calibrating the instrument is simple. Noninductive resistors having standard values from 10 to 1000 Ω are connected across J_2. For each resistor the meter is nulled and

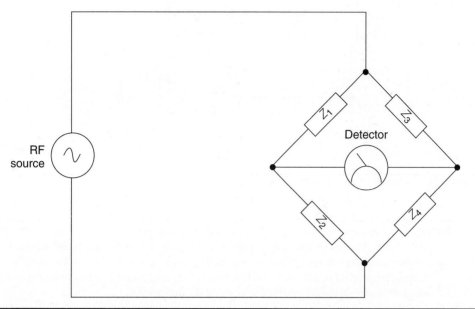

FIGURE 27.2A   Basic Wheatstone bridge.

the dial marking recorded. Alternatively, the value of the load resistor can be inscribed on the panel at each setting of the capacitor index mark.

The basic circuit of Fig. 27.2B is useful only for measuring the resistive component of impedance. By modifying the circuit as shown in Fig. 27.2C we can also measure the reactive component.

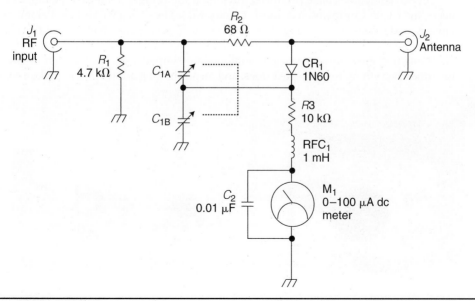

FIGURE 27.2B   SWR bridge.

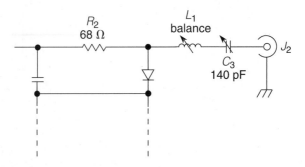

FIGURE 27.2C   Circuit with balance control added.

NOTE *The instrument shown in Fig. 27.2 depends for its operation on an external source of RF, applied at J₁. The power-handling capabilities of this device are limited by the ratings of the resistors and the current through the meter. Never connect this device directly to a transmitter. Instead, use a calibrated low-level RF signal generator or a known attenuator of adequate power rating between the source and the instrument.*

RF Noise Bridges

The *RF noise bridge* was once associated only with engineering laboratories but turns out to have applications in general communications servicing as well. It is one of the most useful low-cost and often overlooked test instruments in our arsenal of devices for troubleshooting antennas.

Several companies have produced low-cost noise bridges: Omega-T, Palomar Engineers, the Heath Company, and, most recently, MFJ. The Omega-T and the Palomar Engineers models are shown in Fig. 27.3. The Omega-T device is a small cube with a single dial and a pair of BNC coax connectors (*antenna* and *receiver*). The dial is calibrated in ohms and measures only the resistive component of impedance. The Palomar Engineers unit does everything the Omega-T does, and, in addition, it allows you to make a rough

FIGURE 27.3   Commercially available amateur noise bridges.

measurement of the reactive component of impedance. The Heath *HD-1422* is, of course, no longer made, but can sometimes be found as used equipment on amateur radio Web sites or at hamfests. MFJ's *204B* antenna bridge is a current production item.

Figure 27.4 shows the circuit of a typical noise bridge. As before, the bridge consists of four arms. An internally generated noise signal is applied to winding T_{1a} of a trifilar wound transformer and coupled into the inductive arms (T_{1b} and T_{1c}) of the bridge circuit. The adjustable arm of the bridge is a series RC circuit consisting of two front panel controls: a 200-Ω potentiometer (R_8) and a 250-pF variable capacitor (C_8). The potentiometer sets the range (0 to 200 Ω) of the resistive component of measured impedance, while the capacitor sets the reactance range (which varies with frequency). Capacitor C_6 in the "unknown" arm of the bridge balances the measurement capacitor. With C_6 in the circuit, the bridge is balanced when C_8 is approximately in the center of its range. This arrangement accommodates both inductive and capacitive reactances, which appear on either side of the "zero" point (i.e., the midrange capacitance of C_8). When the bridge is in balance, the settings of R and C reveal the impedance across the unknown terminal.

An internal source of broadband RF noise uses the spectrum produced by the avalanche process inherent in a reverse-biased zener diode, enhanced by the 1-kHz square-wave modulator (IC_1 and associated components) that chops the noise signal. An amplifier, comprised of transistor stages Q_1 and Q_2, boosts the noise signal to the level needed by the bridge circuit.

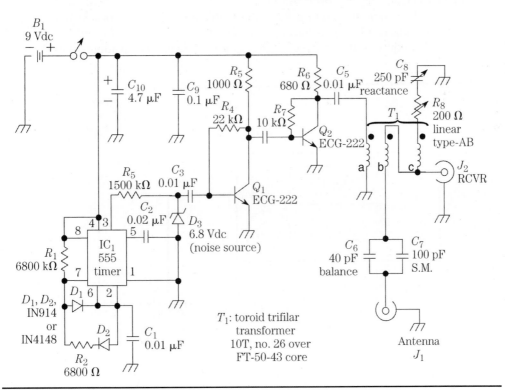

Figure 27.4 Noise bridge circuit.

In use, the output (J_2) of the noise bridge is connected to the antenna input of a tunable receiver covering the frequencies of interest, and R_8 and C_8 are alternately tuned for a null in the received noise at the antenna's design frequency. Preferably, the receiver is set for AM demodulation, but either CW or SSB modes will do in a pinch. Obviously, the accuracy of the test results will be limited by the frequency accuracy and stability of the receiver, as well as the rather coarse readout of R and X provided by most noise bridges commonly available.

Adjusting Antennas with the Noise Bridge

Perhaps the most common use for the antenna noise bridge is finding the impedance and resonant points of an HF antenna. Connect the *Receiver* terminal (J_2 in Fig. 27.4) of the bridge to the *Antenna* input of the HF receiver through a short length of coaxial cable from J_1 on the noise bridge, as shown in Fig. 27.5. Ideally, the cable length should be as short as possible, and its characteristic impedance should match that of the antenna feedline and the receiver input impedance. Next, connect the coaxial feedline from the antenna to the *Antenna* terminal on the bridge. You are now ready to test the antenna.

Set the noise bridge resistance control to the antenna feedline impedance (usually 50 to 75 Ω for most common antennas) and the reactance control to midrange (zero). Next, tune the receiver to the *expected* resonant frequency (f_{EXP}) of the antenna. Turn the noise bridge *on* and look for a noise signal of about S9 (the exact amplitude will vary) on the receiver.

Adjust the *resistance* control R_8 on the bridge for minimum noise signal to the receiver as indicated by the receiver's S-meter. Next, attempt to reduce the signal even further by adjusting the *reactance* control C_8 for an even lower reading on the S-meter.

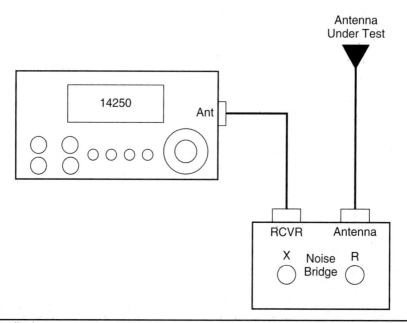

FIGURE 27.5   Connection of noise bridge.

Continue to alternately adjust the two controls for the deepest possible null, as indicated by the lowest reading of the S-meter. (There is usually some interaction between the two, hence the need for the iterative process.)

A perfectly resonant $\lambda/2$ dipole in free space or at least a wavelength above ground should have a reactance reading near 0 Ω and a resistance of 50 to 75 Ω. But "practical" antennas will exhibit some reactance (the less the better) and a resistance that is somewhere between 25 and 100 Ω. Remember, too, that the theoretical impedance of a monopole is ½ that of a corresponding dipole, so don't expect all types of antennas to exhibit the same input impedance.

The impedance-matching methods of Chaps. 4 and 24 can be used to transform the actual resistive component to the 50- or 75-Ω characteristic impedance of the transmission line and eliminate the reactance. Some helpful hints when using a noise bridge:

- If the resistance reading indicates zero or maximum, suspect a short or open circuit somewhere between the bridge and the antenna. But unless you know the exact electrical length of your transmission line (in wavelengths), you can't be sure which it is because odd multiples of $\lambda/4$ will transform one into the other.

- A reactance reading on the X_L side of zero indicates a net inductive reactance at the point on the transmission line where the noise bridge has been inserted, while a reading on the X_C side of zero indicates a net capacitive reactance. Again, without knowing the exact length of the transmission line, there's a limit to what you can do with that information.

As we saw in Chaps. 4 and 26, the impedance of an arbitrary load is repeated every half-wavelength along the line, going back toward the source or transmitter. If we can make the transmission line between the antenna and the noise bridge an exact multiple of $\lambda/2$ at the frequency of interest, we can obtain the input impedance of the antenna directly from the front panel markings on the bridge. (Remember to account for the velocity factor of the transmission line when determining its electrical length.) Otherwise, we will need a Smith chart or other means to translate our readings into something useful.

Suppose, for instance, that we have a 40-m dipole with a full wavelength of transmission line between it and a place where we can insert the noise bridge. Under those circumstances we can treat the noise bridge settings as being roughly the same as the antenna feedpoint resistance and reactance. In particular, if the bridge indicates a net inductive reactance at the design frequency, we can conclude that the dipole is somewhat longer than the resonant length. Similarly, a reading on the X_C side of center implies a length somewhat shorter than resonance.

If we wish to directly connect a transmitter to this feedline without an intermediate ATU in the line, we may choose to adjust the length of the dipole. To determine the correct length, we must find the actual resonant frequency f_{RES}. To do this, set the *reactance* control to zero, and then *slowly* tune the receiver in the proper direction—lower in frequency if we think the antenna is too long or higher if we think it's too short—until the null is found. Call this frequency f_{EXP}. On a high-Q antenna, the null is easy to miss if you tune too fast. Don't be surprised if that null is out of band by quite a bit. The percentage of change is given by

$$\% \text{ Change} = \frac{\left|(f_{RES} - f_{EXP})\right| \times 100}{f_{EXP}} \qquad (27.6)$$

Suppose the change calculated in Eq. (27.6) is –5 percent, i.e., the actual resonance point is 5 percent lower than expected. Shorten the dipole by perhaps 3 or 4 percent (split equally between the two sides) by folding the two ends back on the dipole and remeasure.

It will be rare for a real antenna to have exactly 50- or 75-Ω impedance, so some adjustment of R and C to find the deepest null is in order. You may be surprised how far off some dipoles and other forms of antennas can be if they are not in free space and instead are close to the earth's surface or other conducting objects of comparable dimensions.

Other Jobs for the Noise Bridge

In addition to its antenna-related "chores", the noise bridge can find utility in a variety of jobs around the radio shack. We can find the values of capacitors and inductors, determine the characteristics of series- and parallel-tuned resonant circuits, and find the velocity factor of coaxial cable, to name just a few.

Transmission Line Measurements

Some projects—such as the simple dipole tests and adjustment already described—require a feedline that is either a quarter-wavelength or a half-wavelength at a specific frequency. In other cases, a piece of coaxial cable of specified length is required for other purposes—for instance, the dummy load used to troubleshoot depth sounders is nothing more than a long piece of short-circuited coax that returns the echo at a time interval that corresponds to a specific depth. We can use the bridge to find these lengths as follows:

1. Connect a short circuit across the *Antenna* jack (J_1) and adjust R and X for the best null at the frequency of interest. (*Note*: Both will be near zero.)

2. Remove the short circuit.

3. Connect the length of transmission line to the same jack. (It might be wise to start with a length that is somewhat longer than the expected final length.)

4. For quarter-wavelength lines, cut short lengths from the line until the null is very close to the desired frequency. (A $\lambda/4$ line open-circuited at the far end will appear to be a short circuit at the near end.) For half-wavelength lines, do the same thing, except that the line must be shorted at the far end for each trial length. (A $\lambda/2$ line will repeat the termination of the far end at the near end.)

As we saw in Chap. 4, we need to know the value of v_F, the *velocity factor*, to calculate the *physical* length of a transmission line corresponding to any given *electrical* length. For example, a half-wavelength piece of coax has a physical length in feet of $(492 \times v_F)/f$ when f is given in megahertz. Unfortunately, the real value of v_F is often a bit different from the published value. The noise bridge can be used to find the actual value of v_F for any sample of coaxial cable as follows:

1. Select a convenient length of coax greater than 12 ft in length and install a PL-259 RF connector (or other connector compatible with your instrument) on one end. Short-circuit the other end.

2. Accurately measure the physical length of the coax in feet; convert the "remainder" inches to a decimal fraction of a foot by dividing by 12 (e.g., 32 ft

8 in = 32.67 ft because 8 in/12 in = 0.67). Alternatively, cut off the cable to the nearest foot and reconnect the short circuit.

3. Tune the monitor receiver to the *approximate* frequency corresponding to the length of cable you cut, using the equation $f = 492v_{\text{PUB}}/L$, where v_{PUB} is the published velocity factor (a decimal fraction), f is the frequency in megahertz, and L is the cable length in feet. If f exceeds the tuning range of the monitor receiver, select a longer piece of cable.

4. Set the bridge *resistance* and *reactance* controls to zero.

5. Tune the monitor receiver for deepest null. Use the null frequency to find the velocity factor by rearranging the terms in the preceding equation so that $v_{\text{ACTUAL}} = fL/492$.

As a general rule, conventional RG-8 and RG-213 types of coaxial cable have velocity factors around 0.66, while foam dielectric coaxial cable often has a velocity factor of $v_{\text{F}} = 0.80$.

Tuned Circuit Measurements

A resonant (*LC*) *tank circuit* is the circuit equivalent of a resonant antenna, so there is some similarity between the two measurements. You can measure resonant frequency with the noise bridge to within ±20 percent (or better, if care is taken). This accuracy might seem poor, but it is better than you can usually get with low-cost signal generators, dip meters, absorption wavemeters, and the like.

A *series-tuned circuit* exhibits a low impedance at the resonant frequency and a high impedance at all other frequencies. Conduct the measurement as follows:

1. Connect the series-tuned circuit under test across the *Antenna* terminals of the noise bridge.

2. Set the *resistance* control to a low resistance value, close to 0 Ω. Set the *reactance* control at midscale (zero mark).

3. Tune the receiver to the expected null frequency, and then tune the bridge for the null. Make sure that the null is at its deepest point by rocking the *R* and *X* controls for best null.

4. The receiver is now tuned to the resonant frequency of the tank circuit.

A *parallel-resonant circuit* exhibits a high impedance at resonance and a low impedance at all other frequencies. The measurement is made in exactly the same manner as for the series-resonant circuits, except that the connection is different. Figure 27.6 shows a two-turn link coupling that is needed to inject the noise signal into the parallel-resonant tank circuit. If the inductor is the toroidal

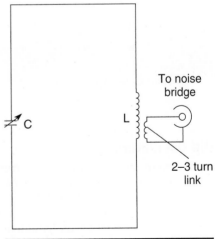

FIGURE 27.6 Noise bridge connection to *LC* tank circuits.

type, the link must go through the hole in the doughnut-shaped core and then connect to the *Antenna* terminals on the bridge. After this, proceed exactly as you would for the series-tuned tank measurement.

Capacitance and Inductance Measurements

With just two additional components—a 100-pF silver mica test capacitor and a 4.7-μH test inductor—the noise bridge is capable of measuring inductance and capacitance, respectively, over a wide range of frequencies. The idea is to use one or the other of the test components to form a series-tuned resonant circuit with an unknown component. If you find a resonant frequency, then you can calculate the value of the unknown component. In both cases, the series-tuned circuit is connected across the *unknown* terminals of the dip meter, and the series tuned procedure is followed.

To measure inductance, connect the 100-pF capacitor in series with the unknown coil across the *Antenna* terminals of the dip meter. When the null frequency is found, find the inductance from

$$L = \frac{253}{f^2} \tag{27.7}$$

where L = inductance, in microhenries
       f = frequency, in megahertz

Similarly, to find the value of an unknown capacitor:

1. Connect the test inductor across the *Antenna* terminals in series with the unknown capacitance.

2. Set the *resistance* control to zero, tune the receiver to 2 MHz, and readjust the *reactance* control for null.

3. Without readjusting the noise bridge control, connect the test inductor in series with the unknown capacitance and retune the receiver for a null.

4. Capacitance can now be calculated from:

$$C = \frac{5389}{f^2} \tag{27.8}$$

where C is in picofarads
       f is in megahertz

Dip Oscillators

One of the most common instruments for determining the resonant frequency of an antenna is the so-called *dip oscillator* or *dip meter*. Originally called the *grid dip meter* or *grid dip oscillator* (GDO) because early circuits were based on the use of a vacuum tube, the principle behind this instrument is that its output energy can be absorbed by a nearby resonant circuit (or antenna, which is an electrically resonant LC circuit). When the inductor of the dip oscillator (see Fig. 27.7) is brought into close proximity to a resonant circuit, a small amount of energy is transferred if the oscillator is tuned to the tank

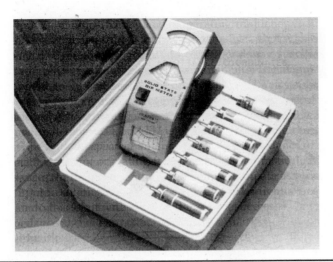

FIGURE 27.7   Dip meter.

circuit resonant frequency. This energy loss shows up on the meter pointer as a slight "dipping" action. The dip is extremely sharp and is easily missed if the meter frequency dial is tuned too rapidly.

When using a dip meter to check lumped-component tank circuits, maximum coupling is usually obtained by orienting the coil of the meter the same as the coil of the circuit. Figure 27.8A shows one way to couple the dip oscillator to a vertical antenna radiator. The coil of the dipper is brought into close proximity to the base of the radiator. Figure 27.8B shows the means for coupling the dip oscillator to systems where the antenna is not easily accessed directly (as when the antenna is at the top of a mast). A small two- or three-turn loop can be connected across the transmitter end of the transmission line, and the coil of the dipper brought close to it.

Dip meters absolutely shine at checking individual traps from multiband Yagi antennas or trap dipoles. A search of the antenna manufacturer's Web site is likely to unearth a service bulletin describing the procedure to be followed for each company's products. The service bulletin dip meter procedure is especially helpful with traps for certain Cushcraft an-

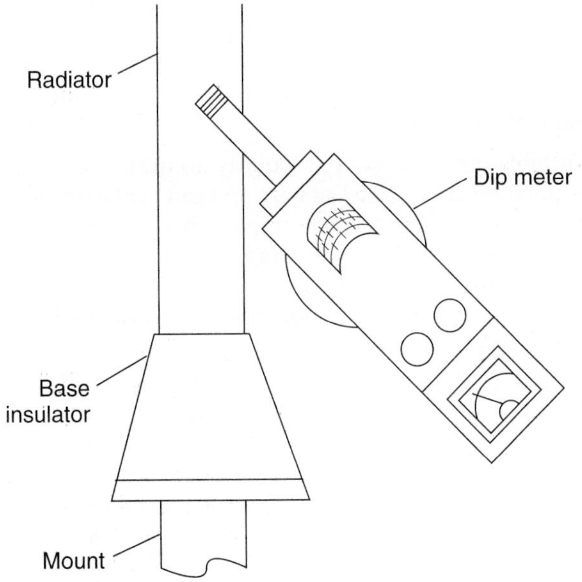

Radiator

Dip meter

Base insulator

Mount

FIGURE 27.8A   Direct coupling the dip meter to antenna.

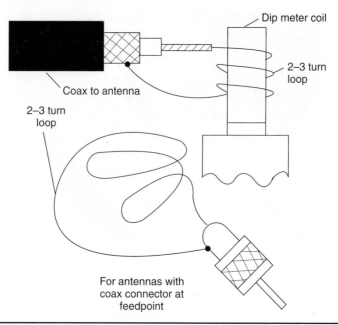

Figure 27.8B   Coupling the dip meter to coaxial cable.

tennas, since weathered traps often lack any positive identification of which end should point toward the boom and whether they're tuned for use in a reflector, director, or driven element.

There are three problems with dip meters that must be recognized in order to best use the instrument:

- The dip is very sharp, and it is easy to tune past it without even seeing it.

- The meter reading tends to drop off gradually from one end of the tuning range to the other, so it's important to tune the meter slowly to avoid confusing the drop-off with an actual resonant circuit dip.

- Frequency readout on affordable meters is not particularly accurate. Dial calibration and frequency interpolation are "coarse" at best, and can lead to erroneous conclusions—especially when working with traps, since the resonant frequencies of traps for the various elements in some Yagis are close together. A "tolerable" work-around is to confirm the exact frequency of the dip meter with a calibrated monitor receiver immediately after conducting each measurement.

Field Strength Meters

A *field strength meter* (FSM) is an instrument that measures the radiated field from an antenna. Commercial engineering-grade instruments are calibrated in terms of either watts per square centimeter or volts per meter, and are used for precise measurements such as those required for broadcast station *proof-of-performance* tests. For adjusting antennas, however, a considerably simpler instrument is sufficient. This section describes two simple passive (which means no dc power is required) field strength meters use-able for adjusting HF radio antennas.

Two circuits are shown in Fig. 27.9; both are basically variations on the old-fashioned "crystal set" theme. Figure 27.9A shows the simplest form of untuned FSM. In this circuit a small whip antenna used for signal pickup is connected to one end of a grounded RF choke (RFC). The RF voltage developed across the RFC by the signal being measured is applied to a germanium diode detector; either a 1N34 or an 1N60 can be used. Silicon diodes are normally preferred in signal applications, but in this case we need the lower contact potential of germanium diodes in order to improve sensitivity (V_g is 0.2 to 0.3 V for Ge and 0.6 to 0.7 V for Si). A potentiometer is used both as the load for the diode and as a sensitivity control to set the meter reading to a convenient level.

The untuned version of the FSM suffers from a lack of sensitivity, which may limit its use, depending on the distances and power levels involved in your tests. Only a certain amount of signal can be developed across the RFC, so this limits the instrument. Also, the RFC does not make a good impedance match to the detector diode D_1. An improvement is possible by adding a tuned circuit and an impedance-matching scheme, as shown in Fig. 27.9B. In this case, a variable capacitor C_3 is used as a tuning control in parallel resonance with inductor L_1. A tapped capacitive voltage divider (C_1/C_2) is used to provide impedance matching to the diode. Figure 27.9C gives values of capacitance and inductance for various bands.

Operation of the tunable FSM is simple and straightforward. Set the sensitivity control to approximately half-scale, and then key the transmitter. Adjust the tuning control (C_3) for maximum deflection of the meter pointer (readjustment of the sensitivity control may be needed). After these adjustments are made, the tunable FSM works just like any other FSM.

As a general rule, the FSM is best utilized at least a few wavelengths from the transmitting antenna.

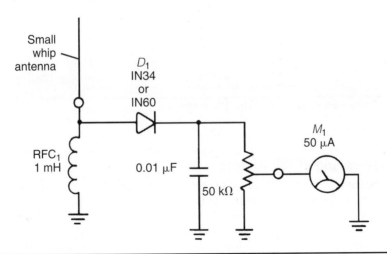

FIGURE 27.9A Simple field strength meter.

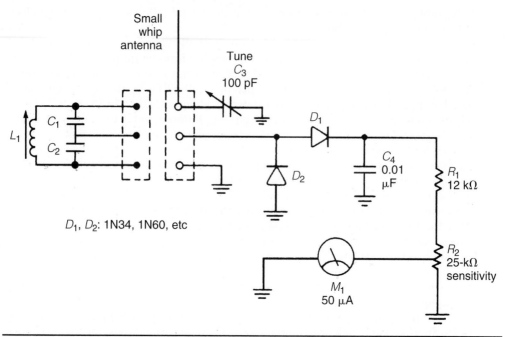

FIGURE 27.9B   Tuned field strength meter.

Band (meters)	L_1 (μH)	C_1 (pF)	C_2 (pF)
75/80	27	25	100
40	10	15	68
30	4.4	15	68
20	2.25	10	68
17	1.7	10	56
15	1.3	10	47
13	0.8	10	39
10	0.5	10	39

FIGURE 27.9C   Component values for field strength meters.

RF Wattmeters and VSWR Meters

Perhaps the single most important instrument to have for day-to-day operation of a transmitter is the RF power meter (or *wattmeter*) capable of reading both *forward (incident)* and *reflected* power. A closely related instrument is the antenna VSWR meter, which gives only a *relative* indication of forward and reflected power. Its meter is calibrated to display the dimensionless units of *voltage standing wave ratio* (VSWR or SWR). Many modern instruments combine both RF power and VSWR measurement capabilities. We will look at two of these as examples in this chapter.

Measuring RF Power

Measuring RF power has traditionally been notoriously difficult, except perhaps in the singular case of continuous-wave (CW) sources that produce pure sine waves. Even in that limited case, however, some measurement methods are distinctly better than others.

Suppose the peak voltage of a waveform is 100 V (i.e., from negative peak to positive peak is 200 V). Since a CW waveform is a pure sine wave, we know that the root mean square (RMS) voltage is 0.707 × peak voltage. Power dissipated in a resistive load is related to the RMS voltage across the load by

$$P = \frac{(V_{RMS})^2}{Z_0} \tag{27.9}$$

where P = power, in watts, delivered to antenna
$\quad V_{RMS}$ = RMS voltage, in volts, measured at feedpoint
$\quad Z_0$ = feedpoint impedance, in ohms

Assuming a load impedance of 50 Ω, the power in our hypothetical illustration waveform is 100 W.

We can measure power on unmodulated sinusoidal waveforms by measuring either the RMS or peak values of either voltage or current, assuming that a constant resistance load is present. But accurate measurement becomes more difficult in the presence of complex waveforms such as modulated signals. For instance, on a Bird model 4311 peak power meter the various power readings—peak(PEP) versus average—vary markedly with modulation type.

One of the earliest practical RF power-measuring devices was the *thermocouple RF ammeter* (see Fig. 27.10). This instrument works by dissipating a small amount of power in a small resistance inside the meter and measuring the resulting heat generated with a thermocouple. A dc ammeter reports thermocouple current. Because it works on the basis of the power dissipated in heating a resistance, a thermocouple RF ammeter is inherently an RMS-reading device. It is thus very useful for making average power measurements. If we know the RMS current into the antenna feedpoint and the resistive component of the load impedance, then we can determine RF power from the familiar expression

$$P = I^2 \times R_L \tag{27.10}$$

provided the reactive component of the load impedance is zero or very low.

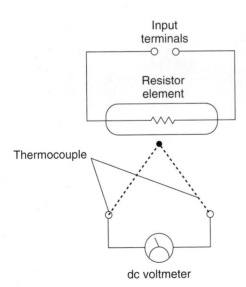

FIGURE 27.10 Circuit for thermocouple RF ammeter.

There is, however, a significant problem that keeps thermocouple RF ammeters from being universally used in RF power measurement: The instruments are highly frequency dependent. Some meters are advertised as operating into the low-VHF region, but the results will have meaning only if a copy of the calibrated frequency response curve for that specific meter is available so that a correction factor can be added to (or subtracted from) the reading. As a general precaution, at 10 MHz and higher, the readings of a thermocouple RF ammeter must be viewed with a certain amount of skepticism unless the original calibration chart is available.

RF power can also be measured by knowing the voltage across the load resistance. In the circuit of Fig. 27.11 the RF voltage appearing across the load is scaled down to a level compatible with the voltmeter by the resistor voltage divider (R_2/R_3). The output of this divider is rectified by CR_1 and filtered to dc by capacitor C_2.

Obtaining the voltage measurement from a simple diode voltmeter is valid only if the RF signal is unmodulated and has a pure sinusoidal waveform. While these criteria are almost always met with CW transmissions, they are not valid for other waveforms, such as those characteristic of voice analog modes. If the voltmeter circuit is peak reading, as in Fig. 27.11, then the peak power is

$$P = \frac{V_0^2}{R_L} \tag{27.11}$$

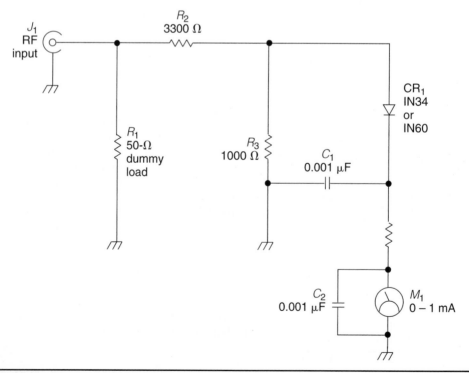

Figure 27.11   RF voltmeter "wattmeter".

The average power is then found by multiplying the peak power by 0.707. Some meter circuits include voltage dividers that precede the meter and thereby convert the reading to RMS, and thus convert the power to average power. Again, it must be stressed that terms like *RMS, average,* and *peak* have meaning only when the input RF signal is both unmodulated and a pure sinusoid. Otherwise, the readings are meaningless unless calibrated against some other source.

A family of RF power meters is based on various bridge methods. Figure 27.12 shows a bridge set up to measure both forward and reverse power. This circuit was once popular for VSWR meters. There are four elements in this quasi-Wheatstone bridge circuit: R_1, R_2, R_3, and the antenna impedance (connected to the bridge at J_2). If R_{ant} is the antenna resistance, then we know that the bridge is in balance (i.e., the null condition) when the ratios R_1/R_2 and R_3/R_{ant} are equal. In an ideal situation, resistor R_3 will have a resistance equal to R_{ant}, but that might overly limit the usefulness of the bridge. In some cases, therefore, the bridge will use a compromise value such as 67 Ω for R_3. Such a resistor will be useable on both 50- and 75-Ω antenna systems with only small errors. Typically, these meters are designed to read relative power level, rather than the actual power.

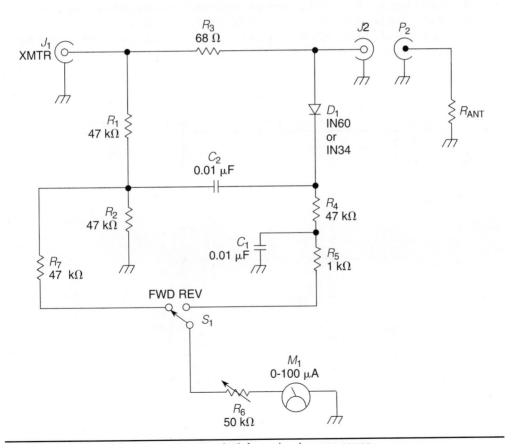

FIGURE 27.12 Bridge arrangement to measure both forward and reverse power.

An advantage of this type of meter is that we can get an accurate measurement of VSWR as long as we properly calibrate the device. With the switch in the *forward* position, and RF power applied to J_1 (XMTR), potentiometer R_6 is adjusted to produce a full-scale deflection on meter M_1. When the switch is then set to the *reverse* position, the meter will read reverse power relative to the VSWR. An appropriate VSWR scale, with calibration specific to the circuit configuration employed, is provided. This design was commercially available from a number of manufacturers in the middle of the twentieth century, most notably being packaged with some of the *E. F. Johnson* Viking Matchboxes of the era.

A significant problem with the bridge of Fig. 27.12 is that it cannot be left in the circuit while transmitting because it dissipates a considerable amount of RF power in the internal resistances. Some of these meters were provided with switches that bypassed the bridge when transmitting so that the bridge was only in the circuit when making a measurement. Others, however, had to be removed from the line before transmitting.

An improved bridge circuit, shown in Fig. 27.13. is the *capacitor/resistor bridge*—also called the *micromatch bridge*. We can immediately see that the micromatch is improved over the conventional bridge because it uses only 1 Ω in series with the line (R_i), thus dissipating considerably less power than the resistance used in the previous example. Because of this, we can leave the micromatch in the line while transmitting. Recall that the ratios of the bridge arms must be equal for the null condition to occur. In this case, the capacitive reactance ratio of C_1/C_2 must match the resistance ratio R_1/R_{ant}. For a 50-Ω antenna the ratio is $\frac{1}{50}$, and for 75-Ω antennas it is $\frac{1}{75}$ (or $\frac{1}{68}$ for a single compro-

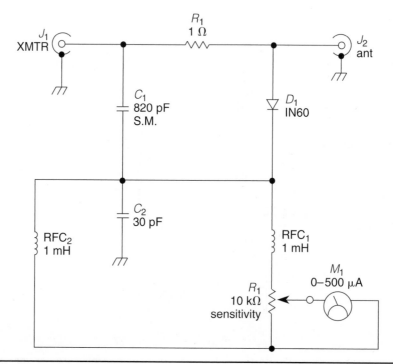

FIGURE 27.13   Micromatch wattmeter.

mise value covering both impedances). The small-value trimmer capacitor (C_2) must be adjusted for a reactance ratio with C_1 of $\frac{1}{50}$, $\frac{1}{75}$, or $\frac{1}{68}$, depending upon how the bridge is set up.

The sensitivity control can be used to calibrate the meter. In one version of the micromatch, there are three power ranges (10, 100, and 1000 W). Each range has its own sensitivity control, and these are switched in and out of the circuit as needed.

The *monomatch bridge* circuit in Fig. 27.14 is the instrument of choice for HF and low-VHF applications. In the monomatch design, first seen by the author in *Collins* accessories, the transmission line is segment B, while RF sampling elements are formed by segments A and C. Although the original designs were based on a coaxial-cable sensor, later versions used printed circuit foil transmission line segments or parallel brass rods for A, B, and C.

The sensor unit is basically a directional coupler with a detector element for both forward and reverse directions. For best accuracy, diodes CR_1 and CR_2 should be matched, as should R_1 and R_2. The resistance of R_1 and R_2 should match the transmission line surge impedance, although in many instruments a 68-Ω compromise resistance is used.

The particular circuit shown in Fig. 27.14 uses a single dc meter movement to monitor the output power. Many recent implementations employ two separate meters (for simultaneous viewing of forward and reverse power, as in the *Kenwood* wattmeters) or a single meter with *crossed-needle* pointers (as in certain Daiwa units).

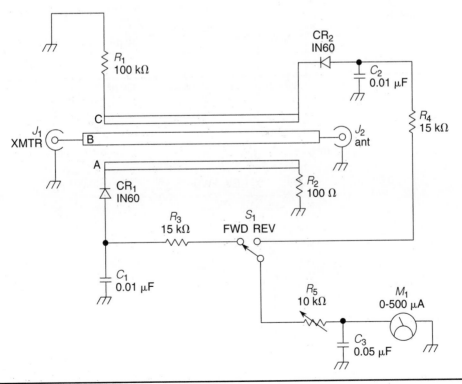

Figure 27.14 Monomatch wattmeter.

One of the latest designs in VSWR meter sensors is the *current transformer* assembly shown in Fig. 27.15. In this instrument, a single-turn ferrite toroid transformer is used as the directional sensor. The transmission line passing through the hole in the toroid forms the primary winding of a broadband RF transformer. The secondary, which consists of 10 to 40 turns of small enamel wire, is connected to a measurement bridge circuit ($C_1 + C_2$ + load) with a rectified dc output.

Figures 27.16 and 27.17 show instruments that can be left in the transmission line for all amateur power levels. The upper unit in Fig. 27.16, from Diamond Antennas, houses the transformer with the metering and control functions. The Kenwood wattmeter below it allows the user to front-panel select any of up to three separate companion transformer units, each capable of being located as much as 18 in from the meter enclosure.

The Bird *Model 43 Thruline* RF wattmeter shown in Fig. 27.17 has for years been one of the industry standards in communications service work. Although it is slightly more expensive than other instruments, it is versatile and has a reputation for being accurate and rugged. The *Thruline* meter can be inserted into the transmission line of an antenna system with so little loss that it may be left permanently in the line during normal operations. The *Model 43 Thruline* is popular with land mobile and marine radio technicians.

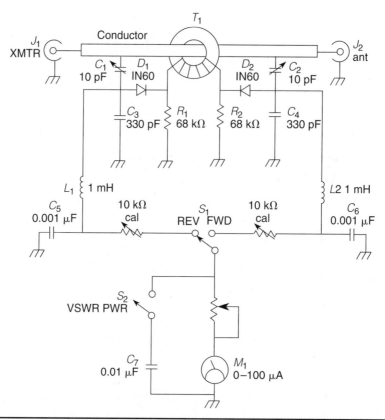

FIGURE 27.15   Current transformer wattmeter.

Figure 27.16 Amateur radio RF wattmeter.

The heart of the *Thruline* meter is the directional coupler transmission line assembly shown in Fig. 27.18A; it is connected in series with the transmission line to the antenna or dummy load. The plug-in directional element can be rotated 180 degrees to measure either forward or reverse power levels. (Some radio repair shops use two meters in series, for simultaneous viewing of the power in both directions.) Each plug-in element contains a sampling loop and diode detector designed to cover a specific range of fre-

Figure 27.17 *Bird model 43 RF wattmeter.*

ELEMENT CIRCUIT

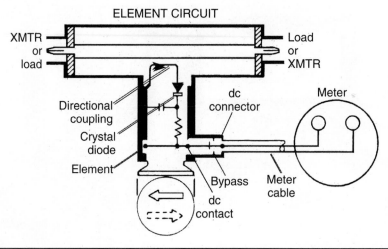

FIGURE 27.18A   *Thruline* sensor circuit.

quencies. The main RF barrel is actually a special coaxial line segment with a 50-Ω or 75-Ω characteristic impedance.

The *Thruline* sensor depends for its operation on the mutual inductance between the sample loop and the center conductor of the coaxial element. Figure 27.18B shows an equivalent circuit. The output voltage from the sampler (e) is the sum of two voltages, e_r and e_m. Voltage e_r is created by the voltage divider action of R and C on transmission line voltage V. If R is much less than X_C, we may write the expression for e_r as

$$e_r = \frac{RV}{X_C} \qquad (27.12)$$

$$= RV(j\omega C)$$

Voltage e_m, on the other hand, is due to mutual inductance, and is expressed by

$$e_m = \pm M(j\omega I) \qquad (27.13)$$

We now have the expression for both contributors to the total voltage e. We know that

$$e = e_r + e_m \qquad (27.14)$$

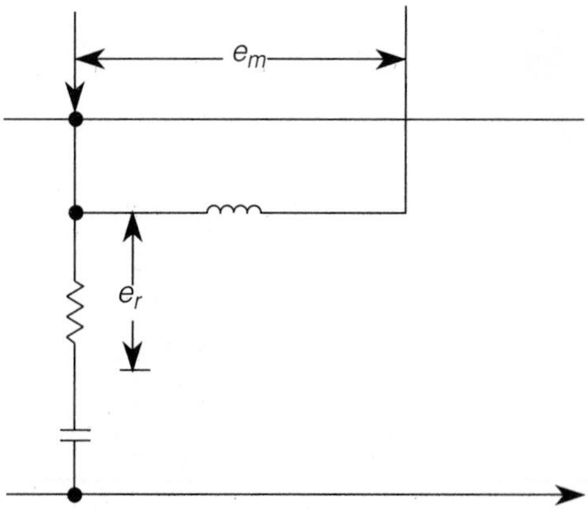

FIGURE 27.18B   Circuit of pick-up element in a *Model 43* wattmeter.

so, by substitution,

$$e = j\omega M\left[\left(\frac{V}{Z_0}\right) \pm I\right] \tag{27.15}$$

At any given point in a transmission line, V is the sum of the forward (V_{FWD}) and reflected (V_{REF}) voltages, and the line current is equal to

$$I = \frac{V_{FWD}}{Z_0} - \frac{V_{REV}}{Z_0} \tag{27.16}$$

where Z_0 is the transmission line impedance.

We may specify e in the forms

$$e = \frac{j\omega M(2V_{FWD})}{Z_0} \tag{27.17}$$

and

$$e = \frac{j\omega M(2V_{REV})}{Z_0} \tag{27.18}$$

The output voltage e of the coupler, then, is proportional to the mutual inductance and frequency (by virtue of $j\omega M$). But the manufacturer terminates R in a capacitive reactance, so the frequency dependence is lessened (see Fig. 27.18C). Each element is

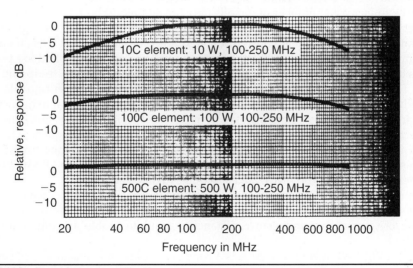

FIGURE 27.18C   Power-frequency calibration.

Figure 27.18D   High-power RF wattmeters.

custom-calibrated, therefore, for specific frequency and power ranges. Beyond the specified ranges for any given element, performance is not guaranteed, but Bird offers many elements to cover most commercial applications. Some of the *Thruline* series intended for very high power (Fig. 27.18D) applications use an in-line coaxial cable coupler (for broadcast-style hardline) and a remote indicator. Since the *Thruline* meter is not a VSWR meter but, rather, a power meter, VSWR can be obtained from the usual formula or by using the nomographs of Fig. 27.19.

SWR Analyzers

A relatively new breed of instrument, often called the SWR analyzer, finds increasing favor with SWL, repair shop technician, and ham radio operator alike. It uses a low-power RF signal generator and modern circuitry to measure the VSWR of the antenna. Lately, an increasing number of models measure and report both the resistive and the reactive parts of the feedpoint impedance—including, in some cases, the *sign* of the reactance. Perhaps the primary breakthrough for these products is that most, if not all, of them are battery-operated and have been kept small enough to be stuffed in a pocket or tote bag and carried to the top of a tower for use right at the antenna feedpoint!

The MFJ *259* VSWR analyzer is shown in Fig. 27.20. This instrument, and its current production version, the *259B*, combine a VSWR analyzer with a digital frequency counter and microcontroller, operating over from 1.8 to 170 MHz. A six-position band switch selects the desired band, and the *tune* control is set to the desired frequency. The meter will then read the VSWR at the that frequency. Alternatively, the tune control can be adjusted until the minimum VSWR is found. The front panel of the *MFJ* unit has two meters, *SWR* and *RESISTANCE,* and a two-line digital readout.

In addition to VSWR, the *259B* can provide the resistance and reactance components (but not the sign) of the feedpoint impedance, and can measure cable length, cable loss, capacitance, inductance, distance to a fault in a cable, and digital frequency readout.

The *VIA* SWR meter from AEA is a compact *graphing* antenna analyzer, successor to their earlier *HF Analyst.* The *VIA* operates in the range from 100 kHz to 54 MHz. Ar-

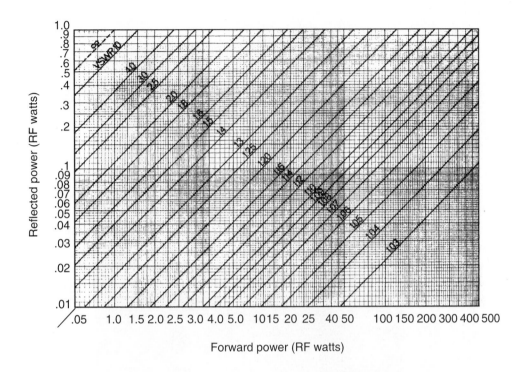

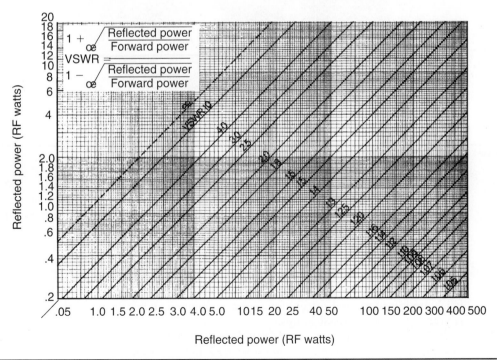

FIGURE 27.19 Forward versus reflected power graphs.

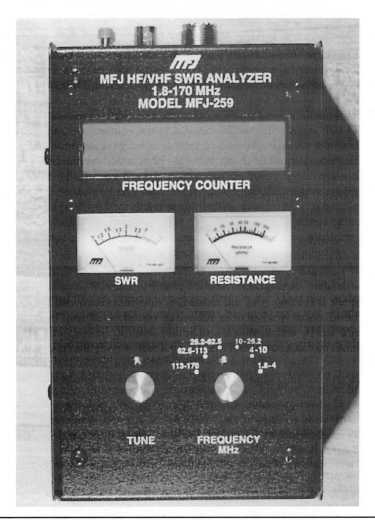

FIGURE 27.20   *MFJ 259 VSWR analyzer.*

guably its most useful feature is its *sweep* display of SWR or *R* or *X* versus frequency, invaluable for quickly determining the effect of an antenna adjustment on its character-istics over the entire band of interest. In addition to being able to provide values of ca-pacitance, inductance, line length, and line loss, the *VIA* includes a serial port for high-resolution data transfer to a PC and/or remote control by a PC. The unit also can store plots in nonvolatile memory, and it has a Smith chart mode as well. Despite its considerable capabilities, the *VIA* is the "low" end of a family of analyzers, the rest of which are intended (and priced) for commercial applications such as cell phone sys-tems and land mobile repair shops.

Because these units employ wideband small-signal detectors to keep costs down, they are susceptible to errors from ambient RF fields within their frequency ranges, such as might be experienced when using the device in the presence of active transmit-ters or in the vicinity of broadcast stations, both AM and FM. Generally, it is not wise to add a notch filter between the analyzer and the device under test because the filter can alter the impedances reported by the instrument.

In addition to antenna measurements, SWR analyzers are equipped to measure other parts of the RF path in a typical installation. In particular, they can measure the velocity factor of a transmission line, help in tuning or adjusting matching stubs or matching networks, measure capacitance or inductance, and determine the resonant frequency of *LC* networks.

Doping Out Coaxial Cable

When you install an antenna, or do a bit of preventive maintenance, or find the antenna is not working properly, one system component it is important to check is the transmission line. Typically, a two-step process is employed. The first step consists of ohmic checks at dc, as shown in Fig. 27.21. This is particularly useful with coaxial cable, where neither the inner conductor and the outer conductor (shield) is available for direct visual inspection. With terminals A and B open there should not be any resistive path across the input terminals (as shown). If a high resistance is seen, there might be some contamination in the system or the dielectric insulation has failed, permitting a *leakage* current to flow. (Or you may have your fingers on the conductive tips of the test leads!) Alternatively, a low resistance indicates a short circuit somewhere along the cable's length. If the cable has been cut, or an object passed through it, or the connector is messed up (commonly because of poor soldering), a short can result.

If terminals A-B are shorted together, a low resistance should be noted. If not, then it is likely that the center conductor is open—typically as the result of some prior excessive stress (such as extreme bending) on the cable. Of course, to determine whether the shield or the center conductor is at fault, you will need to bring *both* ends of the cable

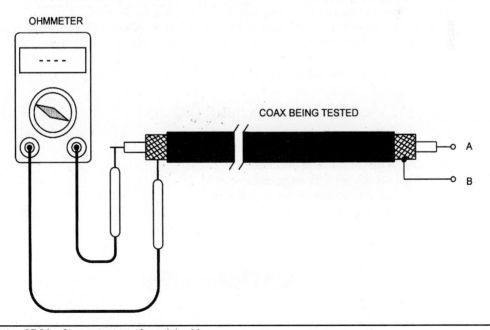

FIGURE 27.21 Ohmmeter test of coaxial cable.

close to the ohmmeter so that you can measure the resistance of the shield separate from that of the center conductor. On the other hand, even if you know which is at fault, it probably isn't going to change the ultimate disposition of the cable!

Once you have completed the basic open-circuit and short-circuit ohmmeter measurements on the cable, your next step should be to measure the loss of the transmission line at one or more frequencies of interest. Although the author owns numerous lengths of RG-8 that exhibit the same losses at HF as his newer stock does, older cables that have not been adequately protected from the elements in prior usage or storage can develop losses that get higher as the cable deterioration progresses, so low loss at the time of initial installation does not guarantee low loss later on. Moisture is the enemy of coaxial cables, so existing cable stock should not be stored outdoors or in damp basements or garages unless all cable ends have been sealed.

Losses in new cables increase with frequency and with the total length of cable used in any particular link between transmitter and antenna. Depending on the specific type of flexible coaxial cable employed, losses can run from as low as 0.2 dB/100 ft at 1.8 MHz (RG-8 with foam dielectric) to as much as 0.5 dB/100 ft (RG-58 and RG-59). At 30 MHz the losses are approximately 1.0 dB/100 ft and 2.5 dB/100 ft, respectively. Arguably the most used type of flexible coaxial cables at MF and HF are members of the RG-8 and RG-213 family. Their loss (with standard dielectric) ranges from 0.25 dB/100 ft at 1.8 MHz to 1.3 dB/100 ft at 30 MHz. The reader is encouraged to consult specific vendors' online catalogs for the correct loss data for cable types of interest, since the number of variations in cable dimensions and materials makes presentation of so many possible choices in a single chart unworkable.

Figure 27.22 shows a test setup for measuring coaxial cable loss. The first step in the procedure is to determine the length of the piece of cable under test. One end of the line should then be terminated in its characteristic impedance—usually 50 or 75 Ω. The termination must be capable of dissipating the output power of the source, so usually a 100W (or higher) dummy load is employed. The simplest RF source is often your own transmitter or transceiver. Two identical RF power meters (M_1 and M_2) are used to simultaneously measure the input power to the line (M_1) and the power delivered to the load (M_2) while the RF source is activated. The difference in the two power readings is then the power lost in the cable itself. To express that loss in decibels, we use

$$\text{Loss}_{dB} = -10 \log\left(\frac{P_{M2}}{P_{M1}}\right) \tag{27.19}$$

where Loss = loss of cable, in decibels
$\quad\quad P_{M1}$ = (cable input) power reading on M_1
$\quad\quad P_{M2}$ = (cable output) power reading on M_2

FIGURE 27.22   RF attenuation test setup for coaxial cable.

If necessary, these tests can be run with a single meter by alternately inserting it in the signal path at each end of the coaxial cable being tested. This has the advantage of eliminating any errors caused by differences in the calibration of two meters but has the disadvantage of changing the test configuration in midmeasurement. When using the single meter approach, it is extremely important that there be a very good match between the cable and the termination—a very low SWR, as close to 1.0:1 as possible, in other words.

In general it is wise to perform these measurements on at least two different frequencies—preferably at the extremes of the frequency range over which the cable will be used. If the cable length is 100 ft, the loss found from Eq. (22.1) is the loss in decibels/100 ft. But if length L is anything other than 100 ft, use the following calculation to obtain the *matched loss per 100 ft*, since it is this number that you will want to compare with the manufacturer's published specifications:

$$\text{Loss}_{\text{dB/100ft}} = \frac{\text{Loss}_{\text{dB}}}{L(\text{ft})} \times 100 \tag{27.20}$$

Of course, the same type of measurement can be performed on other types of coaxial transmission lines as long as a good match between the nominal line impedance and the termination is maintained. Remember, however, that virtually all commonly available *hardline* has a characteristic impedance of 75 Ω, not 50 Ω.

Vector Network Analyzers

The most recent addition to the stable of affordable analyzers for antenna systems is the *vector network analyzer*. Historically, precision antenna analyzers have been quite big and quite expensive laboratory instruments, but VNA designs for the amateur marketplace are now available on a card or in a small enclosure for under $1000—plus the price of the associated PC or Mac, of course! Inherent in its name, a VNA provides resistance, reactance, and the sign of the reactance. VNAs rely on an active USB connection to a PC or Macintosh during their operation, so they are not easily taken to the top of a tower. Instead, VNAs have a procedure for calibrating out the effects of a transmission line between the antenna and the VNA—something not possible with the simpler analyzers described in the previous section. Basic VNA theory can be found on the Internet by googling "vector network analyzer basics" or "vector network analyzer tutorial".

VNAs for the amateur can be built by experimenters with access to surface-mount device (SMD) mounting capabilities, or they can be purchased assembled. The seminal articles on building such a VNA are from Paul Kiciak, N2PK; his Web site is a good starting point for anyone interested in this path. For a while, a VNA kit designed by Tom McDermott, N5EG, and Karl Ireland was available from *TAPR*, the Tucson Amateur Packet Radio group that produced some of the earliest terminal node controllers (TNCs) for packet radio, but the organization has gone out of the VNA manufacturing business, so only used *TAPR* units are available. More recently, rights to the *TAPR* design have been licensed to Ten-Tec, and the company has added it to its product line—fully wired and tested, not as a kit—as the *Model 6000 VNA*. A competing unit, designed by W5BIG, is offered as the *AIM 4170* from Array Solutions, Inc.

VNAs are not just useful for antenna and transmission line analysis, but that is the focus of this book, so we won't expand our discussion to include their more general network analysis capabilities. In addition to swept plots of VSWR, feedpoint $R + jX$, and Smith chart displays of antenna performance across the entire band, VNAs can function as time domain reflectometers (TDRs), providing a detailed graphical look at pulsed waveforms sent down a transmission line. At each point along the line that an outbound pulse encounters an impedance other than the Z_0 of the line, a reflected pulse is created that travels back to the instrument, where it is detected and plotted, very much consistent with the transmission line pulse analysis of Chap. 4. A VNA's ability to locate every connector, every splice, every transition in the feedlines, and—most important—any potential fault (short circuit, open circuit, or other unexpected discontinuity) between the radio room and the antennas hundreds of feet away is amazing!

Figure 27.23 shows a VNA trace of the entire RF path from the radio room to an antenna, obtained while troubleshooting the author's 20-m monoband gamma-matched Yagi. In the first 39 ft of the path, the outbound pulse encounters a series of short lengths of 50-Ω cable and a few inline devices along the path—in particular, connectors, a lightning arrestor, and various relay boxes. While the lumped-component devices and circuits exhibit nominal 50-Ω impedance at HF, the wide bandwidth of the VNA pulses exposes their departure from 50 Ω at other frequencies. The last of the positive-going impedance bumps on the left side of the chart, at 39 ft from the VNA, signals the junction of 50-Ω coaxial cable to 75-Ω hardline. Near the top of the tower, at a total distance of 159 ft from the VNA, the hardline connects to a 39-ft "pigtail" and rotator loop of 50-Ω cable that completes the path to the SO-239 connector at the feedpoint of the four-element beam. (Little additional information can be gleaned from the wiggles and bumps to the right of that spike.)

Because of their circuit implementations, the N2PK and *AIM 4170* VNAs are reportedly less susceptible to measurement errors from ambient RF in the vicinity of the analyzer than the older SWR analyzers described earlier.

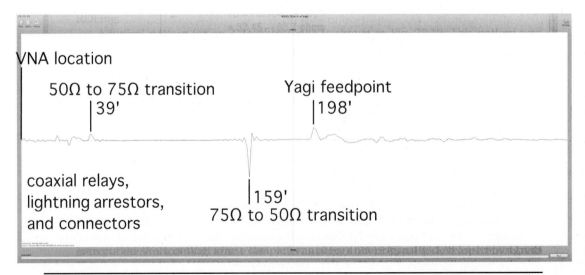

FIGURE 27.23   VNA TDR trace example.

Dummy Loads

A *dummy load* is a nonradiating substitute for an antenna used when measuring or testing any and all parts of the antenna system from the transmitter out, exclusive of the antenna. In fact, British radio engineers often refer to dummy loads as "artificial aerials". There are several uses for these devices. Amateur radio operators should use dummy loads to tune up their transmitters before switching to an actual antenna for final adjustments. Good operating practices require operators to check the frequency for activity by other stations before completing their tune-up procedures on an antenna.

Another use for dummy loads is to aid in troubleshooting antenna systems. Suppose we have a system in which the VSWR has suddenly become very high—high enough to "kick" our transmitter off the air. Starting at the transmitter, we can disconnect successive elements of the RF transmission system and connect the dummy load at the disconnect point. If the VSWR suddenly drops to a normal value, then the difficulty is downstream (i.e., toward the antenna). By repeating this process, we will eventually find the bad element (which is often as likely to be a coaxial cable, a connector, or a switch as it is to be the antenna itself).

Figure 27.24 shows the most elementary form of dummy load, which consists of one or more resistors connected in parallel, series, or series-parallel so that the total resistance is equal to the desired load impedance. The power dissipation is the sum of the individual power dissipations.

NOTE *It is essential that noninductive resistors be used for this application. For this reason, carbon-composition or metal-film resistors are used, although the former are becoming harder and harder to find. For very low frequency (VLF) work, it is permissible to use special low-inductance counterwound wire resistors. These resistors, however, cannot be used for frequencies greater than a few hundred kilohertz. In general, homemade dummy loads should use many resistors in parallel, rather than in series, since it is easier to keep the stray inductance down in the former configuration.*

Several commercial dummy loads are shown in Figs. 27.25A through 27.25C. The load in Fig. 27.25A is a 5W model, typically used for servicing citizens band equipment. The resistor is mounted directly on a PL-259 coaxial connector. These loads typically work to about 300 MHz, although many are not really useful over about 150 MHz. A

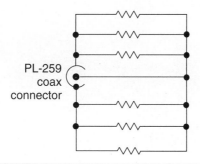

FIGURE 27.24   A very basic dummy load.

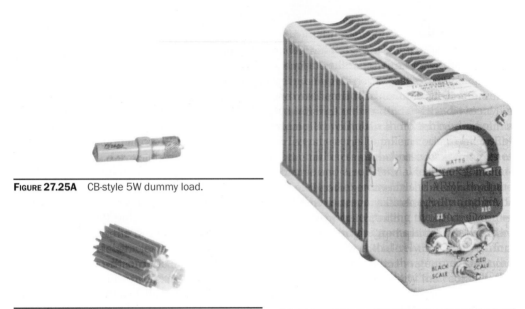

FIGURE 27.25A   CB-style 5W dummy load.

FIGURE 27.25B   15W dummy load.

FIGURE 27.25C   *Termaline* RF wattmeter.

higher-power version of the same type is shown in Fig. 27.25B. This device works to the low VHF region and dissipates up to 50 W in intermittent usage—perfect for servicing VHF land mobile amateur and marine transceivers.

A very high power load is shown in Fig. 27.25C. It is representative of Bird Electronics "coaxial resistors" that operate to power levels up to 10, 30, or 40 kW. These high-power loads are cooled by flowing water through the body of the resistor and then exhausting the heat in an air-cooled radiator.

Our final dummy load resistor is shown in Fig. 27.26. The actual resistor is shown in Fig. 27.26A, and a schematic view is shown in Fig. 27.26B. The long, high-power noninductive resistor element is rated at 50 Ω and can dissipate 1000 W for several min-

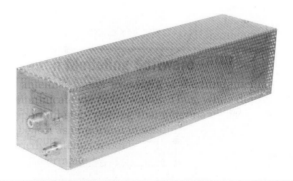

FIGURE 27.26A   Drake *DL-1000* dummy load.

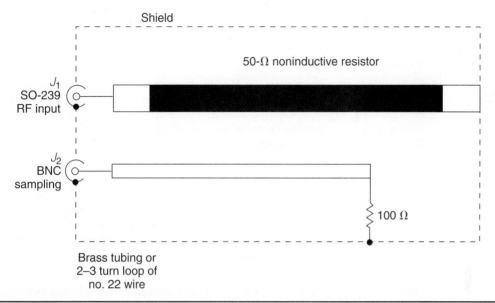

FIGURE 27.26B HF dummy load with pickup loop.

utes. If longer times or higher powers are anticipated, forced air cooling can be applied by adding a blower to one end of the cage. The device has been modified by adding a BNC sampling jack connected internally to either a two-turn loop, made of #22 insulated hookup wire or brass rods that are positioned alongside the resistor element. Thus, it will pick up a sample of the signal so that it can be viewed on an oscilloscope or used as a signal source for other instrumentation.

For intermittent key-down periods at the United States amateur radio service legal limit, the *CanTenna*—originally sold by the Heath Company, now by MFJ—is an inexpensive solution to the dummy load requirement. The internal resistor is cooled with mineral oil. Transformer oils, which historically have contained PCBs, should never be used.

In recent years high-end dummy loads with additional monitoring and metering capabilities have been introduced by a few manufacturers. The *Alpha 2100 Dummy Load* sold by RF Concepts can continuously dissipate a full 1500 W on any frequency from 1.8 to 100 MHz indefinitely! It includes full metering, variable-speed cooling fans, and a price tag comparable to that of a good used legal limit amplifer!

Documenting Antenna System Measurements

Regardless of the instrumentation that may be available, there are certain practices that the owner or installer of antenna systems should follow throughout the life of the system, starting on day one. Here are a few that can save hours and hours of time later on:

- Antenna and transmission line measurements should be made and recorded when the antenna system is first installed and periodically thereafter. If different results are obtained at some time in the future, the same measurements should be repeated and compared with the original data before concluding that something has changed. Measuring and recording the ohmic and RF losses in lengths of coaxial cable and CATV hardline *before* they are installed can pay big dividends later on!

- Many antenna measurements are difficult to make with any degree of accuracy. Nonetheless, some antenna and transmission line parameters can and should be measured, regardless of the difficulty. Probably the simplest are plots of VSWR across the operating band of interest, obtained at a specified point along the feed system. If the appropriate instrumentation is on hand, $R + jX$ impedance measurements are also valuable.

- Measurements of antenna feedpoint impedance versus frequency are best if obtained right at the antenna. Many times, however, this is not accomplished easily, and data taken at the other end of the feedline are also as useful, especially if a VNA is available. Try to also record the "environment" in which the measurements were taken, for instance, length and type of feedline attached to the antenna, heading of the antenna (if it can be rotated), and power level and instrumentation used.

- Dedicate a three-ring binder or a folder in a file cabinet or on your personal computer for collecting measurement data in one place. Memory has a habit of playing tricks on us.

PART VIII

Mechanical Construction and Installation Techniques

CHAPTER 28

Supports for Wires and Verticals

Antenna installation methods vary in complexity from those that can be performed by one person of moderate strength and agility all the way up to large-scale projects that are best left to professional antenna riggers. In this chapter we suggest techniques for erecting wire antennas and lightweight rigid verticals. In Chap. 29 we do the same for full-fledged towers. Keep in mind, however, that the information given herein is merely informal guidance and what you ultimately do should never be in violation of local mechanical, electrical, and other building codes or any zoning ordinances or restrictive covenants pertaining to your property.

Antenna Safety

Before dealing with the radio and performance issues, let's deal first with a few antenna safety matters. You do not want to be hurt either during installation or during the next windstorm. Two areas of potential concern immediately present themselves: *reliable mechanical installation* and *electrical safety*.

Caution No *antenna should ever be erected so the antenna, its feedline, its supports, or any part thereof crosses over or under a power line, transformer, or other utility company equipment—never, ever! Each year we read or hear about colleagues and innocent bystanders being electrocuted while installing or working on an antenna. In virtually all these tragic cases, the antenna came into contact with a power line. Keep in mind one* dictum *and make it an* absolute: *There is* never *a time or situation when any part of an antenna system should be placed near enough to electrical power lines that it can come in contact with them if the antenna, its feedline, or its supports fail or if the power poles or power lines themselves come down! And* never, never *use a utility pole or guy wire as an attachment point for any rope, halyard, tug line, or guy wire—not even temporarily.*

Never rely on the insulation covering antenna wire or the power lines to protect you from high voltages on the utility company wires. Never assume that the power lines are insulated. Old insulation crumbles on contact with even a thin wire antenna, and even new power lines may have small breaks or weakened spots in the insulation, which may, after all, have been lying in an outdoor storage yard for a long time.

Even a simple wire antenna can be dangerous to erect if certain precautions are not followed. It is not possible to foresee all the situations you might face in erecting an antenna. The authors would love to give you a comprehensive list of all possible warnings, but this is just not feasible. You are on your own and must take responsibility for installing your own antenna. We can, however, give you some general safety guidelines and a few tips:

- Prior to starting the project, sketch the installation area to scale, as shown in Fig. 28.1, and draw boundaries of "safe working areas" based on the longest and tallest masts, towers, ladders, cranes, etc., that could fall, break, or tip over. Be sure to include the area around any electric utility *drop wire* from the power pole to the building as part of the "forbidden" zone. Then use stakes and colored ribbons or string to separate the "forbidden" areas from the work area. "Safe" doesn't have to refer only to avoiding power lines. Ravines, structures, and electrified fences are some other common hazards you'd probably prefer not to drop your ladder on or into.

- Make a list of all tools and machinery you expect to need for the project, and make sure you either own them, can rent them, or have friends who will bring or loan them. One of the leading sources of personal injury and property

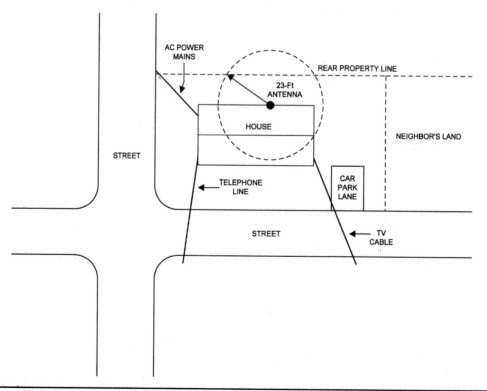

Figure 28.1   Antenna installation on suburban lot.

damage during antenna projects is failure to use the right tool for the job at hand.

- If electricity is required in the work area(s) during the project, make sure that adequate three-wire 115 VAC power with functioning *ground fault interruption* (GFI) circuitry is available at the site. Most municipalities allow GFI protection to be inserted in one of two places: as part of the circuit breaker for the circuit in question or as part of the first outlet on that run from the breaker. If needed, GFI installation should be handled only by an electrician licensed to work in your community.

- Make sure the planned tasks don't require Superman for their successful performance. If climbing is involved, make sure experienced climbers with all the appropriate climbing gear are involved. Any climbing at all, even on ladders, can be taxing. Many common antennas, including small Yagis or quads, are deceptively lightweight on the ground, but when you get up in the air even a short distance they become remarkably difficult to handle, especially in the presence of even light breezes.

- Be aware that wind speeds (and, hence, the effective wind load a given antenna represents) increase with height above ground, especially if there are buildings, trees, and other obstructions to slow the wind below their tops. The wind also has a bad habit of coming up at the worst possible time—usually when you're about to attach the antenna to its support mast. A friend of one of the authors fell from the roof of his two-story beach house, fracturing his pelvis and a leg when an offshore gust of wind came up suddenly and caught a TV antenna. It acted like a hang glider and pulled this very strong man off the roof. Two months of orthopedic casts and a year of physical therapy followed—not to mention lost wages. *Be careful!*

- Never tackle an antenna project with no one else around. Even if it is a project that can easily be handled by one person (stringing a dipole, for instance), make sure a family member, friend, or coworker is on the premises, within earshot, even if he/she is doing other things.

Probably the single best piece of advice we can impart is the time-tested Boy Scout motto: "Be prepared!" There is no substitute for planning ahead. Taking time *before* the day of the project to define and think through the many individual tasks that together result in a successful antenna installation is the single most important thing you can do before you tackle the job or your crew of helpers arrives on the scene. Knowledge of what you face, coupled with hard-nosed, sound judgment and some common sense, are the best tools for *any* antenna job.

One good rule is to always work under the "buddy system". Invite as many friends as are needed to do the job safely, and always have at least one assistant, even when you think you can do it alone.

Install only antennas of the best materials and workmanship in order to minimize the need to repair or replace them. It is not just the electrical or radio reception workings that are important—but also the ability to stay up in the air in an often unfriendly environment.

When planning an antenna job, keep in mind that pedestrian traffic in your yard possibly could affect the antenna system. Wires are difficult to see; if an antenna wire

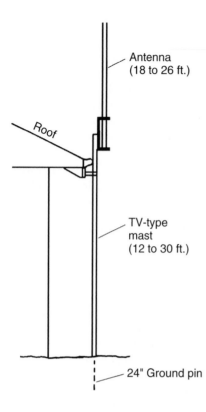

Antenna
(18 to 26 ft.)

Roof

TV-type
mast
(12 to 30 ft.)

24" Ground pin

FIGURE 28.2 Mast-mounted vertical antenna.

is low enough for vehicles, people, or animals to come in contact with it, injury and damage can result. In the United States, even when the person is a trespasser (or even a burglar!), the courts may hold a homeowner liable for injuries caused to the intruder by an inappropriately designed and installed antenna. Take care for not only your own safety but also that of others.

Consider a typical scenario involving a four- or five-band trap vertical antenna (Fig. 28.2). It will be 18 to 26 ft tall, judging from the advertisements in magazines, and will be mounted on a roof or mast 12 to 30 ft off the ground. The total height above ground will be the sum of the two lengths—perhaps 30 to 60 ft overall. You must select a location at which a 30- to 60-ft metal pole can be erected—and can fall—safely. This requirement limits the selection of locations for the antenna. In particular, if the *drop wire* from the power pole to your house attaches to the house high on an end wall rather than coming in underground, you probably will not be able to find a proper location for a house-bracketed or roof-mounted antenna of that overall height.

When installing a vertical antenna, especially one that is not ground-mounted, make sure that you have help. It takes at least two people to safely install a standard HF vertical antenna, and more may be needed for especially tall or heavy designs. Wrenched backs, smashed antennas, crunched house parts, dented vehicles, and other calamities are rare with a well-organized work party that has sufficient hands to do the job safely.

Wire Antenna Construction

Volumes have been written on the electromagnetic design and analysis of wire antennas, which are primarily employed in the MF and HF regions of the radio spectrum. What often seems to be missing, however, are practical details of the mechanical aspects of wire antenna construction and installation, so let's take a look at some of the basics.

Wire antennas come in many forms, but good techniques for construction and support are pretty much the same regardless of the wire configuration, so we shall use our old friend the half-wavelength horizontal dipole for our examples. In its most common configuration (Fig. 28.3), the HF $\lambda/2$ dipole is insulated from other conducting materials at each end and supported by ropes, cables, or other wires attached to trees, buildings, masts, or any other structures of suitable strength and height in fixed positions.

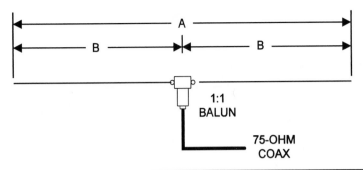

FIGURE 28.3   Typical center-fed wire dipole.

Center Insulators and Feedline Connection

The λ/2 dipole's nominal feedpoint impedance at resonance is 73 Ω in free space and at multiples of a quarter-wavelength (λ/4) above ground, so it is a reasonably good match to standard 50- or 75-Ω coaxial cable. When fed with coax, a 1:1 balun transformer or common-mode RF choke is often inserted at the feedpoint.

Far more important to the long-term success of a horizontal wire than a balun are the quality, strength, and weather resistance of the electromechanical connection between the antenna and the feedline in the face of prolonged exposure to the environment. In virtually all installations (except perhaps for the attic dipole), this junction is subject to temperature and humidity changes, precipitation of all kinds (including ice and snow in many climates), ultraviolet rays from direct sunshine, continual vibration, and abrupt changes in tension resulting from gross motion in the wind.

A common (but poor) practice is to strip the insulation back a few inches at one end of a coaxial cable, part the braid and center conductor, and connect them with ordinary electrical solder to each side of the dipole. We then wonder why the darn thing breaks apart in the next windstorm or why our VSWR measured back at the radio room seems to change over time. Unprotected solder joints exposed to the elements eventually turn gray, brittle, and powdery, and eventually crack under environmental stresses that conventional solder was never designed to resist. Solder for electrical and electronic circuits is not intended to provide a *mechanical* connection—it has little strength—nor is it meant to be exposed to the elements. Equally disastrous, coaxial cable that is open at one end is very hard to protect from moisture ingress; over time, it can become saturated with water. Over the years, there have been many reported cases of water dripping from the indoor end of cables onto the radio desk!

Even if you elect not to use a balun transformer, a ready-made center insulator that accepts a PL-259 or similar connector helps minimize or eliminate many of these problems. Figure 28.4A shows a common form of center insulator for use with dipoles and other wire antennas. At the bottom is an SO-239 coaxial cable receptable, shielded from direct rainfall by the assembly above it. Two eyebolts on the sidewalls provide mechanical strain relief for attaching the two halves of the dipole to the insulator. Although this particular style of center insulator is a compromise because it includes solder connections, if enough slack is provided in the pigtails that attach to the solder lugs the actual solder joint is under little or no mechanical strain and is far less likely to fail from stress-related mechanisms than the old-fashioned junction described above. Figure 28.4B shows a similar center insulator with a self-contained 1:1 balun.

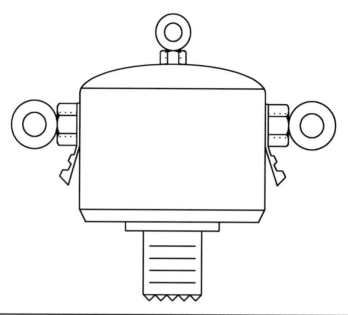

FIGURE 28.4A   Center insulator.

A recommended way to connect a wire dipole to the device is shown in Fig. 28.5. For simplicity, only one side is shown, but the other side is identical. The λ/4 antenna segments (length B in Fig. 28.3) typically are made of #12 or #14 antenna wire, but the authors have used wire as large as #10 and as small as #18 for transmitting applications; the exact gauge is not important except that larger-diameter wire is heavier and sags more, and very thin wire ultimately is not strong enough to support the weight of the center insulator and feedline. In years past, stranded wire was often used, but oxidation of the strands inevitably occurs and may result in some additional loss of transmit signal and the application of large RF voltages to oxidized junctions, so it is no longer favored. Conventional house wiring (THHN) found at electrical supply stores and discount home suppliers can be used (with or without its insulation), but be aware that it is *soft-drawn copper* and can stretch in response to excessive tension such as that encountered during windstorms by dipoles strung from trees without provision for the movement of branches. (Most antenna tuners can compensate for the changing impedance caused by long-term stretching; occasional retensioning of the end support lines may also be required.) To avoid stretch, some recommend copper-clad steel core wire (Copperweld is one well-known trade name), but others have encountered problems with voids in the copper, pitting and rusting, and ultimately failure of the wire from poorer quality "off brands". Copper-clad steel is also quite "springy" and very difficult to work with unless it is always kept under tension once unreeled. Given a choice, the author would rather have a stretched dipole than a snapped dipole!

Pass one end of one side of the dipole through an eyebolt and then double it back and twist it onto itself four to six times to ensure that no slippage of the wire through the eyebolt will occur when the antenna is erected and put under tension. Bare an inch or so of the wire at the end of the pigtail with steel wool, sandpaper, or a utility knife

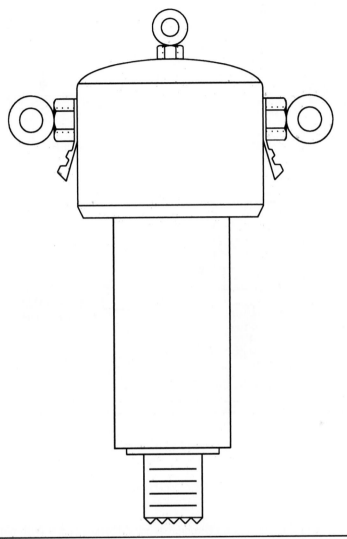

FIGURE 28.4B Center insulator with built-in balun.

until the copper is clean and shiny *all around* the bared section. Form the short section of wire between the eyelet and the solder lug into a *drip loop* and insert the end of the wire into the shank of the solder lug. Be sure to crimp the wings of the solder lug around the wire end with a pair of pliers. Then, while keeping the assembly perfectly still, solder the terminal shank and wire with solder having a *resin core*, not plumber's solder. Some people prefer to use a completely separate wire as the "pigtail", but that doubles the number of solder joints. Whichever approach is used, the intent is that there be no tension at all on the pigtail when the antenna is up in the air. These steps are best performed with the dipole antenna laid out on the ground in a straight line, preferably (weather and tools permitting) in or near the area where it will be hoisted.

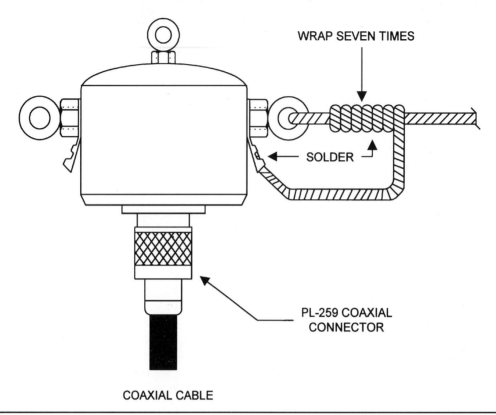

FIGURE 28.5 Center insulator connections.

NOTE *Use only resin-core solder for antennas! Typical solder is 50/50 or 60/40 lead/tin ratio and may be labeled "radio-TV" or "electronic" or something similar. Newer solder alloys, without any lead, are also available. Most plumber's solder is acid-core and will destroy your connection within a short time.*

The life of the solder joint can be extended by painting it with nail polish or spraying it with an acrylic lacquer such as Krylon, after the joint has cooled. As with the soldering process itself, the entire assembly should be not be moved or jostled until the coating has dried.

Keep in mind that even "new" copper wire needs to be sanded or scraped. Bare copper wire may, in fact, have a clear enamel insulation on it that isn't obvious to the eye. But all bare copper oxidizes, forming an insulating layer over the conductor; the early stages of that process result in only a slight dulling of the surface, so cleaning off the wire immediately prior to soldering it is an excellent habit to form. Similarly, when preparing the center connector, it is also good practice to use steel wool or sandpaper on the solder lugs to remove any oxidation that may have formed over time. Oxidized metal simply cannot be soldered!

An SO-239 UHF coaxial receptacle (socket) on the bottom of the center support device shown in Fig. 28.5 accepts any feedline terminated in a standard PL-259 plug. Once that connection is made, wrap high-quality black vinyl electrical tape around the joint, starting on the vinyl cable jacket below the plug and continuing completely over and past the PL-259 until reaching the molded case that supports the SO-239 receptacle. *Half-lapping* the tape like siding or roof shingles should ensure minimal water ingress at the end of the coaxial cable. To minimize water *vapor* ingress caused by changes in humidity, the vinyl tape should also be sprayed with acrylic lacquer or painted with nail polish. Alternatively, use heat-shrink tubing and a heat gun to totally encase the connection from the threads of the SO-239 above the PL-259 to the outer jacket of the cable a half-inch or so below the bottom edge of the PL-259.

In most installations, the weight of the feedline will keep the entire center assembly oriented as shown in Fig. 28.5, thus eliminating the need for a drip loop on the coaxial cable. If, however, a specific installation results in the SO-239 connector being anywhere other than beneath the body of the device, the coaxial line near the connector should be routed with a half-loop that is arranged so as to send any precipitation running along the cable *away* from the connector.

At the top of the center insulator a small eyebolt is provided for strain-relief of the system or for suspension from a center support (such as when the antenna is an inverted-vee). One disadvantage of coaxial cable and commercial center insulator assemblies (as compared to open-wire line directly connected to the two sides of the dipole) is that both add quite a bit of deadweight to the center of the antenna. Unless some kind of auxiliary support (nylon twine, polypropylene rope, or small-diameter aircraft cable pulling up to a high point on a tower, mast, or tree) is available, the sag in a dipole for 80 or 40 m fed with RG-8 or RG-213 cable may be unacceptable to the user. If the VSWR on the line is kept close to 1.0:1 or power to the line is kept substantially below 1000 W, some weight can be removed by using the smaller-diameter RG-58 or RG-59 cables instead of RG-8—especially at 10 MHz and below, where line loss is lowest.

The longer, lower barrel of the center insulator shown in Fig. 28.4B indicates the presence of either a 4:1 or 1:1 impedance ratio balun transformer. The 1:1 size is recommended for ordinary dipoles, whereas a 4:1 unit is a good match for folded dipoles and certain other wire antennas with feedpoint impedances substantially larger than the Z_0 of the feedline.

Some users form the coaxial cable into a 6- to 12-in multiturn loop just below the coaxial connector, as shown in Fig. 28.6. This forms a *common-mode* RF choke intended

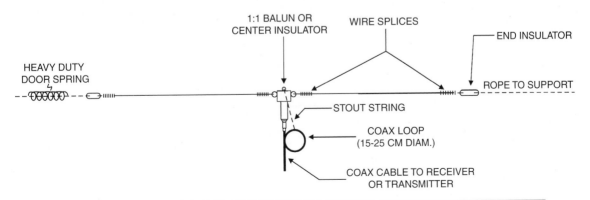

FIGURE 28.6 Dipole and wire antenna installation.

to keep currents from flowing on the outside of the cable shield, thus helping to keep the feedline from becoming an unintended part of the radiating antenna. The number of turns is inversely proportional to the lowest operating frequency. For an 80-m dipole, 16 turns of 8- to 12-in diameter is probably not too few . . . but will be very heavy! Bind the turns together with black electrician's tape or some similar adhesive medium and then fasten the entire loop to the top eyelet with stout string or fishing line for strain relief.

End Insulators

Over the years, end insulators have been made from wood, glass, ceramics, plastics, and composites. In fact, at low power levels (say, 100 W or less), garden-variety polypropylene rope often used for stringing the dipole up is perfectly adequate as its own insulator! Nonetheless, since the wire ends of a dipole are usually high-voltage points, take care to keep the conductors away from flammable surfaces. More than one licensed amateur has set a tree on fire by transmitting into a wire draped across tree limbs!

End insulators come in a large variety of shapes and sizes. Examples of two basic shapes are found in Fig. 28.7. Figure 28.7A depicts the standard *strain insulator*, and Fig. 28.7B is representative of an *egg insulator*. The egg insulator has two sets of wire grooves and through-holes that are orthogonal to each other (only one set is shown). Wire passes through one hole and its grooves, while the supporting rope passes through the other hole/groove set. The purpose of the orthogonal holes is to prevent the antenna from totally separating from the support line if the insulator should fracture.

For higher power levels, the standard strain insulator of Fig. 28.7A is preferable. This style often has a number of grooves or ridges along the body that serve to lengthen

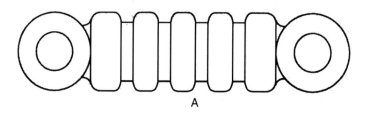

A

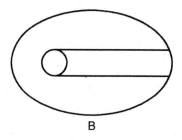

B

FIGURE 28.7   End insulators.

the effective electrical path between the two ends of the insulator, thus reducing losses in the insulator while increasing the breakdown voltage between its two ends. This is particularly important in locations where airborne contaminants gradually collect on the surface of the insulator.

End insulators can be purchased new from many of the radio stores listed at the back of this book; preowned components can be found at hamfests and radio conventions or on Internet sites such as eBay. For the experimenter interested in "rolling his or her own", typical materials for fabricating strain insulators include ceramic, glass, nylon, plastics, and (treated) hardwoods. (Short lengths of hardwood stock can be drilled at each end for wire or rope, then treated with an exterior urethane or other protectant.)

Strictly speaking, the egg insulator of Fig. 28.7B is not designed to be an end insulator for transmitting antennas. The distances between the antenna wire and the support line are not very large and, hence, are subject to breakdown at high power, since at least one end of a wire antenna is a high-voltage point. Egg insulators are okay to use at low power levels or for "receive only" and temporary installations, but so is polypropylene rope! On the other hand, the egg insulator is potentially much stronger than a comparable strain insulator because the interweaving of the wire and/or rope through their respective holes puts the egg in *compression* rather than *tension*. For that reason, the egg insulator and its rectangular sibling seen on power pole guy wires are the only types to use when it is necessary to electrically break up long guy wires into multiple shorter sections.

Figure 28.8 shows how to make connections through an end insulator. The electrical connections to the antenna wire are the same as for the center insulator. The only proviso here is that for transmitting antennas it is wise not to leave any sharp points sticking up. The ends of a λ/2 dipole experience high RF voltages, and sharp points tend to cause corona sparking. Cut off any extraneous leads that form sharp points, and then smooth the whole affair down with tinning (i.e., solder).

The rope is *dressed* in a manner similar to the wire, but it must be knotted with a *self-tightening knot*—i.e., one that tends to clamp down on itself as the support line is tensioned. Some people use the *hangman's knot*, but this seems a bit excessive. The *bowline* is particularly useful, especially if the line is always in tension, and the knot can easily be "broken" and untied later.

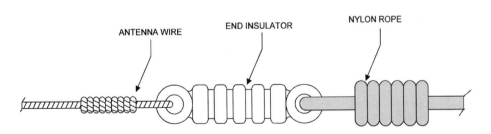

FIGURE 28.8 Connections to end insulators.

Folded Dipole Construction

As discussed in Chap. 6, the folded dipole (Fig. 28.9) consists of two close-spaced conductors shorted together at both ends, but only one of the conductors is split in the center for the feedpoint. The free-space feedpoint impedance at resonance is around 290 Ω, so it makes a good match to 300-Ω, television antenna twin-lead. (Often the antenna itself is cut from the same roll of twin-lead!) Alternatively (as shown in Fig. 28.9), a 4:1 balun transformer at the feedpoint provides a match to 75-Ω, coaxial cable. Despite the fact that 300-Ω, TV twin-lead is disappearing from the market, the folded dipole maintains a following among SWLs and hams. A preassembled folded dipole and feedline rolls up into a small storage space, weighs very little, and can be rapidly deployed in an emergency or portable scenario.

Neither the conductors nor the dielectric of ordinary TV twin-lead are particularly strong, so some ingenuity is required to keep the terribly weak 300-Ω, twin-lead from breaking at the slightest provocation. Figure 28.10 shows a solution developed by one fellow (admittedly a fine worker in plastics and other materials); he fashioned a center insulator and two end insulators from a piece of strong Lucite material.

The 300-Ω TV twin-lead is prepared as shown in Fig. 28.10A. One conductor of the twin-lead is cut at the planned feedpoint, and the insulated wire freed from the center dielectric material a distance of about 0.5 to 0.75 cm in both directions, using a match and/or small diagonal (side) cutters. Use the same tools to strip the insulation from the two wire ends just created. A hand punch—either a large-size leather punch or a paper punch—places two or three holes in the center insulation on either side of the break.

The center insulator is shown in Fig. 28.10B. It is made from a piece of strong plastic, Lucite, or other insulating material. Two identical sections, front and back, are needed. A number of 5-mm holes are drilled into both pieces at the points shown to clamp the twin-lead. A pair of solder lugs provides a mechanical tie-point for connections between the antenna element and the transmission line. The nuts and small bolts that fasten the two halves of the insulator are nylon to eliminate any possibility of short-circuiting the two terminals or the two sides of the twin-lead that form the antenna.

A side view is shown in Fig. 28.10C. Note that the twin-lead causes a gap that can catch water and makes it possible to break one or both insulators by overtightening the

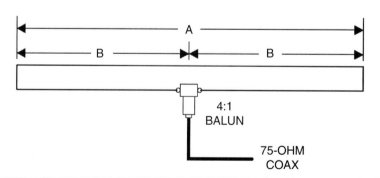

FIGURE 28.9   Folded dipole.

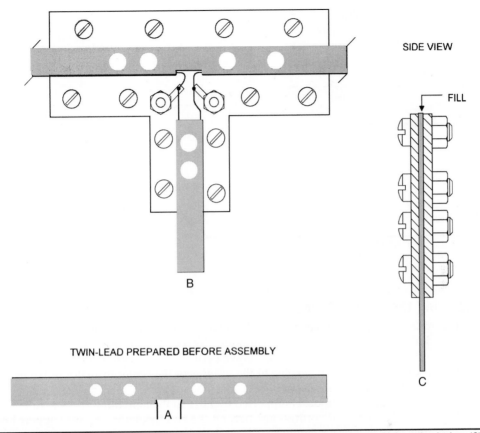

FIGURE 28.10 Center insulator for ribbon cable folded dipole. (A) Cable preparation. (B) front/back view. (C) Edge view.

nuts. To prevent these failures, a gasket of similar material is glued into the space as filler. The gasket material should be the same thickness as the twin-lead.

Rather than burden this antenna with the weight of coaxial cable, the twin-lead feedline should be run to an ATU for balanced lines.

An end insulator for twin-lead folded dipoles is shown in Fig. 28.11. It is constructed in a manner similar to the center insulator. The clamping fixture can be fabricated from metal such as 3- to 6-mm brass or copper stock. The ends of the antenna wires are shorted and soldered together, so use of a conducting material at the ends is not a problem.

An end insulator is then used with a rope in the normal manner. A chain-link section or *carabiner* is used to connect the clamping fixture and the end insulator.

Alternatively, the clamping fixture can do double duty if it is the same material as the center insulator. In that case the other end insulator and chain link can be ignored. If you opt for this approach, it is a good idea to bevel and polish the rope hole on the clamping fixture to prevent chafing of the rope, or insert a guy wire *thimble* of an appropriate size through the hole in the fixture.

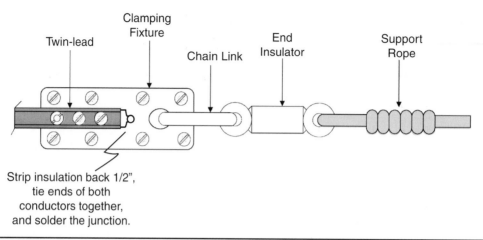

FIGURE 28.11   Twin-lead folded dipole end insulator.

Supporting the Wire Antenna

Figure 28.6 shows the dipole installed. The ends of the two wire elements are connected to end insulators, which are in turn supported by rope to a vertical support (e.g., mast, tree, roofline of the house). Although shown perfectly horizontal here, the actual installation will droop in the center due to the weights of the center insulator or balun, the coaxial cable, and the wire itself.

If either end of the dipole is supported by a tree limb or anything else that can move in the wind, add a door spring—just beyond the end insulator—in one support line. The strength of the spring should be chosen such that the normal tension in the support lines on a calm day causes the spring to be less than halfway extended. For an 80-m dipole with substantial feedline weight, a residential garage door spring is most likely to be appropriate. Most local hardware stores carry an assortment of door springs having various tension ratings.

Trees, Masts, and Other Supports

The end supports of the antenna can be anything that supplies height and is strong enough: a mast, a tree, or the roofline of a building. Figure 28.12 shows the use of a mast, but other forms of support can be used as well. To facilitate easy raising and lowering of the antenna for maintenance and tuning, the support rope should not be tied off at the top, as is true in all too many installations; instead, it should be brought down to ground level, just like the *halyard* of a flagpole. Be sure to provide enough "dead slack" to make lowering the antenna feasible, although once a halyard is created, it can be lengthened temporarily, as needed, to allow the antenna to be pulled sideways (i.e., along its axis) during the raising or lowering process.

The ideal support is arguably a climbable tower (or perhaps a very tall building with an elevator to the roof), but most dipoles and other wire antennas are strung from tall trees. How do we get the antenna up high in the trees? Here are some thoughts about a variety of approaches:

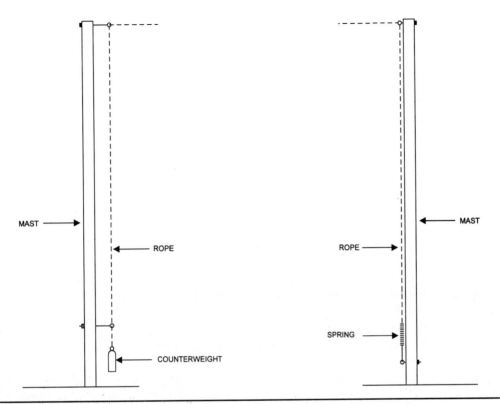

FIGURE 28.12   Mounting mast schemes.

- For those with strong throwing arms whose trees aren't very tall, a baseball with a ³/₁₆-in hole drilled through its center may do the trick. Simply tie one end of some light nylon twine through the center and a half-circumference of the ball. Lay out the twine on clear ground in front of you. (Clearly, this doesn't work very well in the middle of the woods!)

- Many techniques start with a projectile trailing fish line behind it as it is launched, since fish line is lighter than commonly available twine. The fish line should be wound on a *spinning* or *fixed spool* fishing reel. Attach the reel (with duct tape) near one end of a 2- or 3-ft length of wooden dowel, and push the other end into the earth at about a 45-degree angle so it is aimed up high in the targeted tree. There is a tradeoff between the strength of the fish line and how much total weight the projectile can haul up in the air behind it; experiment with different line weights in conjunction with your chosen reel and projectile. Once the fish line is up and over the target tree, attach nylon twine to the far end and reel the fish line back in. Then attach the halyard or an intermediate line to the end of the nylon twine that has been pulled through the tree top. (An assistant to spool out the twine is an extremely helpful addition to this part of the project!)

- One manufacturer sells a *slingshot* in the amateur radio market. The projectile is typically a metal ball with a tab for tying fish line to, or a large metal *hex nut* from the user's parts bin after the metal ball has become lost. Many people have had success with a slingshot, but the author has found its aiming accuracy to be less than that of other approaches, and the elastic sling can deteriorate over time.

- For years the author favored a small *bow and arrow* given to him by a helpful neighbor. Three modifications must be made to conventional arrows:

  ○ Remove the pointed tips (to avoid injury to bystanders, and to avoid having the target tree look like a porcupine).

  ○ Add mass (weight) to the nose of the arrow; a half-dozen framing nails (with large heads aimed forward) can be held in place around the circumference of the arrow at its head long enough to be taped with electrical or duct tape.

  ○ Drill a small hole, large enough to thread fish line through, about an inch or so back from the tail of the arrow.

  Used in conjunction with the spinning reel on a dowel stuck in the ground, this bow-and-arrow outfit allowed the author to put fish line in the tops of trees as tall as 60 ft.

- For greater heights and better accuracy, borrow or purchase a *bowfishing* rig. New sets are expensive, but archery shops often have trade-ins available at a very modest price. The most important element is the compound bow with a tapped hole and matching metal post for direct mounting of a spinning reel. Purchase five or six arrows with removeable points and prepare them as described here. With this setup, the author has sent appropriately sized fish line over the tops of 100-ft trees!

- To get the fish line to drop straight down the back side of the targeted tree (instead of continuing on beyond the tree some distance), halt the flight of the arrow *just after* it passes the top of the tree with a slight turn on the handle of the fishing reel. With practice, you can get the hang of this.

- Never try to retrieve the arrow by reeling it back in, up through the tree. Instead, always untie or cut the fish line at the arrow, rewind the empty fish line, and retie the arrow onto it back at your launch site.

- Wait for a calm day for *all* of these techniques.

- For the ne plus ultra, consider a *potato launcher* with tennis balls. Search the Internet for examples of how enterprising amateurs and others have modified these powerhouses to lay lines over the tops of the very tallest trees.

- If all else fails, pay a tree-climbing professional to place a "permanent" halyard mount near the top of your chosen trees.

CAUTION *Always make sure the area around you and well beyond the targeted tree are clear of humans and animals before launching any projectiles. Even blunt-tipped arrows or hex nuts can cause serious injury!*

Halyards and Other Support Lines

There are many choices for materials to use as a halyard or support line, but not all of them are *good* choices. Of paramount concern is long-term exposure to sun, wind, and moisture. In general, natural cordage (hemp, for instance) is a very poor choice except, perhaps, for a very short period (e.g., a few weeks). Nylon rope—despite having a much longer life—is far too stretchy. Instead, various "poly" compounds, including *polyester* and *polypropylene,* are worth considering, as is braided Dacron. Possible sources include large discount stores, building supply stores, boat dealers, outdoor (mountaineering) stores, your local hardware store, and a number of stores specializing in Internet sales, as listed in App. B. If total cost is an issue, careful shopping is a must; prices for comparable cordage may vary by as much as 4:1, depending on the market the seller is focused on.

Keep in mind that the *safe working range* for rope is typically less than 15 percent of its rated *breaking strength*. Thus, for an antenna that experiences up to 100 lb of tension during windy weather, a 750-lb rated rope should be on your shopping list.

Choosing the best diameter for your cordage is a tradeoff among required strength, size and cost of any pulleys (or *blocks,* in nautical terminology), and what is comfortable for pulling by hand (preferably with reinforced gloves). For supporting a typical dipole, somewhere around $5/16$-in is a good starting point.

Virtually all cordage will eventually unravel at a cut end unless the end is treated. *Laid* or *twisted* three-strand rope can be *eye-spliced* or *back-spliced,* but braided ropes will need to have their ends *fused* (melted), taped, or *whipped* with twine. Fusing can be done with a soldering gun that has a special tip, a propane torch, or even (on a windless day) a simple match. Always fuse cordage outdoors; the fumes from heated synthetics should not be inhaled, and burning poly rope ends often melt and drip on the floor.

The antenna support line is connected to a spring on one end (see Fig. 28.6) and/or a *counterweight* on the other end. As mentioned earlier, the spring should be stout enough so that it is partly extended on a calm day with the antenna support lines normally tensioned. (Ideally, with no wind the spring should be extended less than 50 percent of its total possible expansion.) For an 80-m dipole and about 100 ft of open-wire line, both constructed from #10 enameled copper, a stiff garage door spring seems to work best. The author's 80-m dipole has stayed up for nearly 20 years using this approach!

The counterweight should be just enough to balance the weight of the antenna for a *reasonable* amount of sag along the wire—probably between 15 and 40 lb for the 80-m dipole and OWL described in the previous paragraph. As wind causes the supporting tree limbs to move up and down or back and forth, the counterweight rises and falls, thereby reducing the chances of snapping a support line or the antenna wire itself. The choice of counterweight is limited only by your imagination: drapery cord weights, a small bucket of rocks, a gaggle of fishing weights, a cinder block, and (in one case) a burned-out automobile starter motor. (The mounting hole on the front boss of the motor was ideal for accepting the rope!) Just be sure to locate the counterweight where it cannot harm anyone or anything should the rope break. In that respect, it's helpful to keep the counterweight in balance at a point no more than a few feet above the ground, if possible.

Don't underestimate the value of making the dipole out of ordinary soft-drawn copper (house wiring, in other words). Although you may periodically have to retension the support lines, such wire is more able to withstand the shock of wind gusts snapping the tops of trees around than copper-clad steel is. (Of course, stretch in the antenna wire may change its resonant frequency a little bit, but dipoles that are longer than λ/2 exhibit a tiny bit more gain in their main lobe. If an ATU is used, the stretch is immaterial; if not, perhaps it will be necessary to lop a foot or so off each end of the dipole every five or six years.)

Figure 28.13A and B show two methods for making the connections at the top of a supporting mast. Although egg insulators are shown, any kind can be used. A pulley is mounted at the top of the mast with a link section and a stout eyebolt or screw-eye that passes through the mount. (A simple screw-eye will not harm a tree, but never wrap wire or rope around a tree trunk or branch; it's a sure way to kill the tree.) Make the *downhaul* rope a closed loop so you can raise and lower the antenna just like a flag.

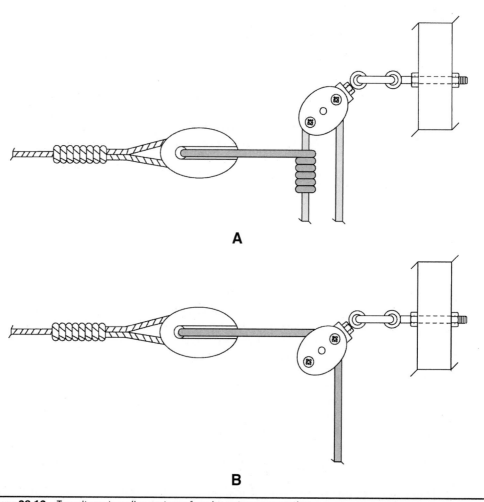

FIGURE 28.13   Two alternate pulley systems for wire antenna mounting.

In theory, a pulley is the best way to change the direction of a support line's path, but few users are willing to pay for the right pulley for the task. The wrong device will tend to rust if made of ferrous metal, or crack if made of plastic. Equally important, the size of the pulley and the diameter of the cordage running through it must be correctly matched to avoid jams. The smaller the diameter of the pulley *sheave,* the shorter the life of the rope that lies in its groove. When selecting a pulley, make sure the groove is neither too wide nor too narrow for the size of line to be run through it. All in all, a pulley is not always the best solution if you have neither the time nor the inclination to select and pay for the right one.

Figure 28.14 shows a "poor man's" pulley, suitable for temporary installations or lightweight (i.e., higher-frequency) dipoles. In this approach, a U-bolt is fastened to the top of the support mast. If the U-bolt is made of brass or galvanized steel, all the better, for it will not corrode. (In general, automobile muffler U-bolts and inexpensive plated U-bolts sold at the local hardware store will rust almost immediately; do not use them!) The rope can be passed through the U-bolt as in the pulley system but without fear of jamming. Be sure to install the U-bolt all the way up to the threads so that the rope has no chance to chafe against them.

The method shown in Fig. 28.15 can only be loved by equestrians! A brass stirrup replaces the U-bolt. If you have switched to cars or trucks and no longer need your

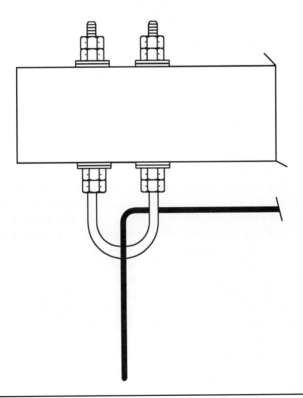

FIGURE 28.14 U-bolt end mounting.

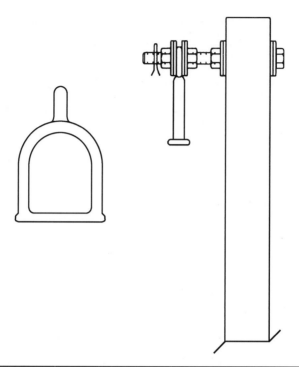

FIGURE 28.15   Stirrup end mounting.

horse stirrups, then perhaps this is a viable means of attaching the support rope to the top of the mast.

Strain Relief of the Antenna

For long dipoles with heavy center insulators and coaxial cable feedlines, the ideal installation includes a high center support capable of supplying strain relief and thus minimizing sag at the dipole center. Such a support also reduces longitudinal forces (tensions) on the dipole wires, end insulators, and halyards or hoist lines at both ends. Most commonly it is used with inverted-vee antennas, where the center of the antenna is pulled up high on the *only* available tall support.

Figure 28.16 shows an alternate method of strain relief for the center insulator or balun to reduce total sag from the height of the dipole ends to its center. A strong *messenger cable* is run between two end supports (which must themselves also be quite strong), and tensioned for minimal sag. By suspending the center of the dipole from the messenger, forces on the various electrical and mechanical components of dipole are substantially lessened. If the messenger cable is a conductor (stranded galvanized steel guy wire material is a favorite), it should be broken up with insulators to prevent detuning the wire antenna. However, there's no reason the messenger cable can't be run at right angles to the antenna wire (using an entirely different pair of end supports), in which case the need for insulators disappears.

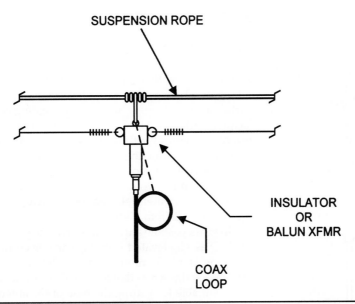

SUSPENSION ROPE

INSULATOR
OR
BALUN XFMR

COAX
LOOP

FIGURE 28.16 Strain relieving the center insulator or balun.

Installing Vertical Antennas

At frequencies above 5 MHz, erecting and supporting HF verticals—both ground-mounted and elevated ground planes—is a relatively easy task for two people, provided the vertical is made of aluminum tubing strong enough to support itself during the brief period when it is being tilted from a horizontal to an upright position. At lower frequencies, verticals should be treated as towers (Chap. 29) or erected with the help of extensive rigging—often including some form of *gin pole*—and a larger crew.

For years the Hy-Gain *14-AVQ* (now sold as the *AV-14AVQ*) has been a popular multiband vertical monopole. By using three parallel-resonant *traps* and a small capacitive *top hat*, the antenna provides a physically short vertical (18 ft overall) that is an electrical quarter-wavelength on 40, 20, 15, and 10 m. (The design is essentially unchanged from the original, pre-WARC band, model, so there is no guarantee of low SWR on the 17- and 12-m bands.) It can be ground-mounted or elevated atop a mast as in Figure 28.17. In either case, the antenna requires radials, but the number and length are different for the two cases. When ground-mounted, at least 16 (nonresonant) radials that are roughly as long as the vertical is tall (18 ft) should be laid out in a circle around the base of the antenna. When elevated, however, three (resonant $\lambda/4$) radials for each band, equally spaced around a 360-degree circle, are sufficient.

Ground-mount installation of the *AV-14AVQ* and similar verticals is simplicity in itself: Drive a 3-ft length of 1½-in OD steel post into the ground, stand the assembled vertical upright, and clamp its base to the post with the U-bolts supplied. To complete the installation, attach radials to the base or add a ring of copper wire or soft copper plumbing tubing around the base as an attachment point. Install a *lightning ground* (see Chap. 30) of multiple 8-ft ground rods in a circle with a radius of about 8 ft centered at the base of the vertical, and connect each ground rod directly and separately to the base

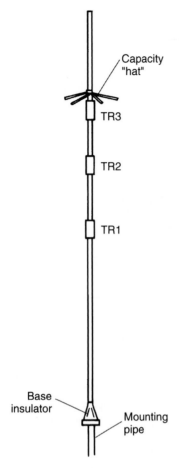

FIGURE 28.17   Mounting commercial trap vertical.

with #6 or heavier gauge wire. Finally, connect 52-Ω coaxial cable at the bottom of the base as shown in Fig. 28.18A or 28.18B and weather-seal the connection with heat-shrink tubing or a half-lap winding of electrical tape (as described earlier in this chapter) and a spray coating of clear acrylic lacquer. Although Hy-Gain does not require or suggest the use of guy wires for the vertical itself, a single set of light-weight, nonconducting guys such as nylon twine or ¼-in poly rope tied just above or just below the middle trap is not a bad idea unless your installation location is such that the antenna is at greater risk from people tripping over the guy ropes than it is from high winds.

In years past the author preferred the performance of the *14-AVQ* as an elevated ground plane, even though that in-stallation is a little more involved. Telescoping *TV antenna mast* material, such as that long sold by RadioShack and oth-ers, is suitable for getting the base of the antenna 15 ft (cor-responding to about λ/8 at the lowest operating frequency) or more up in the air. Because three λ/4 radials are required for each operating band, the radials for the lowest band (40 m) can double as guy wires for the top of the mast, if you so choose. The tradeoff is that adjusting the length of the radials for minimum SWR may be a little more difficult if the radials are doing double duty as guy wires. On the other hand, keeping separate sets of guy wires and radials from getting tangled up during the installation can often be frustrating.

One alternative to a telescoping metal mast is a pivoted wood mast, as shown in Fig. 28.19. At a minimum, one set of guy ropes just below the top of the mast is sufficient, al-though a set of lightweight guy ropes at the midpoint of the vertical eases wind-loading stresses on the antenna.

Although a ground-plane vertical does not require any additional RF ground sys-tem other than its elevated radial system, it does require a lightning ground at the base of the mast. It is a very good idea to also ground the coaxial cable outer conductor there as it comes down from the connector above and heads off to the radio room. A simple way to do that is to use two separate pieces of coaxial cable—one just long enough to run from the SO-239 at the base of the vertical to ground level, and the other to run from the base of the mast to the radio equipment—and join them at the base with a double female (or *barrel*) adapter mounted on a small piece of aluminum mechanically con-nected to the center of the lightning ground system. Alternatively, commercial lightning suppressors that house the barrel, a discharge circuit, and a convenient lug for the ground connection are not very expensive.

Because the normal stresses on guy lines for verticals erected in this fashion are much lower than the tension on halyards and other support lines for horizontal wires, smaller-diameter rope is appropriate. Certainly, ¼-in poly is about the maximum that would ever be needed for verticals of this size.

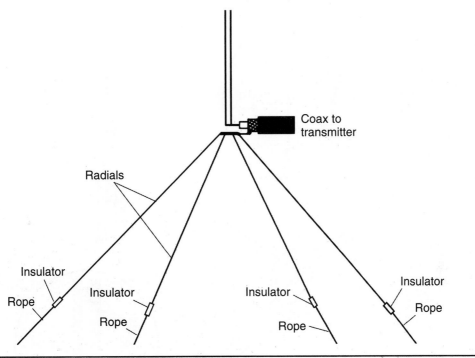

FIGURE 28.18A   Feeding the mast-mounted vertical antenna.

When a vertical is placed atop a mast of any significant height, the support needs to be more substantial than just a pipe driven into the ground. Not only the vertical but the mast itself contributes to the total *moment arm* that exerts a sometimes unexpectedly large sideward or overturning force on the base.

Figure 28.20 shows a standard chain-link fence post used as a support for either a vertical antenna or a mast (which could be metal, PVC, or wood). The typical fence post is 1.25 to 2.00 in OD and made of thick-wall galvanized steel. Thus, such a pipe makes a rugged installation. Some users believe that the pipe will last longer if painted along

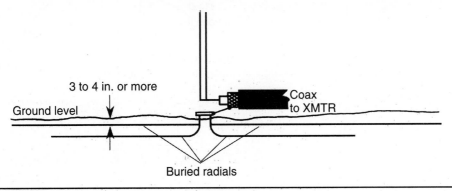

FIGURE 28.18B   Feeding the ground-mounted vertical antenna.

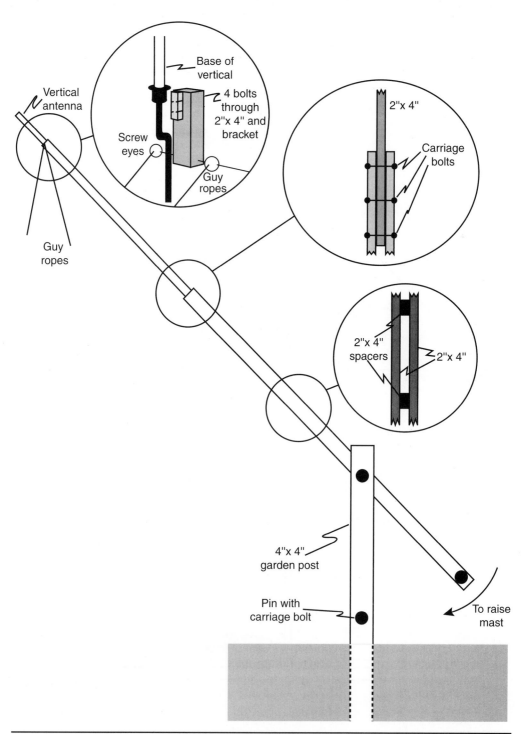

FIGURE 28.19 Pivot mast.

its entire length (especially the underground portion) with a rust-inhibiting paint such as Rustoleum. If your intention is to directly mount the vertical to the metal fence post, make sure the pipe you choose is a smaller outside diameter than the maximum mast size accommodated by the base bracket and U-bolts of the antenna you're putting up. (The maximum for the *AV-14-AVQ* is 1⅝ in, for example.) Otherwise, fabricating a suitable adapter can turn out to be the most time-consuming part of the entire project! On the other hand, if the fence post is supporting an intermediate mast of some height, the larger diameters are more appropriate.

The beauty of the fence post as an antenna support is that it is relatively easy to install, and supplies are obtained easily from hardware stores. The post is mounted in a concrete plug set at depth *D* (in Fig. 28.20) in the ground. To avoid snapped ropes or other damage from *frost heaves*, *D* should be a function of local climate and local building codes and determined by the 100-year depth of the frost line in your area, a figure familiar to local contractors and *codes enforcement* officials. Keep in mind, however, that shallower depths may be legal for fence posts, but installation of a vertical antenna (or mast) on top of the fence post may change the mechanical situation considerably.

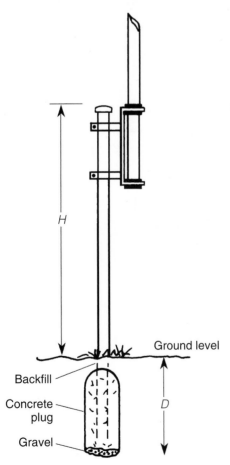

Dig the hole with a post-hole digger or an earth-auger-bit tool. The latter often can be rented in gasoline engine, electric, or manual versions from tool rental stores.

Once the hole is dug, place about 4 in (or local requirement) of clean gravel (available in bags from hardware stores) at the bottom of the hole. Insert the post and *plumb* it (i.e., make it vertical in the hole) by adjusting a single set of guy ropes. Fill with your favorite brand of ready-mix concrete to a level just below the average surface around the hole. Cover with straw or burlap and keep the top of the concrete moist for a few days. Do not subject the post to side loads for at least four days or whatever interval the concrete manufacturer recommends. Concrete needs time to cure; its strength increases exponentially, reaching 99 percent of its ultimate capability in 28 days, but a 7-day hiatus should be adequate before completing the antenna installation. Once the concrete has cured, top it off with soil and sod or garden chips.

Regardless of the height of its base, never attach the antenna base bracket to the mast or other support with just a single mounting point or U-bolt. Always use at least two-point mounting to prevent the antenna from pivoting and shearing off the mounting hardware.

Figure 28.21 shows how to mount a 2 × 4 mast to a fence post. Standard lengths of 2- × 4-in lum-

FIGURE 28.20 Fence post mounting.

ber can be purchased up to 20 ft, although 16- and 20-ft lengths may have to be obtained at professional "contractor" lumberyards rather than local hardware stores. The lumber should be kiln-dried and pressure-treated.

Attach the 2 × 4 to the fence post with U-bolts. A 2 × 4 scrap is used as a wedge or *shim* to take up the difference between the post and the mast. In some installations the U-bolt will go around the perimeter of the 2 × 4, while in others a pair of holes can be drilled in the 2 × 4 to admit the U-bolt arms. The U-bolt must be at least ¼-in #20 thread, and either a ⁵⁄₁₆-in bolt or a ⅜-in bolt is highly recommend if the mast supports any significant weight. Some installers place a cinder block or large brick at the base of the 2 × 4 to bear the static and dynamic loads (i.e., gravity and wind) that eventually can cause the mast to list because of eccentric loading on the footer. An added reason is to keep the butt end of the 2 × 4 off the ground and thereby prevent or slow down rot. Alternatively, a pressure-treated 4 × 4 of the sort used to support decks (Fig. 28.22) can be used.

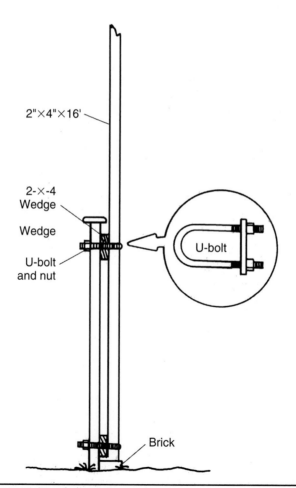

FIGURE 28.21   Mast mounted to a fence post.

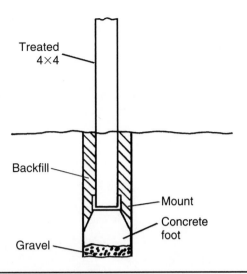

FIGURE **28.22**   4 × 4 antenna support.

Even if ground-mounted, full-length λ/4 verticals (35 ft, more or less) for 40 m require guying unless they are made of extremely strong, self-supporting members. A
full-size 40-m vertical is akin to a very tall flagpole and requires similar construction
unless it is guyed at one or two levels. Even with guying once it is fully vertical, the
tubing must be strong enough to not bend or crease while being pulled or pushed up
from horizontal to vertical. The problem is even more severe for λ/4 80-m ground-
mounted verticals. Irrigation pipe, either butt-joined or telescoped with reducers, is
often employed. Antennas of this length require larger ground crews or a *cherry picker*
for successful installation, and the use of techniques covered in Chap. 29 is appropriate.

CHAPTER 29

Towers

At some point, the height or strength required for a specified antenna installation exceeds the capabilities of trees or cordage, and the methods of the preceding chapter are no longer sufficient. When that point is reached, the focus necessarily shifts to the use of *towers* to safely and reliably maintain antennas at the desired height.

Typically, we call upon towers whenever rotatable HF antennas such as Yagis and quads are to be installed, although UHF, VHF, and even some small HF Yagis are sometimes mounted (if only for temporary events) on well-guyed heavy-duty masts. We even use towers to support very high wires, especially for 160 m and 80 m, since few of us have access to trees in excess of 100 ft tall. The purpose of this chapter is to familiarize the radio enthusiast with fundamental mechanical and safety considerations that are necessarily part of any tower project associated with putting an antenna up high in the air and keeping it there through episodes of severe weather. There are many antenna installation tasks that can be done by amateurs possessing the proper knowledge, the right tools, and a healthy respect for the safety of people and property. But keep in mind when planning an installation that the information given herein is merely informal guidance.

NOTE *The opening paragraphs of Chap. 28 are as valid for tower projects as they are for wires or verticals. Please read or reread the section entitled "Antenna Safety" in that chapter as though it were reprinted here.*

In the United States there is a wide variation from state to state and town to town in how local municipalities deal with towers. Within just a single state the author has found the full gamut of regulation—from detailed zoning ordinance wording and eagle-eyed code enforcement to total disinterest in what happens on private property. Similarly, there is little consistency in the way insurance companies handle the existence of a residential tower or losses resulting from the fall of a tower. But regardless of the stance taken by your local officials or your insurance carrier, you should always have your planned tower installation reviewed by a professional mechanical or structural engineer licensed to practice in your state. Become familiar with—and abide by—any local, state, and federal mechanical, electrical, and building codes that pertain to your project. And always follow the manufacturer's instructions for storing, handling, installing, and using the materials and structures that are part of your project. It is simply not worth the pain or risk to skirt any of these regulations or recommendations.

Antenna support structures (aka *communications towers*) are used extensively in amateur and commercial radio communications systems for a very obvious reason: Antennas tend to work better up in the air than they do nearer the ground. The primary reasons for this are, of course:

- At VHF and above, there is no ionosphere to help extend our transmission and reception range through the reflection (or refraction) of radio waves back to earth. Except for some highly unpredictable propagation modes, most radio communications at these frequencies are limited to line of sight or slightly beyond. Thus, raising the antenna high enough to clear nearby obstacles is an important part of the station design.

- At MF and HF, ground reflection effects can enhance or degrade signal paths by many decibels. Proper *siting* and intelligent choice of antenna height can provide signal enhancements equivalent to a tenfold increase in transmitter power compared to a station assembled with no forethought! In general, at MF and HF, the height required is inversely proportional to the frequency for comparable ground enhancement of the radiation angle.

Because television and FM broadcasting are confined to VHF and above, towers for those services are often 1000 ft tall or more—especially in areas of the United States with flat terrain—to extend line-of-sight reception as far as possible. In metropolitan areas with hilly or sloping terrain or extremely tall downtown buildings, unique siting opportunities for the broadcast transmitter and tower may allow coverage of the primary service area with shorter towers.

In contrast, the AM broadcast band is based on the use of the *ground wave,* so AM broadcast towers *are* the (vertically polarized) radiating elements. Their heights are more likely to be dictated by the exact wavelength of each station and the extent to which it needs to suppress the sky wave component of its signal in certain directions at night, when skip can cause its signal to raise havoc with other stations sharing the same frequency in other regions of the country. Typical heights for AM stations near the top of the broadcast band (1500 to 1700 kHz) might be expected to range from only slightly greater than those of a well-outfitted 160-m amateur station—i.e., around 150 ft for a quarter-wave grounded monopole—to slightly more than twice that figure. At the other end of the AM band (530 to 700 kHz), optimum heights can rise to 400 ft or more.

Because amateurs have multiple bands of frequencies ranging from MF to UHF and beyond, and multiple modes of propagation, including *line of sight* and *ionospheric skip,* their tower requirements are "all over the map". Today, small businesses and active amateurs employ towers as short as 6 or 8 ft (*tripod towers* designed to straddle peaked rooftops) and as tall as 300 ft. But probably the bulk of the nonbroadcast tower installations fall within the 30- to 100-ft range. Given the generalization that "higher is better", why are so many users content with relatively modest towers? Here's a list of possible reasons, in no particular order:

- Cost (a very rough rule of thumb has cost increasing in proportion to the *square* of the height)

- Available space (the taller the tower, the larger the guy wire circle and the *safe fall zone* required)

- Terrain (at some frequencies, properly sloping land or nearby saltwater can provide the same long-haul signal boost as a tall tower on flat land)

- Height optimization (at HF in particular, "higher is *not* always better")
- Intended purpose (e.g., local community coverage for a small business)
- Fear of heights
- Feedline losses
- Proximity to airports and heliports (i.e., FAA limitations)
- Local zoning and building code restrictions

The last two factors are discussed in more detail in Chap. 31, new in this edition.

Types of Towers

Several different forms of tower construction are used by amateurs and other services. By far the vast majority in the United States consist of lattice structures with a triangular cross section, although a few employing a square cross section (called *windmill towers*), such as the Vesto self-supporting towers of the 1950s, are in use. A small but structurally impressive segment of the tower market consists of hollow tube, solid surface poles with square or circular cross section. At least one commercial model, *Big Bertha*—available in heights up to 142 ft—is a thick-walled round pipe that is driven by an electric motor at the base. In this model, the entire tower (called a *rotating pole*) is turned via chain drive from the motor.

Ignoring roof-mount tripods and house-bracketed towers, both of which are severely limited in maximum height, the following list of common tower types is arranged in order of increasing cost for towers in the 70- to 100-ft range:

- Guyed triangular lattice sections
- Guyed triangular lattice section foldover
- Triangular lattice section crank-up (some with tilt-over option)
- Freestanding nonrotating pole (mall parking lot lamppost style)
- Freestanding tapered triangular lattice sections
- Guyed triangular lattice section rotating tower (or a hybrid)
- Freestanding rotating pole

Tower Fundamentals

Before we look at the each type of tower in more detail, a brief review of tower fundamentals is in order.

If the wind never blew here on earth, a tower's role would be limited to supporting the deadweight of antenna(s), rotator, and mast at the desired height. But for most antenna and tower installations, deadweight is a secondary concern compared to the forces that result when strong winds act on the tower, antenna(s), rotator, mast, cables, and other accessories located above ground level.

Consider the freestanding tower of Fig. 29.1. The purpose of this tower is to support and allow rotation of the Yagi beam at a desired height H in the presence of a wind. In

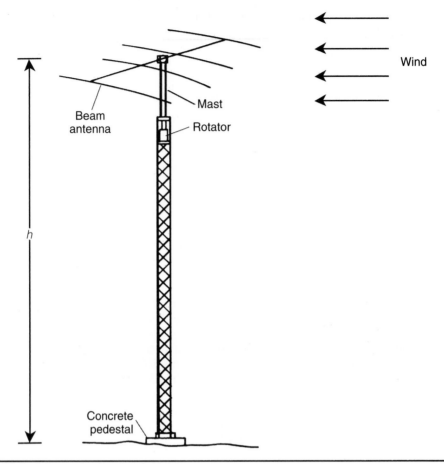

Figure 29.1   Antenna atop freestanding tower.

this highly oversimplified example, the wind is assumed to be applying a perfectly horizontal force to one side of the entire system. In response, the tower and antenna will move sideways, in the same direction as the wind, unless prevented from doing so. Clearly, the first requirement of a tower installation is a strong, solid base!

With the base securely anchored, the tendency of the tower is to *overturn*. Here it is helpful to think of the tall, thin tower as a form of *lever*. To keep the math simple, for the moment assume the tower members are extremely thin so that the force of the wind on them is much, much smaller than the force on the antenna at the top and can be ignored. The force of the wind blowing on the antenna at the top of the tower creates a *moment arm* (*M*) or *torque* on the base of the tower with a magnitude equal to the height (*H*) of the tower multiplied by the horizontal force (*F*) of the wind pushing on the antenna:

$$M_{ANT} = H_{ANT} \times F_{ANT} \tag{29.1}$$

But the force is nothing more than the wind pressure multiplied by the total *effective projected surface area* (A) of the antenna exposed to that pressure:

$$F_{\text{ANT}} = P_{\text{WIND}} \times A_{\text{ANT}} \qquad (29.2)$$

The reason for the term "effective projected" is that different surface contours having the same area develop different resistance, or *drag*, in response to identical wind speeds. It should be obvious, for instance, that the force on a flat plate squarely facing a 30 mph breeze is greater than the force on a long piece of cylindrical tubing in the same breeze. Surface areas for tower-mounted antennas manufactured in the United States are usually given in *equivalent* or *effective square feet*. The "e-word" qualifier is included because most antenna elements and booms are cylindrical and the force on them for a given wind speed is less than if they were flat plates. But read the manufacturer's literature very carefully to be sure you understand how (or "if") the cylindrical nature of the antenna elements and tower members has been accounted for. Common convention is to convert the actual area of one face of each cylinder into an equivalent flat plate area—often by multiplying the cylindrical surface area by 0.6—to account for the reduced drag when air flows past a cylindrical tube, as compared to a flat plate.

For most types of antenna, the equivalent area that directly faces the wind varies with the wind direction (and with the compass orientation of the antenna). In general, the only number of interest is the *worst case*, or maximum equivalent area. For a multi-element Yagi, for instance, that may occur when the beam is facing the wind or when it is broadside to the wind; most likely, however, it will occur when it is turned at some angle in between those two extremes.

The "raw" pressure resulting from wind is given by:

$$P_{\text{WIND}} = K \times v_{\text{WIND}}^2 \qquad (29.3)$$

where (in the U.S., at least) P_{WIND} = pounds per square foot (psf)
v_{WIND} = miles per hour (mph)
K = constant equal to 0.00256 when using mph and psf

Of more interest to us is something called *flat plate impact pressure* (P_{FLAT}). This is the *net* pressure increase experienced by a flat plate squarely facing into a steady wind at a given speed (v). P_{FLAT} is also proportional to the square of wind speed, but the constant is different from the K of Eq. (29.3) and depends on a number of geometrical characteristics—not just of the plate but of the environment around the plate. Thus, the equation for P_{FLAT} may vary with the underlying assumptions embedded in a particular analysis. For some years, curves relating the flat plate impact pressure of flat microwave antenna panels to *true wind speed* often used a $K_{\text{FLAT}} = 0.004$. (See Fig. 29.2.) Today the determination of the K-factor to be used may be conducted on an installation-by-installation basis. (See the nearby Note regarding Revision G of TIA-222.) Nonetheless, the single most important fact to take away from Eq. (29.3) or Fig. 29.2 is that pressure on an antenna—hence, horizontal force on a tower—is directly proportional to the *square* of the wind speed!

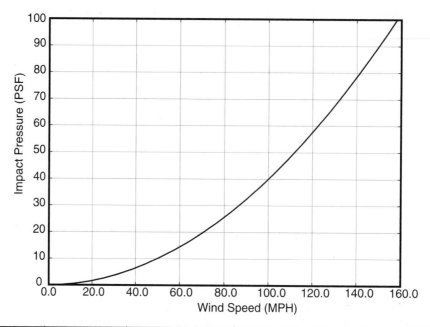

FIGURE 29.2 Pressure versus wind speed.

NOTE *In 2005, the Telecommunications Industry Association (TIA) issued Revision G of TIA-222, their* Structural Standards for Antenna Supporting Structures and Antennas. *Determination of the appropriate wind speed ratings to apply to a tower or antenna in Revision G represents a substantial and significant departure from the methods of earlier revisions. In particular, the focus has shifted from wind speed defined by a* fastest mile *to wind speed defined by a* peak 3-second gust. *Although there is no precise 1:1 equivalence between these numbers (that pesky old "apples and oranges" thing), tower and antenna installations designed for approximately 70 mph using the earlier Revision F standard seem to be comparable to designs based on a 90 mph peak 3-second gust standard. Further, Revision G introduces and requires certain adjustments (usually additions) to peak wind speed numbers because of terrain and/or reliability requirements for the support structure. As before, the standard provides a baseline wind speed specification for each county in the United States. (Most local building codes officials can provide you with the current requirements for their own counties.) As a "glittering generality", the baseline wind speed rating for flat, open countryside across the vast majority of the interior 48 states is 90 mph, but this number must be adjusted upward with icing, with increased tower height, and whenever unusual topographic features (hills, mountains, cliffs, ocean coastlines, etc.) are encountered. Overall, the baseline speeds vary from a low of 85 mph along the Pacific coast states to a high of 150 mph along the coastlines of south Florida and some of the other states bordering the Gulf of Mexico. Although not a universal practice, TIA-222 wind speed ratings already are, or soon will be, made a part of local building codes in many municipalities.*

Now we are in a position to calculate—in very crude fashion, at least—the requirements our tower must meet if it is to survive a high wind. For the purposes of our example, where we are assuming the presence of a steady wind with no gusts, we shall use a K_{FLAT} of 0.004, as previously discussed.

Example 29.1 A typical HF short-boom trap-triband HF Yagi might have a published *effective wind surface area* of between 5 and 6 ft². The Cushcraft A-4S, for instance, is specified at 5.5 ft². This figure is determined by the manufacturer from calculations on both the cylindrical parts (boom, elements, struts, traps, etc.) and the flat plate parts (brackets, end faces of cylindrical items) at different orientations with respect to the wind. Next, we use Fig. 29.2 to obtain the impact pressure for a target wind speed. Let's suppose that we conclude that 100 mph is a conservative value for the highest wind speed we expect our tower installation to survive. From the graph or the equation we obtain $P = 40$ psf.

Suppose we add 0.5 ft² for a rotator housing and other "odds and ends" at the top of the tower. Then our total wind load (ignoring the tower itself) is 40 (psf) × 6.0 (ft²), or 240 lb of horizontal force. If the antenna is mounted at the top of an 80-ft tower, for instance, the total moment arm, measured from the base of the tower, due to the force of the wind on just the antenna and rotator is calculated from Eq. (29.1) as 19,200 lb-ft. Because the bottom of this tower is rigidly held by its base, the tower will resist overturning. To do that, the moment arm is converted back into a *compressive force* in the downwind (or *leeward*) side of the tube if it's a single rigid member or in the downwind leg(s) if it's a triangular lattice structure. That force can be calculated by working backward through a second moment arm equation involving the cross-sectional dimensions of the tower. Specifically, we take the distance of any leg or outer wall from the centerline of the tower and divide that distance into the 19,200 lb-ft. For a tower whose legs form a triangle 18 in on a side, the calculation is:

$$F_{COMP} = \frac{M_{ANT}}{R_{TOWER}}$$

$$= \frac{19,200}{0.866} \tag{29.4}$$

$$= 22,170 \text{ lb}$$

To put it bluntly, mounting a small HF Yagi antenna atop a freestanding 80-ft "stick" has converted a sideward force of 240 lb into a vertical compressive force nearly 100 times greater! Is it any wonder that, while antennas are made of aluminum, towers are almost always made of steel?

For comparison, the "maximum allowable axial compression in a side rail" (or tower leg) for Rohn 45 tower sections, very popular with amateurs and business band users, is slightly less than 8,000 lb. Thus, it is not allowable to put an antenna of this size on top of 80 ft of freestanding Rohn 45 tower sections. In fact, the tower and its accessories all have surface areas, so the wind loads of the tower and accessories on it (rotator, brackets, feedlines, control cables, work platforms, etc.) have to be added to the antenna wind loading to obtain the total "real-world" moment arm and force on the tower legs. When that is done, it can be seen that the maximum allowable height for a freestanding Rohn 45 *with no antennas or accessories* except perhaps a small UHF whip might be more like 50 ft, depending on the maximum wind speed anticipated.

Although the preceding calculation is grossly oversimplified, it provides a basic understanding of the key structural issues germane to communications towers. Clearly, larger antennas and/or taller towers result in larger compressive forces at the base of the tower. Starting from an understanding of the basic *lever* action that is involved, we can now more easily visualize three ways to improve the antenna-carrying capacity or increase the height of our antenna support:

- Use thicker tower legs of the same material, or make the tower members out of stronger materials. (Both approaches are important contributors to the strength of freestanding rotating poles.)

- Increase the distance R from each leg to the centerline of the tower, especially toward the base of the tower. (This is the principle behind tapered self-supporting towers such as the Vesto, the Rohn *SSV* series, some rotating poles, and most crank-up towers such as the EZ-Way products.)

- Reduce the moment resulting from wind pressure on the antenna and tower members with the judicious addition of guy wires that counteract the force of the wind. (This is the principle behind all guyed towers.)

CAUTION *The preceding numerical calculations are overly simplified in order to emphasize basic structural strength issues in towers. No attempt has been made to include the effects of wind gusts, sudden shifts in wind direction, uneven terrain, updrafts and downdrafts, icing, mechanical resonances, and a host of other effects that a thorough design must take into account. Any new tower installation project should be conducted in accordance with the determinations of a professional engineer (PE) licensed to practice in your area. In many municipalities, certain paragraphs of the local building code may apply and periodic inspections may be required during the course of the project. But many, if not most, building code enforcement offices will accept preexisting standard installation drawings and calculations provided by commercial tower manufacturers if they have been signed and stamped by locally licensed PEs, so that avenue should be explored before incurring the expense of designing a tower installation from scratch.*

Guyed Towers

One way to minimize the moment arm at the base of a tower from wind blowing against an antenna at the top of the tower is to directly *resist* the force of that wind through some other physical mechanism. The most efficient way is to *directly oppose* the wind by pushing on the antenna in a direction opposite the wind or by attaching a *guy wire* or *guy line* to a firmly fixed support on the windward side of the antenna. If the guy wire is adjusted to be taut with no wind present, then any increase in force on the antenna when the wind comes up will be met with an increase in tension in the guy wire, with the result that the moment arm at the base of the tower is substantially reduced or eliminated.

In the ideal installation, we would attach the far end of the guy wire to a support at the same height as the antenna so that the guy wire was perfectly horizontal. But if such a support existed, we might as well attach the antenna itself to it and eliminate the tower entirely!

Since our objective in putting an antenna on top of a tower is to get the antenna up above surrounding objects, we must resign ourselves to connecting the far ends of guy wires to attachment points substantially lower than the antenna itself. In fact, in most (but not all) cases, the best we can do is to attach the guy wires to restraints, known as *anchors*, at ground level.

Figure 29.3 is a schematic of a very simple guyed tower. As before, a 6-ft^2 antenna sits atop an 80-ft tower, which is guyed at two levels (40 and 80 ft), but this analysis is going to discuss the effect of only the *top* set of guys on stresses in the tower members. Again, let us assume that the tower members are extremely skinny and thus generate no wind loading of their own. Assume also that the guy wires or guy lines are made of material that does not stretch, or *elongate*, under tensions encountered here, and that they are attached to earth anchors 60 ft out from the base of the tower. From the trigonometry of *3-4-5 right triangles* (see App. A), we know that each top guy wire is 100 ft long.

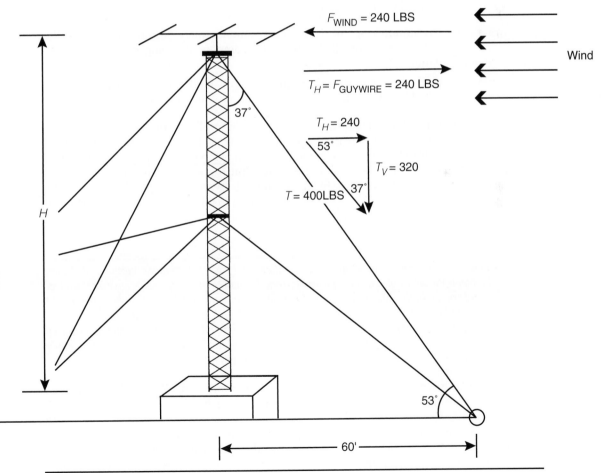

FIGURE 29.3 Forces on a guyed tower.

A little more trigonometry is then all that is necessary to understand the role of the top guy wires. When the wind blows from right to left in Fig. 29.3, it pushes on the antenna and tries to move it to the left. The antenna doesn't move because it is restrained from doing so by the top guy wire that is to windward (i.e., on the right-hand side) of it. For an object to remain motionless in one place, the vector sum of all the forces acting on it must be zero. Thus, if the antenna stays in place, it must mean the horizontal force of the wind is exactly balanced by the horizontal force exerted on it by the guy wire. But the guy wire is not horizontal, and since the tension in the wire has to point along the axis of the wire, we have to break the tension up into two components (trigonometry again): a vertical component (T_V) and a horizontal component (T_H). It is the horizontal component of the tension that must balance the horizontal push of the wind. Thus, if the force on the antenna from the wind is 240 lb, T_H must also be 240 lb, acting on the antenna in the opposite direction.

Again using the trigonometry of right triangles, if T_H = 240 lb, T = 400 lb and T_V = 320 lb. But why are these of interest?

T (400 lb) represents the total force (tension) in the guy wire itself. There are at least two reasons to care how large it gets:

- The maximum T—multiplied by a suitable safety factor—determines the strength required for all the guy wire components being procured for this installation.

- The earth anchor must be designed to withstand a maximum force of T—multiplied by a suitable safety factor—imposed on it by the guy wire.

T_V, on the other hand, is a downward force that is applied to the tower legs. In other words, a diagonal guy wire creates a downward force on the tower that is proportional to the magnitude of the horizontal wind force. That is the penalty we pay for not being able to employ perfectly horizontal guy wires. But the benefits are enormous—now, instead of dealing with compressive forces in the leeward tower leg of over 22,000 lb, we are imposing a paltry 320-lb downward force on that leg! Suddenly our overloaded tower is no longer overloaded and we can consider adding more antennas or going higher.

CAUTION *Once again, this is an oversimplified analysis of a complex mechanical design problem. We have not considered many factors that add to the compressive force on the tower members, including the deadweight of the rotator and antenna, as well as both the weight and wind load of the tower itself, the guy wires, and any cabling or ice buildup, nor have we considered transient loadings from shifts in wind speed or direction, the impact of the initial tensioning of the wires, the interplay between any stretch, however slight, in the guy wire material and the reappearance of a fraction of the original freestanding moment arm, etc.*

Several guyed tower configurations are commonly seen. Figure 29.3 shows the basic form—multiple tower sections (each usually 10 or 20 ft in length) having the same cross-sectional dimensions, braced by multiple levels of guy wires. This type of tower costs the least, and many installers and owners feel it is also the most comfortable platform to climb and work atop. In addition to the standard sections used as building blocks, each tower typically requires a base section and perhaps side-mounting hardware or a special top section designed to accept a thrust bearing for the antenna mast and a rotator mounting platform. The maximum possible height varies with the strength of the sections; Rohn 25, for instance, can be erected to 200 ft, while Rohn 45 is spec'd up to 300 ft. But in areas with higher wind zones, or with antenna loads greater than those assumed by Rohn in its standard configurations, those maximum heights may end up being reduced by 50 percent or more.

Towers are guyed at more than one height to keep side stresses on the tower legs from growing beyond a certain value as distance above or below a *guy station* increases. Rohn 25, for instance should be guyed every 25 to 30 ft, while Rohn 45 can be guyed at 30- to 35-ft intervals and Rohn 55 at 35- to 45-ft intervals for some manufacturer-recommended configurations, dependent on total antenna wind load, maximum wind speed, and other installation factors.

NOTE *These guy spacings are taken from the Rohn data sheets for a sampling of very specific configurations. They may not be appropriate for your installation, antenna loads, and wind speeds. When in doubt, follow the recommendations of the manufacturer to the letter! If you make any changes, make them on the conservative side—but be sure you know what that is—and get them approved by a professional engineer!*

CAUTION *The guy wire spacings previously discussed are for towers that have been completely installed. During the process of erecting a tower, additional temporary guy wires, at closer spacing intervals, should be employed.*

CAUTION *Rope of any kind has too much stretch to be suitable as a guy wire for a steel tower. The only suitable materials for permanent guys are those specified by the manufacturer; typically that will mean only stranded EHS guy wire or Phillystran nonconductive cable.*

Bracketed Towers

Although triangular lattice towers with a constant cross-sectional area (such as Rohn 25 or 45) are designed for guyed installations, very short versions of these towers (say, 20 to 40 ft) can be erected with no guys at all (*freestanding mode*). Further, if bracketed properly to a strong structure such as the gable end of a home, generally at a point 18 ft or more above the base, the allowable unguyed tower height can be increased by nearly the height of the bracket. Example: If a freestanding 30-ft tower comprised of Rohn 45 sections is capable of handling a certain wind load, bracketing the tower at 22 ft may allow the tower to be extended to about the 50-ft level for the same antenna load. But if that specific combination of heights and antenna area is not explicitly given in the manufacturer's data sheets for this tower model, it is imperative that your installation plans be reviewed by a licensed professional engineer.

Bracketed towers share a disadvantage with rooftop (tripod) towers: audible noise inside the structure! In both cases, wind-induced vibrations in the antenna and tower, as well as the "grinding" of any rotator and thrust bearing motion, are transmitted mechanically through the attachment hardware to the framing of the structure. A bracketed 40-ft tower bears an amazing schematic resemblance to a violin or guitar: The tower and antenna at the top are analogous to the taut strings, and the building corresponds to the sounding board! Now throw in the banging of any cables loosely tied to the side of the tower and the clanging of the brake mechanism in some popular antenna rotators. Nonetheless, the author has had a bracketed 40- or 50-ft tower attached to four different residences (often outside a bedroom, no less!) for the past 40 years. Others have reported success in partially damping the vibrations with rubber shims between the tower and the brackets. (A discarded inner tube is an excellent source of rubber stock for this application.)

Often overlooked is the need for strength in the structure the bracket is attached to. Most residential exterior walls are designed primarily to support lots of weight directly above; they have relatively little resistance to horizontal forces. In general, a wall should be reinforced specifically to withstand the sideward force of high winds on a bracketed tower and antenna installation. Obviously this is most easily (and cost-effectively) done at the time the house is constructed! Under no circumstances should you expect attachment to an existing eave or soffit to be adequate; you might as well attach the house bracket to a raincloud. This is another area where expert help is particularly important.

Foldover Towers

A sketch of a foldover tower is shown in Fig. 29.4. This type of tower uses the same triangular lattice construction as the fixed, guyed model but is hinged at a point between one-third and two-thirds the total height. The tower *must* be guyed at the hinge level, and, to avoid interference with the counterbalance boom that goes up in the air as

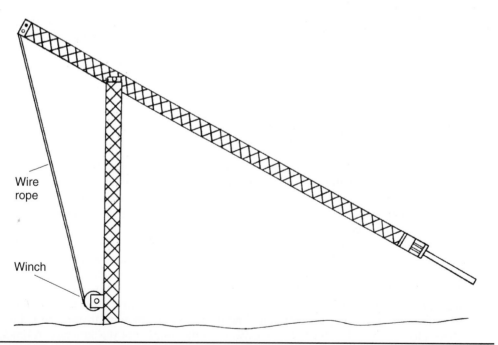

Wire rope

Winch

Figure 29.4 Foldover tower.

the top half of the tower is cranked down, *four* guy wires, uniformly spaced at 90-degree intervals, are required. A waist-high winch spools aircraft cable in or out to raise or lower the upper part of the tower between its fully upright and foldover positions. A safety bolt at the lower end (or farthest point) on the counterbalance boom secures the tilt assembly once erected, thus minimizing end-of-travel "slop" and eliminating the need to rely on the cable to keep the upper portion perfectly vertical when fully erected.

At one time Rohn sold foldover towers in both 25G and 45G ratings, but they are no longer in their catalog. Occasionally one can be found on the used market. Alternatively, if the special hinged section and counterbalance boom can be procured, the remainder of the tower is conventional Rohn product that is still available new.

The advantage of a foldover, of course, is that the top of the tower can be brought close to the ground and all work on antennas, rotator, cables, etc., can be done while standing on the ground or a stepladder. However, there are some disadvantages to this system that are not present in a conventional guyed tower:

- During the process of raising and lowering the tilting portion of the tower, the winch cable and upper tower sections experience maximum stress when the tilt assembly is horizontal. This puts a severe *weight* limit on what can be installed on top (in addition to any wind load limit) because side load limits for triangular lattice tower sections are much lower than their longitudinal compressive and tensile limits. In addition, the winch cable itself has a safe working limit that must not be exceeded.

- The manual winch represents a potential safety hazard. With a full load at the top of the tower, a person of normal build may have difficulty turning the crank smoothly.

- If the winch assembly gets rusty or gummed up such that the safety catch does not properly engage the teeth on the crank assembly, the winch handle can *freewheel* if the operator's hand slips while cranking the antennas up or down. Even if the safety catch does engage, any lag in the catching mechanism can result in a lot of weight on a very large moment arm coming to an abrupt stop, with the possibility of damaging parts of the tower or, worse yet, snapping the aircraft cable and seriously injuring the operator or helpers.

- As marketed, the Rohn foldover had no guy wires at the top. There is merit to adding a set, depending on how many sections have been installed above the hinge and the total wind load of the antennas. However, the owner needs to also factor their added weight into the maximum weight calculations for the tilt process.

- Some antenna rotators—especially those with internal ball bearings—may not take kindly to being tilted partially upside down, depending on internal *play* and retainer spacings.

Crank-up Towers

Figure 29.5 shows a *crank-up* tower. In the example shown, a wide base section supports a smaller moveable upper section. Commercially available models usually have two to five sections. In all of these, a manual or a motorized winch is used to lift the upper section(s) into fully extended position. Typical two-section towers can be raised from a low position of 20 ft to a height of 35 ft or so. When the tower is fully extended, some models incorporate provisions for locking the sections in place with steel bolts, steel bars, or other mechanisms, but these options are by no means universal. A lot of shear force is applied to these bolts, so it is wise to use several very hard stainless steel bolts for the lockdown.

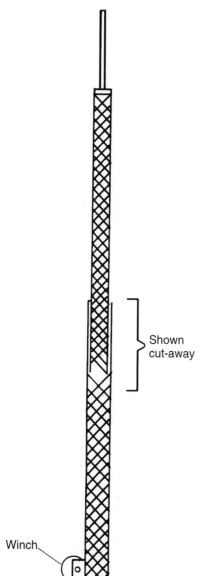

Shown cut-away

Winch

FIGURE 29.5   Crank-up tower.

CAUTION *There is a serious danger inherent in the design of the crank-up tower: the guillotine effect. If a cable supporting extended sections breaks while you are working on or near the tower, one or more sections will come plummeting down the "elevator shaft" formed by the section(s) below it and shear off any arms, legs, or other body parts that get in the way. Whenever you work on this form of tower in its extended position—either fully or partially—use steel fence posts (or similar pieces of metal) as a safety measure (Fig. 29.6); at least two should be used, and both should be attached securely at both ends with rope so they can't be knocked out accidentally. These pipes are used in addition to, not instead of, the normal bolt fasteners that keep the antenna tower erect.*

Some crank-up towers are motorized so that the tower can be raised or lowered without anyone having to stand at the base. Some crank-ups also tilt over, so only a stepladder is needed for installing, removing, or adjusting a typical HF Yagi. Between these two optional features, quite a few of the potential hazards associated with crank-up towers are eliminated or substantially reduced. Of course, crank-up towers with either or both of these features cost quite a bit more.

All towers require periodic maintenance and examination, but crank-up towers as a category require more than most others. In particular, the cable(s) that perform the raising and lowering functions must be frequently inspected and lubricated with compounds specified by the manufacturer of the tower.

In the author's opinion, there is also a major practical inconvenience with crank-up towers. For many models, the specified maximum wind load when the tower is fully extended is so small as to effectively preclude the installation of even a modestly sized HF beam. Some crank-up owners will respond, "No problem! I just crank it down whenever I'm not using it or whenever the wind blows." That may be an acceptable

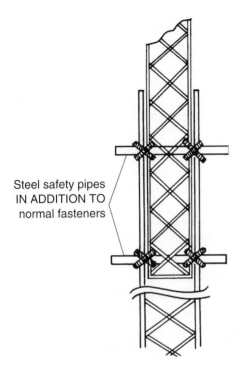

FIGURE 29.6   Safety precaution.

Steel safety pipes
IN ADDITION TO
normal fasteners

solution for some, but prospective owners might wish to consider whether they want to be off the air every time the wind gets above 50 mph or whether they want to go outside and crank it up every time they get the urge to sit down at the rig and operate.

Self-Supporting Towers

As mentioned earlier, self-supporting towers may be designed and constructed with a single fixed diameter or cross section from top to bottom or with a tapered cross section. The greater the taper, the more constant (and smaller) the compressive forces in the downwind leg or side of the tower will be from top to bottom.

Tapered Cross-Section Towers

Old-fashioned windmill towers seen on many farms are an excellent example of the basic design that governs the construction of self-supporting towers. Stated very simply, the slope of the sides of the tower determines the ratio of compressive force in the leeward tower leg(s) at the bottom of the tower to the force exerted on the antenna (or windmill!) by the wind. When there is very little taper (i.e., very little difference in tower cross-sectional area between top and bottom), the multiplier is quite high; when the slope of the sides or legs is pronounced, the multiplier is much lower. The value of sloping the legs of a tower is best understood by considering the value of spreading your feet to brace yourself against a strong wind.

Historically of primary use in commercial applications (and occasionally in residential wind generator systems), the Rohn SSV series of tower sections has found favor with increasing numbers of radio amateurs. The current family of 16 different sections—each 20 ft long—in principle allows for construction of towers up to 320 ft in height. A heavy-duty version of the SSV family replaces the bottom eight sections with beefed-up alternates. A shorter tower can be selected from many different sets of consecutive sections, so if your objective is to build a 100-ft self-supporting tower, the SSV gives you nearly two dozen options(!) to choose from, depending on the maximum wind loading. The beauty of this is that you can customize the strength (and cost!) of your tower based on the maximum wind load you anticipate.

Make no mistake about it—self-supporting towers are expensive! Even though there are no guy wire costs, the superior strength of the sections makes them more expensive than guyed tower sections per foot of height, and the size of the concrete base (which takes the place of the guy wires in resisting any tendency for the tower to tip over in a strong wind) is massive. Further, installation of these 20-ft sections of tower will undoubtedly require professional rigging companies, and the interface between the top of the tower and typical amateur rotators, masts, and thrust bearings will usually entail custom metalworking and welding at a local shop.

One problem with tapered self-supporting towers is the relative difficulty of climbing them. The climbing spikes commonly seen attached to one leg of the tower are usually too small a diameter to be comfortably grasped, and if you've just walked across a muddy field in your climbing boots to get to the tower, you'll find you're trying to hang on to mud-covered spikes when you descend. However, often a matching metal ladder is available for purchase.

AN Wireless Tower Company is another manufacturer of tapered self-supporting towers for the amateur radio community. Like Rohn, their product consists of a family of sections with some heavy-duty options possible. AN Wireless's sections are 10 ft long, and the maximum height currently marketed is 120 ft.

A lighter-duty tapered tower family of 8-ft sections, originally offered by Spaulding and later by Rohn, can occasionally be found on eBay or amateur used equipment bulletin boards. Although inexpensive, these have some disadvantages a prospective buyer needs to be aware of:

- The side rails are made of U-shaped steel channel instead of cylindrical tubes. Consequently, they are very painful to grip while climbing or when working at the top of the tower.

- The cross-bracing is comprised of diagonal steel strips on edge. Since there are no horizontal surfaces to stand on, climbing and working on one of these towers is rough on the feet.

- Invariably the used tower has rust on it. If so, it absolutely must be sandblasted and regalvanized or painted after being inspected for soundness.

- All hardware securing the cross-braces to the side rails must be removed, discarded, and replaced with new high-strength nuts and bolts.

- These tower sections also came in a heavy-duty version (the *HDBX* series). This is the only version that is truly adequate for even small amateur HF Yagis; the standard-duty model is suitable only for VHF and TV antennas.

Rotating Poles

Another form of self-supporting tower is the *rotating pole*. As compared to almost any other kind of tower, the rotating pole provides the greatest degree of flexibility in placing multiple high-gain antennas (notably long-boom Yagis) at specific heights on the tower for optimum stack spacing and full 360-degree (or more) rotation. A typical rotating pole "sits" on a giant ball bearing below ground level and is rotated by chain drive at ground level with a *prop pitch* or similar motor mounted on the same concrete pad that holds the tube in which the bearing is located.

Today the most well-known of the rotating poles is the *Big Bertha*, available through Array Solutions. The name is derived from the original 115-ft rotating pole offered in the 1960s by now-defunct Telrex Antenna Systems. The current *Big Bertha* design is available in heights up to 142 ft, and at least one such tower in the northeastern United States a few years ago sported a bevy of HF Yagis totaling more than 150 ft^2 of equivalent wind load! A typical antenna complement for the "satisfied user" is a three-element 80-m Yagi at the top, a pair of four-element 40-m Yagis at the midpoint and near the top, and three long-boom six-element 20-m Yagis spaced appropriately for maximum stack performance!

No longer in production, the Tri-Ex *Sky Needle* was another rotating pole, available in 70- and 90-ft heights. Sometimes one of these appears on the used market. Of course, dismantling and transporting one of these beasts can cost as much as three or four guyed towers!

A *guyed rotating pole* is substantially less expensive than a *Big Bertha* (but still quite a bit rougher on the wallet than a conventional guyed tower). In some cases, enterprising amateurs erect their own tower(s), using standard Rohn 45, 55, or 65 sections and guy ring assemblies purchased separately. Others buy the complete tower (and often its installation, as well) from Array Solutions, K0XG Systems, and a few other system vendors. Compared with a self-supporting rotating pole, the major advantages of a guyed rotating pole are:

- It is far less expensive.
- It can be erected to greater heights.

The major disadvantages are:

- Guy wire attachment points may interfere with optimum placement of antennas.
- The guy rings can be a "challenging" maintenance item.

A *hybrid guyed rotating pole* consists of a fixed lower tower portion with a rotating portion above. This is sometimes done to reduce the overall cost of the installation when the height of the lowest rotatable antenna on the tower is a substantial percentage of total tower height. The disadvantage of this approach is that the rotator and associated drive system (e.g., chain) now must be located at the junction of the fixed and rotating segments, many feet above ground, making rotator maintenance more difficult.

Tower Bases

Regardless of its exact configuration, a communications tower is typically installed atop a buried pedestal made of concrete (Fig. 29.7). The base serves multiple purposes:

- It keeps the tower from sinking into the ground.
- It keeps the bottom of the tower from "walking" around.
- In conjunction with a layer of stone beneath, it allows proper drainage of tower legs.
- For a self-supporting tower, it resists any overturning moment.
- It provides a physical barrier between the tower legs and corrosive soils.
- It may provide part of the lightning ground system for the tower.

Base construction must follow any relevant local codes, but there are some general considerations to keep in mind. The surface area of the top of the pedestal should be as specified by the manufacturer, but the smallest practical or useful base is probably at least 3 ft by 3 ft. If a backhoe makes the hole, the base is likely to be square, but augers can also be used, especially if a circular hole is desired or necessary.

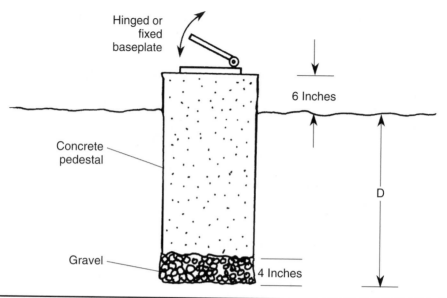

Hinged or
fixed
baseplate

6 Inches

Concrete
pedestal

D

Gravel

4 Inches

Figure 29.7   Ground pedestal.

Caution *If the hole for a tower base or guy anchor is going to be more than 3 ft deep, do not dig it by hand, and do not leave it unattended. The sides of many excavated holes have collapsed unexpectedly. At the very least, if you insist on digging it by hand, make it 6 ft long in one dimension or the other, for it will surely become your grave if a side caves in.*

Regardless of manufacturer specifications, local codes, or other at-a-distance requirements, it is unlikely you will need less than 1.5 cubic yards of concrete for your tower base. Holes in the ground tend to be bigger than planned, forms notwithstanding. (Don't forget to add to that total for any concrete guy anchor stations you may have!) Presuming your site will allow reasonable access to the tower hole by a concrete truck or a backhoe, order no less than 2 yards of premixed concrete from your local supplier. (Around the author's neck of the woods, you can't order fractional yards and 1 yard will not be sufficient for any reasonable base installation.) Further, such a small order will undoubtedly be the last load delivered from that particular truck, and you're apt to get shorted a little bit—it goes with the territory. Most important, you don't really want to mix the equivalent number of bags of Sacrete by hand! You can't keep up, and you will compromise the strength and permanence of your base when its pour is stretched out over the long interval it takes to process that many bags by hand.

By the way, concrete can be ordered and poured at *any* time of year—even in Canada and the northernmost parts of the United States. Additives are part of the secret; long before the big day, talk to your local concrete supplier about what is possible, practical, and reasonable. And when you order the concrete, ask for an extra-strength mix (mention the number "3000"—*they'll* know what you mean, even if *you* don't.)

Regardless of the manufacturer's minimum specifications, the depth of the pedestal and the depth of the gravel or stone base should reflect the local frost line. The concrete pedestal should extend below the local frost line—even if local law permits less; in northern parts of the United States, this will put it at least 4 or 4½ ft below the surface. Beneath the base it is very important to have a layer of stones for proper drainage; layer thickness will be a function of the type of soil encountered. If the tower base consists of a partial section of tower extending all the way through and out the bottom of the concrete, the stones should be large enough so that they can't become wedged in the tower legs and block condensate drainage.

Most manufacturers specify that the concrete pedestal include *rebar* reinforcing steel; refer to the manufacturer's drawings or *Up the Tower* by Steve Morris, K7LXC, before pouring the concrete. Also, make sure you know exactly what needs to be embedded *in* the concrete as part of the tower base before summoning the concrete truck. There are many forms of tower bases and at least an equal number of custom inserts for the concrete. Examples:

- Short section of tower (typically used for constant-diameter guyed triangular lattice towers)
- Pier pin (also used for constant-diameter guyed triangular lattice towers)
- Hinged baseplate (often used for guyed triangular lattice towers up to 40 ft)
- Tilted flanges (for each leg of a tapered self-supporting tower such as Rohn SSV)
- Custom leg brackets (often used for smaller self-supporting towers)
- Single I-beam or mounting post (for crank-up towers)
- Special bolts supplied by the tower vendor

Some professional engineers and tower installers believe the base and its reinforcing metal should be part of the tower grounding system for diverting lightning energy into the earth. Bases designed for this are said to provide a *Ufer ground*. Other professionals prefer the traditional approach of multiple ground rods surrounding the tower base in all directions, with heavy-gauge ground cables running between the bottom of the tower and the rods. See Chap. 30 ("Grounding for Safety and Performance") for additional discussion of grounding techniques.

Standard concrete reaches 99 percent of full strength about a month after it is poured. To the prospective tower owner, that month can seem like an eternity. However, the cure rate is an *exponential* curve, so many experienced contractors will begin to attach hardware within a week of the pour, making sure no extreme stresses are applied to the concrete pad—either directly or through the potent lever arm of installed tower sections. Commonly, the first 20 ft of tower is held *plumb* with temporary guy wires during the pour and cure cycles.

Guy Wires and Anchors

If you are installing a guyed tower, you may also be using concrete pedestals for the guy anchors—a minimum of three, but possibly four (if you're putting up a foldover) or even six (if the tower is very tall and/or you want some redundancy in your guying system). Basically, there are only three reasonable guy anchoring approaches to choose from:

- Conventional concrete block guy anchor
- Earth screw-in anchors (called *screw anchors* for short)
- Special screw eyes held in boulders with equally special expanding cement

Examples of inappropriate and unacceptable guy anchors are:

- Houses, garages, sheds, or other buildings
- Utility poles
- Trees (or any other living things)
- A spike hammered into the top of a tree stump (the author actually has a photo of a real installation where this was done!)
- A steel or iron post anchored in concrete

A special note about the last anchor is in order. Occasionally a steel post anchor appears to be an attractive way to deal with a difficult guying situation, such as when the lower end of a guy wire needs to cross over a driveway, parking area, small shed, or sidewalk. Typically, the plan is to use a 10-ft-long steel tube or I-beam of unknown strength, with 3 or 4 ft of it buried in a concrete pad and 6 or 7 ft sticking up in the air. The guy wire is then attached to a hole drilled near the top of the post. Unfortunately, the very guy wire that is reducing the wind-generated stress on the tower is creating a similar stress on the post. The horizontal component of a 1000-lb tension in the guy wire is 707 lb. This acts with a moment arm of 6 or 7 ft around the point on the I-beam where it enters the concrete. This overturning moment of 4000 lb-ft or greater can become a 20,000-lb compressive force in the post, bending it right at the top of the concrete. The odds are very high that the post is going to bend or break during the first windstorm.

Certainly the easiest, least expensive, and fastest way to create guy anchors in most soils is with screw anchors. Figure 29.8 shows a typical device, available from some tower manufacturers (such as Rohn) and from utility company suppliers such as Graybar and countless regional companies. One excellent anchor is the 6-ft rod with 6-in auger made by A. B. Chance Company. Larger capacities are also available. However, some professional tower installers refuse to climb towers guyed with screw anchors. Compared to concrete pedestals, earth screw anchors have at least two shortcomings:

- The holding power of the soil may be unknown or may vary with weather conditions, flooding, etc.
- The speed and likelihood of corrosion resulting from direct contact with soil of unknown characteristics is uncertain.

A chart relating soil type to screw anchor holding power can be found on the A. B. Chance Web site. Some, leaving nothing to chance (bad pun), have inserted a screw anchor in their soil and measured the tension (force) required to pull it out with a tractor or 4WD SUV. However, holding power is not the only important soil parameter. Soil acidity and the types of trees in the guy anchor area may have a big effect on anchor lifetime. To combat this, some users of screw anchors coat them with a tarlike substance before installing them. Others have recently described electrochemical techniques for neutralizing the effect of corrosive soils. Good news: This is a technical area currently getting a lot of attention as new research and findings are being reported.

FIGURE 29.8 Typical screw anchor.

Screw anchors suitable for modest amateur tower installations can often be installed by hand, by screwing them into the ground with a long *pinch bar* or other strong steel rod threaded through the eye of the rod. But many tool rental outfits have gasoline-powered drivers that make the job easier. Another "glittering generality": The harder it is to insert the screw anchor, the harder it will be for stressed guy wires to pull it out.

Whether screw anchors or concrete pedestals are used, it is critically important to apply *only* longitudinal forces to the anchor rods, which have relatively little strength off-axis. (Some commercially available rods can be bent sideways by hand!) Each guy rod should be installed at an angle that makes the combined force on it from all attached guy wires as close to perfectly in line with the axis of the rod as possible; typically, if the guy wires attached to various heights on the tower are all tensioned to the same specification, the angle of the rod should be equal to the average angle of the guy wires—typically in the 45-degree range. Very little more of the rod than just the knuckle or attachment eyelet should be sticking up beyond the concrete pad or out of the earth. Failure to follow these rules will subject the anchor to the same kind of stresses discussed previously in conjunction with the steel post anchor.

Each and every guy wire should have a *turnbuckle* at its lower end, just above the guy anchor. Two special tools are extremely helpful for proper tensioning of guy wires:

- A *come-along* for maintaining a temporary tension in a guy wire while securing the wire
- A Loos *tension gauge* for tensioning the guy wires following installation

Come-alongs can be purchased at Sears, Home Depot, Lowe's, and local hardware stores. The Loos can be obtained from Champion Radio Products or may be found at sailing accessory stores.

There are only two acceptable types of guy wire material: *EHS stranded galvanized steel* and *Phillystran®*. Either can be obtained from the usual choice of antenna accessory vendors. The advantage of *Phillystran* is that it is nonconductive and long lengths of it do not need to be broken up by insulators to avoid resonances. When *Phillystran* is used, however, the bottom end of each guy wire *must* end in a length of EHS steel to avoid possible breakage from abrasion, vandalism, animals, etc.

Steel guy wires come in many different diameters, each corresponding to a tensile strength rating. Generally, guy wires are tensioned (on a calm day) to some percentage (10 percent is common) of their maximum strength rating. That, of course, adds to the vertical force on the tower legs but ensures that the wires are not slack. (Guy wires that are not properly tensioned can allow tower tops to be snapped back and forth by wind gusts.)

As discussed in other chapters, it's a good idea to insert heavy-duty insulators of the proper compression strength in all guy wires if there's the slightest chance the tower will ever be used as a vertical antenna (typically on 160 or 80 m). Retrofitting a tower installation with insulators is not fun. The tower itself can be directly grounded at its base and fed with shunt-feed techniques described in Chap. 9.

Ideally, all guy wires at a given height (or station) on the tower should come away at the same angle with respect to the tower or ground. When the land surrounding the tower base is not level, or when there is less space on one side of the tower than the other for positioning guy anchors, this isn't always possible. Keep in mind that the top of the tower is in force equilibrium (i.e., stationary and vertical) only when the *horizontal* forces on it are equal and opposing. If guy wires come away at different angles with respect to the tower axis, it will take different tensions, T, in the three guy wires to produce identical horizontal components, T_H. In other words, don't try to make the tensions measured by your Loos gauge in the three guy wires identical if they come away from the same tower attachment height at different vertical angles. Once again, trigonometry is your friend when determining what tension to expect in each guy wire.

Despite being made of rigid, strong steel sections, a tall, slender tower is capable of (often disconcerting) sideward movement far above its base. Blind tensioning of guy wires can lead to incorrect results, since applying too much tension to one guy wire will simply result in pulling the tower sideways at that guy wire's attachment point until the horizontal forces at that guy station are again in equilibrium. Tensioning should always be followed by a check of the entire tower for *trueness*—such as by taking a *carpenter's level* up the tower. However, during the tensioning process, standing or lying at the base of the tower and sighting straight up each face in turn should provide an exaggerated sense of straightness (but *not* whether it's truly vertical!) to guide your subsequent tensioning activities while minimizing the number of trips up and down with the level.

Historically, steel guy wires have been secured with cable clamps, but in recent years *GUY GRIPS®* have become popular, thanks to the speed and ease of their installation. Regardless of which method of securing guy wire ends you choose, there is only one "right way" but many "wrong ways". Follow the manufacturer's recommendations or refer to K7LXC's book (see the end of this chapter).

Erecting the Tower

Raising a triangular lattice tower can be accomplished by any of three methods, although, in the authors' shared opinion, one of the techniques is unsafe. Figure 29.9 shows a method in which the tower is laid out on the ground and then raised with a heavy rope over a high support (such as the peak of the house roof). This method is extremely dangerous unless certain precautions are taken:

- An adjustable-height support must be placed beneath the tower. As the tower is raised, the support is continually moved by a helper so that it is in contact with the underside of the tower until the tower has rotated past the critical angle.

- Temporary guy wires that can be quickly lengthened or shortened by other helpers are needed on each side of the tower (i.e., at the two ends of the back wall of the house).

Towers over 40 ft tall should *not* under any circumstances be erected this way!

The section-by-section method of raising a lattice tower is shown in Fig. 29.10. A "gin pole" is required for this job. The gin pole is a length of thick-walled aluminum pipe—perhaps 2 ft longer than the length of the sections to be installed—fitted with a pulley at the top and a pair of clamps at the bottom. A long piece of suitably strong rope (such as that used by technical rock climbers; in no circumstance should nylon be used!) is run through the pulley. Ideally the line should be twice the height of the tallest tower to be erected with the pole.

Caution Always *wear a* full-body fall-arrest harness *when climbing a tower. At no time should you be* free-climbing *the tower. That is, always have two shock-absorbing safety lines connected to the tower above you while climbing, descending, or working in place, and disconnect only one at a time when changing position. Avoid the possibility of "single-point failures". Carefully inspect your harness and associated hardware prior to each use. The old lineman's climbing belt is no longer considered adequate because you can slip out of the belt and fall to your death if you are thrown upside down by an antenna or tower section that has gotten out of control. Never depend on used equipment. Excellent safety harnesses can be purchased for about $100, while another $100 will procure a pair of shock-absorbing safety lines and perhaps a small tool bucket, as well.*

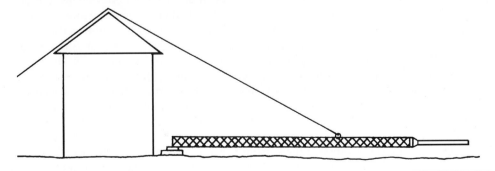

Figure 29.9 Dangerous method for erecting tower.

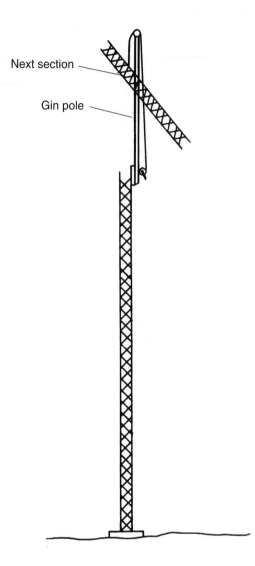

FIGURE 29.10   Proper tower installation using gin pole.

In use, the gin pole is clamped near the top of the highest installed section so that it can be used to raise the next section into place. The ground crew hoists each new section up and holds it in position while the installer (who is belted to the tower just below the top of the old section) maneuvers the bottom of the new section directly over the top of the previous one. Together the installer and the ground crew collaborate to gently lower the upper section onto the lower one so that the corresponding legs of each can be bolted together. The installer then carries the gin pole to the top of the newly installed section and attaches it near the top of this newest section. If this corresponds to a *guy station,* the ground crew uses the gin pole to supply the installer with a guy wire bracket that has the upper end of three guy wires already attached to it. Following installation and tensioning of the three (or four) guy wires, the next section of tower is raised, and the process repeated. Guy wires are typically installed at 30-, 35-, or 40-ft intervals, depending on the strength of the tower sections chosen; with Rohn 25 and 45 or equivalent tower sections, it is appropriate to consider adding temporary guy wires when two sections have been added above the highest existing permanent guy station. Some installers are comfortable with three unsupported sections, but it depends on your own personal "pucker factor" and whether the wind is blowing that day.

The gin pole method is preferred by tower manufacturers and is the method recommended by most of them. For towers constructed from 10-ft sections, installation can be efficiently accomplished with two people—one on the tower and one on the ground. Towers with 20-ft sections and larger should be installed only by professionals.

The third approach to erecting towers is to rent a cherry picker, crane, or helicopter and use it to pick up the entire tower or a major segment of it and hold it in position while it is bolted to the base or segment below it, and guy wires are attached and secured. This should be done only if the crew has at least one person experienced in the proper use and stabilization of the hoisting machine. Many times, cherry pickers can be rented quite reasonably from local equipment rental stores. They are most effective for erecting towers of 80 ft or less. Naturally, your tower site must provide solid and safe access for any machines you plan to bring in.

CAUTION *Always wear a* hard hat *when installing a tower or its antennas. This is especially true for the ground crew because even a small stainless steel bolt or hand tool can cause great injury when dropped from above. But it is also true for any tower climbers, as well, since it's extremely easy to crack your skull on the underside of a guy bracket or boom-to-mast bracket while ascending or while wrestling an antenna into place. Just ask the author, who is a graduate of the school of "hard knocks" (so to speak)!*

At some point you may decide that your best course of action is to have your tower installed by professionals. The boom in cellular tower installations over the past few decades has resulted in no lack of highly competent tower installation companies throughout the world. There are also companies with particular expertise in tower and antenna installations for radio amateurs in addition to the work they do for commercial and institutional customers. An example in the northeastern United States is XX Towers, Inc., headquartered in New Hampshire. Often the tower suppliers listed in App. B can provide a list of reputable installers.

Rotators, Masts, and Thrust Bearings

Many, if not most, antennas placed atop a tower are intended to be rotated because their radiation patterns are highly directional and the desired direction of communication is subject to frequent change. This is true of almost all Yagi and cubical quad installations for not only HF but VHF and UHF as well. The typical installation of such antennas atop a guyed tower is shown in Fig. 29.11: An *antenna rotator* mounted to a shelf a few

feet below the top of the tower turns a strong *mast*, which is often prevented from tilting sideways and binding by a *thrust bearing* at the very top of the tower.

Note particularly the *rotator loop* in the coaxial cable that goes past the rotator. One end attaches to the rotating assembly (usually by being taped to the mast, just above the top of the tower or rotator, whichever is higher) and the other end attaches to the highest practical nonrotating tower surface (usually by being taped to one tower leg). Leave enough slack to allow full rotation of the antenna(s) in both directions without putting any tension on the cable. Dress the loop carefully so that it does not catch on the top of the tower, and check its operation by having a helper rotate the antennas while you stay just below the top of the tower to watch how the cable coils

FIGURE 29.11 Tower top showing rotator on shelf, mast, thrust bearing, and rotator bypass loop in feedlines.

and uncoils. For minimum required loop length, determine where either end of rotation is and attach the cable below the rotator on the leg of the tower that is most nearly directly opposite the end-of-rotation point. An alternative approach involves coiling a turn or two of the cable around the mast just above the top of the tower so that it coils and uncoils just like a spring as the antennas turn. Most hardline and some kinds of coaxial cable are not meant to be flexed frequently; they should not be used to form the rotator loop.

Although some commercial amateur VHF and UHF beams still come with pre-drilled boom-to-mast brackets and mounting hardware for a 1.5-in mast, the standard mast outside diameter (OD) for all but the largest amateur Yagis today is 2 in. Often it is possible to modify older brackets for use with a 2-in mast by simply drilling two new holes in the bracket and replacing the U-bolts.

The best mast material is galvanized seamless steel tubing designed for this purpose. Rohn sells excellent 10-ft masts, and specialty metals suppliers can provide even longer ones. Note that the galvanizing process results in a worst-case OD of slightly more than 2 in, so rotator clamps, thrust bearings, and other support tubes should be sized somewhat larger than 2.0 in. The mast clamp provided with the MFJ/CDE *Ham-M* and *TailTwister* rotators can handle diameters up to $2\frac{1}{8}$ in.

Some rotator manufacturers, including MFJ/CDE, make a lower bracket available for mounting the rotator at the very top of a mast, but this type of installation should be used only for very small VHF and UHF antennas with minimal wind load and torque specifications. Even then, only a short (1 or 2 ft long) mast should be used. Most tower manufacturers sell an accessory rotator shelf to fit inside their towers or tower sections; typically one of these plates is mounted about 2 ft below the top of the tower. A second plate is sometimes used to provide a thrust bearing or clamp support just above the top of the rotator so that if maintenance or repair work on the rotator is required, it can be removed from the tower while the mast and antennas remain clamped in place.

Although some amateurs have successfully used thick-walled aluminum tubing for mast material, this should be limited to Yagis with short booms and elements—i.e., installations that apply only limited torque to the mast and rotator. Aluminum of *any* thickness is no match for steel, and the teeth of a typical rotator mast clamp or the threads of a pinning bolt can eventually chew through even solid aluminum tubing! (Once again, just ask the author!)

For small and medium-size HF Yagis and virtually all VHF and UHF beams, a simple piece of hardwood can serve as a perfectly adequate thrust bearing. Use an inexpensive circle cutter set to the next available diameter greater than 2 in.

The tops of most commercial towers and tower sections sold into the amateur radio market have been predrilled or prepunched at the factory with standard hole patterns for thrust bearings and commonly available rotators.

While some amateurs in areas with low wind loading characteristics have had success using some of the larger *TV rotators*, those devices should not be used to rotate anything larger than small (i.e., short-boom) HF beams—such as a three-element triband Yagi or two-element quad—in a benign environment. Rotators, such as the MFJ/CDE *TR-44*, in this class were originally intended to turn nothing larger than long-boom VHF deep fringe TV antennas; they are sometimes adequate for turning Yagis such as the Cushcraft *A-3S*, Force 12 *C-31*, Mosley *TA-33*, or equivalent.

For larger tribanders (Cushcraft *A-4S*, Force 12 *C-32*) and small monobanders, the venerable MFJ/CDE *Ham-M* or others of equivalent torque and braking capability are appropriate.

The larger MFJ/CDE rotators—notably the *Ham-M* and its big brother, the *TailTwister,* are unique in that the heading of the mast and antennas is locked by a strong brake wedge once the rotation has ended. But this is bane as well as a boon:

- There are only 60 positions of the wedge, so pointing accuracy is never better than within a 6-degree segment. This is not an issue at HF but possibly could be with very high gain, narrow beamwidth VHF or UHF arrays.

- Sometimes the wedge gets jammed by transferred wind load or other sideways forces on the rotator, and the only way to get the rotator turning again is to power the motor in the opposite direction for a few seconds before finally heading off in the desired direction. To minimize this problem, at least one third-party manufacturer (Green Heron) of digital control boxes has incorporated a special start-of-rotation reversing sequence for the *TailTwister* in its controllers' firmware.

As boom lengths increase beyond 26 ft or so, larger rotators are an absolute necessity. In addition to the *TailTwister,* excellent units are available from ProSistel (an Italian company with worldwide distributors), M^2 (pronounced "M squared"), Orion, and Yaesu. Many satisfied amateurs are using heavy-duty rotators from these companies to turn one or more antennas with 40-ft booms on a single high-strength mast. Most of these models have been designed to fit inside any of the top sections in the Rohn 25 product line, and most also use mounting hole locations compatible with those that are prepunched in Rohn's rotator mounting plates—a hole pattern that goes back a half-century or more to the original *CDE Ham-M* rotator.

For the very longest booms, and for rotating the entire tower, modified *prop pitch motors* are popular. Many hams "roll their own", but K7NV sells completely refurbished prop pitch motors with or without a mating controller.

In those rotators that do not have a separate brake, braking is usually accomplished by a combination of gear ratio and *back electromotive force* (emf) generated by the rotator motor.

The single most important specification for a rotator is its braking torque—that is, how well it can withstand the force of the wind on the antennas when they are at rest and not being turned by the controller. Some users eschew rotators that employ a mechanical brake because they prefer the "soft" drift of antennas to the full transmittal of shocks from wind gusts.

Regardless of brand or strength, all rotators require cables to be run from a control box that is usually located indoors, close to the radio equipment, where it can be conveniently accessed by the operator. Today, many of these control boxes include RS-232 Serial or USB ports so that popular logging programs and custom software can access and control them from PCs. (Rotator manufacturers almost always bundle a control box with each tower unit, but increasingly today "naked" tower units can be purchased separately. This trend has undoubtedly been hastened by the near-instant popularity of the Green Heron *RT*-series of third-party controllers that work with virtually every rotator that has been on the market in the past 25 years, including prop pitch motors.)

All rotators have end-of-rotation limits, or *stops*. In general, it is not wise to rotate antennas into either end stop at full speed. Some controllers, such as the Green Heron, include *ramp-up* and *ramp-down* circuitry or software to minimize the overall stresses on the tower and rotator when starting and ending rotation.

Another feature found in certain rotators (and their controllers) is provision for rotation in excess of 360 degrees—in some cases an additional 180 degrees! This is a great convenience for users such as DXers who may be "following propagation" with the heading of their Yagis or quads. However, the owner of one of these devices must be careful to provide an extra-long *rotator loop* in all coaxial cables that pass the rotator on their way up the mast.

Each different type of antenna rotator imposes different requirements on the rotator cable that runs between the control box and the tower unit. For years, the standard in amateur rotators was the CDE *Ham-M*, which requires an eight-wire cable in a conventional installation. However, two of those wires have the high solenoid current of the brake wedge in them and require larger-diameter wire than the other six wires for any given distance between the controller and the top of the tower. One manufacturer, Belden Wire, offers a jacketed cable (their catalog #9405) with two #16 and six #18 conductors that is designed specifically for moderate runs to a *Ham-M* or *TailTwister* rotator. There is nothing magic about the wire, however, so any way the owner of a rotator wants to put together eight cables of sufficient gauge for the length involved is fine. Keep in mind:

- The required wire size depends on the total voltage drop of the rotator control signals over the total length of the run. CDE has converted this to a specification on the maximum allowable resistance for each individual wire; this information can be found in the manuals for their rotators.

- Make sure the cable or individual wires you select are suitable for outdoor and/ or burial applications. Protecting your rotator cables by running them inside PVC or other conduit material (with proper drainage for condensation) is desirable.

- Some brands of wire insulation and cable jackets appear to have a chemical makeup that is more attractive than others to field mice, squirrels, and other "critters". If you feel you have a disproportionately high degree of difficulty with chewed cables, you may wish to try different makes of cable (or route them through conduit with screening at the ends).

One trick for reducing the number of wires running between a CDE rotator and its controller is to replace the *motor start capacitor* in the controller with one of similar capacitance and voltage rating across terminals 4 and 8 on the bottom of the tower unit. This will eliminate the need for two of the eight wires coming from the controller. Keep in mind that motor start capacitors have a pretty sloppy temperature coefficient, so one mounted outside at the rotator or tower base may lose as much as 50 percent of its rated capacitance during cold winters, leading to some potential for sluggishness compared to its summer performance.

Cabling requirements for the Orion and M^2 rotators and controllers are quite different. Beam heading information is obtained from a pulse-counting system in those units, whereas CDE rotator control boxes get headings by reading the dc voltage on the wiper of a wire-wound resistor (*rheostat*) located inside the upper housing of the tower unit. Further, Orion and M^2 don't employ a brake wedge, so there is no brake solenoid to power.

Many users of the Orion and M^2 rotators have reported that it is very important to run a completely separate cable for the pulse-counting position indication system. If not, crosstalk between wires in the same cable can corrupt the beam heading information used by the controller in deciding when to start and stop rotation.

Preowned Towers and Accessories

With respect to obtaining and installing used towers of any type, as well as tower accessories, the author recommends proceeding with extreme caution—if at all! Tower sections (especially those with hollow cylindrical legs) that have been in service or stored outside for long periods may be weakened by hidden rust. However, it may be possible with a sequence of complete sandblasting, inspecting, and *hot-dip regalvanizing* at nearby reputable commercial businesses specializing in those tasks to restore tower sections for your own use for less than the price of new sections plus shipping. Galvanized guy wire and other "mission-critical" hardware that is showing rust or black streaks should simply be discarded. When replacing any original tower bolts, use only the manufacturer's recommended replacement parts. Bolts of a given size come in many different strength ratings, and those you can find at your local hardware store are very likely inadequate for your intended application.

Required Reading

Note that the title of this section is not "Recommended Reading"—it is "*Required* Reading". Installing a tower is no "Saturday afternoon and a keg of beer" picnic—it's a deadly serious project that requires massive amounts of reading and preparation. Failure to do so can lead to death or permanent disability for you, your helpers, your neighbors, or other members of your family.

At the very beginning of any antenna project, there are three kinds of source material that a prospective tower owner absolutely must obtain and read thoroughly and carefully:

- All relevant spec sheets and engineering documents available from the tower and antenna manufacturers.

- The official building code, zoning, and wind speed requirements for your county.

- *Up the Tower: The Complete Guide to Tower Construction,* by Steve Morris, K7LXC, 2009. Available from Champion Radio Products (www.championradio.com).

In addition, someone (professional engineer, local codes enforcement staffer, etc.) associated with the tower project should have access to *TIA Standard RS-222-G* and be skilled in applying it to a specific installation. This document, available from the Telecommunications Industry Association, is expensive and will typically be found only in the offices of professionals who deal with towers and other antenna support structures on a regular basis.

CHAPTER 30

Grounding for Safety and Performance

I n most radio transmitting or receiving installations there will be four different types of ground systems to consider and implement:

- Grounding for lightning protection
- Grounding for power distribution safety
- Grounding all station equipment together
- Grounding for maximum antenna efficiency

The first three are absolutely necessary; the fourth *may* be important, depending on the type(s) of antenna(s) in use.

NOTE *The single most important message to take away from this chapter is this: A good lightning ground is not necessarily a good radio ground, and neither of them provides a safe power distribution or equipment ground in and of itself. In short, for a proper communications installation, all four types of ground systems must be independently considered, designed, and installed. Nonetheless, in many installations, components of any of these ground systems may well function as components of one or more of the other three systems.*

Grounding for Lightning Protection

Although lightning figured prominently in early electricity experiments—especially those conducted (pardon the pun) by Benjamin Franklin—it remains one of the most unpredictable and incompletely understood phenomena in electromagnetics today.

But despite some remaining gaps in our understanding of lightning, the scientific body of knowledge is sufficiently mature today that lightning experts are able to provide valuable guidance to all of us for protection of our lives and our property—usually couched in terms of a statistical "probability". In other words, predicting *exactly* what lightning will do is difficult, if not impossible, but predicting what it is *likely* to do is helpful *most* of the time. The problem is akin to that of weather prediction—the scope of the system is enormous, and the number of variables not under our control is mind-numbing.

Under the right (or "wrong", depending on your point of view) conditions, the earth and its atmosphere in localized regions of the globe can form a power supply of virtually infinite energy. Electrical potential differences of millions of volts build up between objects on the earth and particles in the earth's atmosphere directly above. (Think in terms of an extremely large parallel plate capacitor being charged through a series resistance by an infinite supply voltage, so that there is a time interval associated with how quickly a thunderstorm can "recharge" a given area of a cloud.) At some point, this voltage becomes large enough to jump across air by *ionizing* or breaking down the air molecules. (A typical breakdown figure is something less than 100,000 V/ in of dry air, but the exact voltage depends on the shapes of the two electrodes, and there is much about high-voltage phenomena that is nonlinear.) When there is moisture in the air, the voltage required to ionize the path to earth drops substantially. Any wonder, then, that it is virtually impossible to predict where lightning will strike? Have you ever seen or created a map of humidity versus position in your backyard, for instance? The jagged shape of a lightning bolt is simply the result of the stored energy in our atmospheric system taking the "path of least resistance" . . . literally.

And what could provide less resistance than a wire or a metal tower standing tall in someone's backyard? The very attributes of antennas and their supports that we value are the same characteristics that make them more likely to be targets of frequent lightning strikes. Similarly, tall trees are frequent targets because they exhibit a finite resistance and are capable of electrical conduction, especially when the sap is running. The old adage is a good one: "In a thunderstorm, seek shelter but don't stand under tall trees." To that we can add, "Don't stand under radio towers and utility poles, either."

The second characteristic of lightning that we have to deal with is that a typical stroke passes peak currents of *thousands* of amperes (20,000 A is often cited). The heat generated by such a huge current flowing through even a minuscule resistance is sufficient to blow apart concrete tower bases and vaporize not just solder connections but entire antennas! Thus, in addition to immediate death or permanent disability from the *electromagnetic* effects of being hit by lightning, there is a comparable risk from *proximity* to exploding objects, extreme radiated heat, splashes of boiling water (from the ground) or tree sap, and flash fires (from overheated wires in the walls of wood frame dwellings).

There is no known way to totally prevent a lightning strike from coming to your neighborhood. Thus, the first purpose of a ground system for lightning protection is to *minimize* the damage caused to people and property from a lightning strike that is almost certain to find you someday. To that end, the current professional consensus is to attempt to harmlessly *dissipate* in the loss resistance of the nearby earth as much of the energy in the lightning bolt as possible. The level to which you, the station owner, can do that is determined by the unique characteristics of your site, your tower and cabling configuration, and your wallet.

Bear in mind that lightning does not know the difference between a tower erected for radio communications and a length of house wiring in the attic of a three-story home. The tower may well be somewhat higher than the attic wiring, but both are ultimately grounded and both are "juicy" targets for a lightning bolt looking for a path to earth ground. If conditions are right (such as the air immediately above the house being of higher humidity than the air near the top of the tower), the house may well be more likely to be hit than the tower.

The usual method for dissipating the energy in a lightning surge is to attempt to spread that energy to a large area of ground surrounding the target tower(s) and antennas. This is done most often by connecting the base of the tower to a set of *lightning radials* comprised of straight runs of #6 AWG or larger copper wire or flat copper strap with a cross-sectional resistance no greater than that of #6 wire. Soil conditions permitting, each radial should be connected mechanically or with a brazed joint to multiple 8-ft ground rods of the type found at electrical utility suppliers such as Graybar. As shown in Fig. 30.1, the first ground rod on each radial should be one ground rod length (or 8 ft, in this example) from the base of the tower, and all subsequent rods should be 16 ft (i.e., twice the length of the rod) beyond the previous rod. The minimum practical number of radials is probably 4, with 8 to 16 being a better choice. Two or three rods per radial probably constitute a minimum system.

Since the objective of *lightning radials* is to "dump" the surge currents—whether direct or induced—from the lightning bolt into the soil, there is no reason the radial wires (and the tops of the ground rods) can't be buried many inches under the surface of the ground. However, as we discuss later, if the lightning radials are to be a part of a ground-mounted vertical antenna's radial system, they should not be located more than a few inches below the surface, or the ground losses associated with the vertical's return currents will increase and overall antenna efficiency will suffer. Strange as it may

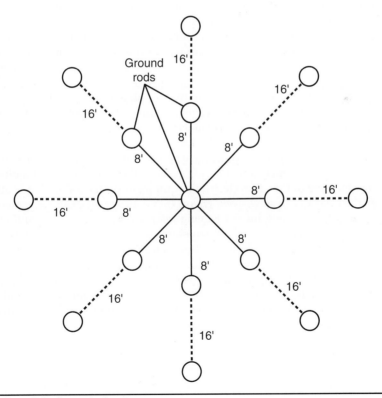

FIGURE 30.1 Tower lightning ground system.

seem, when it comes to lightning surges we want lossy soil but when it comes to radio signals we want lossless soil!

When soil conditions do not allow 8-ft ground rods to be fully pounded in, shorter rods can be used, with some lessening of performance unless the number of radials and rods is increased. As a last resort in very rocky terrain, ground rods can be laid horizontally and buried just under the sod.

Spectral analyses of lightning surges show the bulk of the RF energy to be in the 1- to 3-MHz range. Thus, a key requirement of a *lightning ground* is that it exhibit a very low impedance to lightning—not just at dc and power line frequencies but up through 3 MHz or more. As a practical construction matter, the radial wires and straps being used to dissipate the energy should have no sharp bends or other sources of excessive inductance. In Chap. 3 we noted that a current flowing in a length of wire creates a magnetic field around the wire and that a large enough current will result in a field so strong as to be able to actually move that wire or neighboring wires. So it should not be surprising that large surge currents in conductors from nearby lightning strokes can create mechanical stresses at sharp bends.

The other important thing to remember is Ohm's law: $V = IR$. It does not take very much resistance in a wire or a junction to create extremely high voltage drops (and power dissipation) in the presence of a lightning surge. The heat generated when lightning surges through ground wires and their connections is well above the amount needed to melt any solder you might have available. Consequently, lightning radials should never be soldered; the connections throughout a lightning ground system should be *brazed* or mechanically clamped.

Naturally, our primary concern in grounding for lightning is protection of living beings, followed by protection of our property and possessions. Thus, one reason we attempt to dump the lightning surge current into a properly constructed ground system is to minimize damage to electronics equipment inside the home or business. Many purists feel the only way to be 100 percent sure there is no damage is to totally disconnect all electronic equipment prior to the arrival of a storm or, better yet, immediately following each period of use. This means disconnecting *everything*—power cords, headphones, telephone lines, all audio and control cables, grounding wires, etc. Unfortunately, there have been reports of electronics equipment being damaged during a storm despite being totally disconnected from external wiring, so even this approach is not foolproof.

Today the state of the art in electronic equipment protection for home and business relies on the concept of a *single-point ground* (SPG). In a true SPG arrangement, *all* signal and control cables coming into or leaving the radio room (or entire building, for that matter) are passed through a metal panel or enclosure just *outside* the point of entry (Fig. 30.2). The panel should be ground-mounted or physically as close to earth as can reasonably be attained in a given installation, and solidly attached to its own lightning radial system. It is *imperative* that the telephone company ground and power utility ground be close to the SPG panel and to each other (say, within 1 or 2 ft) and electrically bonded together and to the SPG ground. Any other utilities or quasi-utilities, such as cable or satellite TV system cabling, that enter the building should also pick up their earth ground at the same point. If the building has a protective lightning rod system, it should also be connected here.

The fundamental premise of a single-point ground is that a "rising tide lifts all ships". In most instances of surges from nearby lightning strikes causing damage to electronic equipment, the culprit has been identified as the *difference* in the induced

FIGURE 30.2 Single-point ground (SPG) system brings grounds for all utilities, antenna feedlines, and control lines together at a common entry point for the entire building.

surge voltages on two different cables or within two different systems. As one specific example, the author lost a fax machine, a cordless telephone base station, and a number of PCs—each with an internal fax modem card connected to the telephone line—during thunderstorms on two separate occasions back when his telephone line and ac mains entered—and were grounded—at opposite ends of his home. The author's transceivers, with no connection to the telephone system but with their inputs connected to coaxial cables in the SPG enclosure, sustained no damage during those same storms.

All radio transmission lines, all rotator control cables, all remote switch control lines *must* go through the SPG. At the very least, the GND lead for a set of control wires should be grounded at the SPG. The same is true for the shield braid of each coaxial cable entering or exiting the building. Open-wire transmission lines should, at a minimum, have spark gaps from *both* sides of the line to the SPG ground and ideally would have remotely operated contactors that short out both sides of the transmission line to SPG ground when that line is not in use.

"Hot" leads—the center conductor of coaxial cable, the control lines to remote antenna selectors, rotators, etc.—can be handled either of two ways:

- Use the normally closed contact of relays or contactors to directly ground them all when they're not in use.
- Route each and every hot lead through a lightning protection device.

If you opt for the second approach, you can purchase units manufactured by ICE, Alpha Delta, and others (see Fig. 30.3). Some of these designs employ *metal-oxide varistors* (MOVs) between each control line and ground. Others use gas discharge tubes for protecting any equipment connected to the hot center conductor of coaxial cables. One complaint by some amateurs is that these devices tend to fail "open", making it impossible to know whether they're still operational or not. Further, MOVs have been shown to fail gradually, becoming less and less useful with each nearby lightning surge. The least expensive devices are *shunt* protectors; that is, they operate by shunting the lightning-induced surge to ground. Another category, generally costing more but preferred by some experts, employs *series* protection techniques.

Coaxial signal cables and rotator or relay control cables that have been routed down the legs of metal towers should have their shields and GND leads connected directly to the tower at both the top and bottom. Of course, the bottom connection should be directly tied to a lightning ground radial field emanating from the base of the tower in all compass directions. As mentioned earlier, these connections should be mechanical or brazed, since solder will not survive a lightning surge. Cables dropping from masts and other antenna support structures, including trees, should have their shields and GND leads connected to an earth ground at both the top and the bottom of the vertical span. For maximum protection from lightning surges, the key directive is "Ground everything well and often!"

NOTE *Grounding cables at the tower is not a substitute for a good single-point ground at the point where all the exterior cabling enters the building. At RF, distance is not your friend! A good lightning ground at the tower, while highly desirable, has nothing to do with protecting the electronic equipment inside the building from lightning-induced surges on power lines, telephone lines, or even the cabling between the tower and the building.*

Figure 30.3 SPG enclosure contains surge suppressors on all control lines and antenna feedlines.

Open-wire lines present a special challenge. It is certainly possible to run each side of an open-wire transmission line through its own grounded protective device, but keep in mind that voltages on an open-wire line used in a high-SWR configuration (feeding an 80-m dipole on 40 m, for instance) will be much higher than on a matched OWL or coaxial cable for the same transmitter power levels. Another approach (that can certainly be used in combination with the first) is to pass each side of the open-wire line through a wide-spaced open frame relay with Form C contacts just before the line enters the building. To be sure the relays fail "safe" when power is off, the line connected to the antenna should go to the wiper or common contact and the normally open (NO) contact should connect to the section of OWL that enters the building. The normally closed (NC) contact should be run on as short a path as possible with heavy-gauge wire (#6) to the SPG or lightning ground system, whichever is closer. In an alternate configuration substantially better from a lightning protection viewpoint, the OWL would go to a remotely tuned ATU located in a "doghouse" in the general vicinity of the antenna, and protected coaxial and control cables would run from the doghouse to the SPG. NC contactors and a lightning radial field for the ATU would be located at the doghouse.

Perhaps the best practical approach available to most of us with budget limitations, given the present state of the art, is to employ a three-stage philosophy of equipment protection:

- Ground all transmission and control lines at the top of their respective towers and to lightning ground systems attached to the bases of all towers.

- Use a single-point ground just outside the building that houses the radio room; make sure all incoming utilities are grounded to a suitable lightning ground system at that point and enter the building very close by (i.e., within a few feet).

- At the SPG, mount an enclosure to hold lightning protection devices for all signal and control cables entering and exiting the structure. (See Figs. 30.2 and 30.3.) For an extra measure of protection, include relays or contactors with wide-spaced contacts to disconnect the hot leads of all cables and controls entering the building when they're not in use.

While not a perfect solution, since very large surges and direct hits can still jump these gaps, protection from lesser surges and indirect strikes will be vastly improved.

Grounding for Power Distribution Safety

The second form of ground every radio station installation *must* have is a *power system ground*. The purpose of this ground is to ensure the safe operation of every appliance and every piece of electrical or electronic equipment plugged into a wall outlet.

At least in the United States, current practice is to wire homes with a three-wire system. One wire at each wall outlet is "hot" (usually around 120 V, 60 Hz, and usually with black insulation), one wire (the "neutral", usually white) is a return wire for the current used by the appliance or device, and the third wire (usually green) is the safety "ground". On devices and appliances with a three-wire power cord the green wire is required to be directly connected to the metal chassis of the device.

It is "code" in the United States that the *only* place the white and the green wires are to be connected together is back at the main breaker box, where utility power enters the building. In particular, connecting those two wires together in a subpanel or at an outlet strip alongside your operating desk is strictly and specifically prohibited. There are many reasons for this, all of them good, that are beyond the scope of this antenna book.

Power system ground is a dc and low-frequency ground; circuit impedances can be assumed to be low only for frequencies below 120 Hz or so, and only with respect to the power utility's neutral wire. The earth ground outside your residence does not provide any magic RF grounding capability; instead, it's mostly a 60-Hz ac reference for your site, establishing a defined "zero voltage" point for the utility company at your service entrance. Nonetheless, it is important for the reasons outlined in the preceding "Grounding for Lightning Protection" section to keep the power system ground tied to other system grounds at one specific entry point to the building so that *all* grounds *inside* the building rise and fall with the same surge voltage during a thunderstorm.

Station Grounding

It does no good to provide a topflight ground system outdoors or throughout the structure's wiring if the interconnections between the various pieces of station equipment and the external ground system(s) are substandard. Figure 30.4 shows a method used by one of the authors to good effect. On the back of the operating position is a sheet of copper, 7 in wide, running the length of the equipment platform. This form of copper, weighing 1 lb/ft², is often used on older houses for roofing flashing. An alternative is to run a buss bar of ½-in or ¾-in rigid copper plumbing tubing the length of the equipment table, at the rear. In either approach, each piece of equipment is connected to this *ground*

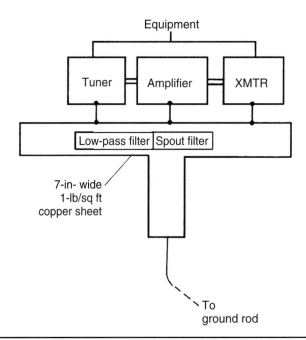

Figure 30.4   Ground system inside shack.

buss or *ground plane* through a short length of braid or #12 wire, secured at the buss with hose clamps or sheet metal screws. Small RF accessories (e.g., a low-pass filter) that have no user controls on them are mounted directly to the copper sheet.

These connections are an important part of all the other grounding systems discussed, and they must therefore be low impedance all the way from dc to many megahertz. Further, the ground plane or ground buss so formed must be connected to the outdoor ground system(s) with as short a run as possible, so that the least possible amounts of both resistance and inductance are introduced into the total ground path. In one installation, the author was able to drop the copper sheet down from the table to connect directly to the ground system outside the building. The run was less than 40 in.

Unfortunately, some amateur radio operators and CBers use the building electrical ground wiring for the RF antenna ground of their station. Neglecting to install an outdoor ground that will properly do the job, they opt instead for a single connection to the grounded "third wire" in a nearby electrical outlet. In addition to being potentially dangerous, this is a very poor RF ground. It is too long for even the lower HF bands, it reradiates RF around the house in large quantity, and there's just not enough of it to serve its intended purpose. Transmitters that depend on the household electrical wiring as the radio ground tend to cause radio and TV broadcast interference, as well as interference to other consumer electronics devices in their own home and in nearby buildings.

Tuned Ground Wire

If you use a power tool or a household appliance on the second or third floor of your home, or at the far end of a long three-conductor extension cord, you still have an effective power safety ground at the point of use, despite the relatively long run from your circuit breaker box in the basement. That's because the wavelength of 60-Hz energy is measured in miles, and the extension cord represents a tiny fraction of a wavelength. However, from a lightning or radio standpoint, second- and third-story grounds are not grounds at all! The reason is because at *radio* frequencies your equipment chassis ground may be a substantial fraction of a wavelength distant from true earth ground. (Remember that typical lightning strikes include large amounts of energy at 1 to 3 MHz and above. At those frequencies, anything longer than a few feet is likely to see a significant induced voltage across its length.)

If you have ever been "bitten" or burned by RF energy when you happened to touch the chassis or a metal knob on your radio equipment while transmitting, you have felt firsthand the effect of having an inadequate RF ground in your radio room. Often the user *thinks* he or she has a ground simply because a wire has been run from the equipment chassis to a ground rod two floors below. That "ground" wire may be 25 or 30 ft in length—long enough to be a quarter-wave vertical on 20 m!

An alternative that some operators use is the *ground wire tuner*. (MFJ Electronics makes such a unit.) These accessories insert an inductor or capacitor (or even a full *LC* network) in series with the ground line. The user adjusts (or "tunes") the device for maximum ground current while transmitting on the desired operating frequency. Although these tuners can bring your radio equipment common chassis connections to a *virtual* RF ground for purposes of providing proper antenna tuning, a low-impedance RF path to a single-point ground located outside the building, at earth level, is still necessary for lightning grounding.

Grounds for Antenna Efficiency

The success or failure of a radio antenna system often hangs on whether it has a good RF ground. Poor grounds cause many antennas to operate at less than best efficiency. In fact, it is possible to waste as much as 50 to 90 percent of your RF power heating the lossy ground under the antenna, instead of sending that RF into the air. This is especially true in the case of ground-mounted monopoles, such as a typical quarter-wave vertical, because the "missing" half of the antenna is the circle of ground lying within roughly $\lambda/2$ of the base of the vertical; RF energy delivered to the feedpoint of the vertical monopole and radiated into space must be balanced by return currents delivered to the transmission line via this ground path. Total effective ground resistance of typical soils in the circle beneath a vertical monopole can vary from a low value of, say, 5 Ω, up to more than 100 Ω, and RF power is dissipated in the ground resistance according to a simple resistive divider equation. The total resistance, R_{LOAD} seen by the transmission line is

$$R_{LOAD} = R_{GND} + R_{RAD} \qquad (30.1)$$

where R_{GND} = effective loss resistance of ground surrounding antenna
$\quad\quad R_{RAD}$ = radiation resistance of antenna

The total power, P_{OUT}, delivered by the transmission line is

$$\begin{aligned} P_{OUT} &= P_{GND} + P_{RAD} \\ &= I^2_{LOAD}(R_{GND} + R_{RAD}) \end{aligned} \qquad (30.2)$$

The power radiated by the antenna is $P_{RAD} = I^2_{LOAD}R_{RAD}$; expressed in terms of the total power delivered by the transmission line, it is

$$P_{RAD} = \frac{I^2_{LOAD}R_{RAD}}{I^2_{LOAD}(R_{GND} + R_{RAD})} \qquad (30.3)$$

$$P_{RAD} = \frac{R_{RAD}}{R_{GND} + R_{RAD}} \qquad (30.4)$$

If, for instance, the effective ground resistance is 30 Ω, and the antenna is a full $\lambda/4$ vertical monopole having a radiation resistance of 30 Ω, also, then half the applied power, or 3 dB, is wasted in ground losses. Any increase in ground losses (usually as a result of an insufficient radial field) or decrease in radiation resistance (a shorter vertical) can conspire to drop the power radiated even further—perhaps to as low as 5 or 10 percent of the power available from the transmission line!

Soil Conductivity

Factors that affect ground resistance include the conductivity of the ground, its composition, and its water content. The effective RF ground depth is rarely right on the surface and—depending on local water table level—might be a few meters or so below the surface.

The conductivity of soil determines how well or how poorly the earth conducts electrical current (Table 30.1). Moist soil over a brackish water dome conducts best (coastal southern swamps make better AM broadcast radio station locations), and the sands of the western deserts make the poorest conductors.

Previous editions of this book described techniques for reducing the electrical resistance of soil through treatment with chemicals. Currently, the author does not recommend the approach. In addition to any possible environmental concerns, any benefit from *salting* will only be temporary and the cost and time required to repetitively salt a useful area surrounding the base of the antenna prohibitive. Salting only the area near ground rods is futile for improving antenna efficiency because the ground currents relevant to antenna efficiency travel *laterally,* near the surface of the ground—from roughly as far away from the base of the vertical as the vertical is tall—back to the antenna feedpoint. In contrast, focusing on deep ground rod conductivity is arguably of interest when attempting to maximize the dissipation of a lightning surge into the ground.

As discussed in Chap. 5, the soil directly underneath a horizontally polarized antenna is important primarily for establishing the effective height of the antenna (and, hence, its feedpoint impedance at its erected height above the ground surface). The soil within, say, a radius of $\lambda/4$ to $\lambda/2$ is important primarily for carrying radial return currents back to vertical monopoles. But with the possible exception of saltwater, any surface normally found under a vertical monopole is not a good enough conductor to seriously consider it as a substitute for copper radials.

Table 30.1 lists the electrical characteristics for a range of commonly encountered soils, along with those of saltwater and freshwater. Note that the highest-conductivity soil ("pastoral hills") in the list is still approximately 200 times lossier than saltwater— and saltwater is nowhere near as good as copper! Bottom line: For superior vertical monopole efficiency, lay at least two dozen radials out under your verticals and quit worrying about your specific soil type!

Type of Soil	Dielectric Constant	Conductivity (siemens / meter)	Relative Quality
Saltwater	81	5	Best
Freshwater	80	0.001	Poor
Pastoral hills	14–20	0.03–0.01	Very good
Marshy, wooded	12	0.0075	Average/poor
Sandy	10	0.002	Poor
Cities	3–5	0.001	Very poor

TABLE 30.1 Sample Soil Conductivity Values

Much farther from the base of your antenna, soil conductivity is important because it affects the first ground reflection of your transmitted signal and the last ground reflection of received signals. But for long-distance (DX) communication, optimum takeoff and arrival angles are usually in the 1- to 5-degree range, so we're talking about bouncing the RF off soil at distances up to a mile away! Odds are high you don't even own the land, much less have the fat wallet needed to salt acres and acres of ground. And for very long DX paths, there are additional ground reflections to consider. Ultimately, it's probably cheaper and a lot less backbreaking labor over a lifetime to buy oceanfront property.

Grounding with Radials

The effectiveness of an RF ground system is enhanced substantially by the use of radials either above ground or buried just below the surface. In Chap. 10 we saw that a vertical monopole antenna is relatively ineffective unless provided with a good RF ground system, and for most installations that requirement is best met through a system of ground radials.

An effective system of radials requires a large number of radials. But how many is enough? Experimental results obtained in 1928 subsequently resulted in new regulations in the United States requiring broadcast stations employing vertical monopoles in the AM band (540 to 1600 kHz at the time) to use a minimum of 120 half-wavelength radials, but 120 was deliberately chosen to be at least twice the number at which the experimenters had found that practical improvement ceased to be meaningful. Recent modeling work has confirmed that nowhere near that number is necessary for non-broadcast services. Installing more than 30 or 40 $\lambda/4$ radials is not only expensive and time-consuming, but totally unnecessary, as well. The author has had a superior signal on 160 m for years using anywhere from 12 to 25 radials of various lengths.

Unlike elevated radials, which should be as long as the electrical height of the vertical, radials on the ground can be just about any length that is convenient, up to perhaps a half-wavelength. A radial in close proximity to earth has lost any pretense of being resonant at a specific length; better to think of it instead as one terminal of a long, skinny capacitor. From modeling and experiments a rough rule of thumb has evolved: Aim for the tips of adjacent radials on the surface of the ground to be about 0.05λ apart. At smaller spacings, no significant reduction in ground losses is noted, and when spacings start to exceed that figure, losses start to grow as an increasing percentage of the vertical's return currents find their way into the earth instead of the copper wires.

A little thought will lead you to the conclusion that for fixed tip-to-tip spacing specified as a fraction of the wavelength in question, the shorter your radials are, the fewer you need!* This seems paradoxical at first, but what's missing is the fact that your antenna's ground return efficiency is not constant during this comparison; truly, shorter radials are not as good as longer ones, no matter how many of them you install. Stated another way, it is a waste of time and good copper to install "additional" short radials in an attempt to compensate for their shortness.

The ideal layout for a system of radials in a vertical antenna system is depicted in a view from above in Fig. 30.5. Here, the radials are laid out in a uniform pattern around the antenna element. This coverage provides both the lowest resistance and the best radiation pattern for the antenna. Tie all radials together at a common point at the base of the vertical, and connect that junction of wires to one side of the feedline—usually the shield braid when coaxial cable is used—and to the ground terminal on any remote

---

*To see this, draw a circle with radius R (the length of a radial). The circumference of the circle is $2\pi R$, or approximately $6.3R$. If $R = \lambda/4$, then $C = 1.6\lambda$. Divide the circumference into little segments, each 0.05λ long. Then N, the number of radials, is $1.6\lambda \div 0.05\lambda = 32$. Now shorten the radials to $\lambda/8$ and redo the calculation. The maximum useful number of radials is now 16.

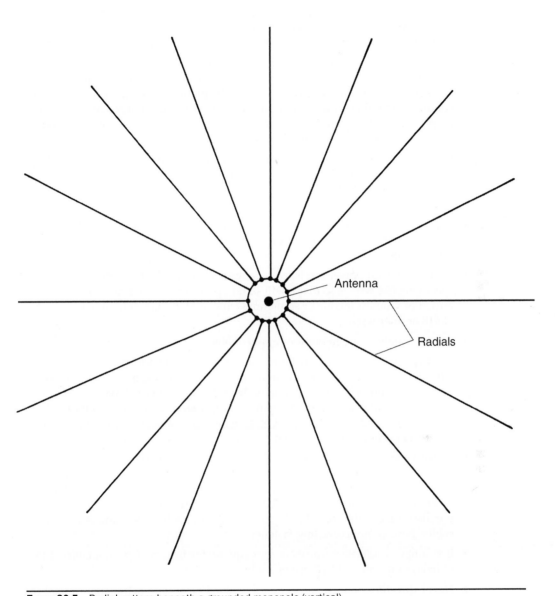

FIGURE 30.5 Radial pattern beneath a grounded monopole (vertical).

matching network you may be using at the base of the vertical. If your radials are going to be part of your lightning surge grounding, use mechanical compression or brazed joints, rather than solder, for the connections.

Here's a Top 10 list (plus one bonus item) of other quick hints about radials:

- For verticals near a property line or a building, make the radials fit the circumstances; if they're short on one side of the antenna and long on the other, no harm done.

- If you need to run a radial slightly "off course" for part of its run to get past a building or other obstruction, do it.

- Don't bother connecting radial tips with circumferential wires; all the return currents from the antenna's fields are *radial* in orientation. Similarly, adding chicken wire mesh under the tower is problematic; in return for lots of added conductor that may or may not be augmenting your original radials, you could be plagued after a couple of winters by rectification from a zillion corroded junctions where the wires of the mesh cross each other.

- Use just about any kind of copper wire you have on hand for your radials. Strip multiconductor cables into their individual wires, if you wish. The primary determinants of what wire to use are cost (if it's free, use it!) and longevity on/ in the ground.

- If you have to purchase any radial wire, get ordinary insulated THHN #14 house wire from a building supply store. The insulation will prolong the life of your radials.

- Even at kilowatt power levels, wire gauge is not critical. The primary reason for specifying #14 wire is so that it will stand up to stresses on it, such as when it is buried or when someone or something trips over it. Many radial fields consist of #18 or finer wire.

- Avoid stranded wire in outdoor applications.

- Ideally, your radials should lie on top of the ground for maximum antenna efficiency, but you can bury them an inch or two below, if necessary, to keep people and animals from tripping on them. Many amateurs have reported that *sod staples,* available at home and garden supply stores, are excellent for holding radials down and out of the reach of lawn mower blades until they sink into the grass and become part of the thatch.

- At least two suppliers of accessories to the amateur market (Array Solutions and DX Engineering) sell radial attachment plates intended to simplify the connection of large numbers of radials to the base of the antenna. Others, including the author, have formed a *grounding ring* from soft copper tubing purchased from the local hardware store and attached the radials to it with solder joints or noncorroding fasteners.

- If you have multiple verticals (a four-square, for instance), do not connect the radials of one vertical to those of another.

- Never directly connect copper to galvanized steel. Use a transition metal in between the two.

Elevated Grounds

Most users of VHF equipment are familiar with the *ground-plane* antenna—a vertical with either three or four radials, mounted atop a mast. The same technique can be used at MF and HF, but the mechanical challenges are not trivial. At VHF, the idea behind the ground-plane antenna is simple: a vertical monopole (whether $\lambda/4$ or $5\lambda/8$ or any length in between) requires a return path for the "missing" half of the charged dipole structure that is necessary for electromagnetic radiation. Starting with a vertical dipole (Fig. 30.6), if we split the wire that is the lower half of the radiating element into four

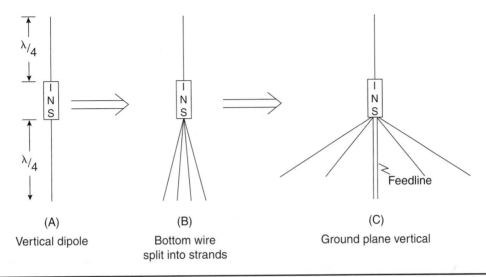

(A)
Vertical dipole

(B)
Bottom wire
split into strands

(C)
Ground plane vertical

Feedline

FIGURE 30.6 Transforming a vertical dipole to a λ/4 ground-plane vertical.

thinner strips, angle them upward by 45 degrees, and space them equally around the azimuth circle, we end up with a *ground-plane vertical.* Thus we have created an *artificial ground plane* high in the air, right where the vertical monopole needs it. A little knowledge of trigonometry will convince the user that *three* radials, spread 120 degrees apart, are sufficient to provide a return current path for all compass directions.

The ground-plane antenna is fairly easy to implement at frequencies as low as 7 MHz. In the 1960s and 1970s a very popular antenna was the Hy-Gain *14-AVQ* (now the *AV-14AVQ*), a trapped vertical for 40, 20, 15, and 10 m. Propped atop a 20- to 30-ft mast, with a handful of radials sloping down from the base of the antenna, the antenna acquitted itself very nicely as a low-angle radiator on 40 m. In some installations the radials did double duty as guy wires for the mast. In such an installation, each combination radial/guy wire must have an insulator in it at a distance of λ/4 *along the radial* from the base feedpoint of the radiating element.

In recent years increasing numbers of hams have installed elevated radials under 80- and 160-m verticals. Unlike VHF ground-plane antennas or even the 14-AVQ, running controlled comparative experiments on 80 m and 160 m is a major undertaking. What has gradually come out of the anecdotes, experiments, and modeling, however, is that a handful of elevated radials (between four and eight) can be as effective as 30 or more radials on the ground—but *only* if the vertical and the elevated radials are well above the earth. Ideally, elevated radials should be at least λ/8 above earth ground, but good results have been reported with lower installed heights. Few amateurs have the resources to put up 160-m verticals whose base is 30 to 70 ft above the earth, so elevated radials are frequently installed as shown in Fig. 30.7.

Compared to earth ground, which is lossy (dissipative), copper radials are essentially lossless. When elevated radials are too close to earth, radial return currents for the vertical are shared between the radials and earth, thus allowing some substantial fraction of the vertical's return currents to be dissipated in the lossy earth. When the vertical and the elevated radials are far enough away from the earth's surface, virtually all

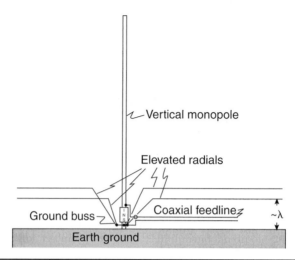

Figure 30.7 Elevated radials for a ground-mounted vertical.

the return currents are collected by the elevated radials, even though there may be only a handful. "Far enough away" turns out to be $\lambda/8$ or more.

Unlike radials on the ground, elevated radials exhibit resonance in the same way that an antenna does. Thus, all radials for a given band should be cut to the same length and should all make approximately the same angle with the ground, insofar as possible. If the vertical (such as the 14-AVQ) is capable of multiband operation, a set of $\lambda/4$ radials should be added for each band to be used. The objective is to make the impedance at the feedpoint end of each active radial as low as possible in order to maximize the return currents it supplies the vertical monopole. One possible simplification might be to add traps so that a single set of four or eight radials serves multiple bands.

CHAPTER 31

Zoning, Restrictive Covenants, and Neighbors

"Beauty is in the eye of the beholder." There is probably no better example of this well-known saying than a tall radio tower topped by a long-boom Yagi antenna. To those of us immersed in the techno-geek world of radio, the sight makes us glow with pride or turn green with envy, depending on whether it is our tower or another's. But to those who are not similarly enamored of the trappings of radio—including, in many cases, our own family members—that same tower and antenna combination may be a blot on the landscape. Never mind that our country is crisscrossed by thousands and thousands of miles of sagging power lines, held up to our view by bent, warped, stained, and rotted utility poles every few hundred feet along the road for as far as the eye can see.

In an ideal world we'd all be able to afford a few hundred acres "in the country", where we could pursue our life's hobby without fear of neighborhood displeasure or municipal sanctions. Even the management of regional cell phone operating companies must feel the same way after being stalled yet another year by an unapologetic town planning board.

In our "civilized" world we all have a mixture of rights and obligations, and a need to balance our own personal interests against those of the larger community in which we live. The difficulty comes when *we* view the value of our towers and antennas differently than our neighbors and local officials do. Often we feel that objections based on personal aesthetics and biases are being cloaked in ill-conceived restrictions in the name of safety.

Challenges to our erecting towers and antennas for the pursuit of our hobbies or our businesses can come from the public sector (governmental bodies, zoning ordinances, and building codes) or the private sector (restrictive covenants and deed restrictions), and it is important not to confuse the two. Let's look at each of those separately.

Public Sector Restraints

Public sector regulations impacting allowable antenna and support structure height may be imposed at any level of government.

Federal

In the United States, federal intervention takes two forms:

- Regulations to minimize hazards to aviation from man-made structures that rise far above the average terrain

- PRB-1, issued by the Federal Communications Commission (FCC) in 1985, preempting certain aspects of local regulation of amateur radio structures

Within the United States the Federal Aviation Agency (FAA) maintains a database of towers and antennas in excess of 200 ft above average terrain and/or close to public airports. It will be necessary for you to notify both the FCC and the FAA *prior* to constructing a tower or other man-made antenna support structure that rises 200 ft or more above the ground around it. Depending on the specifics of your site, your application *may* be acceptable, but it will certainly require FAA-approved lighting in all instances.

Additionally, for potential tower installations in the vicinity of existing airports, the FAA has a published formula that sets a maximum height for towers and any antennas on them as a function of distance to the nearest runway. Actually, there are three individual formulas to be concerned with:

- For airports with one or more runways longer than 3200 ft, the top of your antenna or man-made support cannot exceed 1 ft for every 100 ft between the antenna and the nearest part of any runway at that airport. If you wish to put up a 70-ft tower, for instance, you must be at least 7000 ft (or approximately 1.3 mi) from the nearest part of the runway closest to your tower(s).

- For airports with no runways longer than 3200 ft, the top of your antenna or man-made support cannot exceed 1 ft for every 50 ft between your antenna and the nearest part of any runway at that airport. This effectively doubles your allowable antenna height at a given distance from an airport.

- For heliports, the corresponding limit is 1 ft of antenna height for every 25 ft from your antenna and any portion of the helipad that is used for flight operations.

If the elevation above *mean sea level* (MSL) of the nearby airport is substantially different from that at the base of your proposed tower(s), the calculation may be more complicated than these bullet points indicate. When in doubt, submit an application to the FAA and get a formal ruling for your files.

PRB-1

A quarter-century ago, the FCC issued a document now familiarly called "PRB-1". PRB-1 conveys "limited federal preemption" of local municipal restrictions on amateur towers and antennas. It is potentially useful to you, depending on the specific objections being raised by neighbors or local officials, but it is not a substitute for a clearly worded permissive ordinance and/or a building permit. Occasionally you may encounter a state or locality that has made blanket adjustments in its zoning ordinance in response to PRB-1, but it's primarily a tool for your attorney, should you reach that point in negotiations with your county or town.

State

State-level regulation in the United States is rare. Nonetheless, some states have also preempted certain aspects of local zoning with respect to amateur radio structures. An example is Virginia, where paragraph 15.2-2293.1 of the Code of Virginia establishes minimum heights of 200 ft and 75 ft for populous and rural counties, respectively, below which amateurs may not be unreasonably prevented by local jurisdictions from erecting supports. Note, however, that the VA Code does not give the amateur carte blanche; the local zoning body can still require reasonable safety and screening measures, to name just a couple of factors.

Local

The bulk of zoning restraints on amateur radio is local in scope. Although zoning is defined and regulated at the county level in many states, in other states (especially in the Northeast) it falls under the purview of each individual township or city. While allowing the specific requirements of each ordinance to vary according to local needs, the result is a crazy patchwork of rules that makes one's knowledge of "the way things work" in Podunk of limited usefulness in East Oshkosh.

Nonetheless, some general guidelines and comments are in order:

- There still remain in this country localities with no zoning ordinance at all. For the most part, they are extremely rural and sparsely populated. To install a tower and antennas of virtually any height (under 200 ft, of course), a building permit may or may not be required. This may initially sound like nirvana, but the downside risk is that a restaurant and bar may move in across the street from you next week, complete with neon signs radiating tons of RF energy directly into your specialized receiving antennas.

- Sometimes you will encounter multiple layers of zoning and restrictive regulations collected and administered under the auspices of the local zoning office. State or federal wilderness areas may overlay local zoning districts with other, tighter restrictions. Similarly, a county may invoke a unifying "wrapper" around a collection of unique local ordinances. In many instances, the local ordinance may even list conflicting requirements in different chapters or paragraphs within the same chapter. (A common problem is an overly broad ordinance for cellular and other commercial towers that fails to distinguish between the intended regulation of "for profit" carriers and the unintended inclusion of individual amateurs pursuing their hobby as a permitted accessory use of residential property.)

- Before buying a property upon which you intend to erect towers or antennas, *research* the location. Know with certainty what town or municipality the property is located in, and obtain a *written* copy of the current zoning ordinance and any other relevant documents. Meet with an appropriate zoning official *in person* to confirm that the documents you have are the latest and represent the *totality* of public regulation pertaining to the property you have in mind. If the zoning ordinance is found only in electronic form online, make sure you download, print out, and safely archive the revision that is current as of the exact date you take legal possession of your new property. Depending on the specifics, if the town revises the zoning later on and makes it more restrictive, you may have recourse based on local law with respect to *grandfathering* provisions.

- *Never* depend on oral assurances from *any*one that there is nothing prohibiting you from putting up a tower! Get it all in writing. Do not expect your real estate agent to be a zoning expert, building codes expert, or legal expert—or to do your highly specialized legwork for you. A good agent will do everything he/she can to understand your requirements and help you get in touch with people who *are* experts in the areas you need—but your real estate agent is primarily trained in the mechanics and psychology of helping property owners sell their property. (In fact, in most states, "your" real estate agent is actually a subagent of the *seller* unless you have made specific arrangements for your agent to act as a "buyer's agent" or "dual agent". Note: The author who wrote this section is a licensed real estate broker in one of these 50 United States and has served on his local planning board for the past seven years.)

- If you have located a specific property that seems appropriate for your intended radio uses but the zoning ordinance clearly requires a variance or special use permit for erecting a tower, make the issuance of that permit a *contingency* in your purchase offer and do not let anyone talk you out of it. Once you have a signed purchase contract, immediately set about to *clear* the contingency. This will most likely involve your appearing before the local planning board or zoning board of appeals (exact names of these bodies may vary from state to state) to make a brief presentation in support of your application. Assuming you are successful, before you agree to close on your new property be sure you have in your possession, in writing, all the relevant ordinances, rulings, meeting minutes, and permits to allow you to proceed with your tower(s) and antennas as soon as you become the new owner of the property. Even so, this is not an ideal situation, since you will undoubtedly need to return to the local board for any additional supports you may wish to erect at a later date.

- Finally, it is always better to erect your towers and antennas *sooner* rather than later. Zoning laws do change from time to time, and, depending on how they are worded, construction projects that have not yet been started may well not be protected by any supposed grandfathering provisions. To paraphrase an old adage, "A tower in the yard is worth two in your dreams."

If, when all is said and done, your "dream property" is going to require you to appear in front of a town planning board or zoning board of appeals, or if you already have a tower and unhappy neighbors are complaining about you at monthly town board meetings, you need more help than this chapter can give you in these few pages. By far the dominant reference for zoning matters related to towers and antennas, both amateur and commercial, is authored by Fred Hopengarten, K1VR. Fred is both an active amateur radio operator and a highly skilled lawyer specializing in antenna zoning matters. The cost of either of his books (see the reading list at the end of this chapter) is *de minimus* compared to the costs of a lawsuit or relocation. You can do no better than to follow Fred's recommendations at all stages of the process. In some circumstances, volunteer counsel provided through the good offices of the American Radio Relay League may be available to you and your own attorney—especially if your situation is likely to establish a legal precedent that will be useful in future litigation of antenna support cases.

Private Sector Restraints

In many parts of the United States it is becoming harder and harder to find "affordable" residential properties with enough land to put up a reasonable set of antennas for the high-frequency bands. In many of those same areas, an increasing percentage of homes for sale are located in *planned developments* or other clusters of homes having what are collectively called *restrictive covenants* or *CC&Rs* (codes, covenants, and restrictions). These covenants, by whatever name, differ from the zoning ordinances of the previous section in that they are rules established totally within the *private sector*.

NOTE *If at all possible, avoid purchasing a home or a building lot that is burdened with restrictive covenants!*

If you believe there is a reasonable supply of properties suitable for you in the area you are interested in, you should advise any real estate agents you are working with that under no circumstances are they to waste your time (and theirs) by attempting to show you homes in CC&R developments. Furthermore, proponents of CC&Rs will be quick to remind you that nobody is holding a gun to your head; your purchase of a home is a classic case of an arm's-length contract freely entered into between "a willing buyer and a willing seller".

On rare occasion, the bylaws for a homeowners' association associated with a development may allow for petitioning the association for specific, individual permission to do something generally prohibited. Whether the association directors or the membership as a whole would ever approve your request constitutes a real gamble, and your success will be highly dependent on such specifics as the topography of the development as a whole and your unit specifically, the personalities of the individuals in a position to decide, and your ability as a newcomer to present your case in a collaborative way. In general, the odds of success are extremely low.

If all else fails, and economics or other compromises have led you to a towerless life in a development community, all is not lost. Chapter 15 ("Hidden and Limited-Space Antennas") is written with you in mind!

Final Thoughts

Searching for a home is a complex process. Searching for a home that must also satisfy your antenna requirements is substantially harder. Ask the author—he knows! Here are a few suggestions for those readers who may be in the market for a new location:

- At the very beginning of your search, establish the minimum lot area and dimensions you are willing to tolerate at the new location. Think about what kinds of terrain are acceptable to you. (If your primary interest is VHF, UHF, and/or microwave frequencies, life in a valley is probably not going to be very satisfying.) Recognize also that your interests and requirements may broaden or change with time.

- Do as much of your early searching as possible via the Internet. Insist on obtaining specific street addresses for properties of interest. (Real estate agencies in some areas of the country are stingy with that information until you have registered with them. Further, in many states, vacant land has no numerical street address so it is extremely difficult to pinpoint the exact location of vacant land without the help of the listing agent.)

- Check the topography around each candidate property with the terrain feature (which often is active only for a limited range of mapping scales) or one of the Internet sites specializing in online topographic maps.

- Purchase topographic maps for the area(s) you are considering. Most of these charts explicitly show utility transmission lines; some online topo map sites do not.

- Establish a minimum distance from high-voltage power lines that you are willing to tolerate. (One mile is a reasonably "safe" starting point.)

- When you visit a property, check the electrical noise environment. If you have a mobile rig in your vehicle, great! If not, tune your AM radio to an empty frequency at the top end of the AM broadcast band (which is not far from the 160-m band). Also check into using the (AM) aircraft band of your handheld VHF receiver or transceiver.

- If you have determined where the nearest utility overhead high-voltage transmission lines are, drive toward them, listening to your AM radio as before. Get a sense of how noisy they are when you are right under them. (Keep in mind, however, that high-voltage transmission lines that are "quiet" today may be quite noisy tomorrow when a different piece of heavy machinery somewhere along the line fires up.)

- Use maps, Internet searches, and local residents to determine the location and runway lengths of the nearest airport(s).

- Visit the local zoning office and meet the folks there. Be collaborative, not confrontational. Determine whether the property you are interested in is in a historical or conservation overlay district; if so, your future with antennas may be fraught with disappointment. Confirm that the most up-to-date version of the local zoning ordinance is on the Internet; if not, buy a printed copy and register to receive notice of any future updates. Ask what the procedure is, if any, for obtaining a building permit to put up "residential antenna support structures" (not "towers").

For Additional Reading

Antenna Zoning Book—Professional Edition: Cellular, TV, and Wireless Internet, Fred Hopengarten, www.antennazoning.com.

Antenna Zoning for the Radio Amateur, Fred Hopengarten, K1VR. www.antennazoning.com. (Also available from ARRL, Newington, CT 06111.)

Appendices

Appendices

APPENDIX A

Useful Math

The material in this appendix collects in one place a brief overview of the various math concepts and problem-solving techniques used in this book as aids in discussing and evaluating the performance of a wide variety of antennas, feedlines, and accessories. The overview assumes only that you can perform addition, subtraction, multiplication, and division and have a little familiarity with basic algebra—specifically, methods for finding a single unknown in an equation. If you have taken high school algebra, trigonometry, and geometry courses or their equivalent, this material will be a refresher for you.

A.1 Logarithms

Long before the era of computers and calculators—or even slide rules—*logarithms* were used as a mathematical "shortcut", facilitating the multiplication and division of numbers by simpler addition and subtraction, respectively. In particular, they allow us to convert an extremely wide range of numbers to a compact format that is the foundation of an extremely powerful way to manipulate and display the combined effects of multiple parts of a system.

Our primary interest in logarithms in this book is twofold:

- Logarithms are the basis for *decibels*, a universal means for specifying the *ratio* of two numbers in electronics discussions. We will often discuss antenna and feedline performance in terms of the strength, or amplitude, of a voltage (or a current, a power, a field) *relative to* some other voltage. The "other" voltage may be an industry reference value of some kind or it may be the same voltage measured under a different set of conditions. Decibels make it easier for us to gain insight into the operation and performance of antenna systems without having to deal with extremely large numbers and complex calculations.

- In addition to electronic circuits and systems, many processes in nature respond to input signals with outputs that are logarithmically, rather than linearly, related to their inputs. Our own very subjective sense of *loudness* is one such example; as a result, *psychoacoustics* is a field where the decibel is equally at home as it is in antenna work. Many times the purpose of an antenna is to help us hear a weak signal better; expressing changes in received signal strength in decibels is an excellent match to our internal auditory loudness scale.

Definition   "The *logarithm* of a number x is the power to which we raise a *base number* to obtain x."

What does that mean?

When we talk about the power of 10, or an exponent of 10, we are referring to the number of times we have multiplied the number 10 by itself. Example: "10 raised to the power 2" is the same thing as 10×10, or 10^2 ("10 squared"). We know $10 \times 10 = 100$, so 10^2 must be equal to 100 also. From our definition, therefore, we can say, "The logarithm of 100 using a base number of 10 is 2." In other words, the number "2" is the exponent of the base number 10 that results in the desired number of 100. Similarly, $10 \times 10 \times 10 = 1000 = 10^3$, and the logarithm of 1000 = 3. (It's important to remember that $10^1 = 10$ and $10^0 = 1$.) Thus, the logarithm of 10 is 1 and the logarithm of 1 is 0—as long as we remember we're talking about a base of 10. When speaking, we often abbreviate "logarithm" to "log", so we might say, "Log to the base 10 of 1000 is . . ." or "Log base 10 of 1000 is . . ." as alternatives. In general, if no base number is explicitly stated, it is presumed to be 10.

Here's the part many don't realize: For numbers other than 1, 10, 100, 1000, etc., the relationships are still valid but the resulting exponent (i.e., the logarithm) is not an *integer*. That means we probably can't do the math in our heads, but that's not a problem because centuries ago, long before calculators or slide rules had been invented, mathematicians built *tables of logarithms* for us to use. At the end of this appendix we've included a short table of logarithms for a smattering of numbers having wide-ranging values. Of course, "real" logarithm tables for numbers of 4, 5, 6, 7, etc., digits between 1 and 10 can go on for pages, depending on how fine-grained the numbers need to be, but most of the reasons for having extremely precise log tables have long ago succumbed to the power of the scientific calculator. Since logarithms of numbers get larger as the numbers do (but not in a linearly proportional relationship), we can reasonably expect that the logarithm of 500, say, will be somewhere between the logarithm of 100 (which is 2) and the logarithm of 1000 (which is 3).

The logarithm of a number consists of two parts: a *characteristic*, which is always an integer placed on the left side of the decimal point, and a *mantissa*, which is a decimal number of any length, always less than 1.0 . . ., and placed on the right side of the decimal point. The characteristic is obtained by counting the number of digits in the number to the left of the decimal place and subtracting one (1). The mantissa is obtained from a table of logarithms.

Example A.1.1   Find the logarithm of 250,000.

Solution   There are six digits to the left of the decimal place in the example number, so the *characteristic* in our example is 5. To obtain the *mantissa*, go to the table of logarithms and look down the left-hand column until you find the line for our number's first non-zero digit (which is 2). (In other words, we completely ignore the location of the decimal point in our original number.) Then *interpolate* between the column entries for 2 and 3. The second digit (5) is 5/10, or 50 percent, of the interval between 2 and 3, or 50 percent of the difference between 0.301 and 0.477, so we estimate the mantissa as 0.301 + 0.088 = 0.389. Using the preceding rule, we add the mantissa (0.389) to the characteristic (5) to get 5.389, the logarithm of 250,000. (The exact logarithm is 5.398, but for the type of work we will be using logarithms for, 5.4 is close enough.)

◆

Example A.1.2   Find the value of 100^2 and the value of 100^3 using logarithms.

Solution   A number squared is the same thing as the number multiplied by itself, so $100^2 = 100 \times 100 = 10,000$. A number cubed is the same thing as the number multiplied by itself twice, so $100^3 = 100 \times 100 \times 100 = 1,000,000$. Using the aforementioned rule for creating logarithms to solve these very easy examples, we determine the logarithm of 100 to be 2.0, the logarithm of 10,000 to be 4.0, and the logarithm of 1,000,000 to be 6.0 simply by finding the characteristic for each number. But notice: We could also have found the logarithm of either number by multiplying the logarithm of the original number by the power (or *exponent*) it was raised to. Thus, the logarithm of 100^2 is $2 \times \log_{10} 100 = 2 \times 2.0 = 4.0$, and the logarithm of 100^3 is $3 \times \log_{10} 100 = 3 \times 2.0 = 6.0$.

To convert logarithms back to conventional numbers, use the table of logarithms in reverse. First separate the logarithm of the answer (in the last example, 6.0) to a characteristic (6) and a mantissa (0). From the log table we know that a mantissa of 0 corresponds to the highest-order nonzero digit of the number being a 1. The characteristic of 6 tells us we need to add six zeros to that mantissa to form the answer: 1,000,000.

CAUTION *Logarithms are of no value in adding or subtracting numbers. Adding and subtracting the* logarithms *of numbers is a simpler substitute for multiplying or dividing the numbers themselves. Similarly, multiplication of the logarithm of a number by an exponent is often simpler than calculations directly involving the exponents.*

Admittedly, those were simple examples. But suppose we need to multiply 5,370,000 by 13.7 and then divide the result by 68.3. Today, of course, we can do that quickly with an inexpensive calculator, but up until a half-century ago students and workers alike often used logarithms, especially when the numbers had many more digits than these do.

Common notation for the logarithms described here is "log" or "$\log_{10}$" (spoken and read as "log to the base 10"). Most scientific calculators emboss or print "log" on or above the appropriate key.

CAUTION *Do not confuse these logarithms with another kind, called* natural logarithms, *abbreviated "ln" or "$\log_e$" (read as "log to the base e"). e is a number with special attributes in the eyes of mathematicians. It is known as Euler's number, in honor of the Swiss mathematician, and it has a value of approximately 2.718 in our decimal, or base-10, system. Natural logarithms are of use to us in this book when describing the behavior over time of electronic circuits in response to applied voltages and currents. (See Sec. A.5 for examples.) If we have to, we can always convert between natural logarithms and logs to the base 10 by using the following approximation:*

$$\ln x \approx 2.3 \log_{10} x \qquad \text{(A.1.1)}$$

Occasionally you will run across the word neper. *The neper is to natural logarithms (ln) what the decibel is to base-10 logarithms. Thus, they are related by a fixed ratio:*

$$1\,\text{Np} \approx 8.69\,\text{dB} \quad \text{or} \quad 1\,\text{dB} \approx 0.115\,\text{Np} \tag{A.1.2}$$

At this point you may be thinking, "If we have easy access to calculators these days, why do we care about logarithms at all?" Again, the primary reasons are these:

- Working with logarithms of numbers instead of the numbers themselves allows us to deal with huge spans in the magnitudes of the numbers more easily. The logarithm of 15,000,000,000 is 10.3. Which number would you rather work with?

- Many electronic and auditory phenomena exhibit a logarithmic response to applied signals. In these cases, displaying input/output relationships on graph paper that has a logarithmic scale on one axis (called *semi-log paper*) or on both axes (called *log-log paper*) often enhances our understanding of the physical principles involved.

- When we compare the relative performance of two devices (two different antenna designs, say), or we test a device under two different sets of conditions (one antenna at two different heights above ground, say) we usually express the result as a ratio of two numbers. As we shall see in Section A.2, the use of *decibels* (which are logarithms multiplied by certain standard scale factors) is ideally suited to our purposes.

A.2 Decibels

Throughout this book we shall use *decibels* to help us quantify and compare the performance of antennas and feedlines. The beauty of decibels is that we can multiply and divide what may be very large ratios with simple addition and subtraction of their logarithms, appropriately scaled. This will be especially important whenever we are determining the overall performance of an antenna or an entire RF transmission and reception system consisting of multiple components.

Definition "A *decibel* is the base-10 logarithm of a number (or a *ratio* of two numbers), multiplied by a scale factor." For electronic circuits (including antennas) the formula is:

$$\text{Decibels (dB)} = 10\log\frac{P_2}{P_1} \tag{A.2.1}$$

where P_1 and P_2 are *power* levels at two different points in the circuit or at one specific point under two different sets of test conditions. (If you are not familiar with logarithms, check out Sec. A.1 of this appendix.)

Example A.2.1 An amplifier delivers 1000 W of power (P_2) to a load when driven with an input level of 25 W (P_1). What is the *power gain* of the amplifier, expressed in decibels?

Solution We say the *power gain* of the amplifier is 1000/25, or 40, when expressed as a ratio, and 10 log 40, or 16 dB, when expressed in decibels. (The *characteristic* of the numerical ratio 40, which has two digits to the left of the decimal point, is 2 minus 1, or 1. The *mantissa* for the digit 4 is, from the table at the end of this appendix, 0.602, which we can safely round off to 0.6. The logarithm is then created from the *characteristic* and *mantissa* with a decimal point added between them, so it is 1.6, which, multiplied by 10, becomes 16 dB.)

Returning to our amplifier example, if the impedance of the output circuit is the same as that of the input, Eq. (A.2.1) is identical to

$$\text{Decibels (dB)} = 20\log\frac{V_2}{V_1} \text{ or } 20\log\frac{I_2}{I_1} \qquad\qquad (A.2.2)$$

Equations (A.2.1) and (A.2.2) are equivalent because the power in a circuit is proportional to the *square* of the voltage or current. In our earlier discussion of logarithms we saw that when we square a ratio we *double* its logarithm. Alternatively, from Ohm's law (Eq. [2.1] of Chap. 2) we can see that if we double the voltage in a circuit, we *quadruple* the power in that circuit.

CAUTION *If we are measuring currents or voltages and the impedances at the measuring point(s) are different for the two measurements, the equivalences no longer hold, and the results obtained by using the voltage or current relationships in Eq. (A.2.2) may be erroneous and may lead us to draw the wrong conclusions. When in doubt, it is always safest to use decibels based on* power *ratios.*

Arguably the single most-used term in electronics involves the phrase "3 dB", which corresponds to a *doubling* of power. Once again using the amplifier example: If we double 25 W, we get 50 W, corresponding to a *3-dB gain*. If we double 50 W to 100 W, we add another 3 dB, for a total of 6 dB. Doubling the power three more times brings us to 200, 400, and 800 W, successively, corresponding to overall gains of 9, 12, and 15 dB gain, respectively. If we note that 1000 W (at the output of the amplifier in our example) is a "little bit" more than 800 W, we can very quickly estimate the power gain of the amplifier as a "little bit" more than 15 dB—namely, 16 dB, the same gain we calculated in the example—by working the problem in our head, without needing pencil and paper or a calculator.

Another reason the "3 dB" phrase is frequently seen is that many specifications for electronic circuits refer to a *3-dB bandwidth*. At various chapters in this book we will refer to the *3-dB beamwidth* (usually specified in degrees of a portion of a circle or compass rose) of an antenna as a measure of how narrow, or focused, its main radiation lobe is. In discussion, designers will often use the term *half-power points* interchangeably with *3-dB points*. These are simply different ways of referring to the direction (for an antenna) or the frequency (for an amplifier circuit) at which the output has fallen by 50 percent from its peak value or its design value.

Since decibels are logarithms of ratios, decibels are *dimensionless*; that is, they're not volts or watts or meters or anything else. Sometimes, however, power, signal level, and field strength are stated as being relative to a fixed value of some sort. When this happens, it is standard practice to add a reference identifier to the abbreviation "dB". Here are a few that are often found in conjunction with antenna specifications or testing:

dBm	power level relative to a reference level of 1 milliwatt (1 mW)
dBμv/m	E-field strength relative to a reference level of 1 microvolt/meter (1 μV/m)
dBi	antenna gain in dB relative to the same specification for an *isotropic radiator* (see Chap. 3)
dBd	antenna gain in dB relative to the same specification for a half-wave dipole (dBd is always 2.15 dB less than dBi)

A.3 Sinusoidal Curves

Radio waves of a single frequency are *sine waves*; if we use an *oscilloscope* at some single point along a wire to monitor the effect of a signal of frequency f_c applied to one end of the wire, we might see the voltage there follow the curve of Fig. A.3.1 as time passes. If we measure from one positive peak to the next, we find that this waveform repeats f_c times each second. We define T—the time interval between the same position on two consecutive cycles of the waveform (such as two consecutive positive peaks)—as the *period* of the waveform, and it turns out that

$$T = \frac{1}{f_c} \tag{A.3.1}$$

But why is a sine wave the shape it is? And what physical significance does it have?

If you've ever ridden a Ferris wheel at a carnival, you've created a sine wave yourself—you just didn't realize it at the time! Let's see how you did that.

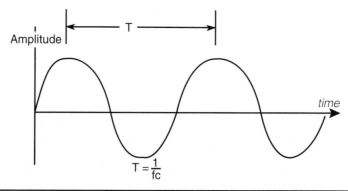

Figure A.3.1   Sine wave.

Suppose the sun is directly overhead at the time when you decide to go for a ride, so that the shadow of the chair you're riding in appears on the ground directly beneath the Ferris wheel. Once your ride has begun, the Ferris wheel rotates at a constant speed, passing the loading gate at the bottom once every 20 seconds, even though each time you go over the top it feels as though you're suddenly going faster.

Your friend, meanwhile, stands on the ground alongside the Ferris wheel, with an extremely accurate clock. Eleven stripes of white paint appear on the ground alongside the carnival ride, dividing the diameter of the wheel into 10 equal segments, as shown in Fig. A.3.2. As you enjoy your ride, your friend marks down the exact time the shadow from your chair passes over each stripe. Of course, every 20 seconds he notes that your chair passes by the loading gate and 10 seconds later it is at the top of the arc, directly above the loading gate.

While you're staggering off the Ferris wheel at the end of your ride, your friend is busy using his scientific calculator or an app on his smartphone to make a graph of the position of your car's shadow as a function of elapsed time. The graph he shows you looks something like Fig. A.3.3, and if you connect the individual data points with a

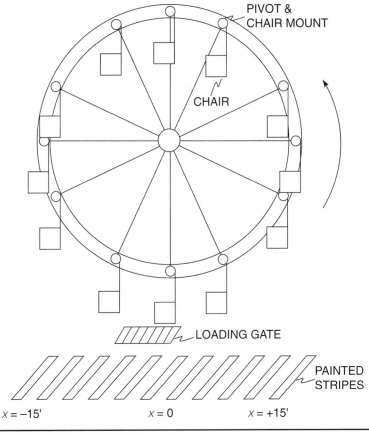

FIGURE A.3.2   Ferris wheel test setup.

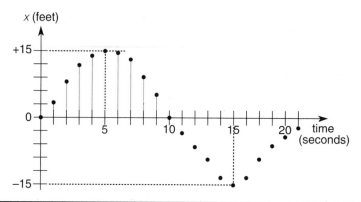

FIGURE A.3.3   Horizontal position of Ferris wheel chair versus elapsed time.

smoothly drawn curve, suddenly you'll have a duplicate of Fig. A.3.1—a sine wave, in other words!

How can this be? You know, from having watched your friends take many rides, that once the ride has started, the Ferris wheel rotates at a constant rate. Yet your shadow has obviously moved along the ground beneath the Ferris wheel at a varying rate. What's going on here?

The answer is this: The shadow on the ground was an exact replica of how the *horizontal* position of your chair varied with time. You, of course, were moving *vertically* as well as horizontally, so what you saw when your eyes followed the car your friends were in was a result of their *total* motion, but what your friend saw when he watched the *shadow* of your car intently was just the horizontal component of your total motion.

Now, it turns out that if you had ridden the Ferris wheel at night and we had shone a very bright light at the Ferris wheel from the side and marked a utility pole or other post on the opposite side of the wheel with the same white stripes as before, with $y = 0$ marked at the midheight of your travel, your friend could collect the same kind of data about your *vertical* motion. If we use the same time intervals and the same exact starting time as before, that curve will look like Figure A.3.4. Note that while it, too, is a sine wave, it "starts" at a different time than the first one. Specifically, when your horizontal position was halfway between the left and right extremes of travel (the first or leftmost data point in Fig. A.3.3), your vertical position was at the very bottom or the very top of its range.

For reasons we don't need to discuss right now, we call the curve of Fig. A.3.1 or Fig. A.3.3 a *sine curve* or a *sine function,* and we call the curve of Fig. A.3.4 a *cosine curve* or a *cosine function.* (This particular cosine function is *negative*—that is, since we assumed the Ferris wheel rotation started when your chair went past the loading gate, the cosine function is starting at $y = -15$.) But *both* are *sine waves* or *sinusoidal functions.* When we travel a complete circle, or one complete rotation, we say we have gone through 360 degrees (often written 360°), just as there are 360 degrees on a compass. Thus, each of our sine waves repeats every 360 degrees, and the sinusoidal wave of your vertical position lags the sinusoidal wave of your horizontal position by a quarter rotation, or 90 degrees. We say these two waves are in *quadrature* with respect to each other.

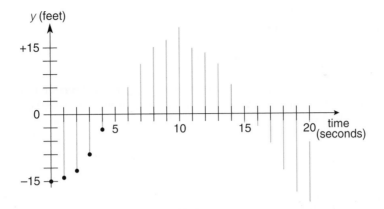

FIGURE A.3.4 Vertical position of Ferris wheel chair versus elapsed time.

NOTE *Sometimes we refer to one complete revolution around a circle or a compass as 2π radians, where a radian is equivalent to 360/2π, or 57.3 degrees. In other words, a radian is the angle formed by laying one radius length along a portion of the circumference of a circle and drawing straight lines from the center of that circle to the two ends of the radius so laid out.*

We also have enough information to determine the frequency, or f_c, of the Ferris wheel rotation. We said earlier it took 20 seconds for your chair to return to the loading gate each time, so we know the lowest part of the vertical (or cosine) curve repeats every 20 seconds. This corresponds to T in Eq. (A.3.1). Thus, $f_c = 1/T = 1/20 = 0.05$ rotations per second. In electronics, we call frequency *cycles per second* (cps), but use the shorthand Hz (or *hertz*) to say the same thing while simultaneously honoring a long-ago radio pioneer. So in engineering lingo, the rotational frequency of your Ferris wheel was 0.05 Hz.

What we have done so far is map the circular motion of your Ferris wheel chair to a sine wave along a horizontal line and to a cosine wave along a vertical line. If we define that horizontal line as the *x axis* and the vertical line as the *y axis*, and if the diameter of the Ferris wheel is 30 ft, we can say

$$x(t) = 15\sin\left(2\pi f_c t\right) \tag{A.3.2}$$

and

$$y(t) = -15\cos\left(2\pi f_c t\right) \tag{A.3.3}$$

NOTE *Because it appears so often, you will frequently see the term 2πf written as ω. Thus, cos (2πf$_c$t) is the same thing as cos (ω$_c$t).*

Velocity is the rate of change of position. Velocity is identical to *speed*, but with the addition of direction information. Equations (A.3.2) and (A.3.3) describe the *position* of a specific chair on the Ferris wheel. While the Ferris wheel is smoothly rotating during your ride, your *angular velocity* is constant; that is, you are always moving through an equal number of degrees of the circle each second. However, both your velocity in the x direction and your velocity in the y direction are constantly changing (and occasionally they have the same magnitude). For example: As you reach the top of the circular path you're traversing, your y (or vertical) speed slows as you approach the top, goes to zero exactly at the top, and then starts to increase in the opposite direction as you head back toward ground level. In other words, your y velocity was positive but is now negative.

Halfway to ground level, when your x position is at its farthest extension in either direction from the loading gate, your y (or vertical) speed is at its greatest, but your velocity along the x axis is now going through zero one-quarter rotation after your y velocity did! So it turns out that while your x position is the sine function of Eq. (A.3.2), your x *velocity* is actually a *cosine* function similar to—but with a different constant or scale factor in front of—Eq. (A.3.3). In calculus terminology, velocity is the *first derivative* of position, so if your position is becoming more negative (i.e., you're heading back down toward the ground), your velocity is a negative number.

Similarly, *acceleration* is the rate of change of velocity. Do you remember the strange feeling you had the first time you went "over the top" on a Ferris wheel—perhaps even a momentary sense of weightlessness? That's because you were reversing your vertical (or y) velocity at that point. Even though your instantaneous y velocity at the top went to zero, you were experiencing maximum acceleration in the (negative) Y direction at that very instant. Through an analysis similar to the one we performed for velocity versus position, a velocity that can be described as a cosine function is the result of an acceleration that is a sine function of the same frequency but with a minus sign. In calculus terminology, acceleration is the *second derivative* of position. In summary:

$$x(t) = k\sin(2\pi ft) \quad \text{and} \quad y(t) = k\cos(2\pi ft) \tag{A.3.4}$$

$$v_X(t) = 2\pi fk\cos(2\pi ft) \quad \text{and} \quad v_Y(t) = -2\pi fk\sin(2\pi ft) \tag{A.3.5}$$

$$a_X(t) = -(2\pi f)^2 k\sin(2\pi ft) \quad \text{and} \quad a_Y(t) = -(2\pi f)^2 k\cos(2\pi ft) \tag{A.3.6}$$

where k = constant proportional to mechanical distance (or electrical amplitude)
f = frequency of driving force (mechanical or electrical)
x and y express instantaneous position along either axis
v_X and v_Y express instantaneous velocity along either axis
a_X and a_Y express instantaneous acceleration along either axis
$x(t)$ means x varies with time t (spoken "x as a function of t" or simply "x of t")

From Eqs. (A.3.4) through (A.3.6) you can see that if an object's position can be described with a sine function of frequency f, its acceleration is also a sine function of frequency f but with the opposite sign (or polarity). This is of particular interest to us in Chap. 3, when we examine the behavior of electrical charges on a very short conductor as they are driven back and forth in sinusoidal motion by an alternating voltage of frequency f instead of the motor and gears of the Ferris wheel.

The other important point to take away from Eq. (A.3.6) is that the higher the frequency f of the sinusoidal driving force, the larger the coefficient of the acceleration terms in front of the sin or cos term. As we discuss in Chap. 3, electromagnetic radiation is caused by *accelerated* charges; it is clear from this equation that the propensity for EM radiation increases with increasing frequency f.

A.4 Triangles

In plane geometry we find three kinds of triangles, as shown in Fig. A.4.1:

- *Equilateral* (where all three sides are of equal length)
- *Isosceles* (where two sides are of equal length)
- *Scalene* (where no two sides are the same length)

All three types share a common trait: The sum of their three interior angles (α, β, and γ) is always 180 degrees.

We can also categorize triangles by their interior angles:

- *Acute* (where no interior angle is 90 degrees or greater)
- *Right* (where one interior angle is exactly 90 degrees)
- *Obtuse* (where one interior angle is greater than 90 degrees)

Right Triangles

In a right triangle, one angle is always 90 degrees, so the sum of the other two angles must also be 90 degrees. The side opposite the right angle is perpendicular to neither of the other two sides, and it is always the longest side of the three; it is called the *hypotenuse*.

If we place a right triangle on an x-y grid so that the two perpendicular sides are on the x and y axes as shown in Fig. A.4.2, the relative lengths of the sides and the three interior angles are related as shown in the figure.

For a right triangle—and *only* for a right triangle—the square of the hypotenuse is equal to the sum of the squares of the other two sides:

$$h^2 = a^2 + b^2 \qquad (A.4.1)$$

(A)
Equilateral

(B)
Isosceles

(C)
Scalene

FIGURE A.4.1 Types of plane triangles.

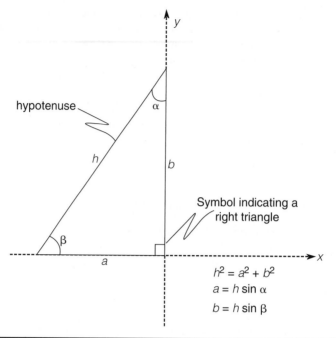

FIGURE A.4.2   Right triangle relationships.

so it follows that

$$h=\sqrt{a^2+b^2} \text{ and } a=\sqrt{h^2-b^2} \text{ and } b=\sqrt{h^2-a^2} \qquad \text{(A.4.2)}$$

Sometimes it is helpful to think of the sides a and b as the *projections* of h onto the x and y axes, respectively. For instance, if the sun is directly overhead, a is the shadow projected by a stick having the length and orientation of h onto the ground directly beneath it.

The sides and interior angles of a right triangle are related by the following equations:

$$\sin\alpha = \frac{a}{h} \quad \text{or} \quad a = h\,\sin\alpha \qquad \text{(A.4.3a)}$$

$$\sin\beta = \frac{b}{h} \quad \text{or} \quad b = h\,\sin\beta \qquad \text{(A.4.3b)}$$

For completeness, we list the following additional relationships in a right triangle:

$$\cos\alpha = \frac{b}{h} \quad \text{or} \quad b = h\,\cos\alpha \qquad \text{(A.4.3c)}$$

$$\cos\beta = \frac{a}{h} \quad \text{or} \quad a = h\,\cos\beta \qquad \text{(A.4.3d)}$$

$$\tan \alpha = {a}/{b} \quad \text{or} \quad a = b \tan \beta \tag{A.4.3e}$$

$$\tan \beta = {b}/{a} \quad \text{or} \quad b = a \tan \beta \tag{A.4.3f}$$

Substituting in Eqs. (A.4.1), (A.4.3a), and (A.4.3.c), we can say:

$$h^2 = a^2 + b^2 = h^2 \sin^2 \alpha + h^2 \cos^2 \alpha \tag{A.4.4}$$

or

$$1 = \sin^2 \alpha + \cos^2 \alpha \tag{A.4.5}$$

Suppose we draw a circle to represent the rotation of a chair on a Ferris wheel. Inside that circle we draw a right triangle whose hypotenuse is equal to the radius of the circle and is connected to the chair at one end, as shown in Fig. A.4.3. As the chair travels the circumference of the circle, the right triangle we draw at each instant is a different shape but we can always draw a right triangle. Because h, the hypotenuse, is equal to the radius of the circle traversed by the chair, it is always constant. But sides a and b are always changing. As an example: When the chair reaches the top of the Ferris wheel, side b of the right triangle goes to 0 and side a is the same length as the hypotenuse h. At that point, $\sin \alpha = 0$ and $\cos \alpha = 1$, agreeing with Eq. (A.4.5).

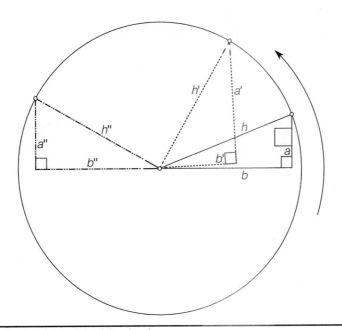

FIGURE A.4.3   Circular motion and the right triangle.

Area of a Triangle

The area inside a triangle is given by

$$A = \frac{1}{2}bh \qquad\qquad\qquad\qquad (A.4.6)$$

where b is the base of a triangle of any shape and h is its height, obtained by creating a right triangle for measurement purposes, as shown in Fig. A.4.4. Start by defining any side of the original triangle as its base, then construct a right triangle inside or outside the original one to reach the highest point of the triangle at the vertex opposite the base. Of course, for a right triangle, the simplest orientation is to select the two sides a and b as the base and height (or vice versa).

In an *equilateral* triangle, there is no way to tell one side from any other (because they are all the same length by definition), and there is no way to distinguish one interior angle from another. Therefore, each interior angle must be exactly 60 degrees. If we draw a line from the junction of any pair of sides to the middle of the third side, we create two right triangles inside the original equilateral triangle, as shown in Fig. A.4.5. If the sides are of length d, for instance, then either right triangle has a hypotenuse of length d and one of the other two sides is $d/2$. From that information, we can find the *height* of an equilateral triangle by using what we know about right triangles:

$$h = \sqrt{d^2 - \left(\frac{d}{2}\right)^2} = \sqrt{\frac{4d^2 - d^2}{4}} = d\sqrt{\frac{3}{4}} = d\sqrt{0.75} = 0.866\,d \qquad (A.4.7)$$

In other words, the height of an equilateral triangle is always 0.866 times the length of any leg. Then the area inside the equilateral triangle is

$$A = \frac{1}{2}bh = \frac{1}{2}d(0.866d) = 0.433d^2 \qquad\qquad (A.4.8)$$

Finally, if we cut a square in half by drawing a straight line between two opposing corners, we create two isosceles triangles. But these are special because one of the inte-

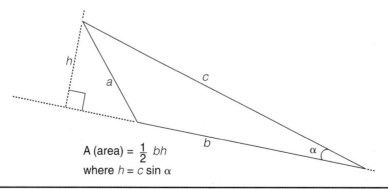

$A \text{ (area)} = \frac{1}{2}\,bh$
where $h = c\sin\alpha$

FIGURE A.4.4   Finding the area inside a triangle.

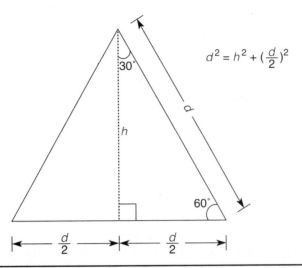

$$d^2 = h^2 + (\tfrac{d}{2})^2$$

FIGURE A.4.5   Height of an equilateral triangle.

rior angles of each is 90 degrees, and we can see (Fig. A.4.6) that, from symmetry, each of the other two interior angles of either isosceles triangle must be 45 degrees. We can find the length of the shared side of the two triangles by recognizing that $a = b$ and proceeding as follows:

$$h^2 = a^2 + b^2 \quad \text{or} \quad h^2 = 2a^2 \qquad (A.4.9)$$

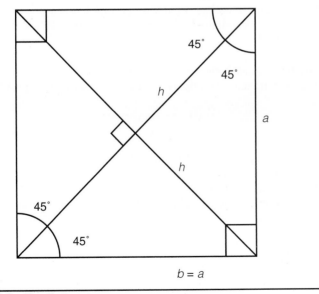

$$h = a\sqrt{2}$$

$$\frac{h}{2} = \frac{a}{\sqrt{2}}$$

FIGURE A.4.6   Bisecting a square.

so

$$h = \sqrt{2a^2} = a\sqrt{2} = 1.414a \qquad \text{(A.4.10)}$$

The ability to calculate lengths and angles in triangles and squares is useful for such diverse antenna tasks as laying out arrays of vertical antennas and obtaining their required phasing line lengths, establishing the locations of guy anchors for a guyed tower, determining the minimum support height needed for a delta loop, quad, or inverted-vee antenna, and estimating the length of strut support cables for a long-boom Yagi, to name but a few.

A.5 Exponentials

As discussed in Chap. 3, if a capacitor has a net surplus of positive charges on one of its plates and a net surplus of negative charges on its other plate, a voltage will exist between the two plates indefinitely. If a resistor is connected across the two plates, however, current will flow in the resistor as long as there is a voltage across it. After an infinite interval of time, the voltage across the capacitor will be zero and there will be equal numbers of positive and negative charges on both plates. As current flows in the resistor, it does so according to Ohm's law:

$$V = IR \qquad \text{(A.5.1)}$$

Because the initial flow of current results in a smaller surplus of charges on the plates of the capacitor for the next short interval, the voltage across the capacitor is slightly less at the end of the first interval. In fact, at the end of any short interval, the voltage across (and, hence, the current through) the resistor is slightly less than at the end of the immediately preceding interval, since R is constant throughout the discharge process. As a result, the current or voltage in any given interval is a percentage (less than 100 percent) of the current or voltage in the preceding interval. When this occurs, the resulting current or voltage waveform is known as an *exponential waveform*, and it has the general shape of the voltage shown in Fig. A.5.1A.

The speed of the discharge (and, hence, the scale of the x axis, which represents time (t) is determined by the capacitance C and the resistance R according to Eq. (A.5.2):

$$V_C(t) = V_0 e^{-\alpha t} \qquad \text{(A.5.2)}$$

where $V_C(t)$ = voltage across capacitor as a function of time
V_0 = initial voltage across capacitor at time $t = 0$
e = Euler's number, 2.7618
α = constant determined by circuit configuration and component values
t = time, in seconds

For this simple circuit,

$$\alpha = \left(\frac{1}{RC}\right) \qquad \text{(A.5.3)}$$

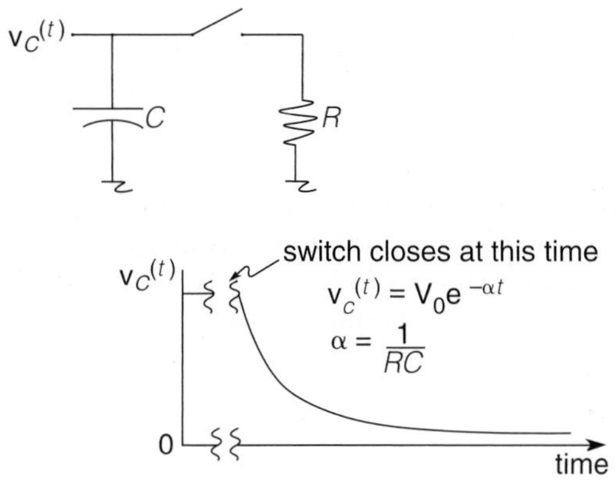

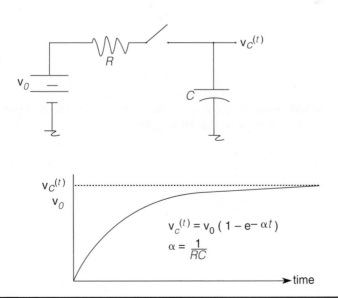

FIGURE A.5.1A   Exponential discharge of a capacitor.

If, alternatively, we had started with no initial voltage on the capacitor and subsequently connected it through a resistor R to a battery of voltage V_0, the voltage $V_C(t)$ across the capacitor would be another exponential; this time, however, it would be

$$V_C(t) = V_0(1 - e^{-\alpha t}) \qquad (A.5.4)$$

This response is shown in Fig. A.5.1B.

FIGURE A.5.1B   Exponential charging of a capacitor.

A.6 Imaginary Numbers and the Complex Plane

Suppose, in the process of solving a math problem, you had to find the square root of −1. What would you do? (Dropping math and enrolling in another field of study is not an option!)

For most of us (and for purposes of this book), the only useful response is to say, "I live in the real world, and I have no idea what the square root of −1 is or even what it means, but I'm going to give it a shorthand notation so that I can easily work with it whenever it comes up. In fact, I'm going to call it j."

So j it is. That is,

$$j = \sqrt{-1} \tag{A.6.1}$$

In some older texts, i is used instead. (That sentence is one of the few known instances of when it is grammatically correct to say "I is . . .")

For clarity, let's reverse the process and obtain the *square* of j:

$$j \times j = j^2 = \sqrt{-1} \times \sqrt{-1} = -1 \tag{A.6.2}$$

To simplify the design and analysis of electronic circuits, scientists and engineers invoke a graphical mathematical tool called the *complex plane*. Because it is a plane, every point on it can be defined in relation to its two orthogonal axes. In particular, we call the x axis the axis of *real numbers* and the y axis the axis of *imaginary numbers*. In common usage, we speak of the *real* axis and the *imaginary* axis.

Thus, a number such as 3 or −7 or 6.14 or −5/8 lies at an appropriate spot on the x or *real* axis. A number such as $+j$ or $−2\pi j$ or $j\lambda/2$ lies at a similarly appropriate point on the y or *imaginary* axis. But what about numbers that lie somewhere in the rest of the space, or plane, between the real and imaginary axes? Just like ordinary graphing techniques, we can represent them as having both an x value and a y value by writing them as (3,2) or (−π,6) or any other appropriate combination, where it's understood that the second number is the imaginary component of the *complex number* being described.

Another way we can format the presentation of complex numbers is as $A + jB$. Thus, (3,2) on the complex plane can be written as $3 + 2j$, and (−π,6) can be written as $−\pi + 6j$. In general, this is the notation used throughout this book.

When we add or subtract complex numbers, we add the real parts together and separately we add the imaginary parts together.

Example A.6.1 What is the sum of $(3 + 2j)$ and $(−\pi + 6j)$?

Solution

$$(3+2j)+(-\pi+6j)=(3-\pi)+j(2+6)=(3-\pi)+8j \tag{A.6.3}$$

◆

When discussing antennas and other electronic circuits, we can't get very far without dealing with inductances and capacitances, introduced in Chap. 3. There we observe that the current and voltage through either of these families of components are in *quadrature*—that is, they are 90 degrees out of phase with each other. We say that capacitors and inductors are *reactive* elements, and it turns out to be very helpful to represent the reactive component of a signal or a circuit parameter such as impedance as imaginary, and to plot it on the y, or imaginary, axis. The *real* part of a signal or circuit parameter is plotted along the x axis.

Throughout this book we will find many antenna and transmission line impedances expressed in the form

$$Z_{IN} = R + jX \tag{A.6.4}$$

Sometimes, however, we need the input admittance, Y_{IN}, corresponding to that impedance. Specifically,

$$Y_{IN} = \frac{1}{Z_{IN}} = \frac{1}{R + jX} \tag{A.6.5}$$

This often turns out to be very inconvenient to work with because we can no longer easily see how to break the complex admittance into its real part (*conductance*) and a separate imaginary part (*susceptance*). However, a little math sleight of hand will help us. If we multiply an equation by 1, we don't change the equation. Furthermore, 1 can be *any* messy number form divided by itself; for instance:

$$1 = \frac{A - jB}{A - jB}$$

So let's multiply both the numerator and denominator of Eq. (A.6.5) by the term $(R - jX)$:

$$\frac{1}{R + jX} \times 1 = \frac{1}{R + jX} \times \frac{R - jX}{R - jX}$$

$$= \frac{R - jX}{(R + jX)(R - jX)}$$

$$= \frac{R - jX}{(R^2 + jXR - jXR - j^2 X^2)}$$

$$= \frac{R - jX}{R^2 + X^2}$$

In other words, for an impedance of the form $Z_{IN} = R + jX$,

$$Y_{IN} = \frac{R - jX}{R^2 + X^2}$$

$$= \frac{R}{R^2 + X^2} \text{ (real part)} - j\frac{X}{R^2 + X^2} \text{ (imaginary part)} \tag{A.6.6}$$

$$= G \text{ (conductance)} + jB \text{ (susceptance)}$$

$(R - jX)$ is called the *complex conjugate* of $(R + jX)$. Multiplication of both the numerator and the denominator of a complex number such as that in Eq. (A.6.5) by the complex conjugate of the denominator (wherein all "j" terms are replaced with "$-j$" terms and vice versa) always results in a pure real denominator, allowing separation of the original complex number into separate real and imaginary components.

Sample Base-10 Logarithms

N	Log
1	0.000
2	0.301
3	0.477
4	0.602
5	0.699
6	0.778
7	0.845
8	0.903
9	0.954
10	1.0
20	1.3
100	2.0
1000	3.0
5000	3.699
10,000	4.000
37,500	4.574
600,000	5.778
1,000,000	6.0

Suppliers and Other Resources

The lists of suppliers and Internet resources that follow are neither comprehensive nor complete. The inclusion of commercial enterprises in these lists does not constitute endorsement of them or their products and services by either the publisher or the author. Every effort has been made to include the most recent contact information publicly available from these entities.

B.1 Manufacturers and Suppliers

B.1.1 Antennas and Repair Parts

Some antenna manufacturers sell antennas "factory direct" only, some sell only through retailers, and others sell both ways. In general, however, antenna repair or replacement parts are obtainable only from the manufacturers themselves.

Alpine Antenna (mobile "screwdriver" HF antennas)
P.O. Box 2062
Alpine, CA 91903
(619) 445-9157
www.alpineantenna.com
E-mail: KA6ELK@cox.net

Arrow Antenna (amateur VHF/UHF Yagis and verticals)
911 East Fox Farm Road
Cheyenne, WY 82007
(307) 638-2369
www.arrowantennas.com
E-mail: info@arrowantennas.com

Barker & Williamson Antennas (wire antennas)
603 Cidco Road
Cocoa, FL 32926
(321) 639-1510
www.bwantennas.com
E-mail: customerservice@bwantennas.com

Bencher Butternut (HF vertical and Yagi antennas)
241 Depot Street
Antioch, IL 60002
(847) 838-3195
www.bencher.com

Bilal Company (*Isotron* antenna)
137 Manchester Drive
Florissant, CO 80816
(719) 687-0650
www.isotronantennas.com
E-mail: WD0EJA@isotronantennas.com

Cal-Av Labs, Inc. (HF beams and rotatable dipoles, baluns, mobile mount springs)
2440 North Coyote Drive, Suite 116
Tucson, AZ 85745
(520) 624-1300
www.cal-av.com
E-mail: info@cal-av.com

Cubex (cubical quad antennas)
P.O. Box 352
Alto, MI 49302
(616) 622-4968
www.cubex.com

Cushcraft (HF/VHF/UHF vertical and Yagi antennas)
308 Industrial Park Road
Starkville, MS 39759
(662) 323-9538
www.cushcraftamateur.com

Diamond Antenna (mobile antennas, wattmeters)
435 South Pacific Street
San Marcos, CA 92078
(770) 614-7443
www.diamondantenna.net
E-mail: diamondantenna@rfparts.com

Directive Systems (UHF and microwave antennas and accessories)
177 Dixon Road
Lebanon, ME 04027
(207) 658-7758
www.directivesystems.com
E-mail: info@directivesystems.com

Dunestar Systems (Portable antennas)
P.O. Box 37
St. Helens, OR 97051
(503) 397-2918
www.dunestar.com
E-mail: dunestar@QTH.com

DWM Communications (wire antennas, small and portable antennas)
P.O. Box 87
Hanover, MI 49241
(517) 315-9892
www.hamradiofun.com
E-mail: tinytenna@hotmail.com

Force 12 (HF/VHF vertical and Yagi antennas)
c/o Texas Antennas
90 Barnett Shale Drive
Bridgeport, TX 76426
(940) 683-8371
www.texasantennas.com
E-mail: force12.sales@mac.com

GAP Antenna Products, Inc. (HF multiband verticals)
99 North Willow Street
Fellsmere, FL 32948
(772) 571-9922
www.gapantenna.com
E-mail: contact@gapantenna.com

Granite State Antenna (HF multiband wire antennas, baluns, rope)
P.O. Box 354
Northwood, NH 03261
(603) 942-7900
www.K1JEK.com
E-mail: K1JEK@K1JEK.com

Ham Radio Deals, LLC (Alpha Delta trap dipoles, arrestors, switches,
    fiberglass masts, rope)
Medina, OH
(330) 730-2598
www.hamradiodeals.com
E-mail: HamRadioDeals@zoominternet.net

High Sierra Antennas (HF/VHF/UHF mobile and fixed antennas, coaxial adapters,
    tools, rope)
P.O. Box 2389
Nevada City, CA 95959
www.hamcq.com

Hi-Q-Antennas (remotely tuned mobile antennas for 160-6 m, accessories)
21085 Cielo Vista Way
Wildomar, CA 92595
(951) 674-4862
www.hiqantennas.com
E-mail: sales@hiqantennas.com

Hy-Gain (HF/VHF/UHF vertical and Yagi antennas)
308 Industrial Park Road
Starkville, MS 39759
(662) 323-9538
www.hy-gain.com
E-mail: hygain@mfjenterprises.com

InnovAntennas Limited (HF/VHF OWA and loop-fed Yagi antennas,
    baluns, accessories)
Antenna House
20 Sanders Road
Canvey Island
Essex, UK SS8 9NY
+44 (0) 800 0124 205
www.innovantennas.com
E-mail: sales@innovantennas.com

KMA Antennas (HF/VHF/UHF log-periodic beams, custom logs and Yagis)
P.O. Box 451
New London, NC 28127
(704) 463-5820
www.kmaantennas.com
E-mail: W4KMA@qsl.net

Light Beam Antenna and Apparatus LLC (HF rotary arrays and accessories)
3613 East Ridge Run
Canandaigua, NY 14424
(585) 698-6310
http://lightbeamantenna.com
E-mail: wfreiert@LightBeamAntenna.com

Loops-N-More (VHF/UHF amateur band aluminum loop antennas)
P.O. Box 586
Augusta, KS 67010
(316) 243-1831
http://loopsnmore.com

Main Trading Company (*Texas Bug-Catcher* mobile HF whips and accessories)
139 Bonham Street
Paris, TX 75460
(903) 737-0773
www.maintradingcompany.com
E-mail: sales@maintradingcompany.com

Mosley Electronics, Inc. (HF/VHF/UHF verticals, dipoles, and Yagi antennas)
1325 Style Master Drive
Union, MO 63084
(636) 583-8595
www.mosley-electronics.com
E-mail: mosley@mosley-electronics.com

New-Tronics Antenna Corporation (*Hustler* mobile antennas)
225 Van Story Street
Mineral Wells, TX 76067
(940) 325-1386
www.new-tronics.com
E-mail: sales@new-tronics.com

NGC Company (*Comet* and *Maldol* HF/VHF/UHF mobile whips)
15036 Sierra Bonita Lane
Chino, CA 91710
(909) 393-6133
www.natcommgroup.com
E-mail: sales@natcommgroup.com

OptiBeam (HF Yagi antennas)
(0049) 07231/45 31 53
www.optibeam.de
info@optibeam.de

Radioware (antenna wire, coaxial cable, connectors, books)
P.O. Box 209
Rindge, NH 03461
(603) 899-6957
www.radio-ware.com
E-mail: radware@radio-ware.com

RadioWavz (custom-built wire antennas)
607 Blue Cove Terace
Lake Saint Louis, MO 63367
(636) 265-0448
www.radiowavz.com
E-mail: sales@radiowavz.com

Radio Works (wire antennas, connectors, baluns, rope)
Box 6159
Portsmouth, VA 23703
(800) 280-8327
www.radioworks.com

SteppIR (motorized vertical and Yagi antennas)
2112 - 116th Avenue NE, Suite 2-5
Bellevue, WA 98004
(425) 453-1910
www.steppir.com
E-mail: support@Steppir.com

SuperBertha.com, LLC (80-6 m long-boom OWA Yagi antennas,
    160-80m nine-circle arrays)
7733 West Ridge Road
Fairview, PA 16415
(814) 881-9258
www.SuperBertha.com
E-mail: sales@SuperBertha.com

S9 Antennas (HF verticals)
c/o LDG Electronics
1445 Parran Road
St. Leonard, MD 20685
(410) 586-2177
LDGelectronics.com

Telewave, Inc. (VHF/UHF antennas)
660 Giguere Court
San Jose, CA 95133
(800) 331-3396
www.telewave.com
E-mail: sales@telewave.com

Tennadyne (log periodic and antennas)
P.O. Box 352
Alto, MI 49302
(616) 622-4968
www.tennadyne.com
E-mail: tennadyne@tennadyne.com

TGM Communications (VHF/UHF/mini-HF Yagi antennas)
121 Devon Street
Stratford, ON, Canada N5A 2Z8
(519) 271-5928
www.tgmcom.com

The Ventenna Company, LLC (portable HF antennas)
P.O. Box 2995
Citrus Heights, CA 95611
(888) 624-7069
www.ventenna.com
E-mail: info@ventenna.com

Ultimax Antennas Inc. (HF/VHF wire antenna kits)
15085 SW 35th Circle
Ocala, FL 34473
(352) 693-3932
www.ultimax-antennas.com
E-mail: ultimaxantennas@hotmail.com

WB4BCR.com (MF/HF receiving four-square controllers and phase switch units)
1001 Angell Road
Mocksville, NC 27028
www.WB4BCR.com

B.1.2 Analyzers, SWR/Wattmeters, Tuners, and Accessories

ABR Industries (coaxial cable)
8561 Rayson Road, Suite A
Houston, TX 77080
(713) 492-2722
www.abrind.com
E-mail: info@abrind.com

AEA Technology, Inc. (HF/VHF/UHF analyzers, TDRs, SWR meters)
5933 Sea Lion Place, Suite 112
Carlsbad, CA 92010
(760) 931-8979
www.aeatechnology.com
E-mail: sales@aeatechnology.com

AIM (Analyzers, VNAs)
www.W5BIG.com
E-mail: Bob@W5BIG.com

Alpha Delta Communications, Inc. (surge protectors)
(606) 598-2029
www.alphadeltacom.com

Ameritron (remote coaxial switches, tuners, wattmeters, mobile screwdriver antennas)
116 Willow Road
Starkville, MS 39759
(662) 323-8211
www.ameritron.com

Autek Research (analyzers, wattmeters, SWR bridges)
P.O. Box 7556
Wesley Chapel, FL 33545
(813) 994-2199
www.autekresearch.com
E-mail: mail@autekresearch.com

Balun Designs LLC (baluns and ununs)
10500 Belvedere
Denton, TX 76207
(817) 832-7197
www.balundesigns.com
E-mail: info@balundesigns.com

Cable X-perts, Inc. (custom and stock coaxial cable assemblies)
540 Zenith Drive
Glenview, IL 60025
(800) 828-3340
www.CableXperts.com
E-mail: info@cablexperts.com

Champion Radio (towers and accessories, climbing equipment, Loos gauges, books)
16541 Redmond Way #281-C
Redmond, WA 98052
(888) 833-3104
www.championradio.com
E-mail: sales@championradio.com

Coaxial Dynamics (HF/microwave wattmeters)
6800 Lake Abram Drive
Middleburg Heights, OH 44130
(440) 243-1100
www.coaxial.com
E-mail: sales@coaxial.com

Elecraft, Inc. (low-power antenna tuner for 160-6 m)
P.O. Box 69
Aptos, CA 95001-0069
www.elecraft.com
E-mail: info@elecraft.com

I.C.E.—Industrial Communication Engineers, Ltd. (lightning suppressors, grounding materials)
www.iceradioproducts.com
E-mail: support@iceradioproducts.com
(Note: I.C.E. is now Morgan Manufacturing, www.morganmfg.us)

LDG Electronics (tuners, wattmeters, SWR meters)
1445 Parran Road
St. Leonard, MD 20685
(410) 586-2177
www.LDGelectronics.com

MFJ Enterprises, Inc. (analyzers, tuners, twin-lead, copper wire,
    window feedthrough panels)
P.O. Box 494
Mississippi State, MS 39762
(662) 323-5869
www.mfjenterprises.com

New Ham Store (coaxial cable tools, connectors, lightning arrestors)
P.O. Box 2389
Nevada City, CA 95959
www.NewHamStore.com

NM3E.com, LLC (wattmeters, coaxial connectors and adapters, dummy loads)
5 Fieldcrest Court
Fleetwood, PA 19522-2009
(610) 207-4865
www.NM3E.com
E-mail: NM3E@verizon.net

PalStar Inc. (antenna analyzer, tuners for 160-6 m, wattmeters, dummy loads)
9676 North Looney Road
Piqua, OH 45356
(937) 773-6255
www.palstar.com
E-mail: sales@palstar.com

RadioCraft LLC (SWR/wattmeter, RF couplers)
P.O. Box 395
Silvana, WA 98287-0395
www.radiocraft-eng.com
E-mail: radiocraft@radiocraft-eng.com

Rig Expert Canada (analyzers for 100 kHz to 500 MHz)
66 Cavell Avenue
Toronto, ON, Canada M8V 1P2
(416) 988-3572
www.rigexpert.net
E-mail: info@rigexpert.net

Ten-Tec (analyzers and HF tuners)
1185 Dolly Parton Parkway
Sevierville, TN 37862
(865) 453-7172
http://tentec.com
E-mail: sales@tentec.com

The Wireman, Inc. (wire, cable, and connectors, baluns, rope)
c/o Clear Signal Products, Inc.
9600 Stevens Avenue
Mustang, OK 73064
(405) 376-9473
www.coaxman.com
E-mail: wire@coaxman.com

Timewave Technology Inc. (analyzers)
27 Empire Drive, Suite 110
St. Paul, MN 55103
(651) 489-5080
www.timewave.com
E-mail: sales@timewave.com

Top Ten Devices, Inc. (A/B and six-way antenna relay switches,
    band reject coaxial stubs)
143 Camp Council Road
Phoenixville, PA 19460
(610) 935-2684
www.QTH.com/topten
E-mail: N3RD@arrl.net

Vectronics (tuners)
300 Industrial Park Road
Starkville, MS 39759
(662) 323-5800
www.vectronics.com

B.1.3 Books

ARRL (antenna handbook, anthologies, tower installation guide, specialties)
225 Main Street
Newington, CT 06111-1494
(888) 277-5289 (U.S.)
www.arrl.org/arrl-store

Champion Radio Products (tower installation techniques and antenna comparisons)
16541 Redmond Way #281-C
Redmond, WA 98052
(888) 833-3104
www.championradio.com
E-mail: sales@championradio.com

CQ Communications, Inc. (antennas, anthologies, specialties)
25 Newbridge Road
Hicksville, NY 11801
516-681-2922
http://store.cq-amateur-radio.com
E-mail: circulation@cq-amateur-radio.com

Radio Book Store (antennas, tower climbing, baluns, lightning protection)
P.O. Box 209
Rindge, NH 03461-0209
(603) 899-6957
www.radiobooks.com
E-mail: radware@radio-ware.com

B.1.4 Components

73CNC.com (*Ladder Snap* open-wire line spacers, center insulator and support, kits)
P.O. Box 249
Caledonia, OH 43314
www.73cnc.com
E-mail: sales@73CNC.com

Allied Electronics, Inc. (wire and cable, connectors, tools, electronic components, enclosures)
7151 Jack Newell Boulevard, South
Fort Worth, TX 76118
(866) 433-5722
www.alliedelec.com

Jameco Electronics (wire and cable, connectors, tools, electronic components, enclosures)
1355 Shoreway Road
Belmont, CA 94002
(800) 831-4242
www.Jameco.com
E-mail: sales@Jameco.com

Mouser Electronics (components, tools, ferrites, wire and cable)
1000 North Main Street
Mansfield, TX 76063
(817) 346-6873
www.mouser.com

Newark (wire and cable, connectors, tools, electronic components, enclosures)
4801 North Ravenswood
Chicago, IL 60640-4496
(773) 784-5100
www.newark.com

Palomar Engineers (toroids)
Box 462222
Escondido, CA 92046
(760) 747-3343
www.Palomar-Engineers.com
E-mail: info@Palomar-Engineers.com

B.1.5 Multiline Dealers

Aero Smith (*Screwdriver* mobile antennas, rotators, feedthroughs, arrestors, analyzers)
325 Netherton Lane
Crossville, TN 38555
http://aero-smith.net
E-mail: info@aero-smith.net

Amateur Electronic Supply (broad line catalog and inventory; 4 locations)
5710 West Good Hope Road
Milwaukee, WI 53223
(414) 358-0333
www.aesham.com
E-mail: milwaukee@aesham.com

Array Solutions (antennas, analyzers, coaxial switches, arrestors, phasing systems)
2611 North Beltline Road, Suite 109
Sunnyvale, TX 75182
(214) 954-7140
www.arraysolutions.com
E-mail: info@arraysolutions.com

DX Engineering (aluminum tubing, coaxial cable, tools, rotators, baluns, antennas)
P.O. Box 1491
Akron, OH 44309-1491
(330) 572-3200
www.DXEngineering.com
E-mail: dxengineering@dxengineering.com

Ham Radio Outlet (broad line catalog and inventory; 12 locations)
933 North Euclid Street
Anaheim, CA 92801
(714) 533-7373
www.hamradio.com
E-mail: Anaheim@hamradio.com

B.1.6 Rotators and Controllers

Alfa Radio Ltd. (*AlfaSpid* rotators)
11211 - 154 Street
Edmonton, Alberta, Canada T5M 1X8
(780) 466-5779
http://alfaradio.ca

Green Heron Engineering LLC (computer-controlled rotator controllers)
1107 Salt Road
Webster, NY 14580
(585) 217-9093
www.greenheronengineering.com
E-mail: info@greenheronengineering.com

Hy-Gain (latest versions of the venerable *Ham-M* and *TailTwister*
    Cornell-Dubiler rotators)
308 Industrial Park Road
Starkville, MS 39759
(800) 973-6572
www.hy-gain.com

Idiom Press (rotator controller enhancements)
P.O. Box 1015
Merlin, OR 97532-1015
(541) 956-1297
www.idiompress.com
E-mail: idiom@idiompress.com

K7NV Prop Pitch Service (prop pitch rotators and controllers, mast clamps,
    YagiStress software)
Minden, NV
www.K7NV.com
E-mail: K7NV@contesting.com

M2 Antenna Systems, Inc. (*Orion* rotators)
4402 North Selland Avenue
Fresno, CA 93722
(559) 432-8873
www.m2inc.com
E-mail: sales@m2inc.com

PRO.SIS.TEL (*Big Boy* rotators and controllers)
C. da Conchia 298
I-70043 Monopoli (Bari) Italy
++ 39 080 8876607
www.prosistel.net
E-mail: prosistel@prosistel.it

B.1.7 Software

Antenna Design Associates, Inc. (PC-aided antenna design program)
55 Teawaddle Hill Road
Leverett, MA 01054
(413) 548-9919
www.antennadesignassociates.com
E-mail: info@antennadesignassociates.com

Kok Chen, W7AY (*cocoaNEC* antenna modeling program for OS X Macs)
Portland, OR
www.W7AY.net
E-mail: W7AY@arrl.net

Nittany Scientific (*NEC-WIN* family of antenna modeling programs for Windows)
1733 West 12600 South, Suite 420
Riverton, UT 84065
(801) 446-1426
www.nittany-scientific.com
E-mail: sales@nittany-scientific.com

Roy Lewallen, W7EL (*EZNEC* family of antenna modeling programs for Windows)
P.O. Box 6658
Beaverton, OR 97007
(503) 646-2885
www.eznec.com
E-mail: W7EL@eznec.com

NEC-4 (computing engine license for U.S. only)
Industrial Partnerships Office
Lawrence Livermore National Laboratory
P.O. Box 808
L-795 Livermore, CA 94551
(925) 422-6416
https://ipo.LLNL.gov/
E-mail: softwarelicensing@lists.LLNL.gov

NEC4Win/VM (antenna modeling program for Windows)
Orion Microsystems
www.orionmicro.com
E-mail: orders@orionmicro.com

4NEC2 (NEC-2 and NEC-4 antenna modeling tools for Windows)
http://home.ict.nl/~arivoors/
E-mail: 4nec2@gmx.net

B.1.8 Towers and Accessories

AN Wireless Tower Company (self-supporting towers, sections, accessories)
1551 Sheep Ridge Road
Somerset, PA 15501
(814) 445-8210
www.anwireless.com
E-mail: info@anwireless.com

Champion Radio Products (towers and accessories, safety and rigging gear,
    Loos gauges)
16541 Redmond Way #281-C
Redmond, WA 98052
(888) 833-3104
www.championradio.com
E-mail: sales@championradio.com

Graybar (screw anchors, aluminum oxide, wire and cable, weathertight fittings and
    enclosures)
34 North Meramec Avenue
St. Louis, MO 63105
(800) 472-9227
www.graybar.com

Hubbell Power Systems (screw anchors)
210 North Allen Street
Centralia, MO 65240
(573) 682-8414
www.abchance.com
E-mail: civilconstruction@hubbell.com

KØXG Systems (rotating towers, guy rings, brackets, rotator shelves, Beverage stands)
1117 Highland Park Drive
Bettendorf, IA 52722
(563) 340-5111
www.KØXG.com
E-mail: sales@KØXG.com

Phillystran, Inc. (nonconducting Kevlar guying material, cable grips, ropes)
151 Commerce Drive
Montgomeryville, PA 18936-9628
(215) 368-6611
www.phillystran.com
E-mail: information@phillystran.com

Rohn Products, LLC (guyed and self-supporting towers, guy wire, accessories, tools, consulting services)
P.O. Box 5999
Peoria, IL 61601-5999
(309) 566-3000
www.rohnnet.com
E-mail: sales@ustower.com

SuperBertha.com, LLC (guyed and self-supporting heavy-duty rotating poles, ring rotators)
7733 West Ridge Road
Fairview, PA 16415
(814) 881-9258
www.SuperBertha.com
E-mail: sales@SuperBertha.com

Texas Towers (towers, masts, tubing, coaxial cable, connectors, safety gear, arrestors)
1108 Summit Avenue, Suite #4
Plano, TX 75074
(800) 272-3467
www.texastowers.com
E-mail: sales@texastowers.com

The Mast Company (stackable aluminum and fiberglass mast sections, HF vertical sections)
P.O. Box 1932
Raleigh, NC 27602-1932
www.tmastco.com

Trylon TSF Inc. (guyed and self-supporting lattice towers and accessories)
21 South Field Drive
Elmira, ON, Canada N3B 0A6
(519) 669-5421
www.trylon.com

Universal Towers (aluminum freestanding towers, masts, and accessories)
43900 Groesbeck Highway
Clinton Township, MI 48036
(586) 463-2560
www.universaltowers.com
E-mail: univ@voyager.net

US Tower Corporation (freestanding crank-up towers, mobile and concealed towers
     and masts)
1099 West Ropes Avenue
Woodlake, CA 93286
(559) 564-6000
www.ustower.com
E-mail: sales@ustower.com

XX Towers, Inc. (towers, tower installation services, antenna replacement and repair)
814 Hurricane Hill Road
Mason, NH 03048
(603) 878-1102
www.xxtowers.com
E-mail: info@xxtowers.com

B.2 Web Resources

B.2.1 Used Antennas, Accessories, and Equipment

http://swap.QTH.com

www.eham.net/classifieds

www.QRZ.com

www.secondhandradio.com

B.2.2 Propagation Forecasts, Beacons, and Archives

http://dx.qsl.net/propagation

http://prop.hfradio.org

www.dxinfocentre.com

www.qth.com/ad5q

www.hamradio-online.com/propagation.html

www.N0HR.com/radio_propagation.htm

www.wcflunatall.com/NZ4O3.htm

www.eham.net/DX/propagation

www.ncdxf.org/beacon/beaconschedule.html

www.swpc.noaa.gov/

www.K7TJR.com/radio_prop1.htm

www.voacap.com

www.solarcycle24.com

www.solen.info/solar/cyclcomp.html (graphical comparison of solar cycles 21–24)

B.2.3 RF Exposure Tutorials and Calculators

http://hintlink.com/power_density.htm (RF near-field calculator)

www.sss-mag.com/rfsafety.html

http://vernon.mauery.com/radio/rfe/rfe_calc.html (N7OH amateur band calculator)

www.wirelessconnections.net/calcs/rfsafety.asp (wireless RF calculator)

B.2.4 Free Online Calculators, Tutorials, and Other Utilities

www.AC6LA.com (Smith chart, transmission line, Moxon rectangles)

http://W8JI.com/antennas.htm (tutorials on a wide variety of antenna topics)

http://tools.rfdude.com (Smith chart)

www.saarsham.net/coax.html (coaxial cable specs, line loss calculator)

www.physicsclassroom.com (tutorials and animations, including waves, optics)

http://www.srrb.noaa.gov/highlights/sunrise/sunrise.html (sunrise/sunset calculator)

www.gpsvisualizer.com/calculators (GPS and great circle calculators)

www.seed-solutions.com/gregordy/Software/SMC.htm (series-section matching calculator)

http://fermi.la.asu.edu/w9cf/tuner/tuner.html (T-network antenna tuner simulator)

www.VK1OD.net/calc/tl/tllc.php (transmission line loss calculator)

http://www.isvr.soton.ac.uk/SPCG/Tutorial/Tutorial/Tutorial_files/Web-basics-frequency.htm (wavelength and frequency visualizations)

http://k7nv.com/notebook/ (antenna, tower, and rotator topics)

Index

Note: Boldface numbers denote definitions and/or primary coverage of a topic.

M

N